An Introduction to Geotechnical Engineering

PRENTICE-HALL CIVIL ENGINEERING AND ENGINEERING
MECHANICS SERIES

N. M. Newmark and W. J. Hall, Editors

An Introduction to Geotechnical Engineering

ROBERT D. HOLTZ, PH.D., P.E.
Purdue University
West Lafayette, IN

WILLIAM D. KOVACS, PH.D., P.E.
National Bureau of Standards
Washington, DC

Prentice-Hall, Inc., Englewood Cliffs, New Jersey 07632

Library of Congress Cataloging in Publication Data

HOLTZ, ROBERT D
    An introduction to geotechnical engineering.

    Includes index.
    1. Soil mechanics. 2. Rock mechanics. I. Kovacs,
William D., joint author. II. Title.
TA710.H564   624.1′513   80-23292
ISBN 0-13-484394-0

DEDICATION: To Our Teachers, Past and Present

Editorial/production supervision
  and interior design by Karen Skrable
Manufacturing buyers: Anthony Caruso and Joyce Levatino
Cover design by Edsal Enterprises

Printed in the United States of America

10 9 8 7 6 5 4 3 2 1

Prentice-Hall International, Inc., *London*
Prentice-Hall of Australia Pty. Limited, *Sydney*
Prentice-Hall of Canada, Ltd., *Toronto*
Prentice-Hall of India Private Limited, *New Delhi*
Prentice-Hall of Japan, Inc., *Tokyo*
Prentice-Hall of Southeast Asia Pte. Ltd., *Singapore*
Whitehall Books Limited, *Wellington, New Zealand*

Contents

7 WATER IN SOILS, II: PERMEABILITY, SEEPAGE, EFFECTIVE STRESS *199*

8 CONSOLIDATION AND CONSOLIDATION SETTLEMENTS *283*

9 TIME RATE OF CONSOLIDATION *376*

Preface

An Introduction to Geotechnical Engineering is intended for use in the first of a two-course sequence in geotechnical engineering usually taught to third- and fourth-year undergraduate Civil Engineering students. We assume the students have a working knowledge of undergraduate mechanics, especially statics and mechanics of materials (including fluids). A knowledge of basic geology, although helpful, is not essential. We introduce the "language" of geotechnical engineering in the first course, that is, the classification and engineering properties of soils. Once the student has a working knowledge of how soil behaves as an engineering material, he/she can begin to predict soil behavior and, in the second course, to carry out the design of simple foundations and earthwork systems.

We feel that there is a need for more detailed and modern coverage of the engineering properties of soils than is found in most undergraduate texts. This applies to both the students "majoring" in geotechnical engineering and the general civil engineering undergraduate student. We find that our students are involved in increasingly more complex projects, especially in transportation, structural, and construction engineering. Environmental, economic, and political constraints demand innovative solutions to civil engineering problems. The availability of modern analytical techniques and the digital computer has had an almost revolutionary effect on engineering design practice. This development demands a better knowledge of site conditions as well as better defined geotechnical engineering design parameters.

We have tried to make the text easily readable by the average undergraduate. To this end, *An Introduction to Geotechnical Engineering* is written at a simple rather than sophisticated level, although the material covered may be rather sophisticated at times. Involved derivations, unread by the typical student, are relegated to appendices where they are available to the interested student.

The emphasis throughout is on the practical, and admittedly empirical, knowledge of soil behavior required by the geotechnical engineer for the design and construction of foundations and embankments. Most of the material in the text is descriptive, since most of the engineering design applications are usually left to the second course in Foundation Engineering. Consequently, in order to interest the student, we have tried to indicate wherever possible the engineering significance and specific design applications of the soil property being discussed. We have tried to emphasize why such a property is needed, how it is determined or measured, and, to some extent, how it is actually used in practice. The only "design"-type problem we cover in a one-semester course (15 weeks) is estimating the settlement of shallow foundations on saturated clays. The text is sufficiently flexible that innovative instructors can add additional design examples should they so desire. It seems that units are always a problem with geotechnical engineers. In line with the trend towards the use of S.I. units encouraged by the American Society of Civil Engineers and American Society for Testing and Materials, we have used this system in the text. The uninitiated may find the discussion of S.I. in Appendix A helpful. In addition to the almost exclusive use of S.I. units in examples and problems, we have been careful to use the latest definition of density (mass/unit volume) in phase relationships as well as in geostatic and hydrostatic pressure computations.

We consider the laboratory portion of the first course to be an essential part of the neophyte engineer's experience with soils as a unique engineering material. How else is the young engineer to begin to develop a "feel" for soils and soil behavior, so essential for the successful practice of geotechnical engineering? Thus, an emphasis on laboratory and field testing is found throughout the text. The order of the laboratory portion of our first course has dictated the organization and development of the material in the text. We begin with phase relations, visual classification of soils, and simple classification tests. Thus, the early chapters introduce the discipline of Geotechnical Engineering, Phase Relationships and Index Properties, Soil Classification, and Clay Minerals and Soil Structure. This material provides the background and terminology for the later chapters. Following a very practical discussion of Compaction in Chapter 5, Chapters 6 and 7 describe how water influences and affects soil behavior. Topics presented include capillarity, shrinkage, swelling, and frost action as well as permeability, seepage, and effective stress. These two chapters again serve as background for the next four chapters which deal with consolidation and shear strength.

The treatment of these latter topics is quite modern and up-to-date. The Schmertmann procedure for determining field compressibility is included as is a modern treatment of secondary compression developed by Prof. Mesri and his co-workers. Prof. Lambe's stress path method is introduced in Chapter 10 and used to advantage in Chapter 11, especially when practical engineering applications of shear strength theory are discussed. The pioneering work of Profs. Seed and Lee on the drained and undrained strength of sands is presented in Chapter 11. Also in this chapter we discuss the stress-deformation and strength characteristics of cohesive soils. Although the treatment is modern, because this is primarily an undergraduate textbook, considerations of strength anisotropy, critical state concepts, the Jürgenson-Rutledge hypothesis, and Hvorslev's strength parameters have been left to more advanced texts.

Even though the book is primarily for the beginning student in geotechnical engineering, advanced students in other disciplines and engineers desiring a refresher in engineering properties may find the book helpful. Because of the many fully-worked example problems, the book is almost "self-teaching." This aspect of the text also frees the instructor in a formal course from the necessity of working example problems during lectures. It allows the instructor to concentrate on explaining basic principles and illustrating specific engineering applications of the points in question. The third group we hope will find this book useful are practicing geotechnical engineers. Typical values are given for all classification and engineering properties for a wide variety of soils; we have found such a compendium very useful in our own engineering practice.

To acknowledge all who have contributed to this book is a formidable task. We have tried whenever possible to indicate by references or quotations, concepts and ideas originating in the literature or with our former teachers, especially Profs. A. Casagrande and H. B. Seed. We apologize for any omissions. We must also mention the students in our beginning geotechnical engineering course at Purdue who have graciously suffered through several versions of *An Introduction to Geotechnical Engineering* in note form. Their criticism and helpful comments on the text have been very valuable. The authors have greatly benefited from discussions with Prof. M. E. Harr of Purdue University regarding the section on the method of fragments in Chapter 7. We hope to bring further attention to the profession of this powerful design method. Dr. E. Simiu of the U.S. National Bureau of Standards critically read a recent version of the manuscript and provided many helpful comments. It should be noted that *An Introduction to Geotechnical Engineering* was written while William D. Kovacs was on the faculty of Purdue University. The text has no connection with Kovacs' present affiliation with the National Bureau of Standards. Our faithful secretaries Mrs. Janice Wait Bollinger, Miss Cathy Minth and Mrs. Edith Vanderwerp deserve special thanks for typing and correcting the several drafts. The first author also wishes to gratefully acknowledge the interest and encouragement of his wife Cricket Morgan. Her work with the proofreading and corrections is especially appreciated. We of course will appreciate any comments and criticism of readers.

R. D. Holtz
W. D. Kovacs

West Lafayette, Indiana

one

Introduction to Geotechnical Engineering

1.1 GEOTECHNICAL ENGINEERING

Geotechnical engineering, as the name implies, concerns the application of civil engineering technology to some aspect of the earth. Usually, the geotechnical engineer is concerned only with the natural materials found at or near the surface of the earth. Civil engineers call these earthen materials *soil* and *rock*. *Soil*, in an engineering sense, is the relatively loose agglomerate of mineral and organic materials and sediments found above the bedrock. Soils can be relatively easily broken down into their constituent mineral or organic particles. *Rocks*, on the other hand, have very strong internal cohesive and molecular forces which hold the constituent mineral grains together. This is true whether the rock is massive bedrock or a piece of gravel found in a clay soil. The dividing line between soil and rock is arbitrary, and many natural materials encountered in engineering practice cannot be easily classified. They may be either a "very soft rock" or a "very hard soil." Other scientific disciplines have different meanings for the terms soil and rock. In geology, for example, *rock* means all the materials found in the earth's crust, independently of how much the mineral particles are bound together. Soils to a geologist are just decomposed and disintegrated rocks generally found in the very thin upper part of the crust and capable of supporting plant life. Similarly, pedology (soil science) and agronomy are concerned with only the very uppermost layers of soil, that is, those materials relating to agriculture and forestry. Geotechnical engineers can learn much from both geology and pedology. Both sciences, especially engineering geology, are important adjuncts to geotechnical engineering and there is considerable overlap between these fields. But differences in terminology, approach, and objectives may cause some confusion, especially for the beginner.

Geotechnical engineering has several different aspects or emphases. *Soil mechanics* is the branch of geotechnical engineering concerned with the engineering mechanics and properties of soil, whereas *rock mechanics* is concerned with the engineering mechanics and properties of rock, usually but not necessarily the bedrock. Soil mechanics applies the basic principles of mechanics including kinematics, dynamics, fluid mechanics, and the mechanics of materials to soils. In other words, soil rather than water or steel or concrete, for example, now becomes the engineering material whose properties and behavior we must understand in order to build with it or upon it. A similar comment could also be made for rock mechanics. It should be noted, however, that there are significant differences between the behavior of soil masses and rock masses, and in principle there is not much overlap between the two disciplines.

Foundation engineering applies geology, soil mechanics, rock mechanics, and structural engineering to the design and construction of foundations for civil engineering and other structures. The foundation engineer must be able to predict the performance or response of the foundation soil or rock to the loads imposed by the structure. Some examples of the kinds of problems faced by the foundation engineer include foundations for industrial, commercial, and residential buildings, and other types of support structures for radar towers, as well as foundations for oil and other kinds of tanks and offshore structures. Even ships must have a dry dock during construction or repairs, and the dry dock must have a foundation. The support of rockets and appurtenant structures during construction and launch have led to very interesting and challenging foundation engineering problems. Related geotechnical engineering problems facing the foundation engineer are the stability of natural and excavated slopes, the stability of permanent and temporary earth-retaining structures, problems of construction, controlling water movement and pressures, and even the maintenance and rehabilitation of old buildings. Not only must the foundation safely support the static structural and construction loads, but it must also adequately resist dynamic loads due to blasting, earthquakes, etc.

If you think about it, it is impossible to design or construct any civil engineering structure without ultimately considering the foundation soils and rocks to some extent, and this is true whether the structure is built on the earth or is extraterrestrial. The performance, economy, and safety of any civil engineering structure ultimately is affected or may even be controlled by its foundation.

Earth materials are often used as a construction material because they are the cheapest possible building material. However, its engineering properties such as strength and compressibility are often naturally poor, and measures must be taken to densify, strengthen, or otherwise stabilize and reinforce soils so that they will perform satisfactorily in service.

Highway and railway embankments, airfields, earth and rock dams, levees, and aqueducts are examples of earth structures, and the geotechnical engineer is responsible for their design and construction. Dam safety and rehabilitation of old dams are important aspects of this phase of geotechnical engineering. Also related, especially for highway and airfield engineers, is the design of the final surface layer on the earth structure, the pavement. Here the overlap between the transportation and geotechnical disciplines is apparent.

Rock engineering, analogous to foundation engineering for soils, is concerned with rock as a foundation and construction material. Because most of the earth's surface is covered with soil (or water), rock engineering usually occurs underground (tunnels, underground power houses, petroleum storage rooms, mines, etc.). But sometimes rock engineering occurs at the surface, such as in the case of building and dam foundations carried to bedrock, deep excavations to bedrock, stability of rock slopes, etc.

In presenting some of the typical problems facing the geotechnical engineer, we wanted you to see, first, how broad the field is and, second, how important it is to the design and construction of civil engineering structures. In a very real sense, geotechnical engineering combines the basic physical sciences, geology and pedology, with hydraulic, structural, transportation, construction, and mining engineering.

1.2 THE UNIQUE NATURE OF SOIL AND ROCK MATERIALS

Geotechnical engineering is highly empirical and is perhaps much more of an "art" than the other disciplines within civil engineering because of the basic nature of soil and rock materials. They are often highly variable, even within a distance of a few millimetres. Another way of saying this is that soils are *heterogeneous* rather than *homogeneous* materials. That is, their material or engineering properties may vary widely from point to point within a soil mass. Furthermore, soils in general are *nonlinear* materials; their stress-strain curves are not straight lines. To further complicate things (as well as to make them interesting!) soils are *nonconservative* materials; that is, they have a fantastic memory—they remember almost everything that ever happened to them, and this fact strongly affects their engineering behavior. Instead of being *isotropic*, soils are typically *anisotropic*, which means that their material or engineering properties are not the same in all directions. Most of the theories we have for the mechanical behavior of engineering materials assume that the materials are homogeneous and isotropic, and that they obey linear stress-strain laws. Common engineering materials such as steel and concrete do not deviate too significantly from these ideals, and consequently we can use, with discretion, simple linear theories to predict their response under engineering loads. With soils and rock, we are not so fortunate. As you shall see in your study of geotechnical engineering, we may assume a linear stress-strain response, but then we must apply large empirical correction or "safety" factors to our designs to account for the real material behavior. Furthermore, the behavior of soil and rock materials in situ is often governed or controlled by joints, fractures, weak layers and zones, and other "defects" in the material; yet our laboratory tests and simplified methods of analysis often do not take into account such real characteristics of the soil and rock. That is why geotechnical engineering is really an "art" rather than an engineering science. Successful geotechnical engineering depends on the good judgment and practical experience of the designer, constructor, or consultant. Put another way, the successful geotechnical engineer must develop a "feel" for soil and rock behavior before a safe and economic foundation design can be made or an engineering structure can be safely built.

1.3 SUGGESTED APPROACH TO THE STUDY OF GEOTECHNICAL ENGINEERING

Because of the nature of soil and rock materials, both laboratory and field testing are very important in geotechnical engineering. One way that student engineers can begin to develop a feel for soil and rock behavior is to get some experience in the laboratory by performing the standard tests for classification and engineering properties on many different types of soils and rocks. In this way the novice begins building up a "mental data bank" of how certain soils and rocks actually look, how they might behave should, for example, the amount of water present change, how they might behave under different kinds of engineering loads, and what the range of probable numerical values is for the different tests. This is sort of a self-calibration process, so that when you are faced with a new soil deposit or rock type, you will in advance have some idea as to the engineering problems you will encounter at that site. You can also begin to judge, at least qualitatively, the validity of laboratory and field test results for the materials at that site. So laboratory as well as field experience is important for you to help develop a "feel" for soil and rock behavior. Of course, just as with any other subject, this exposure in the laboratory to soil and rock properties and behavior must be complemented by a diligent study of the theoretical, empirical, and design components of geotechnical engineering practice.

1.4 SCOPE OF THIS BOOK

Rather than attempt an all-inclusive approach to geotechnical engineering, the primary emphasis in this text will be on the *engineering behavior of soil materials*. Soil mechanics and the analysis and design of foundations and earth structures is generally a fairly straightforward, but creative, application of mechanics, strength of materials, and elementary structural engineering. Often the key in the successful practice and application of geotechnical engineering lies in a sound knowledge and understanding of the engineering properties and behavior of soils in situ, when they are subjected to their engineering loads and environmental conditions. Therefore we feel that the beginning student must first develop an appreciation for the engineering properties of soils as distinct from other common civil engineering materials before proceeding to instruction in the analysis and design phases of foundation and earthwork engineering.

This is an elementary text, and the approach we have tried to follow is to emphasize the fundamentals, with an eye toward the practical applications that you as a practicing civil engineer are likely to encounter in your engineering practice. Finally, we hope you will know enough about soils and soil deposits to avoid serious mistakes or blunders in those aspects of your professional career that involve soil and soil materials.

In the first part of the book, we introduce some of the basic definitions and index properties of soil that are used throughout the book. Then some common soil classification schemes are presented. *Classification* of soils is important because it is the "language" engineers use to communicate certain general knowledge about the engineering behavior of the soils at a particular site. The rest of the book is concerned with the *engineering properties* of soil, properties that are necessary for the design of foundations and earth structures. Topics covered include how water affects soil behavior, their shrinkage and swelling characteristics, and their permeability (how water flows through soils). Then we get into the compressibility of soil, which is the important engineering property one needs to predict the settlement of engineering structures constructed on soil masses. Finally, we describe some of the elementary strength characteristics of both granular and cohesive soils. Soil strength is important, for example, for the design of foundations, retaining walls, and slopes.

Much of the practice of geotechnical engineering depends on topics that include geology and the nature of landforms and soil deposits. You are strongly encouraged to take a physical geology or an engineering geology course in connection with your studies of geotechnical engineering.

It is hoped that with the background of this text, you will be prepared for a follow-up course in foundation and earthwork engineering; you should know how to obtain the soil properties required for most designs, and you should have a pretty good idea as to the probable range of values for a given property if you know the general classification of the soil. Finally, you should have a fairly good idea of what to look for at a site, how to avoid costly and dangerous mistakes, and be aware of your own limitations and knowledge of soils as an engineering material.

1.5 SOIL FORMATION AND THE NATURE OF SOIL CONSTITUENTS

We mentioned earlier that soil from a civil engineering point of view is the relatively loose agglomeration of mineral and organic materials found above the bedrock. In a broader sense, of course, even shallow bedrock is of interest to geotechnical engineers and some of these applications have already been mentioned.

You may remember from your basic science courses that the earth has a crust of granitic and basaltic rocks 10 to 40 km thick. Overlying this more or less solid rock is a relatively thin layer of variable thickness of what geologists call *unconsolidated* materials. These materials can vary in size from sub-microscopic mineral particles to huge boulders. *Weathering* and other geologic processes act on the rocks at or near the earth's surface to form these unconsolidated materials, or soil. Weathering, which usually results from atmospheric processes, alters the composition and structure of these rocks by chemical and physical means. *Physical* or *mechanical weathering* causes disintegration of the rocks into smaller particle sizes. Physical weathering agents include freezing and thawing, temperature changes, erosion, and the activity of plants and animals including man. *Chemical weathering* decomposes the minerals in the rocks by oxidation, reduction, carbonation, and other chemical processes. Generally, chemical weathering is much more important than physical weathering in soil formation. In short then, soils are the products of the weathering of rocks. Soils at a particular site can be *residual* (that is, weathered in place) or *transported* (moved by water, wind, glaciers, etc.), and the geologic history of a particular deposit significantly affects its engineering behavior.

The nature of soil constituents is discussed in greater detail throughout this text. For now, we want to make a few points just to set the stage for what we are about to study. You already have a layman's idea about soil. At least you know in general what *sand* and *gravel* are, and perhaps you even have a general idea about fine-grained soils such as *silts* and *clays*. These terms have quite precise engineering definitions, as we shall later see, but for now the general concept that soils are particles will suffice. Particles of what? Well, usually particles of mineral matter or, more simply, broken up pieces of rock that result from the weathering processes we spoke of previously. If we just talk for now about the size of the particles, gravels are small pieces of rock that typically contain several minerals, whereas sands are even smaller and each grain usually contains only a single mineral. If you cannot see each grain of a soil, then the soil is either a silt or a clay or a mixture of each. In fact, natural soils generally are a mixture of several different particle sizes and may even contain organic matter. Some soils such as *peat* may be almost entirely organic. Futhermore, because soils are a particulate material, they have voids, and the voids are usually filled with water and air. It is the physical and chemical interaction of the water and air in the voids with the particles of soil, as well as the interaction of the particles themselves, that makes soil behavior so complicated and leads to the nonlinear, nonconservative, and anisotropic mechanical behavior we mentioned previously. Now, if you add the variability and heterogeneity of natural soil deposits due to the capriciousness of nature, you probably can begin to see that soils are indeed complex engineering and construction materials. Helping you put some order into this potentially chaotic situation is our primary objective in this book.

1.6 HISTORICAL DEVELOPMENT OF GEOTECHNICAL ENGINEERING

As long as people have been building things, they have used soils as a foundation or construction material. The ancient Egyptians, Babylonians, Chinese, and Indians knew about constructing dikes and levees out of the soils found in river flood plains. Ancient temples and monuments built all around the world involved soil and rock in some way. The Aztecs constructed temples and cities on the very poor soils in the Valley of Mexico long before the Spaniards arrived in the New World. European architects and builders during the Middle Ages learned about the problems of settlements of cathedrals and large buildings. The most noteworthy example is, of course, the Leaning Tower of Pisa. Scandinavians used timber piles to support houses and wharf structures on their soft clays. The "design" of foundations and other constructions involving soil and rock was by rule of thumb, and very little theory as such was developed until the mid-1700's.

Coulomb is the most famous name of that era. He was interested in the problems of earth pressures against retaining walls, and some of his calculation procedures are still in use today. The most common theory for the shear strength of soils is named after him. During the next century, the French engineers Collin and Darcy (D'Arcy) and the Scotsman Rankine made important discoveries. Collin was the first engineer to be concerned with failures in clay slopes as well as the measurement of the shear strength of clays. Darcy established his law for the flow of water through sands. Rankine developed a method for estimating the earth pressure against retaining walls. In England, Gregory utilized horizontal subdrains and compacted earth-fill buttresses to stabilize railroad cut slopes.

By the turn of the century, important developments in the field took place in Scandinavia, primarily in Sweden. Atterberg defined the consistency limits for clays that are still in use today. During the period 1914–1922, in connection with investigations of some important failures in harbors and railroads, the Geotechnical Commission of the Swedish State Railways developed many important concepts and apparatuses in geotechnical engineering. Methods for calculating the stability of slopes were developed. They developed subsurface investigation techniques such as weight sounding and piston and other types of samplers. They understood important concepts such as sensitivity of clays and consolidation, which is the squeezing of water out of the pores of the clay. At that time, clays were thought to be absolutely impervious, but the Swedes made field measurements to show that they weren't. The Commission was the first to use the word *geotechnical* (Swedish: *geotekniska*) in the sense that we know it today: the combination of geology and civil engineering technology.

Even with these early developments in Sweden, the father of soil mechanics is really an Austrian, Prof. Karl Terzaghi. He published in 1925 the first modern textbook on soil mechanics, and in fact the name "soil mechanics" is a direct translation of the German word *erdbaumechanik*, which was part of the title of that book. Terzaghi was an outstanding and very creative engineer. He wrote several important books and over 250 technical papers and articles, and his name will appear many times in this book. He was a professor at Robert College in Istanbul, Technische Hochschule in Vienna, M. I. T., and at Harvard University from 1938 until his retirement in 1956. He continued to be active as a consultant until his death in 1963 at the age of 80.

Another important contributor to the advancement of modern soil mechanics is Prof. Arthur Casagrande, who was at Harvard University from 1932 until 1969. You will see his name often in this book because he made many important contributions to the art and science of soil mechanics and foundation engineering. Other important contributors to the field include Taylor, Peck, Tschebotarioff, Skempton, and Bjerrum. Since the 1950's the field has grown substantially and the names of those responsible for its rapid advancement are too numerous to mention.

Both Terzaghi and Casagrande began the teaching of soil mechanics and engineering geology in the United States. Before the Second World War, the subject was offered only as a graduate course in very few universities. After the war, it became common for at least one course in the subject to be required in most schools of civil engineering. In recent years graduate programs in all phases of geotechnical engineering have been implemented at many universities, and there has been a real information explosion in the number of conferences, technical journals, and textbooks published during the past two decades.

Important recent developments you should know about include developments in earthquake engineering and soil dynamics, the use of digital computers for the solution of complex engineering problems, and the introduction of probability and statistics into geotechnical engineering analysis and design.

1.7 NOTES ON SYMBOLS AND UNITS

At the beginning of each chapter, we list the pertinent symbols introduced in the chapter. As with most disciplines, a standard notation is not universal in geotechnical engineering, so we have tried to adopt the symbols most commonly used. For example, the American Society for Testing and Materials (ASTM, 1979) has a list of Standard Definitions of Terms and Symbols Relating to Soil and Rock Mechanics, Designation D 653, which was prepared jointly some years ago with the American Society of Civil Engineers (ASCE) and the International Society of Rock Mechanics (ISRM). Recently the International Society for Soil Mechanics and Foundation Engineering (ISSMFE, 1977) published an extensive list of symbols. Although there are some deviations from this list because of our personal preference, we have generally tried to follow these recommendations.

Units used in geotechnical engineering can be politely called a mess and, less politely, several worse things. There has developed in practice a jumbled mixture of cgs-metric, Imperial or British Engineering units and hybrid European metric units. With the introduction of the universal and consistent system of units, "Le Système International d'Unités" (SI) in the United States and Canada, we believe it is important that you learn to use those units in geotechnical engineering practice. However since British Engineering units are still commonly used, it is important that you become familiar with the typical values of both sets of units. To assist you with unit conversion where necessary, we have included a brief explanation of SI units as applied to geotechnical engineering in Appendix A.

two

Index and Classification Properties of Soils

2.1 INTRODUCTION

In this chapter we introduce the basic terms and definitions used by geotechnical engineers to index and classify soils. The following notation is used in this chapter.

Symbol	Dimension	Unit	Definition
A	—	—	Activity (Eq. 2-23)
C_c	—	—	Coefficient of curvature (Eq. 2-20)
C_u	—	—	Coefficient of uniformity (Eq. 2-19)
D_{10}	L	mm	Diameter for 10% finer by weight
D_{30}	L	mm	Diameter for 30% finer by weight
D_{60}	L	mm	Diameter for 60% finer by weight
e	—	(decimal)	Void ratio (Eq. 2-1)
LI or I_L	—	—	Liquidity index (Eq. 2-23)
LL or w_L	—	—	Liquid limit
M_t	M	kg	Total mass
M_s	M	kg	Mass of solids
M_w	M	kg	Mass of water
n	—	(%)	Porosity (Eq. 2-2)
PI or I_P	—	—	Plasticity index (Eq. 2-22)
PL or w_P	—	—	Plastic limit
S	—	(%)	Degree of saturation (Eq. 2-4)
SL or w_S	—	—	Shrinkage limit
V_a	L^3	m^3	Volume of air
V_s	L^3	m^3	Volume of solids
V_t	L^3	m^3	Total volume
V_v	L^3	m^3	Volume of voids
w	—	(%)	Water content (Eq. 2-5)
ρ	M/L^3	kg/m^3	Total, wet, or moist density (Eq. 2-6)
ρ'	M/L^3	kg/m^3	Buoyant density (Eq. 2-11)
ρ_d	M/L^3	kg/m^3	Dry density (Eq. 2-9)
ρ_s	M/L^3	kg/m^3	Density of solids (Eq. 2-7)
ρ_{sat}	M/L^3	kg/m^3	Saturated density (Eq. 2-10)
ρ_w	M/L^3	kg/m^3	Density of water (Eq. 2-8)

In this list, L = length and M = mass. When densities of soils and water are expressed in kg/m³, the numbers are rather large. For instance, the density of water ρ_w is 1000 kg/m³. Since 1000 kg = 1 Mg, to make the numbers more manageable, we will usually use Mg/m³ for densities. If you are unfamiliar with SI metric units and their conversion factors, it would be a good idea to read Appendix A before proceeding with the rest of this chapter.

2.2 BASIC DEFINITIONS AND PHASE RELATIONS

In general, any mass of soil consists of a collection of solid particles with voids in between. The soil solids are small grains of different minerals, whereas the voids can be filled either with water, air, or filled partly with both water and air (Fig. 2.1). In other words, the total volume V_t of the soil mass consists of the volume of soil solids V_s and the volume of voids V_v.

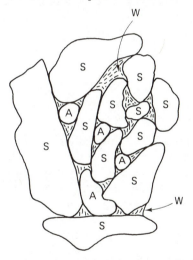

Fig. 2.1   Soil skeleton containing solid particles (S) and voids with air (A) and water (W).

The volume of voids is in general made up of the volume of water V_w and the volume of air V_a. We can schematically represent these three phases in a *phase diagram* (Fig. 2.2) in which each of the three phases is shown separately. On the left side we usually indicate the volumes of the three phases; on the right side we show the corresponding masses of the phases. Even though only two dimensions are shown in the phase diagram, total volume is any convenient unit volume such as m³ or cm³.

In engineering practice, we usually measure the total volume V_t, the mass of water M_w, and the mass of dry solids M_s. Then we calculate the rest of the values and the mass-volume relationships that we need. Most of these relationships are independent of sample size, and they are often dimensionless. They are very simple and easy to remember, especially if

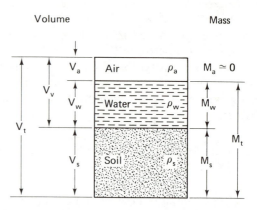

Fig. 2.2   Volumetric and mass relationships for a soil shown in a phase diagram.

you draw the phase diagram. They probably should be memorized, but as you work phase problems memorization will occur almost automatically.

There are three volumetric ratios that are very useful in geotechnical engineering, and these can be determined directly from the phase diagram, Fig. 2.2.

1. The *void ratio, e**, is defined as

$$e = \frac{V_v}{V_s}$$
(2-1)

where V_v = volume of the voids, and
V_s = volume of the solids.

The void ratio e is normally expressed as a *decimal*. The maximum possible range of e is between 0 and ∞. However typical values of void ratios for sands may range from 0.4 to about 1.0; typical values for clays vary from 0.3 to 1.5 and even higher for some organic soils.

2. The *porosity n* is defined as

$$n = \frac{V_v}{V_t} \times 100 \ (\%)$$
(2-2)

where V_v = volume of voids, and
V_t = total volume of soil sample.

Porosity is traditionally expressed as a *percentage*. The maximum range of n is between 0 and 100%. From 2.2 and Eqs. 2-1 and 2-2, it can be shown that

$$n = \frac{e}{1 + e}$$
(2-3a)

and

$$e = \frac{n}{1 - n}$$
(2-3b)

---

*Readers with British backgrounds will note the correct terminology is voids ratio.

3. The *degree of saturation S* is defined as

$$S = \frac{V_w}{V_v} \times 100 \ (\%) \tag{2-4}$$

The degree of saturation tells us what *percentage* of the total volume of voids contains water. If the soil is completely dry then $S = 0\%$, and if the pores are completely full of water, then the soil is fully saturated and $S = 100\%$.

Now let us look at the other side, the mass side, of the phase diagram in Fig. 2.2. First, let us define a mass ratio that is probably the single most important thing we need to know about a soil. We want to know how much water is present in the voids relative to the amount of solids in the soil, so we define a ratio called the *water content w* as

$$w = \frac{M_w}{M_s} \times 100 \ (\%) \tag{2-5}$$

where M_w = mass of water, and
M_s = mass of soil solids.

The ratio of the amount of water present in a soil volume to the amount of soil grains is based on the *dry mass* of the soil and not on the total mass. The water content, which is usually expressed as a *percentage*, can range from zero (dry soil) to several hundred percent. The natural water content for most soils is well under 100%, although it can range up to 500% or higher in some marine and organic soils.

The water content is easily determined in the laboratory. ASTM (1980), Designation D 2216, explains the standard procedure. A representative sample of soil is selected and its total or wet mass is determined. Then the soil sample is dried to constant mass in an oven at 110°C. Normally a constant mass is obtained after the sample is left in the oven overnight. The mass of the drying dish must, of course, be subtracted from both the wet and dry masses. Then the water content is calculated according to Eq. 2-5. Example 2.1 illustrates how the calculations for water content are actually done in practice.

EXAMPLE 2.1

Given:

A sample of wet soil in a drying dish has a mass of 462 g. After drying in an oven at 110°C overnight, the sample and dish have a mass of 364 g. The mass of the dish alone is 39 g.

Required:

Determine the water content of the soil.

Solution:

Set up the following calculation scheme; fill in the "given" or measured quantities (a), (b), and (d), and make the calculations as indicated for (c), (e), and (f).

a. Mass of total (wet) sample + dish $= 462$ g
b.   Mass of dry sample + dish $= 364$ g
c.   Mass of water $(a - b) = \ \ 98$ g
d.   Mass of dish $= \ \ 39$ g
e.   Mass of dry soil $(b - d) = 325$ g
f.   Water content $(c/e) \times 100\% = 30.2\%$

In the laboratory, masses are usually determined in grams (g) on an ordinary chemical balance.

Another very useful concept in geotechnical engineering is density. You know from physics that density is mass per unit volume, so its units are kg/m³. (See Appendix A for the corresponding units in the cgs and British Engineering systems.) The *density* is the ratio that connects the volumetric side of the phase diagram with the mass side. There are several commonly used densities in geotechnical engineering practice. First, we define the total, wet, or moist density ρ, the density of the particles, solid density ρ_s, and the density of water ρ_w. Or, in terms of the basic masses and volumes of Fig. 2.2:

$$\rho = \frac{M_t}{V_t} = \frac{M_s + M_w}{V_t} \tag{2-6}$$

$$\rho_s = \frac{M_s}{V_s} \tag{2-7}$$

$$\rho_w = \frac{M_w}{V_w} \tag{2-8}$$

In natural soils, the magnitude of the total density ρ will depend on how much water happens to be in the voids as well as the density of the mineral grains themselves, but ρ could range from slightly above 1000 kg/m³ to as high as 2400 kg/m³ (1.0 to 2.4 Mg/m³). Typical values of ρ_s for most soils range from 2500 to 2800 kg/m³ (2.5 to 2.8 Mg/m³). Most sands have ρ_s ranging between 2.6 and 2.7 Mg/m³. For example, a common mineral in sands is quartz; its $\rho_s = 2.65$ Mg/m³. Most clay soils have a value of ρ_s between 2.65 and 2.80 Mg/m³, depending on the predominant mineral in the soil, whereas organic soils may have a ρ_s as low as 2.5 Mg/m³. Consequently, it is usually close enough for geotechnical work to *assume* a ρ_s of 2.65 or 2.70 Mg/m³ for most phase problems, unless a specific value of ρ_s is given.

The density of water varies slightly, depending on the temperature. At 4°C, when water is at its densest, ρ_w exactly equals 1000 kg/m³ (1 g/cm³), and this density is sometimes designated by the symbol ρ_o. For ordinary engineering work, it is sufficiently accurate to take $\rho_w \approx \rho_o = 1000$ kg/m³ = 1 Mg/m³.

There are three other useful densities in soils engineering. They are the dry density ρ_d, the saturated density ρ_{sat}, and the submerged or buoyant density ρ'.

$$\rho_d = \frac{M_s}{V_t} \tag{2-9}$$

$$\rho_{sat} = \frac{M_s + M_w}{V_t}(V_a = 0, \ S = 100\%) \tag{2-10}$$

$$\rho' = \rho_{sat} - \rho_w \tag{2-11}$$

Strictly speaking, total ρ should be used instead of ρ_{sat} in Eq. 2-11, but in most cases completely submerged soils are also completely saturated, or at least it is reasonable to assume they are saturated. The dry density ρ_d is a common basis for judging the degree of compaction of earth embankments (Chapter 5). A typical range of values of ρ_d, ρ_{sat}, and ρ' for several soil types is shown in Table 2-1.

From the basic definitions provided in this section, other useful relationships can be derived, as we show in the examples in the next section.

TABLE 2-1 Some Typical Values for Different Densities of Some Common Soil Materials*

Soil Type	Density (Mg/m³)		
	ρ_{sat}	ρ_d	ρ'
Sands and gravels	1.9–2.4	1.5–2.3	1.0–1.3
Silts and clays	1.4–2.1	0.6–1.8	0.4–1.1
Glacial tills	2.1–2.4	1.7–2.3	1.1–1.4
Crushed rock	1.9–2.2	1.5–2.0	0.9–1.2
Peats	1.0–1.1	0.1–0.3	0.0–0.1
Organic silts and clays	1.3–1.8	0.5–1.5	0.3–0.8

*Modified after Hansbo (1975).

2.3 SOLUTION OF PHASE PROBLEMS

Phase problems are very important in soils engineering, and in this section, with the help of some numerical examples, we illustrate how most phase problems can be solved. As is true for many disciplines, practice helps; the more problems you solve, the simpler they become and the more proficient you will become. Also, with practice you soon memorize most of the important definitions and relationships, thus saving the time of looking up formulas later on.

Probably the single most important thing you can do in solving phase problems is *to draw a phase diagram.* This is especially true for the beginner. Don't spend time searching for the right formula to plug into. Instead, always draw a phase diagram and show both the given values and the unknowns of the problem. For some problems, simply doing this leads almost immediately to the solution; at least the correct approach to the problem is usually indicated. Also, you should note that there often are alternative approaches to the solution of the same problem as illustrated in Example 2.2.

EXAMPLE 2.2

Given:

$$\rho = 1.76 \text{ Mg/m}^3 \text{ (total density)}$$
$$w = 10\% \text{ (water content)}$$

Required:

Compute ρ_d (dry density), e (void ratio), n (porosity), S (degree of saturation), and ρ_{sat} (saturated density).

Solution:

Draw the phase diagram (Fig. Ex. 2.2a). Assume that $V_t = 1 \text{ m}^3$.
From the definition of water content (Eq. 2-5) and total density (Eq. 2-6) we can solve for M_s and M_w. Note that in the computations water content is expressed as a decimal.

$$w = 0.10 = \frac{M_w}{M_s}$$

$$\rho = 1.76 \text{ Mg/m}^3 = \frac{M_t}{V_t} = \frac{M_w + M_s}{1.0 \text{ m}^3}$$

Volume (m³)   Mass (Mg)

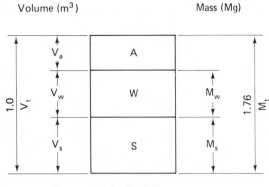

Fig. Ex. 2.2a

Substituting $M_w = 0.10M_s$ we get

$$1.76 \text{ Mg/m}^3 = \frac{0.10M_s + M_s}{1.0 \text{ m}^3}$$

$$M_s = 1.60 \text{ Mg} \quad \text{and} \quad M_w = 0.16 \text{ Mg}$$

These values are now placed on the mass side of the phase diagram (Fig. Ex. 2.2b), and the rest of the desired properties are calculated.

From the definition of ρ_w (Eq. 2-8) we can solve for V_w.

$$\rho_w = \frac{M_w}{V_w},$$

or

$$V_w = \frac{M_w}{\rho_w} = \frac{0.16 \text{ Mg}}{1 \text{ Mg/m}^3} = 0.160 \text{ m}^3$$

Place this numerical value on phase diagram, Fig. Ex. 2.2b.

To calculate V_s, we must assume a value of the density of the solids ρ_s. Here assume $\rho_s = 2.70 \text{ Mg/m}^3$. From the definition of ρ_s (Eq. 2-7) we

Volume (m³)   Mass (Mg)

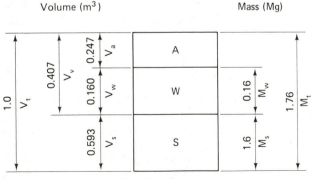

Fig. Ex. 2.2b

can solve for V_s directly, or

$$V_s = \frac{M_s}{\rho_s} = \frac{1.6 \text{ Mg}}{2.70 \text{ Mg/m}^3} = 0.593 \text{ m}^3$$

Since $V_t = V_a + V_w + V_s$, we can solve for V_a, since we know the other terms.

$$V_a = V_t - V_w - V_s = 1.0 - 0.593 - 0.160 = 0.247 \text{ m}^3$$

Once the phase diagram has been filled in, solution of the rest of the problem involves just plugging in the respective numbers into the appropriate definition equations. We recommend that when you make the computations, you write out the equations in symbol form and then insert the numbers in the same order as written in the equation. Also, it is a good idea to have the units accompany the calculations.

Solving for the remainder of the required items is easy.

From Eq. 2-9,

$$\rho_d = \frac{M_s}{V_t} = \frac{1.6 \text{ Mg}}{1 \text{ m}^3} = 1.6 \text{ Mg/m}^3$$

From Eq. 2-1,

$$e = \frac{V_v}{V_s} = \frac{V_a + V_w}{V_s} = \frac{0.247 + 0.160}{0.593} = 0.686$$

From Eq. 2-2,

$$n = \frac{V_v}{V_t} = \frac{V_a + V_w}{V_t} 100 = \frac{0.247 + 0.160}{1.0} 100 = 40.7\%$$

From Eq. 2-4,

$$S = \frac{V_w}{V_v} = \frac{V_w}{V_a + V_w} 100 = \frac{1.160}{0.247 + 0.160} 100 = 39.3\%$$

The saturated density ρ_{sat} is the density when all the voids are filled with water, that is, when $S = 100\%$ (Eq. 2-10). Therefore, if the volume of air V_a were filled with water, it would weigh $0.247 \text{ m}^3 \times 1 \text{ Mg/m}^3$ or 0.247 Mg. Then

$$\rho_{sat} = \frac{M_w + M_s}{V_t} = \frac{(0.247 \text{ Mg} + 0.16 \text{ Mg}) + 1.6 \text{ Mg}}{1 \text{ m}^3} = 2.01 \text{ Mg/m}^3$$

Another, and perhaps even easier way to solve this example problem, is to assume V_s is a unit volume, 1 m^3. Then, by definition, $M_s = \rho_s = 2.7$ (when ρ_s is assumed to be equal to 2.70 Mg/m^3). The completed phase diagram is shown in Fig. Ex. 2.2c.

Since $w = M_w/M_s = 0.10$, $M_w = 0.27$ Mg and $M_t = M_w + M_s = 2.97$ Mg. Also $V_w = M_w$ since $\rho_w = 1 \text{ Mg/m}^3$; that is, 0.27 Mg of water occupies a volume of 0.27 m³. Two unknowns remain to be solved before we can proceed: they are V_a and V_t. To obtain these values, we must use the given information that $\rho = 1.76$ Mg/m³. From the definition of total density (Eq. 2-6),

$$\rho = 1.76 \text{ Mg/m}^3 = \frac{M_t}{V_t} = \frac{2.97 \text{ Mg}}{V_t}$$

Solving for V_t,

$$V_t = \frac{M_t}{\rho} = \frac{2.97 \text{ Mg}}{1.76 \text{ Mg/m}^3} = 1.688 \text{ m}^3$$

Therefore

$$V_a = V_t - V_w - V_s = 1.688 - 0.27 - 1.0 = 0.418 \text{ m}^3$$

You can use Fig. Ex. 2.2c to verify that the remainder of the solution is identical to the one using the data of Fig. Ex. 2.2b.

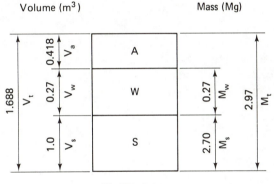

Fig. Ex. 2.2c

EXAMPLE 2.3

Required:

Express the porosity n in terms of the void ratio e (Eq. 2-3a) and the void ratio in terms of the porosity (Eq. 2-3b).

Solution:

Draw a phase diagram (Fig. Ex. 2.3a).

For this problem, assume $V_s = 1$ (units arbitrary). From Eq. 2-1, $V_v = e$ since $V_s = 1$. Therefore $V_t = 1 + e$. From Eq. 2-2, the definition of

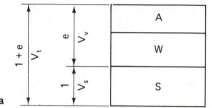

Fig. Ex. 2.3a

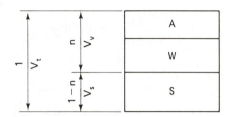

Fig. Ex. 2.3b

n is V_v/V_t, or

$$n = \frac{e}{1 + e} \tag{2-3a}$$

Equation 2-3b can be derived algebraically or from the phase diagram (Fig. Ex. 2.3b). For this case, assume $V_t = 1$.

From Eq. 2-2, $V_v = n$ since $V_t = 1$. Therefore $V_s = 1 - n$. From Eq. 2-1, the definition of $e = V_v/V_s$. So

$$e = \frac{n}{1 - n} \tag{2-3b}$$

EXAMPLE 2.4

Given:

$$e = 0.62, \qquad w = 15\%, \qquad \rho_s = 2.65 \ Mg/m^3.$$

Required:

    a. ρ_d
    b. ρ
    c. w for $S = 100\%$
    d. ρ_{sat} for $S = 100\%$

Solution:

Draw phase diagram (Fig. Ex. 2.4).
    (a) Since no volumes are specified, assume $V_s = 1$ m^3. Just as in

Example 2.3, this makes the $V_v = e = 0.62$ m³ and $V_t = 1 + e = 1.62$ m³. From Eq. 2-9,

$$\rho_d = \frac{M_s}{V_t}$$

and $M_s = \rho_s V_s$ (from Eq. 2-7). So

$$\rho_d = \frac{\rho_s V_s}{V_t} = \frac{\rho_s}{1 + e} \quad \text{since } V_s = 1 \text{ m}^3 \text{ in Fig. Ex.2-4}$$

$$= \frac{2.65}{1 + 0.62} = 1.636 \text{ Mg/m}^3$$

Note: The relationship

$$\rho_d = \frac{\rho_s}{1 + e} \tag{2-12}$$

is often very useful in phase problems.

Fig. Ex. 2.4

(b) Now for ρ:

$$\rho = \frac{M_t}{V_t} = \frac{M_s + M_w}{V_t}$$

We know that

$$M_w = wM_s \text{ (from Eq. 2-5)} \quad \text{and} \quad M_s = \rho_s V_s$$

$$\rho = \frac{\rho_s V_s + w\rho_s V_s}{V_t} = \frac{\rho_s(1 + w)}{1 + e} \quad \text{since } V_s = 1 \text{ m}^3$$

Plug in the numbers.

$$\rho = \frac{2.65(1 + 0.15)}{1 + 0.62} = 1.88 \text{ Mg/m}^3$$

This relationship is often useful to know.

$$\rho = \frac{\rho_s(1 + w)}{(1 + e)} \tag{2-13}$$

Check:

$$\rho_d = \frac{\rho}{1 + w} \tag{2-14}$$

$$= \frac{1.88}{1.15} = 1.636 \text{ Mg/m}^3$$

You should verify that $\rho_d = \rho/(1 + w)$, which is another very useful relationship to remember.

(c) Water content for $S = 100\%$. From Eq. 2-4, we know that $V_w = V_v = 0.62 \text{ m}^3$. From Eq. 2-8, $M_w = V_w \rho_w = 0.62 \text{ m}^3 (1 \text{ Mg/m}^3) = 0.62 \text{ Mg}$. Therefore w for $S = 100\%$ must be

$$w_{(S=100\%)} = \frac{M_w}{M_s} = \frac{0.62}{2.65} = 0.234 \text{ or } 23.4\%$$

(d) ρ_{sat}. From Eq. 2-10, we know $\rho_{\text{sat}} = (M_s + M_w)/V_t$, or

$$\rho_{\text{sat}} = \frac{2.65 + 0.62}{1.62} = 2.019 \text{ or } 2.02 \text{ Mg/m}^3$$

Check, by Eq. 2-13:

$$\rho_{\text{sat}} = \frac{\rho_s(1 + w)}{1 + e} = \frac{2.65(1 + 0.234)}{1.62} = 2.02 \text{ Mg/m}^3$$

EXAMPLE 2.5

Required:

Derive a relationship between S, e, w, and ρ_s.

Solution:

Look at the phase diagram with $V_s = 1$ (Fig. Ex. 2.5). From Eq. 2-4 and Fig. 2.5, we know that $V_w = SV_v = Se$. From the definitions of water content (Eq. 2-5) and ρ_s (Eq. 2-7), we can place the

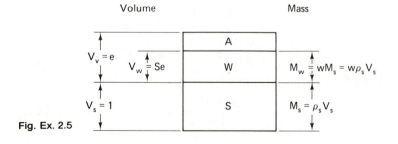

Fig. Ex. 2.5

equivalents for M_s and M_w on the phase diagram. Since from Eq. 2-8, $M_w = \rho_w V_w$, we now can write the following equation:

$$M_w = \rho_w V_w = wM_s = w\rho_s V_s$$

or

$$\rho_w Se = w\rho_s V_s$$

Since $V_s = 1 \text{ m}^3$,

$$\rho_w Se = w\rho_s \qquad (2\text{-}15)$$

Equation 2-15 is one of the most useful of all equations for phase problems. You can also verify its validity from the fundamental definitions of ρ_w, S, e, w, and ρ_s.

Note that using Eq. 2-15 we can write Eq. 2-13 another way:

$$\rho = \frac{\rho_s\left(1 + \dfrac{\rho_w Se}{\rho_s}\right)}{1 + e} = \frac{\rho_s + \rho_w Se}{1 + e} \qquad (2\text{-}16)$$

When $S = 100\%$, Eq. 2-16 becomes

$$\rho_{sat} = \frac{\rho_s + \rho_w e}{1 + e} \qquad (2\text{-}17)$$

EXAMPLE 2.6

Given:

A silty clay soil with $\rho_s = 2700 \text{ kg/m}^3$, $S = 100\%$, and the water content = 46%.

Required:

Compute the void ratio e, the saturated density, and the buoyant or submerged density in kg/m^3.

Solution:

Place given information on a phase diagram (Fig. Ex. 2.6).

Assume $V_s = 1 \text{ m}^3$; therefore $M_s = V_s\rho_s = 2700 \text{ kg}$. From Eq. 2-15, we can solve for e directly:

$$e = \frac{w\rho_s}{\rho_w S} = \frac{0.46 \times 2700}{1000 \times 1.0} = 1.242$$

But e also equals V_v since $V_s = 1.0$; likewise $M_w = 1242 \text{ kg}$ since M_w is

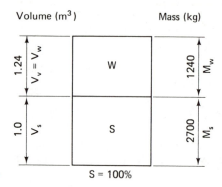

Fig. Ex. 2.6

numerically equal to V_w because $\rho_w = 1000$ kg/m^3. Now that all the unknowns have been found, we may readily calculate the saturated density (Eq. 2-10).

$$\rho_{\text{sat}} = \frac{M_t}{V_t} = \frac{M_w + M_s}{1 + e} = \frac{(1242 + 2700) \text{ kg}}{(1 + 1.24) \text{ m}^3} = 1758 \text{ kg/m}^3$$

We could also use Eq. 2-17 directly.

$$\rho_{\text{sat}} = \frac{\rho_s + \rho_w e}{1 + e} = \frac{2700 + 1000\,(1.242)}{1 + 1.242} = 1758 \text{ kg/m}^3$$

When a soil is submerged, the actual unit weight is reduced by the buoyant effect of the water. The buoyancy effect is equal to the weight of the water displaced. Thus, in terms of densities, (Eqs. 2-11 and 2-17):

$$\rho' = \rho_{\text{sat}} - \rho_w = 1758 \text{ kg/m}^3 - 1000 \text{ kg/m}^3 = 758 \text{ kg/m}^3$$

or

$$\rho' = \frac{\rho_s + \rho_w e}{1 + e} - \rho_w$$

$$= \frac{\rho_s - \rho_w}{1 + e} \qquad (2\text{-}18)$$

$$\rho' = 758 \text{ kg/m}^3$$

In this example, ρ' is less than the density of water. Go back and look at Table 2-1 for typical values of ρ'. The submerged or buoyant density of soil will be found to be very important later on in our discussion of consolidation, settlement, and strength properties of soils.

In summary, for the easy solution of phase problems, you don't have to memorize lots of complicated formulas. Most of them can easily be derived from the phase diagram as was illustrated in the preceding examples. Just remember the following simple rules:

1. Remember the basic definitions of w, e, ρ_s, S, etc.
2. Draw a phase diagram.
3. Assume either $V_s = 1$ or $V_t = 1$, if not given.
4. Often use $\rho_w Se = w\rho_s$.

2.4   SOIL TEXTURE

So far we haven't said very much about what makes up the "solids" part of the soil mass. In Chapter 1 we gave the usual definition of soil from an engineering point of view: the relatively loose agglomeration of mineral and organic materials found above the bedrock. We briefly described how weathering and other geologic processes act on the rocks at or near the earth's surface to form soil. Thus the solid part of the soil mass consists primarily of particles of mineral and organic matter in various sizes and amounts.

The *texture* of a soil is its appearance or "feel," and it depends on the relative sizes and shapes of the particles as well as the range or distribution of those sizes. Thus coarse-grained soils such as *sands* or *gravels* obviously appear coarse textured, while a fine-textured soil might be composed of predominantly very tiny mineral grains which are invisible to the naked eye. *Silts* and *clay* soils are good examples of fine-textured soils.

The soil texture, especially of coarse-grained soils, has some relation to their engineering behavior. In fact, soil texture has been the basis for certain soil classification schemes which are, however, more common in agronomy than in soils engineering. Still, textural classification terms (gravels, sands, silts, and clays) are useful in a general sense in geotechnical engineering practice. For fine-grained soils, the presence of water greatly affects their engineering response—much more so than grain size or texture alone. Water affects the interaction between the mineral grains, and this may affect their *plasticity* and their *cohesiveness*.

Texturally, soils may be divided into coarse-grained versus fine-grained soils. A convenient dividing line is the smallest grain that is visible to the naked eye. Soils with particles larger than this size (about 0.05 mm) are called coarse-grained, while soils finer than the size are (obviously) called fine-grained. Sands and gravels are coarse grained while silts and clays are fine grained. Another convenient way to separate or classify soils is according to their plasticity and cohesion (physics: cohesion—sticking

TABLE 2-2   Textural and Other Characteristics of Soils

Soil name:	Gravels, Sands	Silts	Clays
Grain size:	Coarse grained Can see individ- ual grains by eye	Fine grained Cannot see individual grains	Fine grained Cannot see individual grains
Characteristics:	Cohesionless Nonplastic Granular	Cohesionless Nonplastic Granular	Cohesive Plastic —
Effect of water on engineering behavior:	Relatively unimportant (exception: loose sat- urated granular mater- ials and dynamic loadings)	Important	Very important
Effect of grain size distribution on engineering behavior:	Important	Relatively unimportant	Relatively unimportant

together of like materials). For example, sands are nonplastic and non-cohesive (cohesionless) whereas clays are both plastic and cohesive. Silts fall between clays and sands: they are at the same time fine-grained yet nonplastic and cohesionless. These relationships as well as some general engineering characteristics are presented in Table 2-2. You will need to obtain some practice, best done in the laboratory, in identifying soils according to texture and some of these other general characteristics such as plasticity and cohesiveness. Also you should note that the term *clay* refers both to specific minerals called *clay minerals* (discussed in Chapter 4) and to soils which contain clay minerals. The behavior of some soils is strongly affected by the presence of clay minerals. In geotechnical engineering, for simplicity such soils are usually called *clays*, but we really mean *soils in which the presence of certain clay minerals affects their behavior*.

2.5 GRAIN SIZE AND GRAIN SIZE DISTRIBUTION

As suggested in the preceding section, the size of the soil particle, especially for granular soils, has some effect on engineering behavior. Thus, for classification purposes, we are often interested in the particle or grain sizes present in a particular soil as well as the distribution of those sizes.

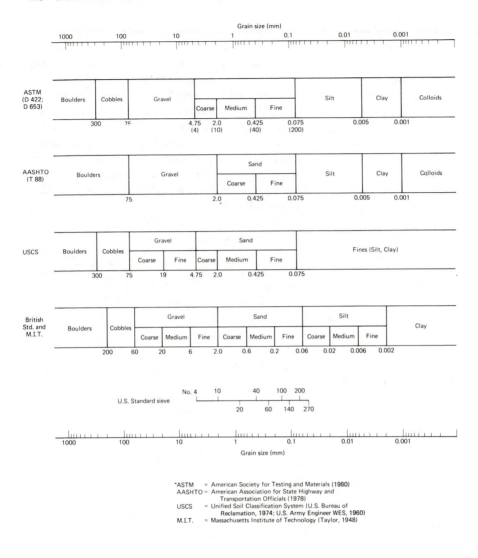

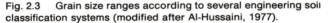

*ASTM   = American Society for Testing and Materials (1980)
AASHTO = American Association for State Highway and
                  Transportation Officials (1978)
USCS   = Unified Soil Classification System (U.S. Bureau of
                  Reclamation, 1974; U.S. Army Engineer WES, 1960)
M.I.T.   = Massachusetts Institute of Technology (Taylor, 1948)

Fig. 2.3   Grain size ranges according to several engineering soil classification systems (modified after Al-Hussaini, 1977).

The range of possible particle sizes in soils is tremendous. Soils can range from boulders or cobbles of several centimetres in diameter down to ultrafine-grained colloidal materials. The maximum possible range is on the order of 10^8, so usually we plot grain size distributions versus the *logarithm* of average grain diameter. Figure 2.3 indicates the divisions between the various textural sizes according to several common engineering classification schemes. It should be noted that traditionally in the

United States the units for the various sizes depend on the grain size. For materials greater than about 5 mm (about 1/4 in.), inches are commonly used, although millimetres could be used just as well. Grain sizes between 5 mm and 0.074 mm are classified according to U.S. Standard sieve number, which of course can be related to a specific grain size as shown in Fig. 2.3. Soils finer than the No. 200 sieve are usually dimensioned in millimetres or, for the very fine-grained colloidal particles, in micrometres.

How is the particle size distribution obtained? The process is called *mechanical analysis* or the *gradation test*. For coarse-grained soils, a *sieve analysis* is performed in which a sample of dry soil is shaken mechanically through a series of woven-wire square-mesh sieves with successively smaller openings. Since the total mass of sample is known, the percentage retained or passing each size sieve can be determined by weighing the amount of soil retained on each sieve after shaking. Detailed procedures for this test are specified by ASTM (1980), Designations C 136 and D 422. The corresponding AASHTO (1978) test standards are T 27 and T 88. The U.S. Standard sieve numbers commonly employed for the particle size analysis of soils are shown in Table 2-3. Since soil particles are rarely perfect spheres, when we speak of particle diameters, we really mean an *equivalent* particle diameter as determined by the sieve analysis.

TABLE 2-3   U.S. Standard Sieve Sizes and Their Corresponding Open Dimension

U.S. Standard Sieve No.	Sieve Opening (mm)
4	4.75
10	2.00
20	0.85
40	0.425
60	0.25
100	0.15
140	0.106
200	0.075

It turns out that the sieve analysis is impractical for sieve openings less than about 0.05 to 0.075 mm (No. 200 U.S. Standard sieve). Thus for the fine-grained soils, silts, and clays, the *hydrometer analysis* is commonly used. The basis for this test is Stoke's law for falling spheres in a viscous fluid in which the terminal velocity of fall depends on the grain diameter and the densities of the grains in suspension and of the fluid. The grain diameter thus can be calculated from a knowledge of the distance and time of fall. The hydrometer also determines the specific gravity (or density) of the suspension, and this enables the percentage of particles of a certain equivalent particle diameter to be calculated. As with the sieve analysis, the percentage of total sample still in suspension (or already out of suspension) can therefore readily be determined. Detailed procedures for the hydrometer test are given by ASTM (1980), Designation D 422, and AASHTO (1978) Standard Method T 88. The U.S.B.R. (1974) and U.S. Army Corps of Engineers (1970) also have similar standardized procedures for this test.

The distribution of the percentage of the total sample less than a certain sieve size or computed grain diameter can be plotted in either a histogram or, more commonly, in a cumulative frequency diagram. The equivalent grain sizes are plotted to a logarithmic scale on the abscissa, whereas the percentage by weight (or mass) of the total sample either passing (finer than) or retained (coarser than) is plotted arithmetically on the ordinate (Fig. 2.4). Note that this figure could just as well be plotted with the smaller grain sizes going towards the right. Some typical grain size distributions are shown in Fig. 2.4. The *well-graded* soil has a good representation of particle sizes over a wide range, and its gradation curve is smooth and generally concave upward. On the other hand, a *poorly graded* soil would be one where there is either an excess or deficiency of certain sizes or if most of the particles are about the same size. The *uniform* gradation shown in Fig. 2.4 is an example of a poorly graded soil. The *gap-graded* or *skip-graded* soil in that figure is also poorly graded; in this case, the proportion of grain sizes between 0.5 and 0.1 mm is relatively low.

We could, of course, obtain the usual statistical parameters (mean, median, standard deviation, etc.) for the grain size distributions, but this is more commonly done in sedimentary petrology than in soil mechanics. Of course the *range* of particle diameters found in the sample is of interest. Besides that, we use certain grain diameters D which correspond to an equivalent "percent passing" on the grain size distribution curve. For example, D_{10} is the grain size that corresponds to 10% of the sample passing by weight. In other words, 10% of the particles are smaller than the diameter D_{10}. This parameter locates the grain size distribution curve (GSD) along the grain size axis, and it is sometimes called the *effective size*. The *coefficient of uniformity* C_u is a crude shape parameter, and it is defined as

$$C_u = \frac{D_{60}}{D_{10}} \qquad (2\text{-}19)$$

where D_{60} = grain diameter (in mm) corresponding to 60% passing, and
D_{10} = grain diameter (in mm) corresponding to 10% passing, by weight (or mass).

Actually, the uniformity coefficient is misnamed since the smaller the number, the more *uniform* the gradation. So it is really a coefficient of

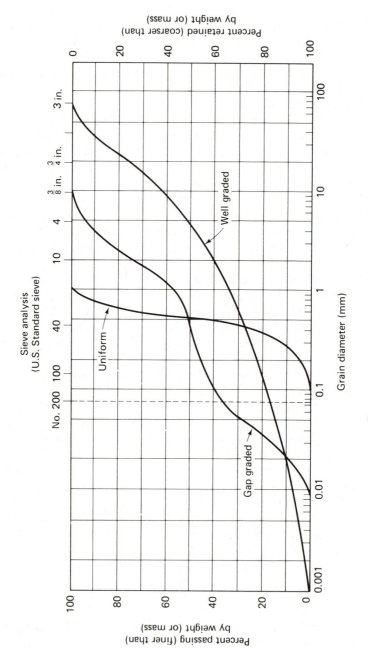

Fig. 2.4 Typical grain size distributions.

"disuniformity." For example, a $C_u = 1$ would be a soil with only one grain size. Very poorly graded soils, for example, beach sands, have C_u's of 2 or 3, whereas very well-graded soils may have a C_u of 15 or greater. Occasionally, the C_u can range up to 1000 or so. As an example, the clay core material for Oroville Dam in California has a C_u of between 400 and 500; the sizes range from large boulders down to very fine-grained clay particles.

Another shape parameter that is sometimes used for soil classification is the *coefficient of curvature* defined as

$$C_c = \frac{(D_{30})^2}{(D_{10})(D_{60})} \tag{2-20}$$

where D_{30} = grain diameter (in mm) corresponding to 30% passing by weight (or mass). The other terms were defined previously.

A soil with a coefficient of curvature between 1 and 3 is considered to be well graded as long as the C_u is also greater than 4 for gravels and 6 for sands.

EXAMPLE 2.7

Given:

The grain size distribution shown in Fig. 2.4.

Required:

Determine D_{10}, C_u, and C_c for each distribution.

Solution:

For Eqs. 2-19 and 2-20 we need D_{10}, D_{30}, and D_{60} for each gradation curve in Fig. 2.4.

    a. Well-graded soil; simply pick off the diameters corresponding to 10%, 30%, and 60% passing.

$$D_{10} = 0.02 \text{ mm}, \qquad D_{30} = 0.6 \text{ mm}, \qquad D_{60} = 9 \text{ mm}$$

From Eq. 2-19,

$$C_u = \frac{D_{60}}{D_{10}} = \frac{9}{0.02} = 450$$

From Eq. 2-20,

$$C_c = \frac{(D_{30})^2}{(D_{10})(D_{60})} = \frac{(0.6)^2}{(0.02)(9)} = 2$$

Since $C_u > 15$ and C_c is between 1 and 3, this soil is indeed well graded.

b. Gap-graded soil; use same procedure as in (a):

$$D_{10} = 0.022, \qquad D_{30} = 0.052, \qquad D_{60} = 1.2$$

From Eq. 2-19,

$$C_u = \frac{D_{60}}{D_{10}} = \frac{1.2}{0.022} = 55$$

From Eq. 2-20,

$$C_c = \frac{(D_{30})^2}{(D_{10})(D_{60})} = \frac{(0.052)^2}{(0.022)(1.2)} = 0.1$$

Even though by the uniformity coefficient criterion, this soil is well graded, it fails the coefficient of curvature criterion. Therefore it is indeed poorly graded.

c. Uniform soil; use same procedure as in (a):

$$D_{10} = 0.3, \qquad D_{30} = 0.43, \qquad D_{60} = 0.55$$

From Eq. 2-19,

$$C_u = \frac{D_{60}}{D_{10}} = \frac{0.55}{0.3} = 1.8$$

$$C_c = \frac{(D_{30})^2}{(D_{10})(D_{60})} = \frac{(0.43)^2}{(0.3)(0.55)} = 1.12$$

This soil is still poorly graded even though the C_c is slightly greater than unity; the C_u is very small.

2.6 PARTICLE SHAPE

The shape of the individual particles is at least as important as the grain size distribution in affecting the engineering response of granular soils. It is possible to quantify shape according to rules developed by sedimentary petrologists, but for geotechnical engineering purposes such refinements are rarely warranted. Only a qualitative shape determination is usually made as part of the visual classification of soils. Coarse-grained soils are commonly classified according to the shapes shown in Fig. 2.5.

Fig. 2.5   Typical shapes of coarse-grained bulky particles. (Photograph by M. Surendra.)

A distinction can also be made between particles that are *bulky* and those which are needlelike or flaky. Mica flakes are an excellent example of the latter, and Ottawa sand is an example of the former. Cylinders of each differ drastically in behavior when compressed by a piston. The bulky grains hardly compress at all, even when in a very loose state, but the mica flakes will compress, even under small pressures, up to about one-half of their original volume. When we discuss the shear strength of sands, you will learn that grain shape is very significant in determining the frictional characteristics of granular soils.

2.7 ATTERBERG LIMITS AND CONSISTENCY INDICES

We mentioned (Table 2-2) that the presence of water in the voids of a soil can especially affect the engineering behavior of fine-grained soils. Not only is it important to know how much water is present in, for example, a natural soil deposit (the water content), but we need to compare or scale this water content against some standard of engineering behavior. This is what the Atterberg limits do—they are important limits of engineering behavior. If we know where the water content of our sample is relative to the Atterberg limits, then we already know a great deal about the engineering response of our sample. The Atterberg limits, then, are water contents at certain limiting or critical stages in soil behavior. They, along with the natural water content, are the most important items in the description of fine-grained soils. They are used in classification of such soils, and they are useful because they correlate with the engineering properties and engineering behavior of fine-grained soils.

The Atterberg limits were developed in the early 1900's by a Swedish soil scientist, A. Atterberg (1911). He was working in the ceramics industry, and at that time they had several simple tests to describe the plasticity of a clay, which was important both in molding clay into bricks, for example, and to avoid shrinkage and cracking when fired. After many experiments, Atterberg came to the realization that at least two parameters were required to define plasticity of clays—the upper and lower limits of plasticity. In fact, he was able to define several limits of consistency or behavior and he developed simple laboratory tests to define these limits. They are:

1. Upper limit of viscous flow.
2. Liquid limit—lower limit of viscous flow.
3. Sticky limit—clay loses its adhesion to a metal blade.
4. Cohesion limit—grains cease to cohere to each other.
5. Plastic limit—lower limit of the plastic state.
6. Shrinkage limit—lower limit of volume change.

He also defined the *plasticity index*, which is range of water content where the soil is plastic, and he was the first to suggest that it could be used for soil classification. Later on, in the late 1920's K. Terzaghi and A. Casagrande (1932b), working for the U.S. Bureau of Public Roads, standardized the Atterberg limits so that they could be readily used for soils classification purposes. In present geotechnical engineering practice we usually use the liquid limit (LL or w_L), the plastic limit (PL or w_P), and sometimes the shrinkage limit (SL or w_S). The sticky and the cohesion limits are more useful in ceramics and agriculture.

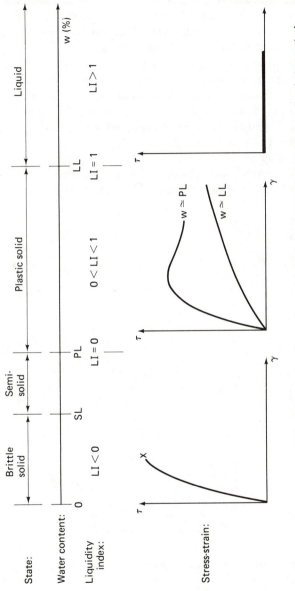

Fig. 2.6   Water content continuum showing the various states of a soil as well as the generalized stress-strain response.

   Since the Atterberg limits are *water contents* where the soil behavior changes, we can show these limits on a water content continuum as in Fig. 2.6. Also shown are the types of soil behavior for the given ranges of water contents. As the water content increases, the state of the soil changes from a brittle solid to a plastic solid and then to a viscous liquid. We can also show on the same water content continuum the generalized material response (stress-strain curves) corresponding to those states.

   You may recall the curves shown in Fig. 2.7 from fluid mechanics, where the shear velocity gradient is plotted versus the shear stress. Depending on the water content, it is possible for soils to have a response represented by all of those curves (except possibly the ideal Newtonian liquid). Note, too, how different this response is from the stress-strain behavior of other engineering materials such as steel, concrete, or wood.

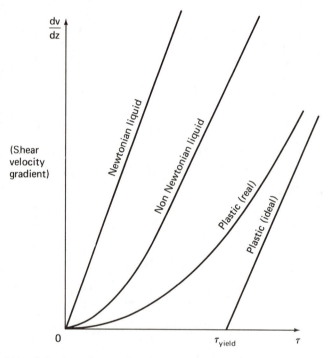

Fig. 2.7   Behavior of several materials including soils over a range of water contents.

   Atterberg's original consistency limit tests were rather arbitrary and not easily reproducible, especially by inexperienced operators. As mentioned, Casagrande (1932b, 1958) worked to standardize the tests, and he developed the liquid limit device (Fig. 2.8) so that the test became more

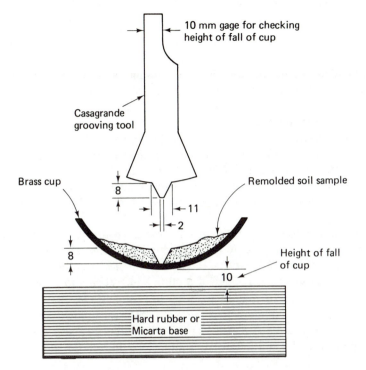

10 mm gage for checking
height of fall of cup

Casagrande
grooving tool

Brass cup

8

←— 11

←— 2

8

Remolded soil sample

Height of fall
of cup

10

Hard rubber or
Micarta base

(a)

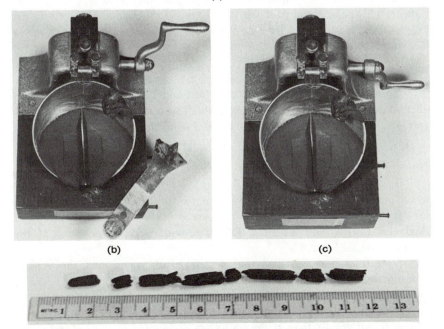

(b)

(c)

(d)

Fig. 2.8 (a) Schematic diagram of the Casagrande liquid limit device and grooving tool; dimensions in millimetres. (b) Cut groove prior to turning the crank. (c) After turning the crank to apply sufficient blows of the cup to close the groove 13 mm. (d) Plastic limit threads. Parts (a) through (c) after Hansbo (1975). (Photographs by M. Surendra.)

operator-independent. He defined the LL as that water content at which a standard groove cut in the remolded soil sample by a grooving tool (Figs. 2.8a, b) will close over a distance of 13 mm ($\frac{1}{2}$ in.) at 25 blows of the LL cup falling 10 mm on a hard rubber or micarta plastic base (Fig. 2.8c). In practice, it is difficult to mix the soil so that the groove closure occurs at exactly 25 blows, but Casagrande found that if you plot the water contents of tests where you get closure at other blow counts versus the logarithm of the number of blows, you get a straight line called the *flow curve*. Where the flow curve crosses 25 blows, that water content is defined as the liquid limit.

The plastic limit test is somewhat more arbitrary, and it requires some practice to get consistent and reproducible results. The PL is defined as the water content at which a thread of soil *just* crumbles when it is carefully rolled out to a diameter of 3 mm ($\frac{1}{8}$ in.). It should break up into segments about 3 to 10 mm ($\frac{1}{8}$ in. to $\frac{3}{8}$ in.) long. If the threads can be rolled to a smaller diameter, then the soil is too wet (above the PL); if it crumbles before you reach 3 mm ($\frac{1}{8}$ in.) in diameter, then you are past the PL. Properly rolled out PL threads should look like those shown in Fig. 2.8d.

Even though the liquid limit and plastic limit tests appear simple, both tests do take some practice to get consistent results. In Sweden, the fall-cone test is used to determine the liquid limit (Hansbo, 1957). It seems to give more consistent results than the Casagrande device, especially for Swedish clays, and it is somewhat simpler and faster to use. Karlsson (1977) presents an excellent discussion of the reliability of both procedures.

Sometimes a *one-point liquid limit* test can be used because, for soils of similar geologic origin, the slopes of the flow curves are similar. Then all you have to do is obtain the water content w_n of the sample with closure of the groove at any blow count n, and use the following relationship

$$\text{LL} = w_n\left(\frac{n}{25}\right)^{\tan \beta} \tag{2-21}$$

where $\tan \beta$ is the slope of the flow curve. For best results the blow count n should be between about 10 and 40. Lambe (1951), U.S. Army Corps of Engineers (1970), and Karlsson (1977) provide good discussions of the one-point liquid limit test.

You may have noticed that we have not mentioned the ASTM procedures for the Atterberg limits tests. We do not recommend the ASTM procedures because, for one thing, they require that the limits be conducted on air-dried specimens. For some soils, such a procedure will give very different results than if the limits are conducted at the natural water content (Karlsson, 1977). The other problem with ASTM is the grooving tool for the liquid limit test. It does not allow for any control of the height of the groove, and therefore it will give inconsistent results. For this reason, we recommend the Casagrande grooving tool (Fig. 2.8) be used.

The range of liquid limits can be from zero to 1000, but most soils have LL's less than 100. The plastic limit can range from zero to 100 or more, with most being less than 40. Even though the Atterberg limits are really water contents, they are also boundaries between different engineering behaviors, and Casagrande (1948) recommends that the values be reported *without* the percent sign. They are *numbers* to be used to classify fine-grained soils, and they *index* soil behavior. You will, however, see the limits reported both ways and using both symbols: LL and PL, and w_L and w_P with a percent sign.

The other Atterberg limit sometimes used in geotechnical engineering practice, the shrinkage limit, is discussed in some detail in Chapter 6.

We mentioned earlier that Atterberg also defined an index called the *plasticity index* to describe the range of water content over which a soil was plastic. The plasticity index, PI or I_p, therefore is numerically equal to the difference between the LL and the PL, or

$$PI = LL - PL \qquad (2\text{-}22)$$

The PI is useful in engineering classification of fine-grained soils, and many engineering properties have been found to empirically correlate with the PI.

When we first started the discussion on the Atterberg limits, we said that we wanted to be able to compare or scale our water content with some defined limits or boundaries or engineering response. In this way, we would know if our sample was likely to behave as a plastic, a brittle solid, or even possibly a liquid. The index for scaling the natural water content of a soil sample is the liquidity index, LI or I_L, is defined as

$$LI = \frac{w_n - PL}{PI} \qquad (2\text{-}23)$$

where w_n is the natural water content of the sample in question. If the LI is less than zero then, from the water content continuum of Fig. 2.6, you would know that the soil will have a brittle fracture if sheared. If the LI is between zero and one, then the soil will behave like a plastic. If LI is greater than one, the soil will be essentially a very viscous liquid when sheared. Such soils can be extremely sensitive to breakdown of the soil structure. As long as they are not disturbed in any way, they can be relatively strong, but if for some reason they are sheared and the structure of the soil breaks down, then they literally can flow like a liquid. There are deposits of *ultra sensitive* (quick) *clays* in Eastern Canada and Scandinavia. Figure 2.9 shows a sample of Leda clay from Ottawa, Ontario, in both the

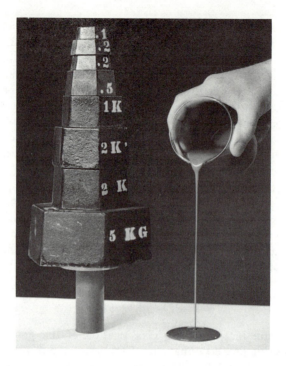

Fig. 2.9   (a) Undisturbed and (b) thoroughly remolded sample of Leda clay from Ottawa, Ontario. (Photograph courtesy of the Division of Building Research, National Research Council of Canada. Hand by D. C. MacMillan.)

undisturbed and remolded states at the same water content. The undisturbed sample can carry a vertical stress of more than 100 kPa; when thoroughly remolded, it behaves like a liquid.

It wasn't emphasized previously, but the limits are conducted on thoroughly *remolded* soils, and when we discuss the structure of clays in Chapter 4, we will see that the natural structure of a soil very strongly governs its engineering behavior. So then how come the Atterberg limits work? They work empirically; that is, they correlate with engineering properties and behavior because *both the Atterberg limits and the engineering properties are affected by the same things.* Some of these "things" include the clay minerals, the ions in the pore water, the stress history of the soil deposit, etc. And these factors are discussed in detail in the chapter on soil structure (Chapter 4). For now, just accept that these very simple, arbitrary, and empirical Atterberg limits are most useful in classifying soils for engineering purposes and that they correlate quite well with the engineering behavior of soils.

2.8 ACTIVITY

In 1953, Skempton defined the *activity A* of a clay as

$$A = \frac{\text{PI}}{\text{clay fraction}} \tag{2-24}$$

where the clay fraction is usually taken as the percentage by weight of the soil less than 2 μm. Clays which have an activity around 1 (0.75 < A < 1.25) are classified as "normal"; A < 0.75 are inactive clays and A > 1.25 are active clays. Activity has been useful for certain classification and engineering property correlations, especially for inactive and active clays. Also, there is fair/good correlation of the activity and the type of clay mineral (Chapter 4). However, the Atterberg limits alone are usually sufficient for these purposes, and the activity provides no really new information.

PROBLEMS

2-1. A water content test was made on a sample of silty clay. The weight of the wet soil plus container was 17.53 g, and the weight of the dry soil plus container was 14.84 g. Weight of the empty container was 7.84 g. Calculate the water content of the sample.

2-2. During a plastic limit test, the following data were obtained for one of the samples:

> Wet weight + container = 22.12 g
> Dry weight + container = 20.42 g
> Weight of container   = 1.50 g

What is the PL of the soil?

2-3. A sample of fully saturated clay weighs 1350 g in its natural state and 975 g after drying. What is the natural water content of the soil?

2-4. For the soil sample of Problem 2-3, compute (a) void ratio and (b) porosity.

2-5. For the soil sample of Problem 2-3, compute (a) the total or wet density and (b) the dry density. Give your answers in Mg/m^3, kg/m^3, and lbf/ft^3.

2-6. A 1 m³ sample of moist soil weighs 2000 kg. The water content is 10%. Assume ρ_s is 2.70 Mg/m³. With this information, fill in all blanks in the phase diagram of Fig. P2-6.

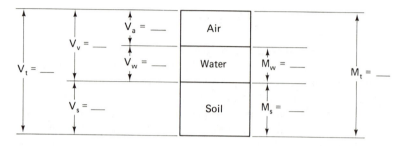

Fig. P2-6

2-7. For the information given in Problem 2-6, calculate (a) the void ratio, (b) the porosity, and (c) the dry density.

2-8. The dry density of a compacted sand is 1.82 Mg/m³ and density of the solids is 2.67 Mg/m³. What is the water content of the material when saturated?

2-9. A 100% saturated soil has a total density of 2050 kg and a water content of 25%. What is the density of the solids? What is the dry density of the soil?

2-10. What is the water content of a fully saturated soil with a dry density of 1.70 Mg/m³? Assume $\rho_s = 2.71$ Mg/m³.

2-11. A dry quartz sand has a density of 1.68 Mg/m³. Determine its density when the degree of saturation is 75%. The density of solids for quartz is 2.65 Mg/m³.

2-12. The dry density of a soil is 1.65 Mg/m³ and the solids have a density of 2.68 Mg/m³. Find the (a) water content, (b) void ratio, and (c) total density when the soil is saturated.

2-13. A natural deposit of soil was found to have a water content of 20% and to be 90% saturated. What is the void ratio of this soil?

2-14. The void ratio of clay soil is 0.5 and the degree of saturation is 70%. Assuming the density of the solids is 2750 kg/m³, compute (a) the water content and (b) dry and wet densities in both SI and British Engineering units.

2-15. The volume of water in a sample of moist soil is 0.056 m³. The volume of solids V_s is 0.28 m³. Given that the density of soil solids ρ_s is 2590 kg/m³, find the water content.

2-16. Verify from first principles that:

(a) $\rho = \rho_s \left(\dfrac{1 + w}{1 + e} \right) = \dfrac{\rho_s + \rho_w Se}{1 + e}$

(b) $\rho = \rho_d(1 + w)$ (c) $w \dfrac{\rho_s}{\rho_w} = Se$

(d) $e = \dfrac{n}{1 - n}$ (e) $n = \dfrac{e}{1 + e}$

2-17. Derive an expression for ρ_s in terms of the porosity n and the water content w for (a) a fully saturated soil and (b) a partially saturated soil.

2-18. Derive an expression for (a) dry density, (b) void ratio, and (c) degree of saturation in terms of ρ, ρ_s, ρ_w, and w.

2-19. Develop a formula for (a) the wet density and (b) the buoyant density in terms of the water content, the density of the soil solids, and the density of water.

2-20. From Archimedes' principle show that Eq. 2-11, $\rho' = \rho_{sat} - \rho_w$, is the same as $(\rho_s - \rho_w)/(1 + e)$.

2-21. The "chunk density" method is often used to determine the unit weight (and other necessary information) of a specimen of irregular shape, especially of friable samples. The specimen at its natural water content is (1) weighed, (2) painted with a thin coat of wax or paraffin (to prevent water from entering the pores), (3) weighed again $(W_t + W_{wax})$, and (4) weighed in water (to get the volume of the sample + wax coating—remember Archimides?). Finally, the natural water content of the specimen is determined. A specimen of silty sand is treated in this way to obtain the "chunk density." From the information given below, determine the (a) wet density, (b) dry density, (c) void ratio, and (d) degree of saturation of the sample. Given:

Weight of specimen at natural water content	= 181.8 g
Weight of specimen + wax coating	= 215.9
Weight of specimen + wax in water	= 58.9
Natural water content	= 2.5%
Soil solid density, ρ_s	= 2700 kg/m³
Wax solid density, ρ_{wax}	= 940 kg/m³

Hint: Use a phase diagram.

2-22. The total volume of a soil specimen is 80 000 mm³ and it weighs 145 g. The dry weight of the specimen is 128 g, and the density of the soil solids is 2.68. Find the (a) water content, (b) void ratio, (c) porosity, (d) degree of saturation, (e) wet density, and (f) dry density. Give the answers to parts (e) and (f) in both SI and British Engineering units.

2-23. The values of minimum e and maximum e for a pure silica sand were found to be 0.46 and 0.66, respectively. What is the corresponding range in the saturated density in kg/m^3?

2-24. A 588 cm^3 volume of moist sand weighs 1010 g. Its dry weight is 918 g and the density of solids is 2670 kg/m^3. Compute the void ratio, the porosity, water content, degree of saturation, and the total density in kg/m^3.

2-25. A sample of saturated glacial clay has a water content of 47%. On the assumption that $\rho_s = 2.70$ Mg/m^3, compute the void ratio, porosity, and saturated density.

2-26. A sensitive volcanic clay soil was tested in the laboratory and found to have the following properties:

(a) $\rho = 1.28$ Mg/m^3 (b) $e = 9.0$ (c) $S = 95\%$
(d) $\rho_s = 2.75$ Mg/m^3 (e) $w = 311\%$

In rechecking the above values one was found to be inconsistent with the rest. Find the inconsistent value and report it correctly.

2-27. The saturated density ρ_{sat} of a soil is 135 lbf/ft^3. Find the buoyant density of this soil in both lbf/ft^3 and kg/m^3.

2-28. A sand is composed of solid constituents having a density of 2.68 Mg/m^3. The void ratio is 0.58. Compute the density of the sand when dry and when saturated and compare it with the density when submerged.

2-29. A sample of natural glacial till was taken from below the ground water table. The water content was found to be 55%. Estimate the wet density, dry density, buoyant density, porosity, and void ratio. Clearly state any necessary assumptions.

2-30. Calculate the maximum possible porosity and void ratio for a collection of (a) ping pong balls (assume they are 30 mm in diameter) and (b) tiny ball bearings 0.3 mm in diameter.

2-31. A cylinder contains 500 cm^3 of loose dry sand which weighs 750 g, and under a static load of 200 kPa the volume is reduced 1%, and then by vibration it is reduced 10% of the original volume. Assume the solid density of the sand grains is 2.65 Mg/m^3. Compute the void ratio, porosity, dry density, and total density corresponding to each of the following cases:

(a) Loose sand.   (b) Under static load.
(c) Vibrated and loaded sand.

2-32. The natural water content of a sample taken from a soil deposit was found to be 11.5%. It has been calculated that the maximum density for the soil will be obtained when the water content reaches 21.5%. Compute how many grams of water must be added to each 1000 g of soil (in its natural state) in order to increase the water content to 21.5%.

2-33. On five-cycle semilogarithmic paper, plot the grain size distribution curves from the following mechanical analysis data on the six soils, A through F. Determine the effective size as well as the uniformity coefficient and the coefficient of curvature for each soil. Determine also the percentages of gravel, sand, silt, and clay according to (a) ASTM, (b) AASHTO, (c) USCS, and (d) the British Standard.

U.S Standard Sieve No.	Percent Passing by Weight					
or Particle Size	Soil A	Soil B	Soil C	Soil D	Soil E	Soil F
75 mm (3 in.)	100		100			
38 $(1\frac{1}{2})$	70		—			
19 $(\frac{3}{4})$	49	100	91			
9.5 $(\frac{3}{8})$	36	—	87			
No. 4	27	88	81		100	
No. 10	20	82	70	100	89	
No. 20	—	80	—	99	—	
No. 40	8	78	49	91	63	
No. 60	—	74	—	37	—	
No. 100	5	—	—	9	—	
No. 140	—	65	35	4	60	
No. 200	4	55	32	—	57	100
40 μm	3	31	27		41	99
20 μm	2	19	22		35	92
10 μm	1	13	18		20	82
5 μm	< 1	10	14		8	71
2 μm	—	—	11		—	52
1 μm	—	2	10		—	39

Note: Missing data is indicated by a dash in the column.

2-34. (a) Explain briefly why it is preferable, in plotting GSD curves, to plot the grain diameter on a logarithmic rather than an arithmetic scale.

(b) Are the shapes of GSD curves comparable (for example, do they have the same C_u and C_c) when plotted arithmetically? Explain.

2-35. The soils in Problem 2-33 have the following Atterberg limits and natural water contents. Determine the PI and LI for each soil and comment on their general activity.

Property	Soil A	Soil B	Soil C	Soil D	Soil E	Soil F
w_n,%	27	14	14	11	8	72
LL	13	35	35	—	28	60
PL	8	29	18	NP	NP	28

2-36. Comment on the validity of the results of Atterberg limits on soils G and H.

	Soil G	Soil H
LL	55	38
PL	20	42
SL	25	—

2-37. The following data were obtained from a liquid limit test on a silty clay.

No. of Blows	Water Content, %
35	41.1
29	41.8
21	43.5
15	44.9

Two plastic limit determinations had water contents of 23.1 and 23.6%. Determine the LL, PI, the flow index, and the toughness index. The flow index is the slope of the water content versus log of number of blows in the liquid limit test, and the toughness index is the PI divided by the flow index.

three

Soil Classification

3.1 INTRODUCTION

From the discussion in Chapter 2 on soil texture and grain size distributions, you should have at least a general idea about how soils are classified. For example, in Sec. 2.4 we described sands and gravels as coarse-grained soils, whereas silts and clays were fine grained. In Sec. 2.5, we showed the specific size ranges on a grain size scale (Fig. 2.3) for these soils according to the standards of ASTM, AASHTO, etc. Usually, however, general terms such as sand or clay include such a wide range of engineering characteristics that additional subdivisions or modifiers are required to make the terms more useful in engineering practice. These terms are collected into *soil classification systems*, usually with some specific engineering purpose in mind.

A soil classification system represents, in effect, a language of communication between engineers. It provides a systematic method of categorizing soils according to their probable engineering behavior, and allows engineers access to the accumulated experience of other engineers. A classification system does not eliminate the need for detailed soils investigations or for testing for engineering properties. However, the engineering properties have been found to correlate quite well with the index and classification properties of a given soil deposit. Thus, by knowing the soil classification, the engineer already has a fairly good general idea of the way the soil will behave in the engineering situation, during construction, under structural loads, etc. Figure 3.1 illustrates the role of the classification system in geotechnical engineering practice.

Many soil classification systems have been proposed during the past 50 years or so. As Casagrande (1948) pointed out, most systems used in

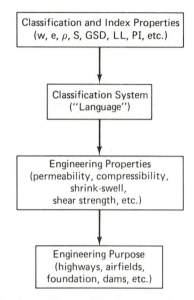

Fig. 3.1   Role of classification system in geotechnical engineering practice.

civil engineering had their roots in agricultural soil science. This is why the first systems used by civil engineers classified soil by grain size or soil texture. Atterberg (1905) apparently was the first to suggest that something other than grain size could be used for soil classification. To this end, in 1911 he developed his consistency limits for the behavior of fine-grained soils (Sec. 2.7), although at that time for agricultural purposes. Later the U.S. Bureau of Public Roads based the classification of fine-grained soils almost entirely on the Atterberg limits and other simple tests. Casagrande (1948) describes several other systems that have been used in highway engineering, airfield construction, agriculture, geology, and soil science.

Today, only the Unified Soil Classification System (USCS) and the American Association of State Highway and Transportation Officials (AASHTO) system are commonly used in civil engineering practice. The Unified Soil Classification System is used mostly by engineering agencies of the U.S. Government (U.S. Army Corps of Engineers and U.S. Department of the Interior, Water and Power Resources Service*) and many geotechnical engineering consulting firms and soil testing laboratories. With slight modification this system is also in fairly common use in Great Britain and elsewhere outside the United States. Nearly all of the state Departments of Transportation and Highways in the United States use the AASHTO system, which is based upon the observed behavior of soils

*Formerly the U.S. Bureau of Reclamation.

under highway pavements. The Federal Aviation Administration (FAA) of the U.S. Department of Transportation had its own soil classification system for the design of airport pavements, but it now uses the Unified Soil Classification System.

Once you become familiar with the details, both the USCS and AASHTO systems are easy to use in engineering practice.

3.2 THE UNIFIED SOIL CLASSIFICATION SYSTEM (USCS)

This system was originally developed by Professor A. Casagrande (1948) for use in airfield construction during World War II. It was modified in 1952 by Professor Casagrande, the U.S. Bureau of Reclamation, and the U.S. Army Corps of Engineers to make the system also applicable to dams, foundations, and other construction (U.S. Army Engineer Waterways Experiment Station, 1960). The basis for the USCS is that coarse-grained soils can be classified according to their grain size distributions, whereas the engineering behavior of fine-grained soils is primarily related to their plasticity. In other words, soils in which "fines" (silts and clays) do not affect the engineering performance are classified according to their grain size characteristics, and soils in which fines do control the engineering behavior are classified according to their plasticity characteristics. Therefore, only a sieve analysis and the Atterberg limits are necessary to completely classify a soil in this system.

The four major divisions in the USCS are indicated in Table 3-1. They are (1) *coarse-grained*, (2) *fine-grained*, (3) *organic soils*, and (4) *peat*. Classification is performed on the material passing the 75 mm sieve, and the amount of "oversize" material is noted on the drill logs or data sheets. Particles greater than 300 mm equivalent diameter are termed *boulders*, while materials between 75 mm and 300 mm are called *cobbles*. Coarse-grained soils, sands, and gravels are those having 50% or more material retained on the No. 200 sieve. These fractions have been arbitrarily but conveniently subdivided as shown in Table 3-1. Fine-grained soils are those having more than 50% passing the No. 200 sieve. The highly organic soils and peat can generally be identified visually.

The symbols in Table 3-1 are combined to form soil group symbols which correspond to the names of typical soils as shown in Table 3-2.

The coarse-grained soils are subdivided into gravels and gravelly soils (G) and sands and sandy soils (S). The gravels are those having the greater percentage of the coarse fraction (particles larger than 4.75 mm diameter) retained on the No. 4 sieve, and the sands are those having the greater portion passing the No. 4 sieve. Both the gravel (G) and the sand (S)

TABLE 3-1 USCS Definitions of Particle Size, Size Ranges, and Symbols

	Soil Fraction or Component	Symbol	Size Range
	Boulders	None	Greater than 300 mm
	Cobbles	None	75 mm to 300 mm
(1)	Coarse-grained soils:		
	Gravel	G	75 mm to No. 4 sieve (4.75 mm)
	Coarse		75 mm to 19 mm
	Fine		19 mm to No. 4 sieve (4.75 mm)
	Sand	S	No. 4 (4.75 mm) to No. 200 (0.075 mm)
	Coarse		No. 4 (4.75 mm) to No. 10 (2.0 mm)
	Medium		No. 10 (2.0 mm) to No. 40 (0.425 mm)
	Fine		No. 40 (0.425 mm) to No. 200 (0.075 mm)
(2)	Fine-grained soils:		
	Fines		Less than No. 200 sieve (0.075 mm)
	Silt	M	(No specific grain size— use Atterberg limits)
	Clay	C	(No specific grain size— use Atterberg limits)
(3)	Organic Soils:	O	(No specific grain size)
(4)	Peat:	Pt	(No specific grain size)
	Gradation Symbols Well-graded, W Poorly-graded, P		*Liquid Limit Symbols* High LL, H Low LL, L

groups are divided into four secondary groups, GW and SW, GP and SP, GM and SM, GC and SC, depending on the grain size distribution and nature of fines in the soils. Well-graded (W) soils have a good representation of all particle sizes whereas the poorly graded (P) soils are either uniform or skip- or gap-graded (Fig. 2.4). Whether a gravel or sandy soil is well graded can be determined by plotting the grain size distribution curve and computing the coefficient of uniformity C_u and the coefficient of curvature C_c. These coefficients are defined in Chapter 2 as

$$C_u = \frac{D_{60}}{D_{10}} \qquad (2\text{-}19)$$

and the coefficient of curvature is

$$C_c = \frac{D_{30}^2}{D_{10} \times D_{60}} \qquad (2\text{-}20)$$

where D_{60} = grain diameter at 60% passing,
  D_{30} = grain diameter at 30% passing, and
  D_{10} = grain diameter at 10% passing by weight (or mass).

Gradation criteria for gravelly and sandy soils are shown in Table 3-2 (column 6). The GW and SW groups are well-graded gravelly and sandy soils with less than 5% passing the No. 200 sieve. The GP and SP groups are poorly graded gravels and sands with little or no nonplastic fines.

The fine-grained soils, those having *more* than 50% passing the No. 200 sieve, are subdivided into silts [M for the Swedish terms *mo* (= very fine sand) and *mjäla* (= silt)] and clays (C) based on their liquid limit and plasticity index. Organic soils (O) and peat (Pt) are also included in this fraction although, as shown in Table 3-1, no grain size range is specified. Fine-grained soils are silts (M) if their liquid limits and plasticity indices plot *below* the A-line on Casagrande's (1948) plasticity chart (Fig. 3.2). The fines are clays (C) if the LL and PI plot *above* the A-line. The A-line generally separates the more claylike materials from those that are silty and also the organics from the inorganics. The exception is organic clays (OL and OH) which plot below the A-line. However, these soils do behave similarly to soils of lower plasticity. The silt, clay, and organic fractions are further subdivided on the basis of relatively low (L) or high (H) liquid limits. The dividing line between the low and high liquid limits has been arbitrarily set at 50. Representative soil types for fine-grained soils are also shown in Fig. 3.2. This figure, columns 4 and 5 of Table 3-2, and Table 3-3, will be helpful in the visual identification and classification of fine-grained soils. You can see from Fig. 3.2 that several different soil types tend to plot in approximately the same area on the LL-PI chart, which means that these soils tend to have about the *same engineering behavior*. This is why the Casagrande chart is so useful in the engineering classification of soils. For example, Casagrande (1948) observed the behavior of soils at the same liquid limit with plasticity index as compared with their behavior at the same plasticity index but with an increasing liquid limit, and he obtained the following results:

Characteristic	Soils at Equal LL with Increasing PI	Soils at Equal PI with Increasing LL
Dry strength	Increases	Decreases
Toughness near PL	Increases	Decreases
Permeability	Decreases	Increases
Compressibility	About the same	Increases
Rate of volume change	Decreases	—

Toughness near the PL and dry strength are very useful visual classification properties, and they are defined in Table 3-3. The other characteristics are engineering properties, and they are discussed in great

TABLE 3-2   Unified Soil Classification System*

Major Divisions			Group Symbols (†)	Typical Names	Field Identification Procedures (excluding particles larger than 75 mm and basing fractions on estimated weights)		
1	2		3	4	5		
Coarse-grained Soils — More than half of material is larger than No. 200 (†) (75 μm) sieve size.	Gravels — More than half of gravel fraction is larger than No. 4 sieve size. (4.75 mm)	Clean Gravels (little or no fines)	GW	Well-graded gravels, gravel sand mixtures, little or no fines.	Wide range in grain sizes and substantial amounts of all intermediate particle sizes.		
			GP	Poorly graded gravels, gravel-sand mixture, little or no fines.	Predominantly one size or a range of sizes with some intermediate sizes missing.		
		Gravels with Fines (appreciable amount of fines)	GM	Silty gravels, gravel-sand-silt mixtures.	Nonplastic fines or fines with low plasticity (for identification procedures see ML belo		
			GC	Clayey gravels, gravel-sand-clay mixtures.	Plastic fines (for identification procedures see CL below).		
	Sands — More than half of coarse fraction is smaller than No. 4 sieve size. (4.75 mm)	Clean Sands (little or no fines)	SW	Well-graded sands, gravelly sands, little or no fines.	Wide range in grain sizes and substantial amounts of all intermediate particle sizes.		
			SP	Poorly graded sands, gravelly sands, little or no fines.	Predominantly one size or a range of sizes with some intermediate sizes missing.		
		Sands with Fines (appreciable amount of fines)	SM	Silty sands, sand-silt mixtures.	Nonplastic fines or fines with low plasticity (for identification procedures see ML below		
			SC	Clayey sands, sand-clay mixtures.	Plastic fines (for identification procedures see CL below).		
Fine-grained Soils — More than half of material is smaller than No. 200 (75 μm) sieve size.	Silts and Clays — Liquid limit less than 50				Identification Procedures on Fraction Smaller than No. 40 Sieve Size		
					Dry Strength (crushing characteristics)	Dilatancy (reaction to shaking)	Toughness (consistency near PL)
			ML	Inorganic silts and very fine sands, rock flour, silty or clayey fine sands or clayey silts with slight plasticity.	None to slight	Quick to slow	None
			CL	Inorganic clays of low to medium plasticity, gravelly clays, sandy clays, silty clays, lean clays.	Medium to high	None to very slow	Medium
			OL	Organic silts and organic silty clays of low plasticity.	Slight to medium	Slow	Slight
	Silts and Clays — Liquid limit greater than 50		MH	Inorganic silts, micaceous or diatomaceous fine sandy or silty soils, elastic silts.	Slight to medium	Slow to none	Slight to medium
			CH	Inorganic clays of high plasticity, fat clays.	High to very high	None	High
			OH	Organic clays of medium to high plasticity, organic silts.	Medium to high	None to very slow	Slight to medium
Highly Organic Soils			Pt	Peat and other highly organic soils.	Readily identified by color, odor, spongy feel and frequently by fibrous texture.		

Note: "The No. 200 sieve size is about the smallest particle visible to the naked eye." appears in the leftmost column of the coarse-grained section. "For visual classification, 5 mm may be used as equivalent to the No. 4 sieve size" appears within the Gravels/Sands division column.

† Boundary classifications: soil, possessing characteristics of two groups are designated by combinations of group symbols. For example: GW-GC, well-graded gra

  sands mixture with clay binder.

‡ All sieve sizes on this chart are U.S. Standard.

*After U.S. Army Engineer Waterways Experiment Station (1960) and Howard (1977).

TABLE 3-2 Continued

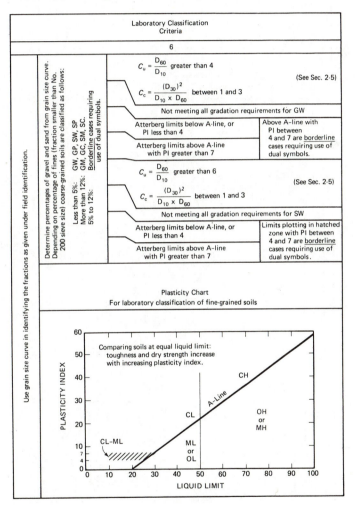

Laboratory Classification Criteria

6

Use grain size curve in identifying the fractions as given under field identification.

Determine percentages of gravel and sand from grain size curve. Depending on percentage of fines (fraction smaller than No. 200 sieve size) coarse-grained soils are classified as follows:

Less than 5%: GW, GP, SW, SP
More than 12%: GM, GC, SM, SC.
5% to 12%: Borderline cases requiring use of dual symbols.

$C_u = \dfrac{D_{60}}{D_{10}}$ greater than 4

$C_c = \dfrac{(D_{30})^2}{D_{10} \times D_{60}}$ between 1 and 3

(See Sec. 2-5)

Not meeting all gradation requirements for GW

Atterberg limits below A-line, or PI less than 4

Atterberg limits above A-line with PI greater than 7

Above A-line with PI between 4 and 7 are borderline cases requiring use of dual symbols.

$C_u = \dfrac{D_{60}}{D_{10}}$ greater than 6

$C_c = \dfrac{(D_{30})^2}{D_{10} \times D_{60}}$ between 1 and 3

(See Sec. 2-5)

Not meeting all gradation requirements for SW

Atterberg limits below A-line, or PI less than 4

Atterberg limits above A-line with PI greater than 7

Limits plotting in hatched zone with PI between 4 and 7 are borderline cases requiring use of dual symbols.

Plasticity Chart
For laboratory classification of fine-grained soils

Comparing soils at equal liquid limit: toughness and dry strength increase with increasing plasticity index.

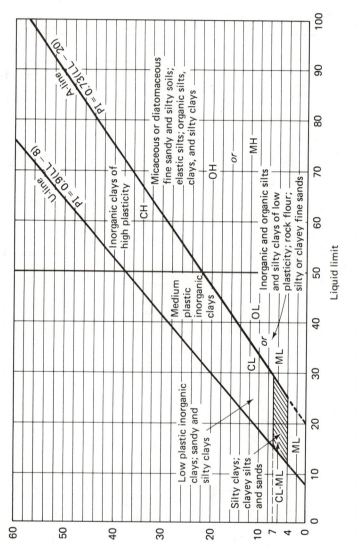

Fig. 3.2 Casagrande's plasticity chart, showing several representative soil types (developed from Casa-grande, 1948, and Howard, 1977).

TABLE 3-3   Field Identification Procedures for Fine-Grained Soils or Fractions*

These procedures are to be performed on the minus No. 40 sieve size particles, approximately 0.4 mm. For field classification purposes, screening is not intended; simply remove by hand the coarse particles that interfere with the tests.

Dilatancy (reaction to shaking):	Dry Strength (crushing characteristics):
After removing particles larger than No. 40 sieve size, prepare a pat of moist soil with a volume of about 5 cm³. Add enough water if necessary to make the soil soft but not sticky. Place the pat in the open palm of one hand and shake vigorously against the other hand several times. A positive reaction consists of the appearance of water on the surface of the pat, which changes to a livery consistency and becomes glossy. When the sample is squeezed between the fingers, the water and gloss disappear from the surface, the pat stiffens, and finally it cracks or crumbles. The rapidity of appearance of water during shaking and of its disappearance during squeezing assist in identifying the character of the fines in a soil.	After removing particles larger than No. 40 sieve size, mold a pat of soil to the consistency of putty, adding water if necessary. Allow the pat to dry completely by oven, sun, or air, and then test its strength by breaking and crumbling between the fingers. This strength is a measure of the character and quantity of the colloidal fraction contained in the soil. The dry strength increases with increasing plasticity.
Very fine clean sands give the quickest and most distinct reaction, whereas a plastic clay has no reaction. Inorganic silts such as a typical rock flour show a moderately quick reaction.	High dry strength is characteristic for clays of the CH group. A typical inorganic silt possesses only very slight dry strength. Silty fine sands and silts have about the same slight dry strength but can be distinguished by the feel when powdering the dried specimen. Fine sand feels gritty, whereas a typical silt has the smooth feel of flour.

Toughness
(consistency near plastic limit):

After removing particles larger than the No. 40 sieve size, a specimen of soil about $\frac{1}{2}$ in. cube in size is molded to the consistency of putty. If too dry, water must be added and, if sticky, the specimen should be spread out in a thin layer and allowed to lose some moisture by evaporation. Then the specimen is rolled out by hand on a smooth surface or between the palms into a thread about 3 mm in diameter. The thread is then folded and refolded repeatedly. During this manipulation the moisture content is gradually reduced and the specimen stiffens, finally loses its plasticity, and crumbles when the plastic limit is reached.

After the thread crumbles, the pieces should be lumped together and a slight kneading action continued until the lump crumbles.

The tougher the thread near the plastic limit and the stiffer the lump when it finally crumbles, the more potent the colloidal clay fraction in the soil. Weakness of the thread at the plastic limit and quick loss of coherence of the lump below the plastic limit indicate either inorganic clay of low plasticity or materials such as kaolin-type clays and organic clays which occur below the A-line.

Highly organic clays have a very weak and spongy feel at the plastic limit.

*After U.S. Army Engineer Waterways Experiment Station (1960) and Howard (1977).

detail later in this book. For now, just rely on your general knowledge and ingenuity to figure out what those words mean.

The upper limit line (U-line) shown in Fig. 3.2 indicates the upper range of plasticity index and liquid limit coordinates found thus far for soils (A. Casagrande, personal communication). Where the limits of any soil plot to the left of the U-line, they should be rechecked. Some highly active clays such as bentonite may plot high above the A-line and close to the U-line. It is shown in Chapter 4 that Casagrande's plasticity chart can even be used to identify qualitatively the predominant clay minerals in a soil.

Coarse-grained soils with *more than 12% fines* are classified as GM and SM if the fines are silty (limits plot below the A-line on the plasticity chart) and GC and SC if the fines are clayey (limits plot above the A-line). Both well-graded and poorly graded materials are included in these two groups.

Soils having *between 5% and 12%* passing the No. 200 sieve are classed as "borderline" and have a dual symbol. The first part of the dual symbol indicates whether the coarse fraction is well graded or poorly graded. The second part describes the nature of the fines. For example, a soil classified as a SP-SM means that it is a poorly graded sand with between 5% and 12% silty fines. Similarly a GW-GC is a well-graded gravel with some clayey fines that plot above the A-line.

Fine-grained soils can also have dual symbols. Obviously, if the limits plot within the shaded zone on Fig. 3.2 (PI between 4 and 7 and LL between about 12 and 25), then the soil classifies as a CL-ML. Howard (1977) makes the practical suggestion that if the LL and PI values fall near the A-line or near the LL = 50 line, then dual symbols should be used. Possible dual symbols then are:

ML-MH
CL-CH
OL-OH

CL-ML
CL-OL

CH-MH
CH-OH

Borderline symbols can also be used for soils with about 50% fines and coarse-grained fractions. In this case possible dual symbols are:

GM-ML
GM-MH

GC-CL
GC-CH

SM-ML
SM-MH

SC-CL
SC-CH

Figure 3.3 is a practical guide for borderline cases of soil classification.

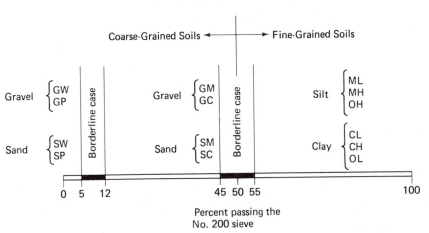

UNIFIED SOIL CLASSIFICATION SYSTEM
(Borderline Classifications)

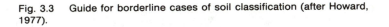

Percent passing the
No. 200 sieve

Note: Only two group symbols may be used to describe a soil.
Borderline classifications can exist within each of the above groups.

Fig. 3.3   Guide for borderline cases of soil classification (after Howard, 1977).

A step-by-step procedure for USCS classification of soils, conveniently presented in Fig. 3.4, shows a process of elimination of all the possibilities until the only one left indicates the specific classification. The following steps, adapted from the Corps of Engineers, may help in this process (U.S. Army Engineer Waterways Experiment Station, 1960). Classification should be done in conjunction with Table 3-2 and Fig. 3.4:

1. Determine if the soil is coarse grained, fine grained, or highly organic. This is done by visual inspection and/or by determining the amount of soil passing the No. 200 sieve.
2. If *coarse grained*:
   a. Perform a sieve analysis and plot the grain size distribution curve. Determine the percentage passing the No. 4 sieve and classify the soil as *gravel* (greater percentage retained on No. 4) or *sand* (greater percentage passing No. 4).
   b. Determine the amount of material passing the No. 200 sieve. If less than 5% passes the No. 200 sieve, examine the shape of the grain size curve; if well graded, classify as GW or SW; if poorly graded, classify as GP or SP.
   c. If between 5% and 12% of the material passes the No. 200 sieve, it is a borderline case, and the classification should have dual

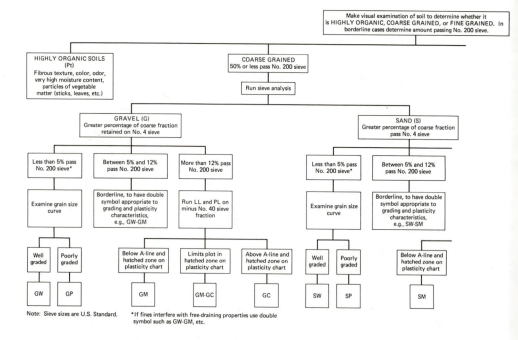

Fig. 3.4   Auxiliary laboratory identification procedure (after USAEWES,

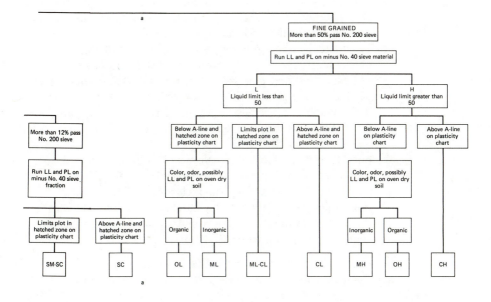

Fig. 3.4   Continued.

symbols appropriate to grading and plasticity characteristics (GW-GM, SW-SM, etc.).

   d. If more than 12% passes the No. 200 sieve, perform the Atterberg limits on the minus No. 40 sieve fraction. Use the plasticity chart to determine the correct classification (GM, SM, GC, SC, GM-GC, or SM-SC).

3. If *fine grained*:

   a. Perform Atterberg limits tests on minus No. 40 sieve material. If the liquid limit is less than 50, classify as L, and if the liquid limit is greater than 50, classify as H.

   b. For L: If the limits plot below the A-line and the hatched zone on the plasticity chart, determine by color, odor, or the change in liquid limit and plastic limit caused by oven-drying the soil, whether it is organic (OL) or inorganic (ML). If the limits plot in the hatched zone, classify as CL-ML. If the limits plot above the A-line and the hatched zone on the plasticity chart (Fig. 3.2), classify as CL.

   c. For H: If the limits plot below the A-line on the plasticity chart, determine whether organic (OH) or inorganic (MH). If the limits plot above the A-line, classify as CH.

   d. For limits which plot in the hatched zone on the plasticity chart, close to the A-line or around LL = 50, use dual (borderline) symbols as shown in Fig. 3.3.

Although the letter symbols in the USCS are convenient, they do not completely describe a soil or soil deposit. For this reason, descriptive terms should also be used along with the letter symbols for a complete soil classification. Table 3-4 from U.S. Army Engineer Waterways Experiment Station (1960) provides some useful information for describing soils.

In the case of all soils, such characteristics as color, odor, and homogeneity of the deposit should be observed and included in the sample description.

For coarse-grained soils such items as grain shape, mineralogical content, degree of weathering, in situ density and degree of compaction, and presence or absence of fines should be noted and included. Adjectives such as rounded, angular, and subangular are commonly used to describe grain shape (see Fig. 2.5). The in situ density and degree of compaction is normally obtained indirectly by observing how difficult the material is to excavate or to penetrate with devices called *penetrometers*. Terms such as *very loose*, *loose*, *medium*, *dense*, and *very dense* are used to describe in situ density. A granular deposit which can, for example, be excavated readily

TABLE 3-4 Information Required for Describing Soils*

Coarse-grained soils:	Fine-grained soils:

Coarse-grained soils:

For undisturbed soils add information on stratification, degree of compactness, cementation, moisture conditions and drainage characteristics.

Give typical name. Indicate approximate percentages of sand and gravel, maximum size, angularity, surface condition, and hardness of the coarse grains, local or geologic name and other pertinent descriptive information, and symbol in parentheses.

Example:

Silty sand, gravelly. About 20% hard, angular gravel particles 12 mm maximum size, rounded and subangular sand grains coarse to fine, about 15% nonplastic fines with low dry strength, well compacted and moist in place, alluvial sand, (SM).

Fine-grained soils:

Give typical name. Indicate degree and character of plasticity, amount and maximum size of coarse grains, color in wet condition, odor if any, local or geologic name, and other pertinent descriptive information, and symbol in parentheses.

For undisturbed soils add information on structure, stratification, consistency in undisturbed and remolded states, moisture and drainage conditions.

Example:

Clayey silt, brown, slightly plastic, small percentage of fine sand, numerous vertical root holes, firm and dry in place, loess, (ML).

Note: Be prepared for wide variations in soil description among organizations and testing laboratories. They all have their own ways of doing things.

*After U.S. Army Engineer Waterways Experiment Station (1960).

by hand would be considered very loose, whereas a deposit of the same material which requires power tools for excavation would be described as very dense or perhaps cemented.

For the fine-grained fraction, natural water content, consistency, and remolded consistency should be noted in the sample description. *Consistency* in the natural state corresponds in some respects to degree of compaction in coarse-grained soils and is usually evaluated by noting the ease by which the deposit can be excavated or penetrated. Such terms as *very soft*, *soft*, *medium*, *stiff*, *very stiff*, and *hard* are employed to describe consistency. (Sometimes the word *firm* is used synonymously with the term *stiff*.) Fine-grained soils may be additionally described by using the tests explained in Table 3-3 for dilatancy, toughness, and dry strength. Other techniques for visual classification of soils should be learned and practiced in the laboratory. Excellent descriptions of visual classification and identification procedures are found in the U.S.B.R. (1974) *Earth Manual*, Appendix E-3, and ASTM (1980) Designation D 2488.

EXAMPLE 3.1

Given:

Sieve analysis and plasticity data for the following three soils.

Sieve Size	Soil 1, % Finer	Soil 2, % Finer	Soil 3, % Finer
No. 4	99	97	100
No. 10	92	90	100
No. 40	86	40	100
No. 100	78	8	99
No. 200	60	5	97
LL	20	—	124
PL	15	—	47
PI	5	NP*	77

*Nonplastic.

Required:

Classify the three soils according to the Unified Soil Classification System.

Solution:

Use Table 3-2 and Fig. 3.4.

1. Plot the grain size distribution curves for the three soils (shown in Fig. Ex. 3.1).
2. For soil 1, we see from the curve that more than 50% passes the No. 200 sieve (60%); thus the soil is a fine-grained soil and the Atterberg limits are required to further classify the soil. With LL = 20 and PI = 5, the soil plots in the hatched zone on the plasticity chart. Therefore the soil is a CL-ML.
3. Soil 2 is immediately seen to be a coarse-grained soil since only 5% passes the No. 200 sieve. Since 97% passes the No. 4 sieve, the soil is a sand rather than a gravel. Next, note the amount of material passing the No. 200 sieve (5%). From Table 3-2 and Fig. 3.4, the soil is "borderline" and therefore has a dual symbol such as SP-SM or SW-SM depending on the values of C_u and C_c. From the grain size distribution curve, Fig. Ex. 3.1, we find that $D_{60} = 0.71$ mm, $D_{30} = 0.34$ mm, and $D_{10} = 0.18$ mm. The coefficient of

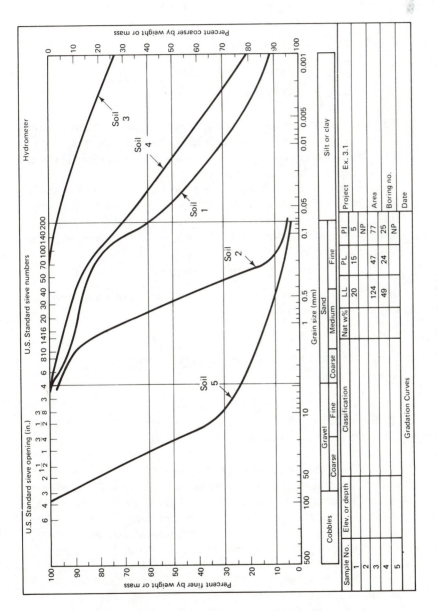

Fig. Ex. 3.1

uniformity C_u is

$$C_u = \frac{D_{60}}{D_{10}} = \frac{0.71}{0.18} = 3.9 < 6$$

and the coefficient of curvature C_c is

$$C_c = \frac{(D_{30})^2}{D_{10} \times D_{60}} = \frac{(0.34)^2}{0.18 \times 0.71} = 0.91 \simeq 1$$

For a soil to be considered well graded, it must meet the criteria shown in column 6 of Table 3-2; it does not, so the soil is considered poorly graded and its classification is SP-SM. The soil is SM because the fines are silty (nonplastic).

4. A quick glance at the characteristics for soil 3 indicates the soil is fine grained (97% passes the No. 200 sieve). Since the LL is greater than 100 we cannot directly use the plasticity chart (Fig. 3.2). Use instead the equation for the A-line on Fig. 3.2 to determine if the soil is a CH or MH.

$$PI = 0.73(LL - 20) = 0.73(124 - 20) = 75.9$$

Since the PI is 78 for soil 3, it lies *above* the A-line and thus the soil is classified as a CH.

3.3 THE AASHTO SOIL CLASSIFICATION SYSTEM

In the late 1920's the U.S. Bureau of Public Roads (now the Federal Highway Administration) conducted extensive research on the use of soils especially in local or secondary road construction, the so-called "farm-to-market" roads. From that research the Public Roads Classification System was developed by Hogentogler and Terzaghi (1929). The original system was based on the stability characteristics of soils when used as a road surface or with a thin asphalt pavement. There were several revisions since 1929, and the latest in 1945 is essentially the present AASHTO (1978) system. The applicability of the system has been extended considerably; AASHTO states that the system should be useful for determining the relative quality of soils for use in embankments, subgrades, subbases, and bases. But you might keep in mind its original purpose when using the system in your engineering practice. (See Casagrande, 1948, for some comments on this point.)

TABLE 3-5 AASHTO Definitions of Gravel, Sand, and Silt-Clay

Soil Fraction	Size Range
Boulders	Above 75 mm
Gravel	75 mm to No. 10 sieve (2.0 mm)
Coarse sand	No. 10 (2.0 mm) to No. 40 (0.425 mm)
Fine sand	No. 40 (0.425 mm) to No. 200 (0.075 mm)
Silt-clay (combined silt and clay)	Material passing the 0.075 mm (No. 200) sieve

Soil fractions recognized in the AASHTO system are listed in Table 3-5. Boulders should be excluded from the sample to be classified, but as with the USCS the amount of boulders present should be noted. Fines are silty if they have a PI less than 10 and clayey if the PI is greater than 10.

The AASHTO system classifies soils into eight groups, A-1 through A-8, and it includes several subgroups. Soils within each group are evaluated according to the *group index*, which is calculated by an empirical formula. The only tests required are the sieve analysis and the Atterberg limits. Table 3-6 illustrates the current AASHTO (1978) soil classification.

Granular materials fall into classes A-1 to A-3. A-1 materials are well graded, whereas A-3 soils are clean, poorly graded sands. A-2 materials are also granular (less than 35% passing the No. 200 sieve), but they contain a significant amount of silts and clays. A-4 to A-7 are fine-grained soils, the silt-clay materials. They are differentiated on the basis of their Atterberg limits. Figure 3.5 can be used to obtain the ranges of LL and PI for groups A-4 to A-7 and for the subgroups in A-2. Highly organic soils including peats and mucks may be placed in group A-8. As with the USCS, classification of A-8 soils is made visually.

The *group index* is used to further evaluate soils within a group. It is based on the service performance of many soils, especially when used as pavement subgrades. It may be determined from the empirical formula given at the top of Fig. 3.6, or you may use the nomograph directly.

Using the AASHTO system to classify soils is not difficult. Once you have the required test data, proceed from left to right in the chart of Table 3-6, and find the correct group by the process of elimination. The first group from the left to fit the test data is the correct AASHTO classification. A complete classification includes the group index to the nearest whole number, in parentheses, after the AASHTO symbol. Examples are A-2-6(3), A-4(5), A-6(12), A-7-5(17), etc.

Figure 3.7 will be helpful in classifying soils according to the AASHTO system.

TABLE 3-6 Classification of Soils and Soil-Aggregate Mixtures*

General Classification	Granular Materials (35% or less passing 0.075 mm)							Silt-Clay Materials (More than 35% passing 0.075 mm)			
	A-1		A-3	A-2				A-4	A-5	A-6	A-7 A-7-5 A-7-6
Group classification	A-1-a	A-1-b		A-2-4	A-2-5	A-2-6	A-2-7				
Sieve analysis, percent passing:											
2.00 mm (No. 10)	50 max.	—	—	—	—	—	—	—	—	—	—
0.425 mm (No. 40)	30 max.	50 max.	51 min.	—	—	—	—	—	—	—	—
0.075 mm (No. 200)	15 max.	25 max.	10 max.	35 max.	35 max.	35 max.	35 max.	36 min.	36 min.	36 min.	36 m
Characteristics of fraction passing 0.425 mm (No. 40):											
Liquid limit	—		—	40 max.	41 min.	40 max.	41 min.	40 max.	41min.	40 max.	41 m
Plasticity index	6 max.		NP	10 max.	10 max.	11 min.	11 min.	10 max.	10 max.	11 min.	11 m
Usual types of significant constituent materials	Stone fragments, gravel, and sand		Fine sand	Silty or clayey gravel and sand				Silty soils		Clayey soils	
General rating as subgrade	Excellent to good							Fair to Poor			

*©American Association of State Highway and Transportation Officials, 1978. Used by permission.
†Plasticity index of A-7-5 subgroup is equal to or less than LL minus 30. Plasticity index of A-7-6 subgroup is greater than LL minus 30 (see Fig. 3.5).

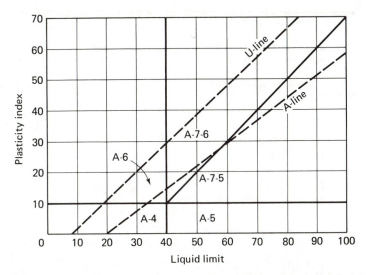

Fig. 3.5   Atterberg limits ranges for subgrade subgroups A-4, A-5, A-6, and A-7. Note that Casagrande's A-line and U-line have been superimposed on the chart (after Liu, 1970, and Al-Hussaini, 1977).

EXAMPLE 3.2

Given:

The following data for soils 4 and 5. See Fig. Ex. 3.1 for the grain size distribution curves.

Sieve Size	Soil 4, % Finer	Soil 5, % Finer
No. 4	99	23
No. 10	96	18
No. 40	89	9
No. 100	79	5
No. 200	70	4
LL	49	—
PL	24	—
PI	25	NP

Required:

Classify the soils according to the AASHTO Soil Classification System.

Group index (GI) = (F − 35)[0.2 + 0.005(LL − 40)] + 0.01(F − 15)(PI − 10), where F = % passing 0.075 mm sieve, LL = liquid limit, and PI = plasticity index.

When working with A-2-6 and A-2-7 subgroups the Partial Group Index (PGI) is determined from the PI only.

When the combined partial group indices are negative, the group index should be reported as zero.

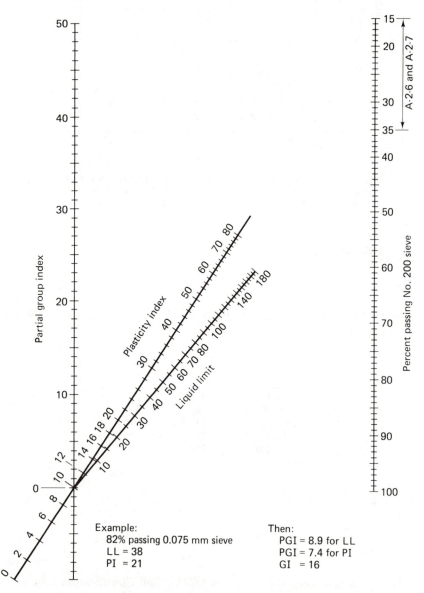

Example:
  82% passing 0.075 mm sieve
  LL = 38
  PI = 21

Then:
  PGI = 8.9 for LL
  PGI = 7.4 for PI
  GI = 16

Fig. 3.6   Group index chart (after AASHTO, 1978). ©American Association of State Highway and Transportation Officials, 1978. Used by permission.

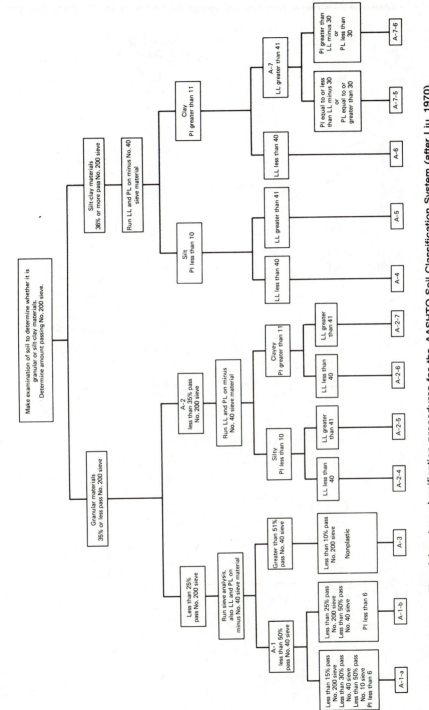

Fig. 3.7 Chart for auxiliary laboratory classification procedures for the AASHTO Soil Classification System (after Liu, 1970).

TABLE 3-7 Comparable Soil Groups in the AASHTO and USCS Systems*

Soil Group in Unified System	Comparable Soil Groups in AASHTO System		
	Most Probable	Possible	Possible but Improbable
GW	A-1-a	—	A-2-4, A-2-5, A-2-6, A-2-7
GP	A-1-a	A-1-b	A-3, A-2-4, A-2-5, A-2-6, A-2-7
GM	A-1-b, A-2-4, A-2-5, A-2-7	A-2-6	A-4, A-5, A-6, A-7-5, A-7-6, A-1-a
GC	A-2-6, A-2-7	A-2-4, A-6	A-4, A-7-6, A-7-5
SW	A-1-b	A-1-a	A-3, A-2-4, A-2-5, A-2-6, A-2-7
SP	A-3, A-1-b	A-1-a	A-2-4, A-2-5, A-2-6, A-2-7
SM	A-1-b, A-2-4, A-2-5, A-2-7	A-2-6, A-4, A-5	A-6, A-7-5, A-7-6, A-1-a
SC	A-2-6, A-2-7	A-2-4, A-6, A-4, A-7-6	A-7-5
ML	A-4, A-5	A-6, A-7-5	—
CL	A-6, A-7-6	A-4	—
OL	A-4, A-5	A-6, A-7-5, A-7-6	—
MH	A-7-5, A-5	—	A-7-6
CH	A-7-6	A-7-5	—
OH	A-7-5, A-5	—	A-7-6
Pt	—	—	—

*After Liu (1970).

TABLE 3-7 Continued

Soil Group in AASHTO System	Comparable Soil Groups in Unified System		
	Most Probable	Possible	Possible but Improbable
A-1-a	GW, GP	SW, SP	GM, SM
A-1-b	SW, SP, GM, SM	GP	—
A-3	SP	—	SW, GP
A-2-4	GM, SM	GC, SC	GW, GP SW, SP
A-2-5	GM, SM	—	GW, GP, SW, SP
A-2-6	GC, SC	GM, SM	GW, GP SW, SP
A-2-7	GM, GC, SM, SC	—	GW, GP, SW, SP
A-4	ML, OL	CL, SM, SC	GM, GC
A-5	OH, MH, ML, OL	—	SM, GM
A-6	CL	ML, OL, SC	GC, GM, SM
A-7-5	OH, MH	ML, OL, CH	GM, SM, GC, SC
A-7-6	CH, CL	ML, OL, SC	OH, MH, GC, GM, SM

Solution:

1. Because more than 35% of soil 4 passes the No. 200 sieve, from Table 3-6 we see that the soil is an A-4 or higher. Since the LL is 49, the soil is either A-5 or A-7. with a PI of 25, the soil is an A-7. A check of Fig. 3.5 shows the soil is classified as an A-7-6.
2. Because soil 5 has less than 35% passing the No. 200 sieve, it is granular. (A glance at Fig. Ex. 3.1 provides the same information!) Proceeding from left to right in Table 3-6, we see that the first group from the left that meets the criterion is A-1-a.

3.4 COMPARISON OF THE USCS AND AASHTO CLASSIFICATION SYSTEMS

There are several significant differences between the USCS and AASHTO soil classification systems, which is not surprising considering the differences in their history and purpose. You can see the differences in the treatment of coarse-grained soils by comparing Table 3-1 with 3-5. The major differences in the fine-grained soils are shown in Fig. 3.5, where we have superimposed both the A-line and the U-line on the LL-PI chart. AASHTO (1978) actually plots LL versus PI, but we have turned the chart 90° for easy comparison with the Casagrande plasticity chart (Fig. 3.2). The differences are significant. Also, use of PI = 10 as the dividing line between silty and clayey soils seems rather arbitrary and probably does not realistically relate to the engineering properties of fine-grained soils. Al-Hussaini (1977) discusses several other significant differences between the two systems.

Table 3-7 shows a comparison of the two systems in terms of the probable corresponding soil groups.

PROBLEMS

3-1. Classify soils 4 and 5 in Example 3.2 according to the Unified Soil Classification System. Explain your steps. Compare with Table 3-5 and Example 3.1.

3-2. Classify soils 1, 2, and 3 in Example 3.1 according to the AASHTO Soil Classification System. Compare with Table 3-5 and Example 3.2.

3-3. Given the grain size distribution curves of Problem 2-12 and the Atterberg limits data of Problem 2-14, classify soils A through F using the USCS and AASHTO soil classification systems.

3-4. For the data below, classify the soils according to the USCS.

(a) 100% material passed No. 4 sieve, 25% retained on No. 200 sieve.
  Fines exhibited:
  Medium to low plasticity.
  Dilatancy—none to very slow.
  Dry strength—medium to high.

(b) 65% material retained on No. 4 sieve, 32% material retained on No. 200 sieve.

$C_u = 3$, $C_c = 1$.

(c) 100% passed No. 4 sieve, 90% passed No. 200 sieve.

Dry strength—low to medium.
Dilatancy—moderately quick.
LL = 23, PL = 17.

(d) 5% retained on No. 4 sieve, 70% passed No. 4 and retained on No. 200 sieve.

Fines exhibited low plasticity and high dilatancy.

(e) 100% material passed No. 4 sieve, 20% retained on No. 200 sieve.

Fines exhibit high plasticity.
Dilatancy—none.
Dry strength—high.

(f) 90% material passed No. 4 sieve, and retained on No. 200 sieve.

$C_u = 3$, $C_c = 1$.
10% passed No. 200 sieve.
Fines exhibited high dilatancy.

(g) 5% material retained on No. 4 sieve, 70% passed No. 4 sieve and retained on No. 200 sieve.

Fines exhibited medium dry strength, medium toughness.
Dilatancy—none.
LL = 25, PL = 15.

(h) 70% material retained on No. 4 sieve, 27% retained on No. 200 sieve.

$C_u = 5$, $C_c = 1.5$.

3-5. For the soils of Problem 3-4, estimate the compressibility, permeability, and toughness.

3-6. Grain size distributions and Atterberg limits are given for 16 soils in the six graphs comprising Fig. P3.6. Classify the soils according to (a) USCS and (b) AASHTO systems. (All data from USAEWES, 1960.)

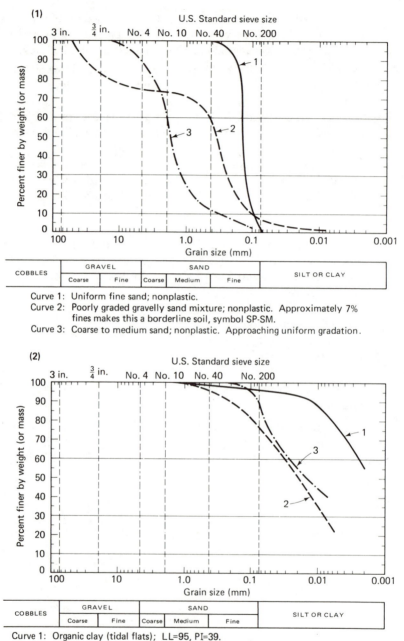

(1)

Curve 1: Uniform fine sand; nonplastic.
Curve 2: Poorly graded gravelly sand mixture; nonplastic. Approximately 7% fines makes this a borderline soil, symbol SP-SM.
Curve 3: Coarse to medium sand; nonplastic. Approaching uniform gradation.

(2)

Curve 1: Organic clay (tidal flats); LL=95, PI=39.
Curve 2: Alkali clay with organic matter; LL=66, PI=27.
Curve 3: Organic silt; LL=70, PI=33 (natural water content); LL=53, PI=19 (oven dried).

Fig. P-3.6

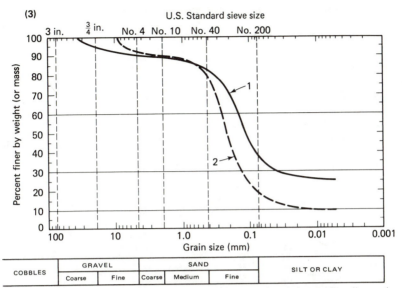

(3)

Curve 1: Clayey sand; LL=23, PI=10. Poorly graded mixture of sand-clay and fine silty sand.
Curve 2: Limerock and sand mixture; LL=23, PI=8. Poorly graded.

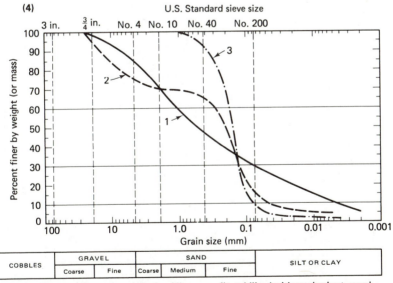

(4)

Curve 1: Silty gravelly sand; nonplastic. Micaceous silt stabilized with sandy chert gravel.
Curve 2: Mixture of gravel-sand and fine silty sand; nonplastic. Poorly graded mixture; note absence of coarse and medium sand.
Curve 3: Silty fine sand; LL=22, PI=5.

Fig. P-3.6   Continued.

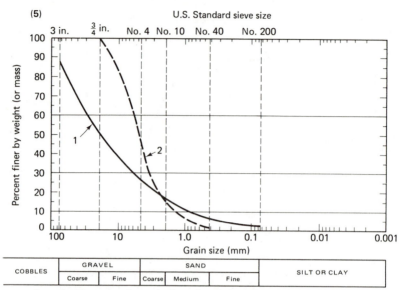

(5)

Curve 1: Pit run gravel; nonplastic; well graded; small percentage of fines.
Curve 2: Sandy gravel; nonplastic; no fines.

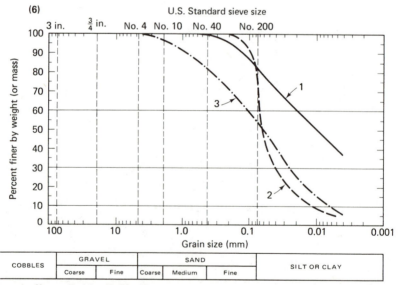

(6)

Curve 1: Clayey silt; LL = 46, PI = 16.
Curve 2: Uniform sandy silt; LL = 30, PI = 3.
Curve 3: Sandy silt; LL = 34, PI = 3
General: Note curves 2 and 3 have about the same plasticity but vary in grain size distribution.

Fig. P-3.6   Continued.

four

Clay Minerals and
Soil Structure

4.1 INTRODUCTION

At this stage it is useful to again define the term *clay*. Clay can refer to specific minerals such as kaolinite or illite, as are discussed in detail in this chapter. However, in civil engineering, *clay* often means a *clay soil*—a soil which contains some clay minerals as well as other mineral constituents, has plasticity, and is "cohesive." Clay soils are fine grained, as indicated in Chapter 2 (Table 2-2), but not all fine-grained soils are cohesive or clays. Silts are both granular and fine grained. The individual silt grains, like clays, are invisible to the naked eye, but silts are noncohesive and nonplastic. Rock flour is another example of a very fine-grained cohesionless soil.

Also remember that certain characteristics of granular soils such as the grain size distribution and the grain shape affect the engineering behavior of these soils. On the other hand, the presence of water, with a few important exceptions, is relatively unimportant in their behavior. In contrast, for clay soils the grain size distribution has relatively little influence on the engineering behavior, but water markedly affects their behavior. Silts are an "in between" material. Water affects their behavior—they are *dilatant*—yet they have little or no plasticity ($PI \simeq 0$), and their strengths, like sands, are essentially independent of water content.

As we indicate in this chapter, clay minerals are very small particles which are very active electrochemically. The presence of even a small amount of clay minerals in a soil mass can markedly affect the engineering properties of that mass. As the amount of clay increases, the behavior of the soil is increasingly governed by the properties of the clay. When the clay content is about 50%, the sand and silt grains are essentially floating in a clay matrix and have little effect on the engineering behavior.

In this chapter, we briefly describe the important clay minerals, how they are identified, and how they interact with water and with each other. We also describe some of the latest thinking about soil fabric and structure, concepts which are fundamentally important for a good understanding of cohesive soil behavior. Finally, cohesionless soil structures and the concept of relative density are discussed.

Only one new symbol is introduced in this chapter.

Symbol	Dimension	Unit	Definition
D_r or I_D	—	(%)	Relative density or density index

4.2 CLAY MINERALS

Clay minerals are very tiny crystalline substances evolved primarily from chemical weathering of certain rock-forming minerals. Chemically, they are *hydrous aluminosilicates* plus other metallic ions. All clay minerals are very small, colloidal-sized crystals (diameter less than 1 μm), and they can only be seen with an electron microscope. The individual crystals look like tiny plates or flakes, and from X-ray diffraction studies scientists have determined that these flakes consist of many crystal sheets which have a repeating atomic structure. In fact, there are only two fundamental crystal sheets, the *tetrahedral* or *silica*, and the *octahedral* or *alumina*, sheets. The particular way in which these sheets are stacked, together with different bonding and different metallic ions in the crystal lattice, constitute the different clay minerals.

The tetrahedral sheet is basically a combination of silica tetrahedral units which consist of four oxygen atoms at the corners, surrounding a single silicon atom. Figure 4.1a shows a single silica tetrahedron; Fig. 4.1b shows how the oxygen atoms at the base of each tetrahedron are combined to form a sheet structure. The oxygens at the bases of each tetrahedron are in one plane, and the unjoined oxygen corners all point in the same direction. A common schematic representation of the tetrahedral sheet which is used later is shown in Fig. 4.1c. A top view of the silica sheet showing how the oxygen atoms at the base of each tetrahedron belong to two tetrahedrons and how adjacent silicon atoms are bonded is shown in Fig. 4.1d. Note the hexagonal "holes" in the sheet.

The octahedral sheet is basically a combination of octahedral units consisting of six oxygen or hydroxyls enclosing an aluminum, magnesium, iron, or other atom. A single octahedron is shown in Fig. 4.2a, while Fig. 4.2b shows how the octahedrons combine to form a sheet structure. The rows of oxygens or hydroxyls in the sheet are in two planes. Figure 4.2c is

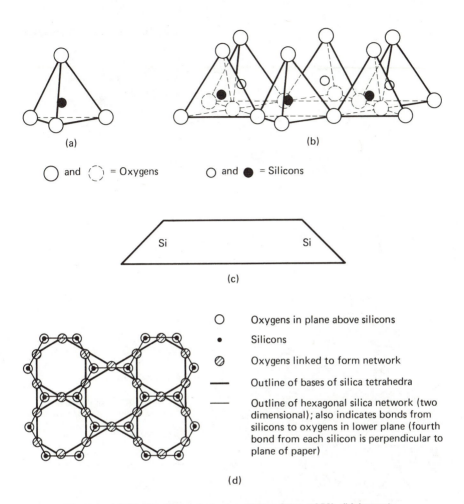

(a)

(b)

○ and ⌐◯⌐ = Oxygens   ○ and ● = Silicons

Si   Si

(c)

○   Oxygens in plane above silicons

•   Silicons

⊘   Oxygens linked to form network

——   Outline of bases of silica tetrahedra

——   Outline of hexagonal silica network (two dimensional); also indicates bonds from silicons to oxygens in lower plane (fourth bond from each silicon is perpendicular to plane of paper)

(d)

Fig. 4.1   (a) Single silica tetrahedron (after Grim, 1959). (b) Isometric view of the tetrahedral or silica sheet (after Grim, 1959). (c) Schematic representation of the silica sheet (after Lambe, 1953). (d) Top view of the silica sheet (after Warshaw and Roy, 1961).

a schematic representation of the octahedral sheet which we use later. For a top view of the octahedral sheet showing how the different atoms are shared and bonded see Fig. 4.2d.

Substitution of different cations in the octahedral sheet is rather common and leads to different clay minerals. Since the ions substituted are approximately the same physical size, such substitution is called *isomorphous*. Sometimes not all the octahedrons contain a cation, which results in a somewhat different crystalline structure with slightly different physical properties and a different clay mineral. If all the anions of the octahedral sheet are hydroxyls and two-thirds of the cation positions are filled with aluminum, then the mineral is called *gibbsite*. If magnesium is substituted

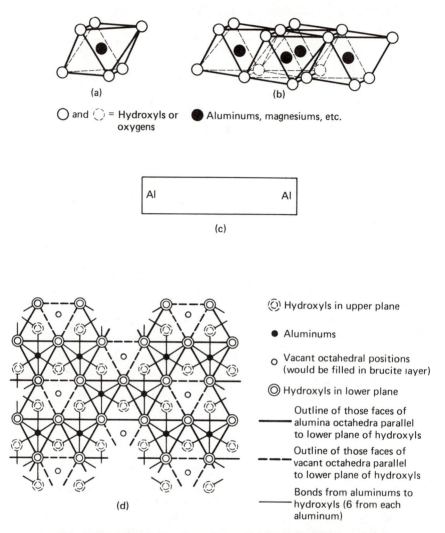

Fig. 4.2 (a) Single aluminum (or magnesium) octahedron (after Grim, 1959). (b) Isometric view of the octahedral sheet (after Grim, 1959). (c) Schematic representation of the octahedral or alumina (or magnesia) sheet (after Lambe, 1953). (d) Top view of the octahedral sheet (after Warshaw and Roy, 1961).

for the aluminum in the sheet and it fills all the cation positions, then the mineral is called *brucite*. The variations in the basic sheet structures make up the dozens of clay minerals which have been identified. All clay minerals consist of the two basic sheets which are stacked together in certain unique ways and with certain cations present in the octahedral and tetrahedral sheets. For engineering purposes it is usually sufficient to describe only a few of the more common clay minerals which are found in clay soils.

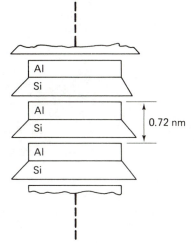

Fig. 4.3 Schematic diagram of the structure of kaolinite (after Lambe, 1953).

Kaolinite consists basically of repeating layers of one tetrahedral (silica) sheet and one octahedral (alumina or gibbsite) sheet. Because of the stacking of one layer of each of the two basic sheets, kaolinite is called a 1:1 clay mineral (Fig. 4.3). The two sheets are held together in such a way that the tips of the silica sheet and one of the layers of the octahedral sheet form a single layer, as shown in Fig. 4.4. This layer is about 0.72 nm thick

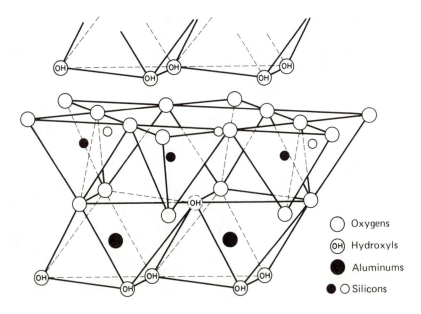

Fig. 4.4 Atomic structure of kaolinite (after Grim, 1959).

Fig. 4.5 Scanning electron
micrograph of a well-crystallized
kaolinite from Georgia. The
length of the light bar is 5 μm
(photograph by R. D. Holtz).

and extends indefinitely in the other two directions. A kaolinite crystal, then, consists of a stack of several layers of the basic 0.72 nm layer. Successive layers of the basic layer are held together by hydrogen bonds between the hydroxyls of the octahedral sheet and the oxygens of the tetrahedral sheet. Since the hydrogen bond is very strong, it prevents hydration and allows the layers to stack up to make a rather large crystal. A typical kaolin crystal can be 70 to 100 layers thick. Figure 4.5 is a scanning electron micrograph (SEM) of kaolinite.

Kaolinite is the primary constituent in china clay; it is also used in the paper, paint, and pharmaceutical industries. For example, as a pharmaceutical it is used in *Kao*pectate and Rolaids.

Another 1:1 mineral related to kaolinite is *halloysite*. It differs from kaolinite in that when it was formed it somehow became hydrated between the layers, causing a distortion or random stacking in the crystal lattice so that it is tubular in shape (Fig. 4.6). The water can easily be driven out from between the layers by heating or even air drying, and the process is irreversible. That is, the halloysite will not rehydrate when water is added. Halloysite, although not very common, occasionally plays an important

Fig. 4.6 Scanning electron micrograph of halloysite from Colorado.
The length of the light bar is 5 μm
(photograph by R. D. Holtz).

engineering role. Classification and compaction tests made on air-dried samples can give markedly different results than tests on samples at their natural water content. If the soil will not be air dried in the field, it can be extremely important that laboratory tests be carried out at the field water contents so that the results will have some validity.

Montmorillonite, also sometimes called *smectite*, is an important mineral composed of two silica sheets and one alumina (gibbsite) sheet (Fig. 4.7). Thus montmorillonite is called a 2:1 mineral. The octahedral sheet is between the two silica sheets with the tips of the tetrahedrons combining with the hydroxyls of the octahedral sheet to form a single layer, as shown in Fig. 4.8. The thickness of each 2:1 layer is about 0.96 nm, and like kaolinite the layers extend indefinitely in the other two directions. Because the bonding by van der Waals' forces between the tops of the silica sheets is weak and there is a net negative charge deficiency in the octahedral sheet, water and exchangeable ions can enter and separate

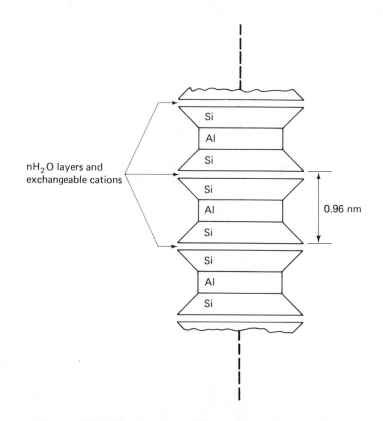

Fig. 4.7 Schematic diagram of the structure of montmorillonite (after Lambe, 1953).

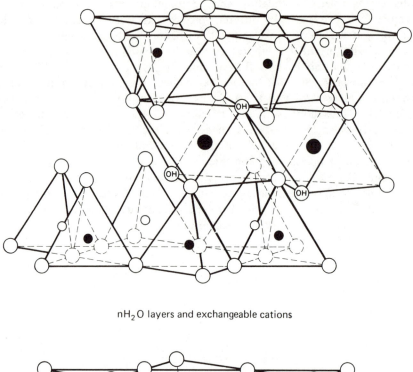

nH$_2$O layers and exchangeable cations

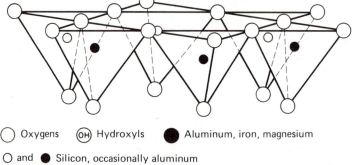

◯ Oxygens   (OH) Hydroxyls   ● Aluminum, iron, magnesium

○ and ● Silicon, occasionally aluminum

Fig. 4.8 Atomic structure of montmorillonite (after Grim, 1959).

the layers. Thus montmorillonite crystals can be very small (Fig. 4.9), but at the same time they have a very strong attraction for water. Soils containing montmorillonite are very susceptible to swelling as they change (increase) water content, and the swelling pressures developed can easily damage light structures and highway pavements. Montmorillonite is also the primary constituent in drilling mud and kitty litter, and it has many

Fig. 4.9 Scanning electron micrograph of Na-montmorillonite from Wyoming. The length of the light bar is 5 μm (photograph by R. D. Holtz).

other important industrial and pharmaceutical applications. It is even used in chocolate bars!

Illite, discovered by Prof. R. E. Grim of the University of Illinois, is another important constituent of clay soils. It also has a 2:1 structure similar to montmorillonite, but the interlayers are bonded together with a potassium atom. Remember the hexagonal hole in the silica sheet (Fig. 4.1d)? It has almost exactly the right diameter so that a potassium atom just fills that hexagonal hole and rather strongly bonds the layers together (Fig. 4.10). In addition, there is some isomorphous substitution of aluminium for silicon in the silica sheet.

Illites have a crystal structure similar to the mica minerals but with less potassium and less isomorphous substitution; thus they are chemically much more active than the other micas. Figure 4.11 is a SEM of illite.

Chlorite, relatively common in clay soils, is made of repeating layers of a silica sheet, an alumina sheet, another silica, and then either a gibbsite (Al) or brucite (Mg) sheet (Fig. 4.12). It could be called a 2:1:1 mineral. Chlorite can also have considerable isomorphous substitution and be missing an occasional brucite or gibbsite layer; thus it may be susceptible to swelling because water can enter between the sheets. Generally, however, it is significantly less active than montmorillonite.

As mentioned previously, there are literally dozens of clay minerals, with virtually every conceivable combination of substituted ions, interlayer water, and exchangeable cations. Some of the more important from an engineering viewpoint include *vermiculite*, which is similar to montmorillonite, a 2:1 mineral, but it has only two interlayers of water. After it is dried at high temperature, which removes the interlayer water, "expanded" vermiculite makes an excellent insulation material. Another clay mineral, *attapulgite*, Fig. 4.13, does not have a sheet structure but is a chain silicate; it consequently has a needle or rodlike appearance. *Mixed-layer* minerals

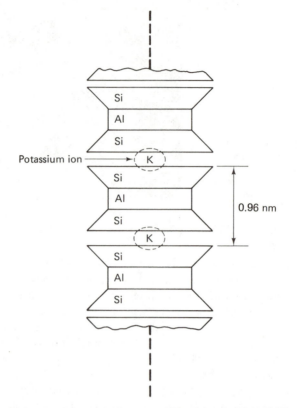

Fig. 4.10   Schematic diagram of illite (after Lambe, 1953).

Fig. 4.11   Scanning electron micrograph of illite from Fithian, Illinois. The length of the light bar is 5 μm (photograph by R. D. Holtz).

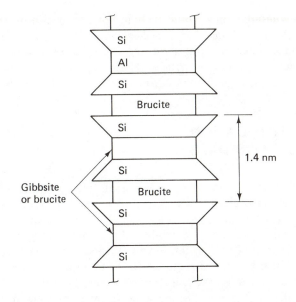

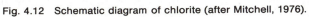

Fig. 4.12 Schematic diagram of chlorite (after Mitchell, 1976).

Fig. 4.13 Scanning electron micro graph of attapulgite from Florida. The length of the space between the bars of light is 0.5 μm (photograph by R. D. Holtz).

are relatively common; they would include, for example, montmorillonite mixed with chlorite or illite. Because *allophane* is an alumino-silicate, it is often classified as a clay mineral. However, it is *amorphous*, which means it has no regular crystalline structure. Under specialized conditions of weathering, it may be a locally important constituent of clay soils.

4.3 IDENTIFICATION OF CLAY MINERALS

Since the clay minerals are so very small, their identification by the usual optical mineralogical techniques used in geology is not possible so other means must be employed to identify them. From your engineering materials courses, you may remember that materials with regular or repeating patterns of crystal structure will diffract X-rays. Different minerals with different crystalline structures will have different *X-ray diffraction* patterns, and in fact these different patterns were how the minerals were identified in the first place. The patterns for the common minerals are published, and it is relatively simple to compare the diffraction pattern of your unknown with the patterns of known minerals. There is a problem, however, with soils which are mixtures of clay minerals, soils which contain organics and other non-clay mineral constituents, and soils with mixed-layer minerals. Usually a detailed quantitative analysis is impossible —about all that one can tell is which minerals are present and roughly how much of each.

Another technique that is sometimes used to identify clay minerals is *differential thermal analysis* (DTA). A specimen of the unknown soil along with an inert control substance is continuously heated to several hundred degrees in an electric furnace, and certain changes in temperature occur because of the particular structure of the clay minerals. The changes occur at specific temperatures for specific minerals, and the record of these changes may be compared with those of known minerals.

Electron microscopy, both transmission and scanning, can be used to identify clay minerals in a soil sample, but the process is not easy and/or quantitative.

A simple approach suggested by Prof. Casagrande is to use the Atterberg limits. It was mentioned (Sec. 2.8) that the activity has been related to specific active or inactive clays. Montmorillonites will be highly active since they are very small and have large plasticity indices. Use of Casagrande's plasticity chart (Fig. 3.2) can also tell you just about as much, at least from an engineering point of view, as the more sophisticated diffraction and DTA analyses. The procedure is shown in Fig. 4.14. You simply locate your sample on the LL-PI chart, and compare its location

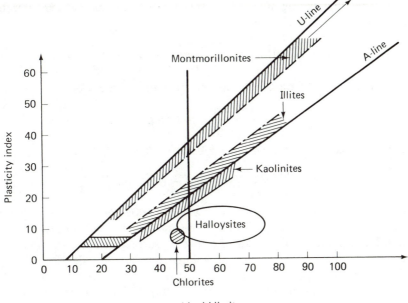

Fig. 4.14 Location of common clay minerals on Casagrande's plasticity chart (developed from Casagrande, 1948, and data in Mitchell, 1976).

with those of known minerals. If your sample has Atterberg limits that plot high above the A-line near the U-line, then chances are that it contains a lot of active clay minerals such as montmorillonite. Even if the soil is classified as a CL, for example a sandy clay (CL), and still plots near the U-line, the clay portion of the soil is predominantly montmorillonite. The glacial lake clays from around the Great Lakes region in the United States and Canada are predominantly illitic and they plot right above the A-line. Scandinavian marine clays which are illitic also plot in this region. Kaolinites, which are relatively inactive minerals, plot right below the A-line. Even though they are technically clays, they behave like ML-MH materials.

4.4 SPECIFIC SURFACE

Specific surface is the ratio of the surface area of a material to either its mass or volume. In terms of volume

$$\text{specific surface} = \text{surface area}/\text{volume} \tag{4-1}$$

The physical significance of specific surface can be demonstrated using a $1 \times 1 \times 1$ cm cube.

$$\text{specific surface} = \frac{6(1 \text{ cm}^2)}{1 \text{ cm}^3} = 6/\text{cm} = 0.6/\text{mm}$$

If the cube is 1 mm on a side, the specific surface would be

$$\frac{6(1 \text{ mm}^2)}{1 \text{ mm}^3} = 6/\text{mm}$$

If the cube is 1 μm on a side, the specific surface would be

$$\frac{6(1 \ \mu\text{m}^2)}{1 \ \mu\text{m}^3} = 6/\mu\text{m} = 6000/\text{mm}$$

This illustrates that large particles, whether cubes or soil particles, have smaller surface areas per unit of volume and thus smaller specific surfaces than small particles. To obtain the specific surface in terms of mass, you just divide the value in terms of volume by the mass density ρ_s; units would then be m^2/g or m^2/kg.

Now, if sufficient water was present to just dampen the surface area of the cubes in the above example, it would take ten times as much water to wet the surface of all the grains when the cubes were 1 mm on a side than when the same volume occupied a single cube of 1 cm^3. Note also that if one were trying to remove water from the surface wet soil, there would be ten times as much water to remove from the smaller grains.

Specific surface is inversely proportional to the grain size of a soil. We generally do not compute the specific surface for practical cases since the soil grains are too irregular in shape to do so. But it should be clear that a soil mass made up of many small particles will have on the average a larger specific surface than the same mass made up of large particles.

From the concept of specific surface, we would expect larger moisture contents for fine-grained soils than for coarse-grained soils, all other things such as void ratio and soil structure being equal.

You may recall from your materials courses that specific surface is a primary factor in concrete and asphalt mix design. In both cases it is necessary to provide sufficient cement paste or asphalt to coat the aggregate surfaces.

4.5 INTERACTION BETWEEN WATER AND CLAY MINERALS

As mentioned previously, water usually doesn't have much effect on the behavior of granular soils. For example, the shear strength of a sand is approximately the same whether it is dry or saturated. An important exception to this fact is the case of water present in loose deposits of sand subjected to dynamic loadings such as earthquakes or blasts.

On the other hand, fine-grained soils, especially clay soils, are strongly influenced by the presence of water. The variation of water content gives rise to plasticity, and the Atterberg limits are an indication of this influence. Grain size distribution only rarely is a governing factor in the behavior of fine-grained soils.

Why is water important in fine-grained soils? Recall the discussion of specific surface, in the previous section, where the smaller the particle, the larger the specific surface. Clay minerals, being relatively small particles, have large specific surfaces, and everything else being equal, you might expect that they would have very active surfaces.

The relative sizes of four common clay minerals and their specific surfaces are shown in Fig. 4.15. Kaolinite, the largest clay mineral, has a

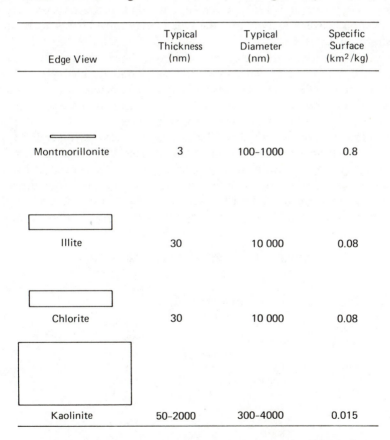

Edge View	Typical Thickness (nm)	Typical Diameter (nm)	Specific Surface (km^2/kg)
Montmorillonite	3	100–1000	0.8
Illite	30	10 000	0.08
Chlorite	30	10 000	0.08
Kaolinite	50–2000	300–4000	0.015

Fig. 4.15 Average values of relative sizes, thicknesses, and specific surfaces of the common clay minerals (after Yong and Warkentin, 1975).

thickness or edge dimension of about 1 μm, while montmorillonite, the smallest clay mineral, has a thickness of only a few nanometres. Since the crystals have roughly the same average "diameter," at least within an order of magnitude, it is not surprising that the specific surfaces are so different. Of course, there are rather wide variations in the sizes of the crystals depending on weathering and other factors, but the values given are average values. Since surface activity is related to the particle size, you can see why montmorillonite, for example, is more "active" than kaolinite. Similarly, the surface activity of a sand or silt grain is practically zero.

In Sec. 2.8, we defined the *activity* of a clay as

$$A = \frac{\text{PI}}{\text{clay fraction}} \tag{2-24}$$

where the clay fraction is usually taken as the percentage of the sample less than 2 μm (Skempton, 1953). We mentioned that there was a pretty good correlation between activity and the type of clay mineral. This correlation is shown in Table 4-1.

Now, it seems that clay particles are almost always hydrated in nature; that is, there are layers of water surrounding each crystal of clay. This water is called *adsorbed water*. As discussed in the next section, the structure of clay soils and thus their engineering properties ultimately depend on the nature of this adsorbed water layer.

How is water adsorbed on the surface of a clay particle? First, you may recall from chemistry or materials courses that water is a *dipolar* molecule (Fig. 4.16). Even though water is electrically neutral, it has two separate centers of charge, one positive and one negative. Thus the water molecule is electrostatically attracted to the surface of the clay crystal. Secondly, water is held to the clay crystal by *hydrogen bonding* (hydrogen

TABLE 4-1 Activities of Various Minerals*

Mineral	Activity
Na-montmorillonite	4–7
Ca-montmorillonite	1.5
Illite	0.5–1.3
Kaolinite	0.3–0.5
Halloysite (dehydrated)	0.5
Halloysite (hydrated)	0.1
Attapulgite	0.5–1.2
Allophane	0.5–1.2
Mica (muscovite)	0.2
Calcite	0.2
Quartz	0

*After Skempton (1953) and Mitchell (1976).

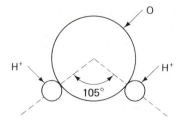

Fig. 4.16 Schematic diagram of a water molecule (after Lambe, 1953).

of the water is attracted to the oxygens or hydroxyls on the surface of the clay). The third factor is that the negatively charged clay surface also attracts cations present in the water. Since all cations are hydrated to some extent, depending on the ion, cations also contribute to the attraction of water to the clay surface. Of these three factors, hydrogen bonding is probably the most important factor.

The attraction of water to the clay surface is very strong near the surface and diminishes with distance from that surface. It seems that the water molecules right at the surface are very tightly held and strongly oriented. Measurements show that some thermodynamic and electrical properties of the water next to the clay surface are different than that of "free water" (Mitchell, 1976).

The source of the negative charge at the surface of the clay crystal results from both isomorphous substitution, mentioned earlier, and imperfections in the crystal lattice, especially at the surface. "Broken" edges contribute greatly to unsatisfied valence charges at the edges of the crystal. Since the crystal wants to be electrically neutral, cations in the water may be strongly attracted to the clay, depending on the amount of negative charge present. Different clays have different charge deficiencies and thus have different tendencies to attract the exchangeable cations. They are called *exchangeable* since one cation can easily be exchanged with one of the same valence or by two of one-half the valence of the original cation. As might be expected from their relative sizes and specific surfaces, montmorillonite has a much greater charge deficiency and thus a much greater attraction for exchangeable cations than kaolinite. Illite and chlorite are intermediate in this respect.

Calcium and magnesium are the predominant exchangeable cations in soils, whereas potassium and sodium are less common. Aluminium and hydrogen are common in acidic soils. The depositional environment as well as subsequent weathering and leaching will govern what ions are present in a particular soil deposit. As might be expected, marine clays are predominately sodium and magnesium since these are the most common cations in sea water. Cation exchange or replacement is further complicated by the presence of organic matter.

The ease of replacement or exchange of cations depends on several factors, primarily the valence of the cation. Higher valence cations easily replace cations of lower valence. For ions of the same valence, the size of the hydrated ion becomes important; the larger the ion, the greater the replacement power. A further complication is the fact that potassium, even though it is monovalent, fits into the hexagonal holes in the silica sheet. Thus it will be very strongly held on the clay surface, and it will have a greater replacement power than sodium, for example, which is also monovalent. The cations can be listed in *approximate* order of their replacement ability. The specific order depends on the type of clay, which ion is being replaced, and the concentration of the various ions in the water. In order of increasing replacement power the ions are

$$Li^+ < Na^+ < H^+ < K^+ < NH_4^+ \ll Mg^{++} < Ca^{++} \ll Al^{+++}$$

There are several practical consequences of ion exchange. The use of chemicals to stabilize or strengthen soils is possible because of ion exchange. For example, lime (CaOH) stabilizes a sodium clay soil by replacing the sodium ions in the clay since calcium has a greater replacing power than sodium. The swelling of sodium montmorillonitic clays can be significantly reduced by the addition of lime.

What does a clay particle look like with adsorbed water on it? Figure 4.17 shows a sodium montmorillonite and kaolinite crystal with layers of adsorbed water. Note that the thickness of the adsorbed water is approximately the same, but because of the size differences the montmorillonite will have much greater activity, higher plasticity, and greater swelling, shrinkage, and volume change due to loading.

In this section only a brief overview of the very complex subject of the interaction between water and clay minerals has been presented. For additional information, you should consult Yong and Warkentin (1975) and Mitchell (1976) and references included therein.

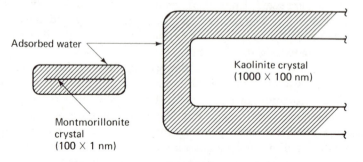

Fig. 4.17 Relative sizes of adsorbed water layers on sodium montmorillonite and sodium kaolinite (after Lambe, 1958a).

4.6 INTERACTION OF CLAY PARTICLES

The association of clay minerals and their adsorbed water layer provides the physical basis for soil structure. The individual clay particles interact through their adsorbed water layers, and thus the presence of different ions, organic materials, different concentrations, etc., affect or contribute to the multitude of soil structures found in natural soil deposits. Clay particles can repulse each other electrostatically, but the process depends on the ion concentration, interparticle spacing, and other factors. Similarly, there can be attraction of the individual particles due to the tendency for hydrogen bonding, van der Waals' forces, and other types of chemical and organic bonds. The interparticle force or potential fields decrease with increasing distance from the mineral surface, as shown in Fig. 4.18. The actual shape of the potential curve will depend on the valence and concentration of the dissolved ion and on the nature of the bonding forces.

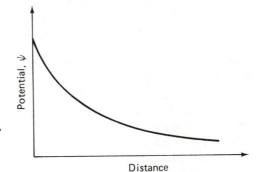

Fig. 4.18 Chemical, electrostatic, etc., potential versus distance from the clay mineral surface.

Particles can flocculate or be repelled (disperse or separate). They can flocculate in several possible configurations; edge-to-face is the most common, but edge-to-edge and face-to-face flocculation are also possible. The tendency towards flocculation will depend on *increasing* one or more of the following (Lambe, 1958a):

Concentration of the electrolyte
Valence of the ion
Temperature

or *decreasing* one or more of the following:
Dielectric constant of the pore fluid
Size of the hydrated ion
pH
Anion adsorption

Just about all natural clay soils are flocculated to some extent. Only in very dilute solutions (at very high water content) is dispersion of clay particles possible, and this might occur in a sedimentary deposit during deposition.

4.7 SOIL STRUCTURE AND FABRIC

In geotechnical engineering practice, the *structure* of a soil is taken to mean both the geometric arrangement of the particles or mineral grains as well as the interparticle forces which may act between them. Soil *fabric* refers only to the geometric arrangement of the particles. In granular or cohesionless soils, the interparticle forces are very small, so both the fabric and structure of gravels, sands, and to some extent silts are the same. On the contrary, however, interparticle forces are relatively large in fine-grained cohesive soils, and thus both these forces and the fabric of such soils must be considered as the structure of the soil. The structure strongly affects or, some would say, governs the engineering behavior of a particular soil. All the clay structures found in nature and described in the next section result from some combination of these factors, the geologic environment at deposition, the subsequent geologic and engineering stress history, and the nature of the clay mineral. We study these very complicated factors because they fundamentally affect soil behavior and the engineering properties of soil. Geotechnical engineers must consider the soil structure and fabric at least qualitatively when cohesive soils are encountered in engineering practice.

A complete description of the structure of a fine-grained cohesive soil requires a knowledge of both the interparticle forces as well as the geometrical arrangement (fabric) of the particles. Since it is extremely difficult, if not impossible, to directly measure the interparticle force fields surrounding clay particles, most studies of cohesive soil structure involve only the fabric of these soils, and from the fabric certain inferences are made about the interparticle forces.

4.8 COHESIVE SOIL FABRICS

Classification of cohesive soil fabrics into simple systems involving only a few clay particles is not really possible. Single grain or single particle units occur only rarely in nature and then in only very dilute clay-water systems under special environmental conditions. From recent studies of real clay soils with the scanning electron microscope (SEM), the individual clay particles seem to always be aggregated or flocculated together in submicroscopic fabric units called *domains*. Domains then in turn group together to form *clusters*, which are large enough to be seen with a visible light microscope. Clusters group together to form *peds* and even groups of peds. Peds can be seen without a microscope, and they and other macrostructural features such as joints and fissures constitute the macrofabric system. A schematic sketch of this system proposed by Yong and Sheeran (1973) is shown in Fig. 4.19; a microscopic view of a marine clay is also included (Pusch, 1973). Collins and McGown (1974) suggest a somewhat more elaborate system for describing microfabric features in natural soils. They propose three types of features:

1. *Elementary particle arrangements*, which consist of single forms of particle interaction at the level of individual clay, silt, or sand particles (Figs. 4.20a and b) or interaction between small groups of clay platelets (Fig. 4.20c) or clothed silt and sand particles (Fig. 4.20d).
2. *Particle assemblages*, which are units of particle organization having definable physical boundaries and a specific mechanical function. Particle assemblages consist of one or more forms of elementary particle arrangements or smaller particle assemblages, and they are shown in Fig. 4.21.
3. *Pore spaces* within and between elementary particle arrangements and particle assemblages.

Collins and McGown (1974) show microphotographs of several natural soils which illustrate their proposed system.

A SEM photograph of a silty clay ped from Norway is shown in Fig. 4.22. Note how complex the structure appears, which suggests that the engineering behavior is also probably quite complex.

Macrostructure, including the stratigraphy of fine-grained soil deposits, has an important influence on soil behavior in engineering practice. Joints, fissures, silt and sand seams, root holes, varves, and other "defects" often control the engineering behavior of the entire soil mass. Usually, the strength of a soil mass is significantly less along a crack or fissure than through the intact material, and thus if the defect happens to be unfavorably oriented with respect to the applied engineering stresses, instability or failure may occur. As another example, the drainage of a clay layer can be markedly affected by the presence of a silt or sand layer or seam. Consequently, in any engineering problem involving stability, settlements, or drainage, the geotechnical engineer must investigate carefully the clay macrostructure.

Microstructure is more important from a fundamental than an engineering viewpoint, although an understanding of the microstructure aids

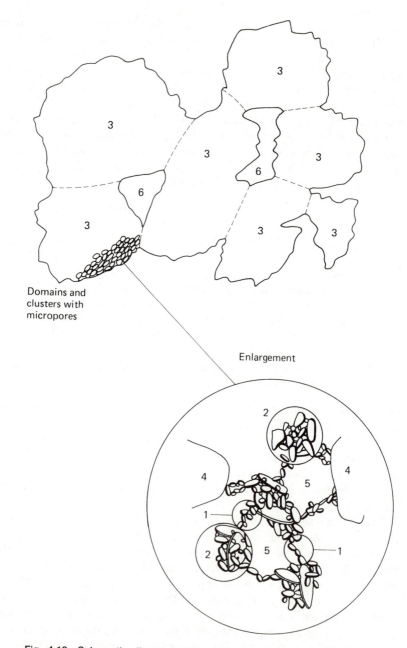

Domains and
clusters with
micropores

Enlargement

Fig. 4.19 Schematic diagram of the soil microfabric and macrofabric system proposed by Yong and Sheeran (1973) and Pusch
(1973): 1, domain; 2, cluster; 3, ped; 4, silt grain; 5, micropore; and
6, macropore.

(a)

(b)

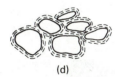

(c)

(d)   (e)

Fig. 4.20 Schematic representations of elementary particle arrangements: (a) individual clay platelet interaction; (b) individual silt or sand particle interaction; (c) clay platelet group interaction; (d) clothed silt or sand particle interaction; (e) partly discernible particle interaction (after Collins and McGown, 1974).

in a general understanding of soil behavior. The microstructure of a clay reflects the entire geologic and stress history of that deposit. Virtually everything that ever happened to that soil which will affect the engineering response of the clay is imprinted in some manner on the microstructure. The microstructure reflects the depositional history and environment of the deposit, its weathering history, both chemical and physical: in effect its stress history, that is, all changes caused both geologically and by man.

Recent research on clay microstructure suggests that the greatest single factor influencing the final structure of a clay is the electrochemical environment existing at the time of sedimentation. Flocculated structures or aggregations can result during sedimentation in virtually all depositional environments, whether marine, brackish, or in fresh water. The degree of

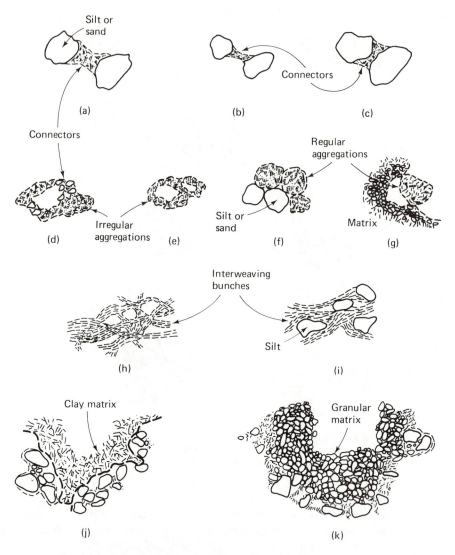

Fig. 4.21 Schematic representations of particle assemblages: (a), (b), and (c) connectors; (d) irregular aggregations linked by connector assemblages; (e) irregular aggregations forming a honeycomb arrangement; (f) regular aggregations interacting with silt or sand grains; (g) regular aggregation interacting with particle matrix; (h) interweaving bunches of clay; (i) interweaving bunches of clay with silt inclusions; (j) clay particle matrix; (k) granular particle matrix (after Collins and McGown, 1974).

5 μ

Fig. 4.22 Drammen silty clay: large ped of silt and clay with weak links to other peds (from Barden and McGown, 1973; photograph courtesy of A. McGown).

openness of the structure is apparently influenced to a large degree by the clay mineralogy as well as the amount and angularity of silt grains present. Silt particles have been observed to have a thin skin of apparently strongly oriented clay particles or even amorphous materials parallel to their surfaces. Some grain-to-grain contacts of silt particles have been observed (see Fig. 4.23), but at present it is difficult to ascertain whether actual mineral contact occurs in clays.

In summary, the structure of most naturally occurring clay deposits is highly complex. The engineering behavior of these deposits is strongly influenced by both the macro- and the microstructure. At present, no quantitative connection exists between microstructure and the engineering

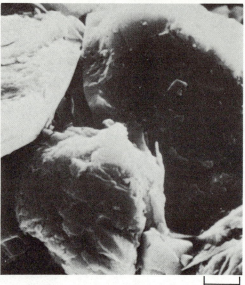

Fig. 4.23 SEM of silt grain-to-grain contact in Swedish till (from McGown, 1973).

⊢——⊣
10 μm

properties, but it is important for the engineer to have an appreciation of the complexity of the structure of cohesive soils and their relation to engineering behavior.

4.9 COHESIONLESS SOIL FABRICS

Grains of soil which can settle out of a soil-fluid suspension independently of other grains (generally larger than 0.01 to 0.02 mm) will form what is called a *single grained* structure. This is the structure of, for example, a sand or gravel pile, and some sand-silt mixtures. The weight of the grains causes them to settle and come to equilibrium in the bottom of the fluid as soon as the velocity can no longer support the particles in suspension.

Deposition media include both air (loess deposits, sand dunes; grain size generally < 0.05 mm) and water (rivers, beaches, etc.).

Single grained structures, shown in Fig. 4.24, may be "loose" (high void ratio or low density) or "dense" (low void ratio or high density). Depending on the grain size distribution as well as the packing or arrangement of the grains, a wide range of void ratios is possible. Table 4-2 lists some typical values for a variety of granular soils. It is possible, under some conditions of deposition, for a granular material to achieve a honeycombed structure (Fig. 4.25) which can have a very high void ratio. Such a structure is meta-stable. The grain arches can support static loads, but the

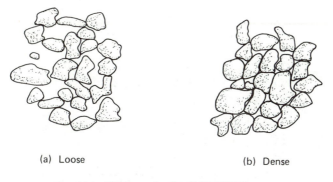

(a) Loose (b) Dense

Fig. 4.24 Single grained soil structures.

structure is very sensitive to collapse when vibrated or loaded dynamically. The presence of water in very loose grain structures also can alter their engineering behavior. Phenomena typical of loose grain structures, such as *bulking*, a capillary phenomenon, and *quicksand* are discussed in Chapters 6 and 7.

The greatest possible void ratio or loosest possible condition of a soil is called the *maximum void ratio* (e_{max}). It is determined in the laboratory by pouring dry sand very carefully with no vibration into a calibrated mold of known volume. From the weight of sand in the mold, e_{max} can be calculated. Similarly, the *minimum void ratio* (e_{min}) is the densest possible condition that a given soil can attain. The value of e_{min} is determined by vibrating a known weight of dry sand into a known volume and calculating the void ratio. The range of possible void ratios for some typical granular soils are shown in Table 4-2.

The relative density D_r, also called the density index I_D, is used to compare the void ratio e of a given soil with the maximum and minimum void ratios. Relative density is defined as

$$D_r = I_D = \frac{e_{max} - e}{e_{max} - e_{min}} \times 100(\%) \qquad (4\text{-}2)$$

and is usually expressed as a percentage. Relative density can also be stated in terms of maximum and minimum dry densities as

$$D_r = I_D = \frac{1/\rho_{d\,min} - 1/\rho_d}{1/\rho_{d\,min} - 1/\rho_{d\,max}} \times 100(\%) \qquad (4\text{-}3)$$

where ρ_d = dry density of the soil with void ratio e,

$\rho_{d\,min}$ = minimum dry density of the soil with the void ratio e_{max}, and

$\rho_{d\,max}$ = maximum dry density of the soil with the void ratio e_{min}.

The relative density of a natural soil deposit very strongly affects its engineering behavior. Consequently, it is important to conduct laboratory

TABLE 4-2 Typical Index Properties for Granular Soils*

	Particle Size and Gradation				Voids			
	Approx. Size Range (mm)		Approx. D_{10} (mm)	Approx. Range C_u	Void Ratio		Porosity (%)	
	D_{max}	D_{min}			e_{max} (loose)	e_{min} (dense)	n_{max} (loose)	n_{min} (dense)
1. Uniform materials:								
(a) Equal spheres	—	—	—	1.0	0.92	0.35	48	26
(b) Standard Ottawa sand	0.84	0.59	0.67	1.1	0.80	0.50	44	33
(c) Clean, uniform sand (fine or medium)	—	—	—	1.2 to 2.0	1.0	0.40	50	29
(d) Uniform, inorganic silt	0.05	0.005	0.012	1.2 to 2.0	1.1	0.40	52	29
2. Well-graded materials:								
(a) Silty sand	2.0	0.005	0.02	5 to 10	0.90	0.30	47	23
(b) Clean, fine to coarse sand	2.0	0.05	0.09	4 to 6	0.95	0.20	49	17
(c) Micaceous sand	—	—	—	—	1.2	0.40	55	29
(d) Silty sand and gravel	100	0.005	0.02	15 to 300	0.85	0.14	46	12

*Modified after B. K. Hough (1969), *Basic Soils Engineering*, © 1969 by the Ronald Press, Co. Reprinted by permission of John Wiley & Sons, Inc.

TABLE 4-2 Continued

	Density (Mg/m³)†						
	Dry Density, ρ_d			Wet Density, ρ		Submerged Density, ρ'	
	Min.	100% Mod. Proctor	Max.	Min.	Max.	Min.	Max.
	(loose)		(dense)	(loose)	(dense)	(loose)	(dense)
1. Uniform materials:							
(a) Equal spheres (theoretical values)	—	—	—	—	—	—	1.12
(b) Standard Ottawa sand	1.49	—	1.78	1.51	2.12	0.93	1.18
(c) Clean, uniform sand (fine or medium)	1.35	1.86	1.92	1.37	2.20	0.85	1.18
(d) Uniform, inorganic silt	1.29	—	1.92	1.31	2.20	0.83	1.28
2. Well-graded materials:							
(a) Silty sand	1.41	1.98	2.06	1.43	2.30	0.88	1.40
(b) Clean, fine to course sand	1.38	2.14	2.23	1.40	2.39	0.86	1.23
(c) Micaceous sand	1.23	—	1.95	1.24	2.23	0.77	1.49
(d) Silty sand and gravel	1.44	—	2.36	1.46	2.51	0.91	

†Tabulation is based on $\rho_s = 2.65$ Mg/m³. Multiply by 62.4 to obtain lbf/ft³.

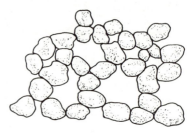

Fig. 4.25 Honeycomb structure.

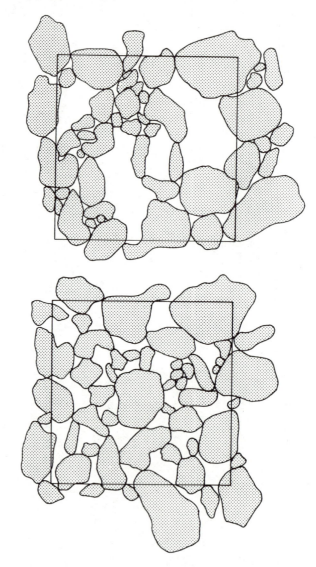

Fig. 4.26 Potential ranges in packing of identical particles at the same relative density (G. A. Leonards, 1976, personal communication).

tests on samples of the sand at the same relative density as in the field. Sampling of loose granular materials, especially at depths greater than a few metres, is very difficult. Since the materials are very sensitive to even the slightest vibration, one is never sure the sample has the same density as the natural soil deposit. Therefore different kinds of penetrometers are used in engineering practice, and the penetration resistance values are roughly correlated with relative density. For deposits at shallow depths where direct access is possible, other techniques have been developed to measure the in-place density of compacted soils. These techniques are discussed in detail in Chapter 5.

Finally, it should be noted in this discussion of the structure of granular soils that relative density alone is not sufficient to characterize their engineering properties. It is possible for two sands, for example, to have identical void ratios and relative densities but significantly different fabrics and thus significantly different engineering behaviors. Figure 4.26 is a two-dimensional example of such a fabric. Both "sands" are identical — they have the same grain size distributions and the same void ratios. But their fabrics are obviously very different. Stress history is another factor that must be considered when dealing with sands and gravels in engineering practice. Deposits of granular materials which have been preloaded by nature or man will have very different stress-strain properties and therefore very different settlement responses (Lambrechts and Leonards, 1978).

PROBLEMS

4-1. Calculate the specific surface of a cube (a) 10 mm, (b) 1 mm, (c) 1 μm, and (d) 1 nm on a side. Calculate the specific surface in terms of both areas and m^2/kg. Assume for the latter case that $\rho_s = 2.65$ Mg/m^3.

4-2. Calculate the specific surface of (a) tennis balls, (b) ping pong balls, (c) ball bearings 1 mm in diameter, and (d) fly ash with approximately spherical particles of 50 μm in diameter.

4-3. The values of e_{min} and e_{max} for a pure silica sand ($\rho_s = 2.65$ Mg/m^3) were found to be 0.46 and 0.66, respectively. (a) What is the corresponding range in dry density? (b) If the in situ void ratio is 0.63, what is the density index?

4-4. Describe briefly the crystalline or atomic structure of the following ten minerals. Also list any important distinguishing characteristics.

(a) Smectite   (b) Brucite   (c) Gibbsite
(d) Attapulgite   (e) Bentonite   (f) Allophane
(g) Halloysite   (h) Illite   (i) Mica
(j) Chlorite

4-5. Describe the following types of bonding agents found with clay minerals.

(a) Hydrogen bond   (b) Covalent bond
(c) van der Waals' forces   (d) James bond

4-6. The wet density of a sand in an embankment was found to be 1.9 Mg/m^3 and the field water content was 10%. In the laboratory, the density of the solids was found to be 2.66 Mg/m^3, and the maximum and minimum void ratios were 0.62 and 0.44, respectively. Calculate the relative density of the sand in the field.

4-7. Which sheet, silica or alumina, would you wear to a toga party? Why?

4-8. Given the particles in Fig. 4.26, is it realistic to show that all the particles are in contact with each other for this given plane? Any given plane? Why?

five

Compaction

5.1 INTRODUCTION

In geotechnical engineering practice the soils at a given site are often less than ideal for the intended purpose. They may be weak, highly compressible, or have a higher permeability than desirable from an engineering or economic point of view. It would seem reasonable in such instances to simply relocate the structure or facility. However, considerations other than geotechnical often govern the location of a structure, and the engineer is forced to design for the site at hand. One possibility is to adapt the foundation to the geotechnical conditions at the site. Another possibility is to try to *stabilize* or improve the engineering properties of the soils at the site. Depending on the circumstances, this second approach may be the most economical solution to the problem. Stabilization is usually *mechanical* or *chemical*, but even thermal and electrical stabilization have occasionally been used or considered.

In this chapter we are primarily concerned with mechanical stabilization or densification, also called *compaction*. Chemical stabilization includes the mixing or injecting of chemical substances into the soil. Portland cement, lime, asphalt, calcium chloride, sodium chloride, and paper mill wastes are common chemical stabilization agents.

Other methods for stabilizing unsuitable foundation soils include *dewatering*, which is the removal or reduction of unwanted excess ground water pressures, and *preloading*, in which the foundation soils are *surcharged* with a temporary overload so as to increase the strength and decrease anticipated settlement. The details of these and other methods are described in textbooks on foundation and highway engineering. A good state of the art discussion and reference to methods for improving the engineering characteristics of soils is provided in the ASCE (1978) committee report "Soil Improvement—History, Capabilities, and Outlook."

Compaction and stabilization are very important when soil is used as an engineering material; that is, the structure itself is made of soil. Earth dams and highway embankments are examples of *earth structures*. If soils are dumped or otherwise placed at random in a fill, the result will be an embankment with low stability and high settlement. In fact, prior to the 1930's, highway and railroad fills were usually constructed by end-dumping soils from wagons or trucks. There was very little attempt to compact or densify the soils, and failures of even moderately high embankments were common. Of course earthworks such as levees are almost as old as man, but these structures, for example in ancient China or India, were constructed by people carrying small baskets of earth and dumping them in the embankment. People walking over the dumped materials compacted and thus strengthened the soils. Even elephants have been used in some countries to compact soils, but research has shown that they are not very good at it (Meehan, 1967).

The following symbols are introduced in this chapter.

Symbol	Dimension	Unit	Definition
D	L	m	Depth of influence (Eq. 5-2)
g	L/T^2	m/s^2	Acceleration of gravity, 9.80665m/s^2
R.C.	—	(%)	Relative compaction (Eq. 5-3)
W	M	t	Mass of falling weight (Eq. 5-2)
w_{opt} or OMC	—	(%)	Optimum water content; sometimes called the optimum moisture content (OMC)
$\rho_{d\,max}$	M/L^3	Mg/m^3	Maximum dry density
$\rho_{d\,field}$	M/L^3	Mg/m^3	Field dry density

5.2 COMPACTION

Compaction is the densification of soils by the application of mechanical energy. It may also involve a modification of the water content as well as the gradation of the soil. Cohesionless soils are efficiently compacted by vibration. In the field, hand-operated vibrating plates and motorized vibratory rollers of various sizes are very efficient for compacting sand and gravel soils. Rubber-tired equipment can also be used efficiently to compact sands. Even large free-falling weights have been used to dynamically compact loose granular fills. Some of these techniques are discussed later in this chapter.

Fine-grained and cohesive soils may be compacted in the laboratory by falling weights and hammers, by special "kneading" compactors, and even statically in a common loading machine or press. In the field, common compaction equipment includes hand-operated tampers, sheepsfoot rollers, rubber-tired rollers, and other types of heavy compaction equipment (Sec. 5.5). Considerable compaction can also be obtained by proper routing of the hauling equipment over the embankment during construction.

The objective of compaction is the improvement of the engineering properties of the soil mass. There are several advantages which occur through compaction:

Detrimental settlements can be reduced or prevented.

Soil strength increases and slope stability can be improved.

Bearing capacity of pavement subgrades can be improved.

Undesirable volume changes, for example, caused by frost action, swelling, and shrinkage may be controlled.

5.3   THEORY OF COMPACTION

The fundamentals of compaction of cohesive soils are relatively new. R. R. Proctor in the early 1930's was building dams for the old Bureau of Waterworks and Supply in Los Angeles, and he developed the principles of compaction in a series of articles in *Engineering News-Record* (Proctor, 1933). In his honor, the standard laboratory compaction test which he developed is commonly called the *Proctor test*.

Proctor established that compaction is a function of four variables: (1) dry density ρ_d, (2) water content w, (3) compactive effort, and (4) soil type (gradation, presence of clay minerals, etc). *Compactive effort* is a measure of the mechanical energy applied to a soil mass. In countries where British Engineering units are used, compactive effort is usually reported in ft·lbf/ft^3, whereas the SI units are J/m^3 (J = joules). Since 1 J = 1 N·m, and using the conversion factors in Appendix A, we can determine that 1 ft·lbf/ft^3 = 47.88 J/m^3. In the field, compactive effort is the number of passes or "coverages" of the roller of a certain type and weight on a given volume of soil. In the laboratory, *impact* or *dynamic*, *kneading*, and *static compaction* are usually employed. During *impact compaction*, which is the most common type, a hammer is dropped several times on a soil sample in a mold. The mass of the hammer, height of drop, number of drops, number of layers of soil, and the volume of the mold are specified. For example, in the *standard Proctor test* [also standard AASHTO

(1978), Designation T 99, and ASTM (1980), Designation D 698], the mass of the hammer is 2.495 kg (5.5 lb) and the height of fall is 304.88 mm (1 ft). The soil is placed in three layers in an approximately 1 litre (0.944 $\times$ 10^{-3} m³ or 1/30 ft³) mold, and each layer is tamped 25 times. Compactive effort can then be calculated as shown in Example 5.1.

EXAMPLE 5.1

Given:

Standard Proctor test hammer and mold.

Required:

Calculate the compactive effort in both SI and British Engineering units.

Solution:

a. *SI units*:

$$\frac{\text{compactive}}{\text{effort}} = \frac{2.495 \text{ kg}(9.81 \text{ m/s}^2)(0.3048 \text{ m})(3 \text{ layers})(25 \text{ blows/layer})}{0.944 \times 10^{-3} \text{ m}^3}$$

$$= 592.7 \text{ kJ/m}^3$$

If exact values of g and the volume are used, the standard Proctor compactive effort is 592.576 kJ/m³.

b. *British Engineering units*:

$$\text{compactive effort} = \frac{5.5 \text{ lbf}(1 \text{ ft})(3)(25)}{\frac{1}{30} \text{ ft}^3}$$

$$= 12,375 \frac{\text{ft} \cdot \text{lbf}}{\text{ft}^3}$$

This calculation is, strictly speaking, incorrect since the 5.5 lb hammer is really a mass not a weight. However, the differences are negligible.

For other types of compaction, the calculation of compactive effort is not so simple. In kneading compaction, for example, the tamper kneads the soil by applying a given pressure for a fraction of a second. The kneading action is supposed to simulate the compaction produced by a sheepsfoot roller and other types of field compaction equipment. In *static compaction*, the soil is simply pressed into a mold under a constant static stress in a laboratory testing machine.

The process of compaction for cohesive soils can best be illustrated by considering the common laboratory compaction or Proctor test. Several samples of the same soil, but at different water contents, are compacted according to the standard Proctor compaction test specifications given previously. Typically, the total or wet density and the actual water content of each compacted sample are measured. Then the dry density for each sample can be calculated from phase relationships we developed in Chapter 2.

$$\rho = \frac{M_t}{V_t} \qquad (2\text{-}6)$$

$$\rho_d = \frac{\rho}{1 + w} \qquad (2\text{-}14)$$

When the dry densities of each sample are determined and plotted versus the water contents for each sample, then a curve called a *compaction curve* for standard Proctor compaction is obtained (Fig. 5.1, curve *A*).

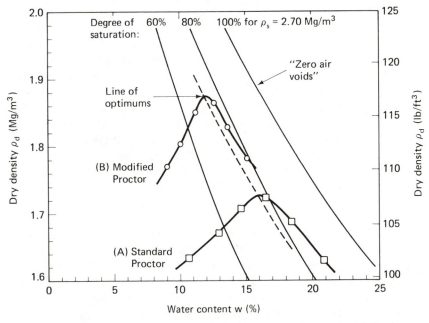

Fig. 5.1 Standard and modified Proctor compaction curves for Crosby B till.

Each data point on the curve represents a single compaction test, and usually four or five individual compaction tests are required to completely determine the compaction curve. This curve is unique for a given soil type, method of compaction and (constant) compactive effort. The peak point of the curve is an important point. Corresponding to the *maximum dry density* $\rho_{d \; max}$ is a water content known as the *optimum water content* w_{opt} (also known as the optimum moisture content, OMC). Note that the maximum dry density is only a maximum for a specific compactive effort and method of compaction. This does not necessarily reflect the maximum dry density that can be obtained in the field.

Typical values of maximum dry density are around 1.6 to 2.0 Mg/m^3 (100 to 125 lbf/ft^3) with the maximum range from about 1.3 to 2.4 Mg/m^3 (80 to 150 lbf/ft^3). (Densities are also given in British Engineering units because you are likely to encounter them in practice.) Typical optimum water contents are between 10% and 20%, with an outside maximum range of about 5% to 40%. Also shown on Fig. 5.1 are curves representing different degrees of saturation of the soil. From Eqs. 2-12 and 2-15, we can derive the equation for these theoretical curves.

$$\rho_d = \frac{\rho_w S}{w + \dfrac{\rho_w}{\rho_s} S} \tag{5-1}$$

The exact position of the degree of saturation curves depends only on the value of the density of the soil solids ρ_s. Note that at optimum water content for this particular soil, S is about 75%. Note too that the compaction curve, even at high water contents, never actually reaches the curve for "100% saturation" (traditionally called the *zero air voids* curve). And this is true even for higher compactive efforts, for example, curve B of Fig. 5.1. Curve B is the compaction curve obtained by the *modified Proctor compaction test* [modified AASHTO (1978), Designation T 180, and ASTM (1980), Designation D 1557]. This test utilizes a heavier hammer (4.536 kg or 10 lb), a greater height of fall (457 mm or 1.5 ft), and 5 layers tamped 25 times into a standard Proctor mold. You should verify that the compactive effort is 2693 kJ/m^3 or 56,250 ft·lbf/ft^3. The modified test was developed during World War II by the U.S. Army Corps of Engineers to better represent the compaction required for airfields to support heavy aircraft. The point is that increasing the compactive effort tends to increase the maximum dry density, as expected, but also decreases the optimum water content. A line drawn through the peak points of several compaction curves at different compactive efforts for the same soil will be almost parallel to a 100% S curve. It is called the *line of optimums*.

Typical compaction curves for different types of soils are illustrated in Fig. 5.2. Notice how sands that are well graded (SW soils, top curve) have a higher dry density than more uniform soils (SP soils, bottom curve). For clay soils, the maximum dry density tends to decrease as plasticity increases.

Why do we get compaction curves such as those shown in Fig. 5.1 and 5.2? Starting at low water contents, as the water content increases, the particles develop larger and larger water films around them, which tend to "lubricate" the particles and make them easier to be moved about and reoriented into a denser configuration. However we eventually reach a water content where the density does not increase any further. At this point, water starts to replace soil particles in the mold, and since $\rho_w \ll \rho_s$ the dry density curve starts to fall off, as is shown in Fig. 5.3. Note that no matter how much water is added, the soil never becomes completely saturated by compaction.

Compaction behavior of cohesive soils as described above is typical for both field and laboratory compaction. The curves obtained will vary

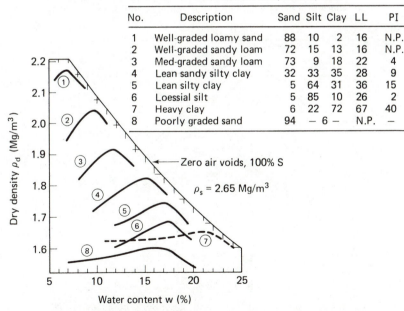

No.	Description	Sand	Silt	Clay	LL	PI
1	Well-graded loamy sand	88	10	2	16	N.P.
2	Well-graded sandy loam	72	15	13	16	N.P.
3	Med-graded sandy loam	73	9	18	22	4
4	Lean sandy silty clay	32	33	35	28	9
5	Lean silty clay	5	64	31	36	15
6	Loessial silt	5	85	10	26	2
7	Heavy clay	6	22	72	67	40
8	Poorly graded sand	94	– 6 –		N.P.	–

Soil texture and plasticity data

Zero air voids, 100% S

$\rho_s = 2.65$ Mg/m³

Fig. 5.2 Water content-dry density relationships for eight soils compacted according to the standard Proctor method (after Johnson and Sallberg, 1960).

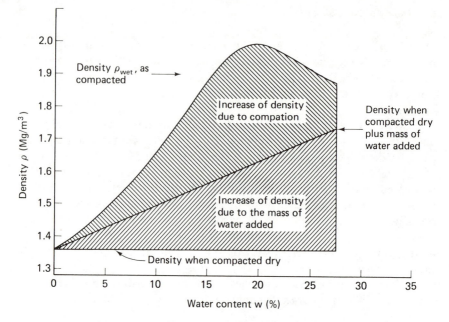

Fig. 5.3 The water content-density relationship indicating the increased density resulting from the addition of water and that due to the applied compaction effort. Soil is a silty clay, LL = 37, PI = 14, standard Proctor compaction (after Johnson and Sallberg, 1960).

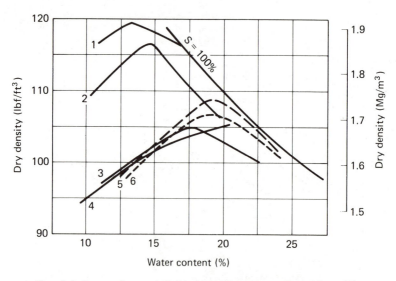

Fig. 5.4 Comparison of field and laboratory compaction. (1) Laboratory static compaction, 2000 psi; (2) modified Proctor; (3) standard Proctor; (4) laboratory static compaction, 200 psi; (5) field compaction, rubber-tired load, 6 coverages; (6) field compaction, sheepsfoot roller, 6 passes. Note: Static compaction from top and bottom of soil sample. (After Turnbull, 1950, and as cited by Lambe and Whitman, 1969.)

somewhat, that is, have different shapes and positions on the ρ_d versus w plot, but in general the response will be similar to that shown in Fig. 5.4, where the same soil is compacted under different conditions. The standard and modified Proctor laboratory tests were developed as a standard of comparison for field compaction, that is, to see if the rolling or compaction was sufficient. The approximation to field compaction is not exact, as mentioned, since the standard laboratory compaction is a dynamic-impact type, whereas field compaction is essentially a kneading-type compaction. This difference led to the development of the Harvard miniature compactor (Wilson, 1970) as well as larger kneading compactors. Field compaction control procedures are described in Sec. 5.6.

5.4 PROPERTIES AND STRUCTURE OF COMPACTED COHESIVE SOILS

The structure and thus the engineering properties of compacted cohesive soils will depend greatly on the method or type of compaction, the compactive effort applied, the soil type, and on the molding water content. Usually the water content of compacted soils is referenced to the optimum water content for a given type of compaction. Depending on their position, soils are called *dry of optimum, near or at optimum*, or *wet of optimum*. Research on compacted clays has shown that when they are compacted dry of optimum, the structure of the soils is essentially independent of the type of compaction (Seed and Chan, 1959). Wet of optimum, however, the type of compaction has a significant effect on the soil structure and thus on the strength, compressibility, etc., of the soil.

The comments in this section are very general, and you should keep in mind that the real fabric of compacted clays is about as complex as the fabric of natural clays described in Chapter 4. At the same compactive effort, with increasing water content, the soil fabric becomes increasingly oriented. Dry of optimum the soils are always flocculated, whereas wet of optimum the fabric becomes more oriented or dispersed. In Fig. 5.5, for example, the fabric at point C is more oriented than at point A. Now, if the compactive effort is increased, the soil tends to become more oriented, even dry of optimum. Again, referring to Fig. 5.5, a sample at point E is more oriented than at point A. Wet of optimum, the fabric at point D will be somewhat more oriented than at point B, although the effect is less significant than dry of optimum.

Permeability (Chapter 7) at constant compactive effort decreases with increasing water content and reaches a minimum at about the optimum. If the compactive effort is increased, the coefficient of permeability decreases because the void ratio decreases (increasing dry unit weight). This change in permeability with molding water content is shown in Fig. 5.6a, where it can be seen that the permeability is about an order of magnitude higher when this soil is compacted dry of optimum than when it is compacted wet of optimum.

Compressibility (Chapter 8) of compacted clays is a function of the stress level imposed on the soil mass. At relatively low stress levels, clays compacted wet of optimum are more compressible. At high stress levels, the opposite is true. In Fig. 5.6b it can be seen that a larger change in void ratio (a decrease) takes place in the soil compacted wet of optimum for a given change (increase) in applied pressure.

Swelling of compacted clays is greater for those compacted dry of optimum. They have a relatively greater deficiency of water and therefore have a greater tendency to adsorb water and thus swell more. Soils dry of optimum are in general more sensitive to environmental changes such as changes in water content. This is just the opposite for shrinkage as shown in Fig. 5.7a, where samples compacted wet of optimum have the highest shrinkage. Also illustrated in the upper part of this figure is the effect of different methods of compacting the samples.

The strength of compacted clays is rather complex. However, for now, just remember that samples compacted dry of optimum have higher strengths than those compacted wet of optimum. The strength wet of optimum also depends somewhat on the type of compaction because of differences in soil structure. If the samples are soaked, the picture changes

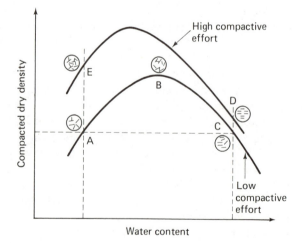

Fig. 5.5 Effect of compaction on soil structure (after Lambe, 1958a).

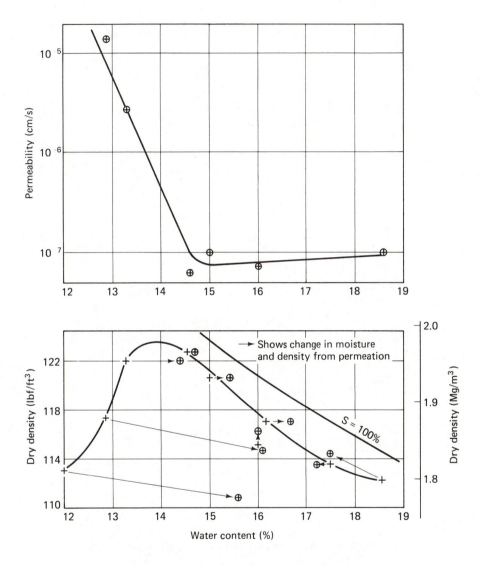

(a) Compaction-permeability tests on Jamaica sandy clay.

Fig. 5.6(a) Change in permeability with molding water content (after Lambe, 1958b).

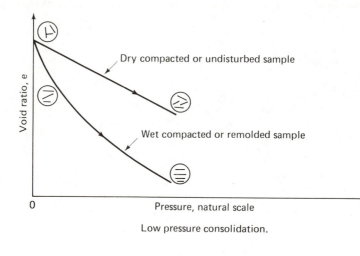

Low pressure consolidation.

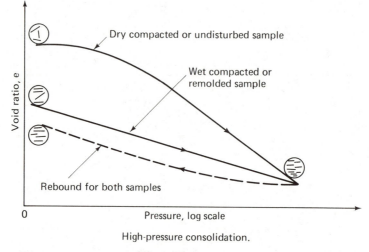

High-pressure consolidation.

(b)

Fig. 5.6(b) Change in compressibility with molding water content (after Lambe, 1958b).

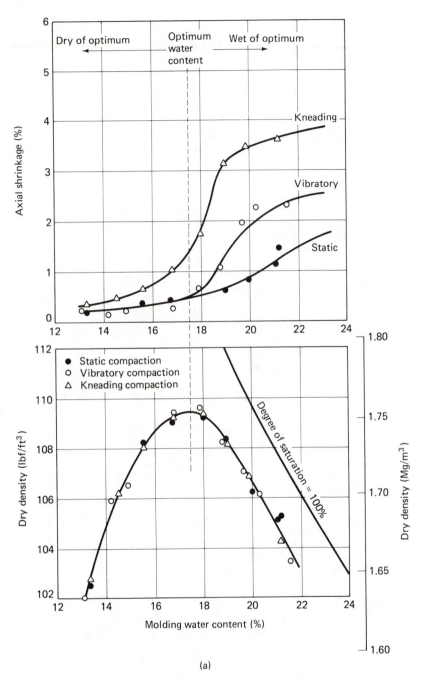

(a)

Fig. 5.7(a) Shrinkage as a function of water content and type of compaction (after Seed and Chan, 1959).

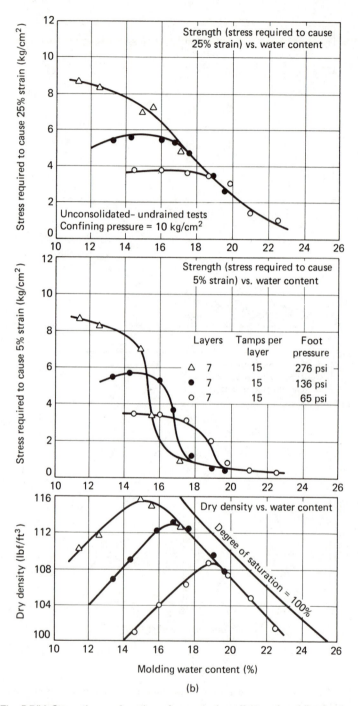

Fig. 5.7(b) Strength as a function of compactive effort and molding water content (after Seed and Chan, 1959).

TABLE 5-1 Comparison of Soil Properties between Dry of Optimum and Wet of Optimum Compaction*

Property	Comparison
1. Structure:	
A. Particle arrangement	Dry side more random
B. Water deficiency	Dry side more deficient; thus imbibes more water, swells more, has lower pore pressure
C. Permanence	Dry side structure sensitive to change
2. Permeability:	
A. Magnitude	Dry side more permeable
B. Permanence	Dry side permeability reduced much more by permeation
3. Compressibility:	
A. Magnitude	Wet side more compressible in low pressure range, dry side in high pressure range
B. Rate	Dry side consolidates more rapidly
4. Strength:	
A. As molded:	
(a) Undrained	Dry side much higher
(b) Drained	Dry side somewhat higher
B. After saturation:	
(a) Undrained	Dry side somewhat higher if swelling prevented; wet side can be higher if swelling permitted
(b) Drained	Dry side about the same or slightly greater
C. Pore water pressure at failure	Wet side higher
D. Stress-strain modulus	Dry side much greater
E. Sensitivity	Dry side more apt to be sensitive

*After Lambe (1958b).

123

due to swelling, especially dry of optimum. The strength curves for a silty clay compacted by kneading compaction for three different compactive efforts are shown in Fig. 5.7b. They show the stress required to cause 25% strain (upper) and 5% strain (middle) for the three compactive efforts. The strengths are about the same wet of optimum and increase significantly on the dry side of optimum.

Note too that at a given water content wet of optimum, the stress at 5% strain is actually less for the higher compaction energies. This fact is also shown in Fig. 5.8, where strength is measured by the CBR (California bearing ratio) test. In this test, the resistance to penetration of a 3 in.^{2} piston developed in a compacted specimen is compared to that developed by a standard sample of densely compacted crushed rock. The CBR is a common pavement design test. In Fig. 5.8 a greater compactive effort produces a greater CBR dry of optimum, as you would expect. But notice how the CBR is actually less wet of optimum for the higher compaction energies. This fact is important in the proper design and management of a compacted earth fill; we shall discuss its implication later in this chapter.

Table 5-1 from Lambe (1958b) is a summary of the effects of wet versus dry of optimum compaction on several engineering properties.

5.5 FIELD COMPACTION EQUIPMENT AND PROCEDURES

Soil to be used in a compacted fill is excavated from a *borrow area*. *Power shovels*, *draglines* and self-propelled *scrapers* or *"pans"* are used to excavate the borrow material. A self-loading scraper is shown in Fig. 5.9a and an elevating scraper in Fig. 5.9b. Sometimes "dozers" are necessary to help load the scraper. Scrapers may cut through layers of different materials, allowing several grain sizes to be mixed, for example. The power shovel mixes the soil by digging along a vertical surface, whereas the scraper mixes the soil by cutting across a sloping surface where different layers may be exposed.

The borrow area may be on site or several kilometres away. Scrapers, off the road vehicles, are often used to transport and spread the soil in *lifts* on the fill area. Trucks may be used as well, on or off the highway, and they may *end dump*, *side dump*, or *bottom dump* the fill material (Fig. 5.10a). For economic reasons, the hauling contractor usually tries to spread the fill material when dumping in order to reduce spreading time. However, unless the borrow materials are already within the desired water content range, the soil may need to be wetted, dried, or otherwise reworked. Where possible, the contractor directs his earth-moving equipment over previously

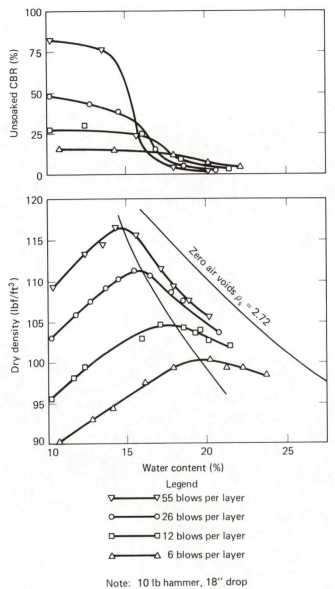

Fig. 5.8 Strength as measured by the CBR and the dry density versus water content for laboratory impact compaction (after Turnbull and Foster, 1956).

125

(a)

(b)

Fig. 5.9 Two types of scrapers: (a) Conventional or self-loading scraper. Sometimes a "dozer" or two helps push the "pan" to load up. Here the two scrapers are assisting each other load up in push-pull. (b) Elevating scraper where the elevating machine loads by itself and eliminates the need for a pusher. (Photographs courtesy of Caterpillar Tractor Co.)

(a)

(b)

Fig. 5.10 Examples of equipment used for hauling and spreading fill materials: (a) fill material being hauled by end dump truck; (b) motor grader spreading and preparing fill subgrade (photographs courtesy of Caterpillar Tractor Co.).

uncompacted soil, thereby reducing the amount of compactive effort required later.

Once borrow material has been transported to the fill area, *bull-dozers*,[*] front loaders, and *motor graders*, called *blades* (Fig. 5.10b), spread the material to the desired layer or *lift* thickness. Lift thickness may range from 150 to 500 mm (6 to 18 in.) or so, depending on the size and type of compaction equipment and on the maximum grain size of the fill.

The kind of compacting equipment or *rollers* used on a job will depend on the type of soil to be compacted. Equipment is available to apply pressure, impact, vibration, and kneading. Figure 5.11 shows two types of rollers.

A *smooth wheel*, or *drum*, roller (Fig. 5.11a) supplies 100% coverage under the wheel, with ground contact pressures up to 380 kPa (55 psi) and may be used on all soil types except rocky soils. The most common use for

*Genus bovinas masculinus sonambulorum.

(a)

(b)

Fig. 5.11 Types of rollers: (a) smooth-wheel roller; (b) rubber-tired roller (self-propelled) (photographs courtesy of Hyster Company, Construction Equipment Division).

large smooth wheel rollers is for *proofrolling* subgrades and compacting asphalt pavements. The *pneumatic*, or *rubber-tired*, *roller* (Fig. 5.11b) has about 80% coverage (80% of the total area is covered by tires) and tire pressures may be up to about 700 kPa (100 psi). A heavily loaded wagon with several rows of four to six closely spaced tires is self-propelled or towed over the soil to be compacted. Like the smooth wheel roller, the rubber-tired roller may be used for both granular and cohesive highway fills, as well as for earth dam construction.

Probably the first roller developed and perhaps the most common type of compactor used today is the *sheepsfoot roller*. This roller, as its name implies, has many round or rectangular shaped protrusions or "feet" (Fig. 5.12a) attached to a steel drum. The area of these protrusions ranges from 30 to 80 cm² (5 to 12 in.²). Because of the 8% to 12% coverage, very high contact pressures are possible, ranging from 1400 to 7000 kPa (200 to 1000 psi) depending on the drum size and whether the drum is filled with water. The drums come in several diameters. Surprisingly enough, a "4 by 4" (which means 4 ft long and 4 ft in diameter) roller provides a higher strength compacted fill in clay soils than a heavier, higher pressure "5 by 5" roller because there is less kneading or shearing action with the "4 by 4" than the "5 by 5" roller, which produces a different soil structure (see Fig. 5.8). Sheepsfoot rollers are usually towed in tandem by crawler tractors or are self-propelled, as shown in Fig. 5.12b.

The sheepsfoot roller starts compacting the soil below the bottom of the foot (projecting about 150 to 250 mm from the drum) and works its way up the lift as the number of passes increases. Eventually the roller "walks out" of the fill as the upper part of the lift is compacted. The sheepsfoot roller is best suited for cohesive soils.

Other rollers with protrusions have also been developed to obtain high contact pressures for better crushing, kneading, and compacting of a rather wide variety of soils. These rollers can either be towed or self-propelled. *Tamping foot rollers* (Fig. 5.13) have approximately 40% coverage and generate high contact pressures from 1400 to 8400 kPa (200 to 1200 psi), depending on the size of the roller and whether the drum is filled for added weight. The special hinged feet (Fig. 5.13a) of the tamping foot roller apply a kneading action to the soil. These rollers compact similarly to the sheepsfoot in that the roller eventually "walks out" of a well-compacted lift. Tamping foot rollers are best for compacting fine-grained soils.

Still another kind of roller is the *mesh*, or *grid pattern*, *roller* with about 50% coverage and pressures from 1400 to 6200 kPa (200 to 900 psi) (Fig. 5.14). The mesh roller is ideally suited for compacting rocky soils, gravels, and sands. With high towing speed, the material is vibrated, crushed, and impacted.

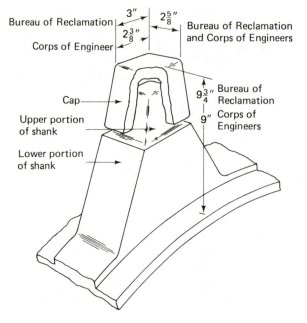

(a)

(b)

Fig. 5.12 Sheepsfoot rollers: (a) detail of a rectangular sheepsfoot (drawing provided by Hyster Company, Construction Equipment Division); (b) self-propelled sheepsfoot roller in foreground ("pan" in background) (photograph courtesy of Caterpillar Tractor Co.).

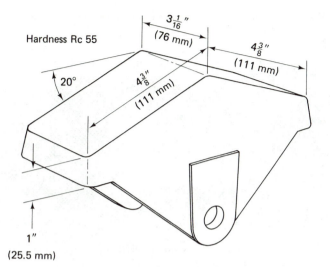

Hardness Rc 55

$3\frac{1}{16}''$
(76 mm)

$4\frac{3}{8}''$
(111 mm)

20°

$4\frac{3}{8}''$
(111 mm)

1"
(25.5 mm)

(a)

(b)

Fig. 5.13 Tamping foot roller: (a) details of a tamping foot; (b) self-propelled tamping foot compactor. Note how the blade is used to spread the material before compaction by the rollers (drawing and photograph courtesy of Caterpillar Tractor Co.).

Fig. 5.14 Grid roller (photograph courtesy of Hyster Company, Construction Equipment Division).

Several compaction equipment manufacturers have attached vertical vibrators to the smooth wheel and tamping foot rollers so as to more efficiently densify granular soils. Figure 5.15 shows a vibrating drum on a smooth wheel roller compacting a gravelly material. Also available are *vibrating plates* and *rammers* that range in size from 230 to 122 mm^2 (9 to 45 in.^{2}) and weigh from 50 to 3000 kg (100 to 6000 lb). Compaction depth for even the larger plates is less than 1 metre. These devices are used in

Fig. 5.15 Vibrating drum on smooth-wheel roller (photograph courtesy of the Hyster Company, Construction Equipment Division).

areas where the larger rollers cannot operate. Broms and Forssblad (1969) have listed the different types of vibratory soil compactors (see Table 5-2) and also indicate the common frequency of operation. In Table 5-3, they illustrate the practical application of these machines.

Probably the best explanation of why roller vibration causes densification of granular soils is that particle rearrangement occurs due to cyclic deformation of the soil produced by the oscillations of the roller. In addition, vibratory compaction can work in materials with some cohesion (Selig and Yoo, 1977). When oscillation is added to a static component, compaction is significantly increased, as shown in Fig. 5.16. For soils compacted on the dry side of optimum, adding the dynamic component results in increased density.

There are many variables which control the vibratory compaction or densification of soils. Some are compactor dependent and some depend on the soil being compacted. The list of variables would include:

Characteristics of the compactor:
  Mass, size
  Operating frequency and frequency range

Characteristics of the soil:
  Initial density
  Grain size and shape
  Water content

Construction procedures:
  Number of passes of the roller
  Lift thickness
  Frequency of operation of vibrator
  Towing speed

The compactor characteristics influence the stress level and depth of influence of the dynamic force, and the initial density strongly influences the final density. For example, the upper 30 cm of medium dense sand may never become compacted higher than the initial density, whereas dense sands will be vibrated loose in the top 30 cm. Once the compactor is chosen, the actual construction procedures essentially govern the results. The influence of operating frequency for various soil types is shown in Fig. 5.17. Note how a peak in the density-frequency curve develops for most of the soils, even clays. The frequency at which a maximum density is achieved is called the *optimum frequency*. It is a function of the compactor-soil system, and it changes as the density increases during the process of compaction. Clearly, it is desirable for a compactor to have the capability to vary its operating frequency and have the range required to obtain

TABLE 5-2 Different Types of Vibratory Soil Compactors*

Surface Vibrators			Internal Vibrators		
Type of Machine	Mass	Frequency	Type of Machine	Diameter	Frequency
Vibrating tampers (rammers):			*Concrete vibrators:*		
Hand-guided	50–150 kg (100–300 lb)	About 10 Hz	Manually operated or tractor-mounted	5–15 cm (2–6 in.)	100–200 Hz
Vibrating plate compactors:			*Vibroflotation equip.:*		
Self-propelled, hand-guided	50–3000 kg (100–6000 lb)	12–80 Hz	Crane-mounted	23–38 cm (9–15 in.)	About 30 Hz
Multiple-type, mounted on tractors, etc.	200–300 kg (400–600 lb)	30–70 Hz			
Crane-mounted†	Up to 20 tons	10–15 Hz			
Vibrating rollers:					
Self-propelled, hand-guided (one or two drums)	250–1500 kg (500–3000 lb)	40–80 Hz			
Self-propelled, tandem-type	0.7–10 tons	30–80 Hz			
Self-propelled, rubber tires	4–25 tons	20–40 Hz			
Tractor-drawn	1.5–15 tons	20–50 Hz			

*After Broms and Forssblad (1969).
†Used on a limited scale.

TABLE 5-3 Applications of Vibratory Soil Compaction*

Type of Machine	Applications
Vibrating tampers (rammers):	Street repair. Fills behind bridge abutments, retaining and basement walls, etc. Trench fills.
Vibrating plate compactors:	
Self-propelled, hand-guided	Base and subbase compaction for streets, sidewalks, etc. Street repair. Fills behind bridge abutments, retaining and basement walls, etc. Fills below floors. Trench fills.
Multiple-type	Base and subbase compaction for highways.
Vibrating rollers:	
Self-propelled, hand-guided	Base, subbase, and asphalt compaction for streets, sidewalks, parking areas, garage driveways, etc. Fills behind bridge abutments and retaining walls. Fills below floors. Trench fills.
Self-propelled, tandem type	Base, subbase, and asphalt compaction for highways, streets, sidewalks, parking areas, garage driveways, etc. Fills below floors.
Self-propelled, rubber tires	Base, subbase, and embankment compaction for highways, streets, parking areas, airfields, etc. Rock-fill dams. Fills (soil or rock) used as foundations for residential and industrial buildings.
Tractor-drawn	Base, subbase, and embankment compaction on highways, streets, parking areas, airfields, etc. Earth- and rock-fill dams. Fills (soil or rock) used as foundations for residential and industrial buildings. Deep compaction of natural deposits of sand.

*After Broms and Forssblad (1969).

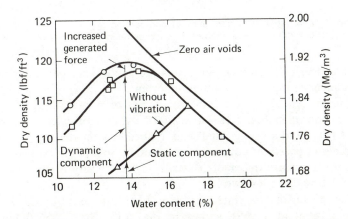

Fig. 5.16 Compaction results on 30 cm (12 in.) layers of silty sand, with and without vibration, using a 7700 kg (17,000 lb) towed vibratory roller (after Parsons, et al., 1962, as cited by Selig and Yoo, 1977).

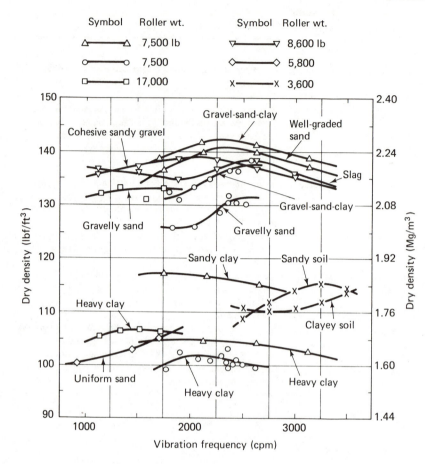

Fig. 5.17 Variation with frequency of compaction by smooth-drum vibratory rollers (after several sources as cited by Selig and Yoo, 1977).

maximum density. However, the peaks are gentle and, percentagewise, a wide frequency range is not all that important.

The influence of the number of passes of a roller and the towing speed are shown in Fig. 5.18 for a 7700 kg roller compacting a "heavy" (high LL) clay and a well-graded sand. Notice how the density increases as the number of passes or coverages increases, up to a point. Not so obvious is that, for a given number of passes, a higher density is obtained if the vibrator is towed more slowly!

The effect of lift thickness may be illustrated by the work of D'Appolonia, et al. (1969), shown in Fig. 5.19. Here a 5670 kg roller operating at a frequency of 27.5 Hz is used to compact a 240 cm thick layer of northern Indiana dune sand. The initial relative density was about

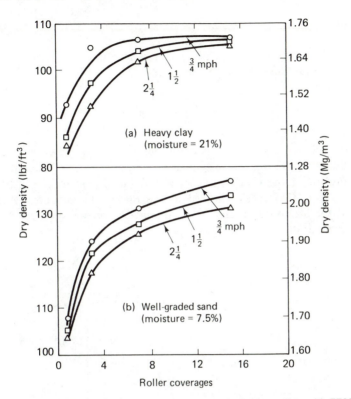

Fig. 5.18 Effect of roller travel speed on amount of compaction with 7700 kg (17,000 lb) towed vibratory roller (after Parsons, et al., 1962, as cited by Selig and Yoo, 1977).

50% to 60%. Field density tests were made in test pits before and after compaction. Note how the density varies with depth. In the upper 15 cm (6 in.), the soil is vibrated loose, whereas the soil reaches its maximum density for a given number of coverages at about 45 cm; thereafter the increase in density tapers off. When compacting past five or so coverages, there is not a great increase in density.

EXAMPLE 5.2

Given:

It is decided, based on economics, that five coverages of a certain roller and operating frequency shall be used.

Required:

What would be the maximum lift thickness to obtain a minimum relative density? Use the data shown in Fig. 5.19.

Solution:

Trace the relative density versus depth curve for five passes on onionskin paper. Superimpose that drawing over the original one, and slide it up and down until the desired relative density is obtained (shown in Fig. Ex. 5.2). About 45 cm (18 in.) is indicated as the maximum thickness. Actually, however, the lift could be thicker as compaction of the top layer densifies the lower layer the second time around.

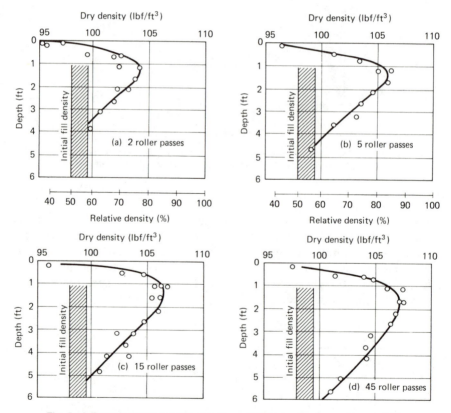

Fig. 5.19 Density-depth relationship for a 5670 kg roller operating at 27.5 Hz for a 240 cm lift height (after D'Appolonia, et al., 1969).

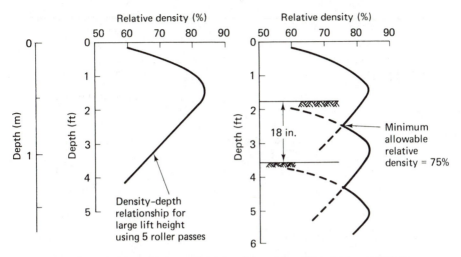

Fig. Ex. 5.2 Approximate method for determining lift height required to achieve a minimum compacted relative density of 75% with five roller passes, using data for a large lift height (after D'Appolonia, et al., 1969).

Figure 5.20 summarizes the applicability of various types of compaction equipment as a function of soil type, expressed as a percentage of sand to clay. These "zones" are not absolute, and it is possible for a given piece of equipment to compact satisfactorily outside the given zone.

When structures are to be founded on relatively deep deposits of loose granular materials, densification by even heavy surface vibratory rollers is usually insufficient, and other techniques must be employed. *Excavation and replacement* of the soil in compacted layers may be economical under certain conditions. Blasting has also been used at some sites (Mitchell, 1970). *Vibro-flotation* (Mitchell, 1970) is often used to increase the density of building foundations on loose sand. Another technique which is gaining in popularity is *dynamic compaction*. Basically, the method consists of repeatedly dropping a very heavy weight (10 to 40 tons mass) some height (10 to 40 m) over the site. The impact produces shock waves that cause densification of unsaturated granular soils. In saturated granular soils, the shock waves can produce partial liquefaction of the sand, a condition similar to quicksand (discussed in Chapter 7), followed by consolidation (discussed in Chapter 8) and rapid densification. The variables include energy (drop height and weight of pounder), the number of drops at a single point (3 to 10), and the pattern of the drops at the surface (5 to 15 m center-to-center). Figure 5.21 shows a pounder just impacting the surface of a loose sand layer. Eventually this site will look like a set of

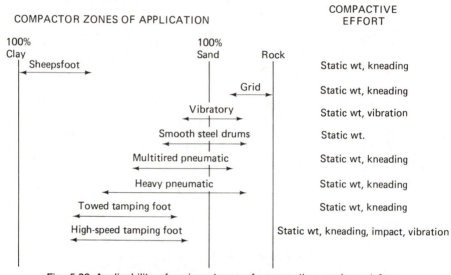

Fig. 5.20 Applicability of various types of compaction equipment for a given soil type (modified after Caterpillar Tractor Co., 1977).

organized moon craters. The craters can be filled with sand and additionally tamped or the area between them smoothed out by the pounder itself.

Dynamic compaction was apparently first used in Germany in the mid-1930's during construction of the Autobahns (Loos, 1936). It has also been used in the USSR to compact loessial soils up to 5 m deep (Abelev, 1957). The technique was further refined and promoted in France and elsewhere by Louis Ménard (Ménard and Broise, 1975), who pioneered in the development of very heavy pounders (up to 200 metric tons mass) and massive cranes and tripods for lifting them to drop heights up to 40 m. Improvement is claimed to depths down to 40 m. In the United States, dynamic compaction has been used on a more modest scale by contractors using ordinary equipment (Leonards, Cutter, and Holtz, 1980; Lukas, 1980).

The depth of influence D, in metres, of the soil undergoing compaction is conservatively given by Leonards, et al. (1980) as

$$D \simeq \tfrac{1}{2}(\text{Wh})^{1/2} \tag{5-2}$$

where W = mass of falling weight in metric tons, and
    h = drop height in metres.

The heavier the weight and/or the higher the drop height, the greater the depth of compaction. Leonards, et al. (1980) also found that the amount of improvement due to compaction in the zone of maximum improvement correlates best with the product of the energy per drop times the total energy applied per unit of surface area.

Fig. 5.21 Dynamic compaction at a site in Bangladesh. The 100 ton crane is dropping a 16 metric ton weight 30 m (courtesy of S. Varaksin, Techniques Louis Ménard, Longjumeau, France).

5.6 FIELD COMPACTION CONTROL AND SPECIFICATIONS

Since the objective of compaction is to stabilize soils and improve their engineering behavior, it is important to keep in mind the desired *engineering properties* of the fill, not just its dry density and water content. This point is often lost in earthwork construction control. Major emphasis is usually placed on achieving the specified dry density, and little consideration is given to the engineering properties desired of the compacted fill. Dry density and water content correlate well with the engineering properties (Sec. 5.4), and thus they are convenient construction control parameters.

The usual design-construct procedure is as follows. Laboratory tests are conducted on samples of the proposed borrow materials to define the properties required for design. After the earth structure is designed, the compaction specifications are written. Field compaction *control tests* are specified, and the results of these become the standard for controlling the project. Construction control inspectors then conduct these tests to see that the specifications are met by the contractor.

There are basically two categories of earthwork specifications: (1) *end-product specifications* and (2) *method specifications*. With the first type, a certain *relative compaction*, or *percent compaction*, is specified. Relative compaction is defined as the ratio of the field dry density $\rho_{d\,field}$ to the laboratory maximum dry density $\rho_{d\,max}$, according to some specified standard test, for example, the standard Proctor or the modified Proctor test; or

$$\text{relative compaction (R.C.)} = \frac{\rho_{d\,field}}{\rho_{d\,max}} \times 100(\%) \tag{5-3}$$

You should note the difference between relative compaction and *relative density* D_r or *density index* I_D, defined in Chapter 4. Relative density, of course, applies only to granular soils. If some fines are present, it is difficult to decide which type of test is applicable as a standard test. ASTM (1980), Designation D 2049, suggests that the relative density is applicable if the soil contains less than 12% fines (passing the No. 200 sieve); otherwise the compaction test should be used. A relationship between relative density and relative compaction is shown in Fig. 5.22. A statistical study of published data on 47 different granular soils indicated

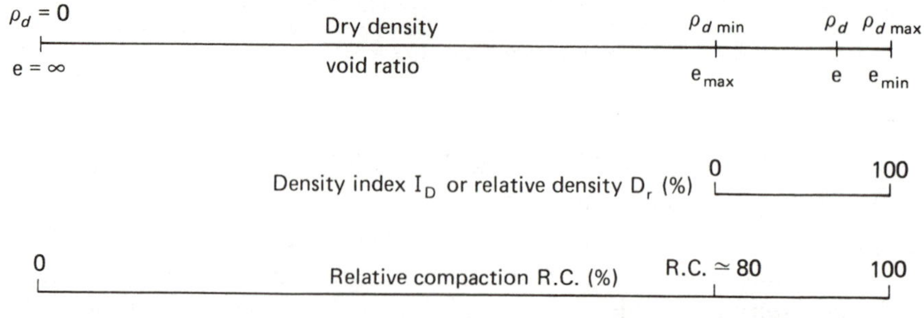

Fig. 5.22 Relative density and relative compaction concepts (after Lee and Singh, 1971).

that the relative compaction corresponding to zero relative density is about 80%.

With *end-product specifications*, which are used for most highways and building foundations, as long as the contractor is able to obtain the specified relative compaction, how he obtains it doesn't matter, nor does the equipment he uses. The economics of the project supposedly ensure that the contractor will utilize the most efficient compaction procedures. The most economical compaction conditions are illustrated in Fig. 5.23, showing three hypothetical field compaction curves of the same soil but at different compactive efforts. Assume that curve 1 represents a compactive effort that can easily be obtained by existing compaction equipment. Then

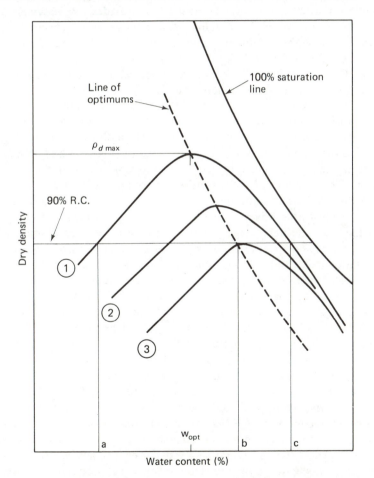

Fig. 5.23 Dry density versus water content, illustrating the most efficient conditions for field compaction (after Seed, 1964).

to achieve, say, 90% relative compaction, the placement water content of the compacted fill must be greater than water content a and less than water content c. These points are found where the 90% R.C. line intersects compaction curve 1. If the placement water content is outside of the range a to c, then it will be difficult, if not impossible, to achieve the required percentage of relative compaction called for, no matter how much the contractor compacts that lift. This is why it may be necessary at times to wet or dry (rework) the soil prior to rolling in the field.

Now that we have established the range of placement water contents, the contractor might ask: "What is the best placement water content to use?" From a purely economical viewpoint, the most efficient water content would be at b, where the contractor provides the *minimum* compactive effort to attain the required 90% relative compaction. To consistently achieve the minimum relative compaction for the project, the contractor will usually use a slightly higher compactive effort, as shown by curve 2 of Fig. 5.23. Thus the most efficient placement water contents exist between the optimum water content and b.

However what may be best from the contractor's viewpoint may not provide a fill with the desired engineering properties. Compacting a soil on the wet side generally results in a lower shear strength than compacting the soil on the dry side of the optimum water content (Figs. 5.7 and 5.8). Other properties such as permeability and shrink-swell potential will also be different. Thus a *range* of placement water contents should also be specified by the designer in addition to the percentage of relative compaction. This point illustrates why the desired engineering performance of the fill rather than just the percentage of compaction must be kept in mind when writing compaction specifications and designing field control procedures.

Figures 5.23 and 5.1 also illustrate that specified densities can be achieved at higher water contents if more compactive effort is applied, either by using heavier rollers or more passes of the same roller. But, as shown in Fig. 5.8, at higher water contents the strength measured by the CBR test curves cross, and a lower strength will be obtained with higher compaction energies wet of optimum. This effect is known as *overcompaction*. Overcompaction can occur in the field when wet of optimum soils are *proofrolled* with very heavy, smooth wheeled rollers (Fig. 5.11a) or an excessive number of passes are applied to the lift (Mills and DeSalvo, 1978). Otherwise even good material can become weaker. You can also detect overcompaction in the field by careful observation of the behavior of the soil immediately under the compactor or the wheel of a heavily loaded scraper. If the soil is too wet and the energy applied is too great, *pumping* or *weaving* of the fill will result as the wheel shoves the wet weaker fill ahead of itself. Also, sheepsfoot rollers won't be able to "walk out."

In *method specifications*, the second general category, the type and weight of roller, the number of passes of that roller, as well as the lift thicknesses are specified by the engineer. A maximum allowable size of material may also be specified. In contrast to the end-product specifications, where the contractor is responsible for proper compaction, with method specifications the responsibility rests with the owner or owner's engineer as to the quality of the earthwork. If compaction control tests performed by the engineer fail to meet a certain standard, then the contractor will be paid extra for additional rolling. This specification requires prior knowledge of the borrow soils so as to be able to predict in advance how many passes of, for example, a certain type of roller will produce adequate compaction performance. This means that during design, test fills must be constructed with different equipment, compactive efforts, etc., in order to determine which equipment and procedures will be the most efficient. Since test fill programs are expensive, method specifications can only be justified for very large compaction projects such as earth dams. However, considerable savings in earthwork construction unit costs are possible because a major part of the uncertainty associated with compaction will be eliminated for the contractor. He can estimate quite well in advance just how much construction will cost. The contractor also knows that if extra rolling is required he will be adequately compensated.

How is relative compaction determined? First, the test site is selected. It should be representative or typical of the compacted lift and borrow material. Typical specifications call for a new field test for every 1000 to 3000 m^3 or so, or when the borrow material changes significantly. It is also advisable to make the field test at least one or maybe two compacted lifts below the already compacted ground surface, especially when sheepsfoot rollers are used or in granular soils.

Field control tests can either be *destructive* or *nondestructive*. Destructive tests involve excavation and removal of some of the fill material, whereas nondestructive tests determine the density and water content of the fill indirectly. The steps required for the common destructive field tests are:

1. Excavate a hole in the compacted fill at the desired sampling elevation (the size will depend on the maximum size of material in the fill). Determine the mass of the excavated material.
2. Take a water content sample and determine the water content.
3. Measure the volume of the excavated material. Techniques commonly employed for this include the sand cone, the balloon method, or pouring water or oil of known density into the hole (Fig. 5.24). In the sand cone method, dry sand of known dry

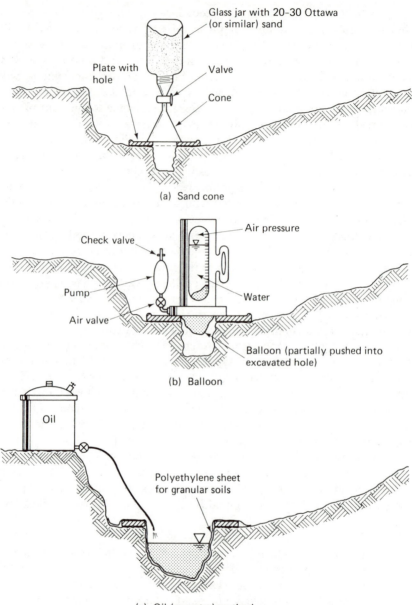

(a) Sand cone

(b) Balloon

(c) Oil (or water) method

Fig. 5.24 Some methods for determining density in the field.

density is allowed to flow through a cone-pouring device into the hole. The volume of the hole can then easily be determined from the weight of sand in the hole and its dry density. (It is necessary that the contractor stop earth-moving equipment from vibrating and densifying the sand in the hole during the sand cone test; otherwise the measured percent relative compaction will be lower than actual.) In the balloon method, the volume is determined directly by the expansion of a balloon directly in the hole.

4. Compute the total density ρ. Knowing M_t, the total mass of the material excavated from the hole, and the volume of the hole, we can compute ρ. Since we also know the water content, we can obtain the dry density of the fill, $\rho_{d\ field}$.

5. Compare $\rho_{d\ field}$ with $\rho_{d\ max}$ and calculate relative compaction (Eq. 5-3).

EXAMPLE 5.3

Given:

A field density test is performed by the balloon method (Fig. 5.24b). The following data were obtained from the test:

$$
\begin{array}{ll}
\text{Mass of soil removed + pan} = & 1590 \text{ g} \\
\text{Mass of pan} = & 125 \text{ gm} \\
\text{Balloon readings:} & \\
\quad \text{Final} = & 1288 \text{ cm}^3 \\
\quad \text{Initial} = & 538 \text{ cm}^3 \\
\end{array}
$$

$$
\begin{array}{ll}
\text{Water content information:} & \\
\quad \text{Mass of wet soil + pan} = & 404.9 \text{ g} \\
\quad \text{Mass of dry soil + pan} = & 365.9 \text{ g} \\
\quad \text{Mass of pan} = & 122.0 \text{ g} \\
\end{array}
$$

Required:

a. Compute the dry density and water content of the soil.

b. Using curve B of Fig. 5.1 as the laboratory standard, compute the relative compaction.

Solution:

a. Compute the wet density, $\rho = \dfrac{M_t}{V_t}$

$$\rho = \frac{1590 - 125 \text{ g}}{1288 - 538 \text{ cm}^3} = \frac{1465 \text{ g}}{750 \text{ cm}^3} = 1.95 \text{ g/cm}^3 = 1.95 \text{ Mg/m}^3$$

Water content determination:

1. Mass of wet soil + pan = 404.9 g
2. Mass of dry soil + pan = 365.9 g
3. Mass of water M_w (1 − 2) = 39.0 g
4. Mass of pan = 122.0 g
5. Mass of dry soil M_s (2 − 4) = 243.9 g
6. Water content $(M_w/M_s) \times 100$ (3 ÷ 5) = 16%

For calculation of dry density, use Eq. 2-14:

$$\rho_d = \frac{\rho}{1 + w} = \frac{1.95 \text{ Mg/m}^3}{1 + 0.16} = 1.68 \text{ Mg/m}^3$$

b. For calculation of relative compaction, use Eq. 5-3:

$$\text{R.C.} = \frac{\rho_{d \text{ field}}}{\rho_{d \text{ max}}} = \frac{1.68}{1.86} \times 100 = 90.3\%$$

There are several problems associated with the common destructive field density test. First, the laboratory maximum density may not be known exactly. It is not uncommon, especially in highway construction, for a series of laboratory compaction tests to be conducted on "representative" samples of the borrow materials for the highway. Then, when the field test is conducted, its result is compared with the results of one or more of these job "standard" soils. If the soils at the site are highly variable, this is a poor procedure. Another alternative is to determine the complete compaction curve for each field test—a time-consuming and expensive proposition.

A second alternative is to perform a *field check point*, or 1 point Proctor test. When the field engineer knows in advance that the soil in which he is performing a field density test does not exactly visually match one of the borrow soils, an extra amount of soil is removed from the compacted fill during the test. The total amount of soil removed should be sufficient to perform a single laboratory compaction test. The only restrictions necessary for the performance of the field check point are that:

1. During compaction, the mold must be placed on a smooth solid mass of at least 100 kg, a requirement which may be difficult to achieve in the field. Asphalt pavement or compacted soil should *not* be used.

2. The soil to be compacted must be dry of optimum for the compactive effort used, and to know when the soil is dry of optimum takes some field experience.

The reason for this second requirement may be apparent from Fig. 5.25. Three compaction curves are shown for soils *A*, *B*, and *C* from a given construction job borrow area. The soil just tested for density, as identified by the field engineer, does not match any soils for which curves exist. The field check point is plotted as point *X* on the graph. By drawing a line parallel to the dry side of optimum of curves *A*, *B*, and *C* and reaching a maximum at the "line of optimums," a reasonable approximation of the maximum dry density may be obtained. If the soil was not adequately dried out before the compaction test, a point such as *Y* would be obtained. Then it would be difficult to distinguish which laboratory curve the soil belonged to, and therefore an estimate of the maximum dry density would be impossible. Some experience is required to "feel" when the soil is dried out enough for the field check point water content to be less than the OMC.

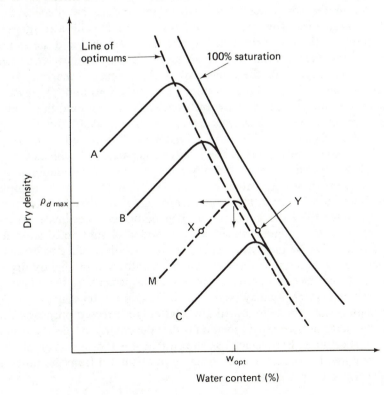

Fig. 5.25 Principle of the check point test.

The second major problem with the common destructive density test procedure is that the determination of the water content takes time (several hours or overnight according to ASTM, 1980). Time is often of the utmost value on a compaction job, and if it takes a day or even several hours before the results are available, several lifts of fill may have been placed and compacted over the "bad" or "failing" test area. Then the engineer has a difficult decision to make: should the contractor be required to tear out a lot of possibly good fill just to improve the relative compaction of that one "bad" lift? Contractors understandably are very hesitant to do that, and yet how many zones of "bad" compaction are allowable in an embankment? Of course, the problem is statistical and again, on a typical job, it is difficult and expensive to conduct sufficient tests for a statistical analysis of the compaction results.

Since determining the water content takes the most time, several methods have been proposed to obtain a more rapid water content. Pan drying or "frying" the sample over an open flame is commonly used, but since it is difficult to control the temperature, it gives poor results, especially for fat clay (CH) soils. The "speedy" moisture meter, in which the water in the soil reacts with carbide to produce acetylene gas, is another alternative. The gas pressure shown on a calibrated gage is proportional to the water content. Burning with methanol and the special alcohol-hydrometer method are also sometimes used. The correlation with standard oven drying for these methods is approximate—generally satisfactory for silts and lean clays but poor for organic soils and fat clays.

Another method for quickly and efficiently determining the relative compaction of cohesive soils was developed in the 1950's by the U.S. Bureau of Reclamation (1974, and Hilf, 1961). The procedure makes it possible to determine accurately the relative compaction of a fill as well as a very close approximation of the difference between the optimum water content and the fill water content without oven drying the sample. Samples of the fill materials are compacted according to the desired laboratory standard at the fill water content and, depending on an estimate of how close the fill is to optimum, water is either added or subtracted from the sample (Fig. 5.26). With a little experience it is relatively easy to estimate whether the fill material is about optimum, slightly wet, or slightly dry of optimum. From the wet density curve, the exact percent relative compaction based on dry density may be obtained. Only one water content, the fill water content, need be determined and that only for record purposes. The main advantage of the "rapid" method is that the contractor has the results in a very short time. Experience has shown that it is possible to obtain the values required for control of construction in about 1 h from the time the field density test is performed.

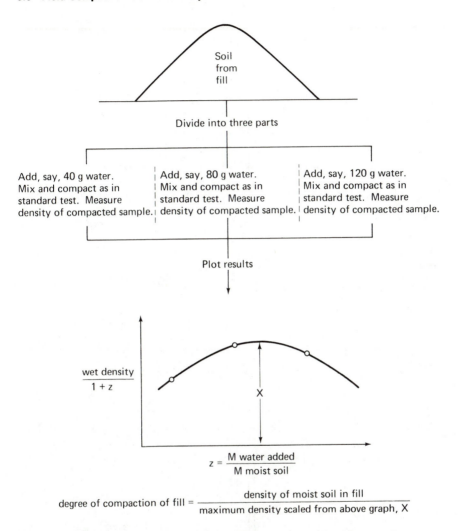

Divide into three parts

Add, say, 40 g water.	Add, say, 80 g water.	Add, say, 120 g water.
Mix and compact as in	Mix and compact as in	Mix and compact as in
standard test. Measure	standard test. Measure	standard test. Measure
density of compacted sample.	density of compacted sample.	density of compacted sample.

Plot results

$$\frac{\text{wet density}}{1 + z}$$

X

$$z = \frac{M \text{ water added}}{M \text{ moist soil}}$$

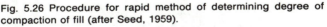

$$\text{degree of compaction of fill} = \frac{\text{density of moist soil in fill}}{\text{maximum density scaled from above graph, X}}$$

Fig. 5.26 Procedure for rapid method of determining degree of compaction of fill (after Seed, 1959).

Other problems with destructive field tests are associated with the determination of the volume of the excavated material. The sand cone, often taken as the "standard," is subject to errors. For example, vibration from nearby working equipment will increase the density of the sand in the hole, which gives a larger hole volume than it should have; this results in a lower field density. All of the common volumetric methods are subject to error if the compacted fill is gravel or contains large gravel particles. Any

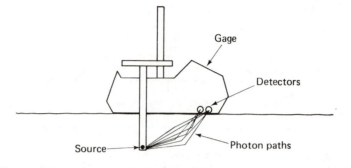

(a)

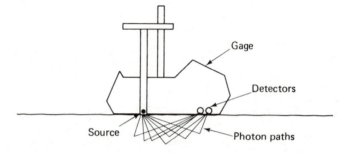

(b)

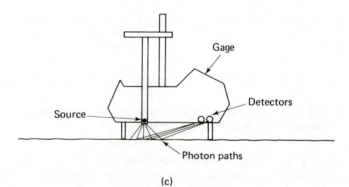

(c)

Fig. 5.27 Nuclear density and water content determination: (a) direct transmission; (b) backscatter; (c) air gap (after Troxler Electronic Laboratories, Inc., Research Triangle Park, North Carolina).

kind of unevenness in the walls of the hole causes a significant error in the balloon method. If the soil is coarse sand or gravel, none of the liquid methods works well, unless the hole is very large and a polyethylene sheet is used to contain the water or oil.

Because of some of the problems with destructive field tests, nondestructive density and water content testing using radioactive isotopes has increased in popularity during the past few years. Nuclear methods have several advantages over the traditional techniques. The tests can be conducted rapidly and results obtained within minutes. Therefore the contractor and engineer know the results quickly, and corrective action can be taken before too much additional fill has been placed. Since more tests can be conducted, a better statistical control of the fill is provided. An average value of the density and water content is obtained over a significant volume of fill, and therefore the natural variability of compacted soils can be considered. Disadvantages of nuclear methods include their relatively high initial cost and the potential danger of radioactive exposure to field personnel. Strict radiation safety standards must be enforced when nuclear devices are used.

Basically, two types of sources or emitters are necessary to determine both the density and the water content. Gamma radiation, as provided by radium or a radioactive isotope of cesium, is scattered by the soil particles; the amount of scatter is proportional to the total density of the material. The spacing between the source and pickup, usually a scintillation or Geiger counter, is constant. Hydrogen atoms in water scatter neutrons, and this provides a means whereby water content can be determined. Typical neutron sources are americium-beryllium isotopes. Calibration against compacted materials of known density is necessary, and for instruments operating on the surface the presence of an uncontrolled air gap can significantly affect the measurements.

Three nuclear techniques are in common use. The *direct transmission* method is illustrated schematically in Fig. 5.27a, and the *backscatter* technique is shown in Fig. 5.27b. The less common *air-gap* method (Fig. 5.27c) is sometimes used when the composition of the near-surface materials adversely affects the density measurement.

5.7 ESTIMATING PERFORMANCE OF COMPACTED SOILS

How will a given soil behave in a fill, supporting a foundation, holding back water, or under a pavement? Will frost action be a critical factor? For future reference, we present the experience of the U.S. Army

Corps of Engineers on compaction characteristics applicable to roads and airfields (Table 5-4) and the experience of the U.S. Department of the Interior, Water and Power Resources Service for several types of earth structures (Table 5-5).

In Table 5-4, the terms *base*, *subbase*, and *subgrade* (columns 7, 8, and 9) refer to components of a pavement system, and they are defined in Fig. 5.28. In column 16, the term CBR represents the California bearing ratio. The CBR is used by the Corps of Engineers for the design of *flexible* pavements. The modulus of subgrade reaction (column 17) is used by them for *rigid* pavement design. Though the difference is rather arbitrary, the upper layers of flexible pavements usually are constructed of asphaltic concrete, whereas rigid pavements are made of Portland cement concrete. A good reference for the design of pavements is the book by Yoder and Witczak (1975).

The use of these tables in engineering practice is best shown by an example. They are very helpful for preliminary design purposes, choosing the most suitable compaction equipment, and for rapid checking of field and laboratory test results.

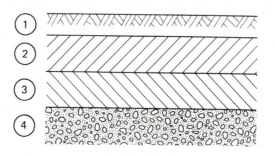

(1) Wearing surface: 20-25 cm Portland cement or 2-8 cm asphaltic concrete.

(2) Base: 5-10 cm asphaltic concrete, 15-30 cm sand-gravel base, 20-30 cm soil-cement, or 15-20 cm asphalt stabilized sand.

(3) Subbase material (this layer may be omitted): 15-30 cm sand-gravel.

(4) Subgrade: The natural soil at the site. The top 0.15-0.5 m is usually compacted prior to the placement of the other layers of the pavement.

Fig. 5.28 Definitions of terms relating to pavement systems, with typical dimensions and materials for each component.

TABLE 5-4 Characteristics Pertinent to Roads and Airfields*

Major Divisions (1) (2)		Letter (3)	Hatching (4)	Color (5)	Name (6)	Value as Subgrade When not Subject to Frost Action (7)	Value as Subbase When not Subject to Frost Action (8)	Value as Base When not Subject to Frost Action (9)	Potential Frost Action (10)	Compressibility and Expansion (11)
COARSE-GRAINED SOILS	GRAVEL AND GRAVELLY SOILS	GW		Red	Well-graded gravels or gravel-sand mixtures, little or no fines	Excellent	Excellent	Good	None to very slight	Almost none
		GP		Red	Poorly graded gravels or gravel-sand mixtures, little or no fines	Good to excellent	Good	Fair to good	None to very slight	Almost none
		GM d / u		Yellow	Silty gravels, gravel-sand-silt mixtures	Good to excellent / Good	Good / Fair	Fair to good / Poor to not suitable	Slight to medium / Slight to medium	Very slight / Slight
		GC		Yellow	Clayey gravels, gravel-sand-clay mixtures	Good	Fair	Poor to not suitable	Slight to medium	Slight
	SAND AND SANDY SOILS	SW		Red	Well-graded sands or gravelly sands, little or no fines	Good	Fair to good	Poor	None to very slight	Almost none
		SP		Red	Poorly graded sands or gravelly sands, little or no fines	Fair to good	Fair	Poor to not suitable	None to very slight	Almost none
		SM d / u		Yellow	Silty sands, sand-silt mixtures	Fair to good / Fair	Fair to good / Poor to fair	Poor / Not suitable	Slight to high / Slight to high	Very slight / Slight to medium
		SC		Yellow	Clayey sands, sand-clay mixtures	Poor to fair	Poor	Not suitable	Slight to high	Slight to medium
FINE-GRAINED SOILS	SILTS AND CLAYS LL IS LESS THAN 50	ML		Green	Inorganic silts and very fine sands, rock flour, silty or clayey fine sands or clayey silts with slight plasticity	Poor to fair	Not suitable	Not suitable	Medium to very high	Slight to medium
		CL		Green	Inorganic clays of low to medium plasticity, gravelly clays, sandy clays, silty clays, lean clays	Poor to fair	Not suitable	Not suitable	Medium to high	Medium
		OL		Green	Organic silts and organic silt-clays of low plasticity	Poor	Not suitable	Not suitable	Medium to high	Medium to high
	SILTS AND CLAYS LL IS GREATER THAN 50	MH		Blue	Inorganic silts, micaceous or diatomaceous fine sandy or silty soils, elastic silts	Poor	Not suitable	Not suitable	Medium to very high	High
		CH		Blue	Inorganic clays of high plasticity, fat clays	Poor to fair	Not suitable	Not suitable	Medium	High
		OH		Blue	Organic clays of medium to high plasticity, organic silts	Poor to very poor	Not suitable	Not suitable	Medium	High
HIGHLY ORGANIC SOILS		Pt		Orange	Peat and other highly organic soils	Not suitable	Not suitable	Not suitable	Slight	Very high

*After U.S.Army Waterways Experiment Station (1960).

155

TABLE 5-4 Continued

Major Divisions (1)	(2)	Symbol Letter (3)	Symbol Hatching (4)	Color (5)	Drainage Characteristics (12)	Compaction Equipment (13)	Unit Dry Densities lbf/ft³ (14)	Mg/m³ (15)	CBR (16)	Subgrade Modulus k (lbf/in.³) (17)
COARSE-GRAINED SOILS	GRAVEL AND GRAVELLY SOILS	GW		Red	Excellent	Crawler-type tractor, rubber-tired roller, steel-wheeled roller	125-140	2.00-2.24	40-80	300-500
		GP		Red	Excellent	Crawler-type tractor, rubber-tired roller, steel-wheeled roller	110-140	1.76-2.24	30-60	300-500
		GM d			Fair to poor	Rubber-tired roller, sheepsfoot roller; close control of moisture	125-145	2.00-2.32	40-60	300-500
		GM u		Yellow	Poor to practically impervious	Rubber-tired roller, sheepsfoot roller	115-135	1.84-2.16	20-30	200-500
		GC		Yellow	Poor to practically impervious	Rubber-tired roller, sheepsfoot roller	130-145	2.08-2.32	20-40	200-500
	SAND AND SANDY SOILS	SW		Red	Excellent	Crawler-type tractor, rubber-tired roller	110-130	1.76-2.08	20-40	200-400
		SP		Red	Excellent	Crawler-type tractor, rubber-tired roller	105-135	1.68-2.16	10-40	150-400
		SM d			Fair to poor	Rubber-tired roller, sheepsfoot roller; close control of moisture	120-135	1.92-2.16	15-40	150-400
		SM u		Yellow	Poor to practically impervious	Rubber-tired roller, sheepsfoot roller	100-130	1.60-2.08	10-20	100-300
		SC		Yellow	Poor to practically impervious	Rubber-tired roller, sheepsfoot roller	100-135	1.60-2.16	5-20	100-300
FINE-GRAINED SOILS	SILTS AND CLAYS LL IS LESS THAN 50	ML		Green	Fair to poor	Rubber-tired roller, sheepsfoot roller; close control of moisture	90-130	1.44-2.08	15 or less	100-200
		CL		Green	Practically impervious	Rubber-tired roller, sheepsfoot roller	90-130	1.44-2.08	15 or less	50-150
		OL		Green	Poor	Rubber-tired roller, sheepsfoot roller	90-105	1.44-1.68	5 or less	50-100
	SILTS AND CLAYS LL IS GREATER THAN 50	MH		Blue	Fair to poor	Sheepsfoot roller, rubber-tired roller	80-105	1.28-1.68	10 or less	50-100
		CH		Blue	Practically impervious	Sheepsfoot roller, rubber-tired roller	90-115	1.44-1.84	15 or less	50-150
		OH		Blue	Practically impervious	Sheepsfoot roller, rubber-tired roller	80-110	1.28-1.76	5 or less	25-100
HIGHLY ORGANIC SOILS		Pt		Orange	Fair to poor	Compaction not practical				

156

TABLE 5-4 Continued

Notes:

1. In column 3, division of GM and SM groups into subdivisions of d and u are for roads and airfields only. Subdivision is on basis of Atterberg limits; suffix d (e.g., GMd) will be used when the liquid limit is 25 or less and the plasticity index is 5 or less; the suffix u will be used otherwise.

2. In column 13, the equipment listed will usually produce the required densities with a reasonable number of passes when moisture conditions and thickness of lift are properly controlled. In some instances, several types of equipment are listed because variable soil characteristics within a given soil group may require different equipment. In some instances, a combination of two types may be necessary.

   a. *Processed base materials and other angular materials.* Steel-wheeled and rubber-tired rollers are recommended for hard, angular materials with limited fines or screenings. Rubber-tired equipment is recommended for softer materials subject to degradation.

   b. *Finishing.* Rubber-tired equipment is recommended for rolling during final shaping operations for most soils and processed materials.

   c. *Equipment size.* The following sizes of equipment are necessary to assure the high densities required for airfield construction:

   Crawler-type tractor—total weight in excess of 30,000 lb (14 000 kg).

   Rubber-tired equipment—wheel load in excess of 15,000 lb (7000 kg), wheel loads as high as 40,000 lb (18 000 kg) may be necessary to obtain the required densities for some materials (based on contact pressure of approximately 65 to 150 psi or 450 kPa to 1000 kPa).

   Sheepsfoot roller—unit pressure (on 6 to 12 in.^{2} or 40 to 80 cm^2 foot) to be in excess of 250 psi (1750 kPa) and unit pressures as high as 650 psi (4500 kPa) may be necessary to obtain the required densities for some materials. The area of the feet should be at least 5% of the total peripheral area of the drum, using the diameter measured to the faces of the feet.

3. In columns 14 and 15, densities are for compacted soil at optimum water content for modified AASHTO compaction effort.

4. In column 16, the maximum value that can be used in design of airfields is, in some cases, limited by gradation and plasticity requirements.

TABLE 5-5 Tabulations of Engineering Properties of Compacted Soils Used in Earth Structures*

Typical names of Soil Groups	Group Symbols	Important Engineering Properties				Relative Desirability for Various Uses (No. 1 is considered the best)									
		Permeability When Compacted	Shear Strength When Compacted and Saturated	Compressibility When Compacted and Saturated	Workability as a Construction Material	Rolled Earthfill Dams			Canal Sections		Foundations		Fills		Roadways
						Homogeneous Embankment	Core	Shell	Erosion Resistance	Compacted Earth Lining	Seepage Important	Seepage Not Important	Frost Heave Not Possible	Frost Heave Possible	Surfacing
Well-graded gravels, gravel-sand mixtures, little or no fines	GW	Pervious	Excellent	Negligible	Excellent	—	—	1	1	—	—	1	1	1	3
Poorly graded gravels, gravel-sand mixtures, little or no fines	GP	Very pervious	Good	Negligible	Good	—	—	2	2	—	—	3	3	3	—
Silty gravels, poorly graded gravel-sand-silt mixtures	GM	Semipervious to impervious	Good	Negligible	Good	2	4	—	4	4	1	4	4	9	5
Clayey gravels, poorly graded gravel-sand-clay mixtures	GC	Impervious	Good to fair	Very low	Good	1	1	—	3	1	2	6	5	5	1
Well-graded sands, gravelly sands, little or no fines	SW	Pervious	Excellent	Negligible	Excellent	—	—	3 if gravelly	6	—	—	2	2	2	4
Poorly graded sands, gravelly sands, little or no fines	SP	Pervious	Good	Very low	Fair	—	—	4 if gravelly	7 if gravelly	—	—	5	6	4	—
Silty sands, poorly graded sand-silt mixtures	SM	Semipervious to impervious	Good	Low	Fair	4	5	—	8 if gravelly	5 erosion critical	3	7	8	10	6

Description	Symbol														
Clayey sands, poorly graded sand-clay mixtures	SC	Impervious	Good to fair	Low	Good	3	2	—	5	2	4	8	7	6	2
Inorganic silts and very fine sands, rock flour, silty or clayey fine sands with slight plasticity	ML	Semipervious to impervious	Fair	Medium	Fair	6	6	—	—	6 erosion critical	6	9	10	11	—
Inorganic clays of low to medium plasticity, gravelly clays, sandy clays, silty clays, lean clays	CL	Impervious	Fair	Medium	Good to fair	5	3	—	9	3	5	10	9	7	7
Organic silts and organic silt-clays of low plasticity	OL	Semipervious to impervious	Poor	Medium	Fair	8	8	—	—	7 erosion critical	7	11	11	12	—
Inorganic silts, micaceous or diatomaceous fine sandy or silty soils, elastic silts	MH	Semipervious to impervious	Fair to poor	High	Poor	9	9	—	—	—	8	12	12	13	—
Inorganic clays of high plasticity, fat clays	CH	Impervious	Poor	High	Poor	7	7	—	10	8 volume change critical	9	13	13	8	—
Organic clays of medium to high plasticity	OH	Impervious	Poor	High	Poor	10	10	—	—	—	10	14	14	14	—
Peat and other highly organic soils	Pt	—	—	—	—	—	—	—	—	—	—	—	—	—	—

*After USBR (1974).

EXAMPLE 5.4

Given:

A soil, classified as a CL according to the USCS, is proposed for a compacted fill.

Required:

Consider the soil to be used as:

    a. Subgrade
    b. Earth dam
    c. Foundation support for a structure

Use Tables 5-4 and 5-5 and comment on:

    1. The overall suitability of the soil
    2. Potential problems of frost
    3. Significant engineering properties
    4. Appropriate compaction equipment to use

Solution:

Prepare a table for soil type CL.

Item Use:	Subgrade	Earth Dam	Structural Foundation
1. Applicability	Poor to fair	Useful as central core	Acceptable if compacted dry of optimum and if not saturated during service life
2. Frost potential	Medium to high	Low if covered by nonfrost heaving soil of sufficient depth	Medium to high if not controlled by temperature and water availability
3. Engineering properties	Medium compressibility fair strength CBR $\leq$ 15	Low permeability, compact for low permeability and high strength but also for flexibility	Potential for poor strength and therefore poor performance
4. Appropriate compaction equipment	Sheepsfoot and/or rubber-tired roller	Sheepsfoot and/or rubber-tired roller	Sheepsfoot and/or rubber-tired roller

Note: Once you have finished with this book and a course in foundation engineering, you could readily expand the information in this table.

PROBLEMS

5-1. For the data in Fig. 5.1:

(a) Estimate the maximum dry density and optimum water content for both the standard curve and the modified Proctor curve.

(b) What is the placement water content range for 90% relative compaction for the modified Proctor curve and 95% relative compaction for the standard Proctor curve?

(c) For both curves, estimate the maximum placement water content for the minimum compactive effort to achieve the percent relative compaction in part (b).

5-2. The natural water content of a borrow material is known to be 10%. Assuming 6000 g of *wet* soil is used for laboratory compaction test points, compute how much water is to be added to other 6000 g samples to bring their water contents up to 13, 17, 20, 24, and 28%.

5-3. For the soil shown in Fig. 5.1, a field density test provided the following information:

Water content = 14%
Wet density = 1.89 Mg/m³ (118 lbf/ft³)

Compute the percent relative compaction based on the modified Proctor and the standard Proctor curves.

5-4. For the data given below ($\rho_s = 2.64$ Mg/m³):

(a) Plot the compaction curves.

(b) Establish the maximum dry density and optimum water content for each test.

(c) Compute the degree of saturation at the optimum point for data in column A.

(d) Plot the 100% saturation (zero air voids) curve. Also plot the 70, 80, and 90% saturation curves. Plot the line of optimums.

A		B		C	
(modified)		(standard)		(low energy)	
ρ_d (Mg/m³)	$w/c\%$	ρ_d (Mg/m³)	$w/c\%$	ρ_d (Mg/m³)	$w/c\%$
1.873	9.3	1.691	9.3	1.627	10.9
1.910	12.8	1.715	11.8	1.639	12.3
1.803	15.5	1.755	14.3	1.740	16.3
1.699	18.7	1.747	17.6	1.707	20.1
1.641	21.1	1.685	20.8	1.647	27.4
		1.619	23.0		

5-5. A field compaction control test was conducted on a compacted lift. The mass of the material removed from the hole was 1814 g and the volume of the hole was found to be 944 cm^3. A small sample of the soil lost 15 g in the drying test and the mass remaining after drying was 100 g. The laboratory control test results are shown in Fig. P5-5.
   (a) If end-product specification requires 100% relative compaction and w = (optimum $-$ 3%) to (optimum $+$ 1%), determine the acceptability of the field compaction and state why this is so.
   (b) If it is not acceptable, what should be done to improve the compaction so that it will meet the specification?

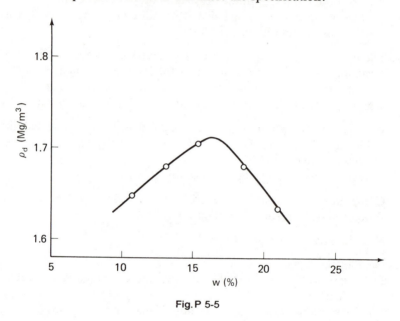

Fig. P 5-5

5-6. Calculate the compactive effort of the modified Proctor test in both (a) SI and (b) British Engineering units.

5-7. Why does the relative compaction decrease if there is vibration during the sand cone test?

5-8. In a field density test, using the oil method, the wet mass of soil removed from a small hole in the fill was 1.59 kg. The mass of oil (sp. gr. = 0.90) required to fill the hole was 0.81 kg, and the field water content was found to be 25%. If the ρ_s of the soil solids is 2700 kg/m^3, what is the dry density and degree of saturation of the fill?

5-9. Let's pretend you are an earthwork construction control inspector and you are checking the field compaction of a layer of soil. The laboratory compaction curve for the soil is shown in Fig. P5-9. Specifications call for the compacted density to be at least 95% of the maximum laboratory value and within ± 2% of the optimum water content. When you did the sand cone test, the volume of soil excavated was 1153 cm³. It weighed 2209 g wet and 1879 g dry.

(a) What is the compacted dry density?
(b) What is the field water content?
(c) What is the relative compaction?
(d) Does the test meet specifications?
(e) What is the degree of saturation of the field sample?
(f) If the sample were saturated at constant density, what would be the water content?

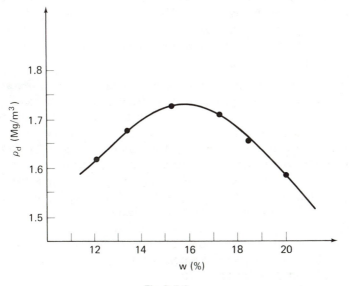

Fig. P 5-9

5-10. You are checking a field compacted layer of soil. The laboratory control curve has the following values:

ρ_d (lb/ft³)	w (%)
104	14
105.5	16
106	18
105	20
103.5	22
101	24

The specification for compaction states that the compacted field soil must be at least 95% of the maximum control density and within $\pm 2\%$ of the optimum moisture for the control curve. You dig a hole of $\frac{1}{30}$ ft^3 in the compacted layer and extract a sample that weighs 4 lb wet and 3.4 lb dry.

(a) What is the compacted ρ_d? The compaction w? The percent compaction? Does the sample meet the specifications?

(b) If the density of solids is 2.70 Mg/m^3, what is the compacted degree of saturation? If the sample were saturated at constant density, what would be the water content? (After C. W. Lovell.)

5-11. A mixture contains 30% by *dry weight* fines and 70% coarse. When the coarse material has a $w = 2\%$, its affinity for water is completely satisfied. The fines have a PL = 20 and an LL = 40. This mixture is *compacted* by rolling to $\rho_d = 130$ pcf and $w_{mix} = 15\%$. What is the water content of the *fines* in the compacted mass? What is the liquidity index of the fines in the compacted mass? (After C. W. Lovell.)

5-12. A soil proposed for a compacted fill contains 40% fines and 60% coarse material by *dry weight*. When the coarse fraction has $w = 1.5\%$, its affinity for water is completely satisfied (that is, it is saturated but surface dry). The Atterberg limits of the fines are LL = 27 and PL = 12. The soil is compacted by rolling to a $\rho_d = 2.0$ Mg/m^3 at $w = 13\%$. Note: This is the water content of the entire soil mixture.

(a) What is the water content of the *fines* in the compacted mass?

(b) What is the likely classification of the soil? (Give both the Unified and the AASHTO classifications).

(c) What is the liquidity index of the fines?

(d) What can you say about the susceptibility of the fill to
(1) shrinkage-swelling potential?
(2) potential for frost action?

(e) Is there a certain type of compaction equipment you would especially recommend for this job? Why?

5-13. A fine sand with poor gradation is to be used as a subgrade for a flexible pavement. Give as much information as you can about the suitability of this soil as a pavement subgrade.

5-14. What soils, if properly compacted, would make the best foundation material for a structure? Give your answers in terms of the Unified Soil Classification System symbols.

5-15. The same as problem 5-14, for an earth dam.

5-16. Given: The data shown in Fig. 5.2. Soil types 3 and 4 are mixed in the borrow area to some unknown extent. After air drying a representative sample of the combined material to a uniform water content (hopefully on the dry side of optimum), a compaction test is performed and a value of 1.8 Mg/m^3 dry density at 11% water content is obtained. (a) Estimate the maximum dry density of the combined soils. (b) If a field dry density of 1.58 Mg/m^3 were obtained after compaction by a sheepsfoot roller, compute the relative compaction.

5-17. The core of an earth dam is to be compacted on the wet side of optimum so as to ensure low permeability and flexibility (a nonbrittle stress-strain relationship). You have the choice of using a sheepsfoot roller or a smooth wheel roller to compact the soil. To reduce potential shrinkage of the dam core, which is the best piece of equipment to use? If the soil were to be compacted dry of optimum, would it matter?

six

Water in Soils, I: Capillarity, Shrinkage, Swelling, Frost Action

6.1 INTRODUCTION

From previous discussions on the Atterberg limits, classification of soils, and soil structure, you should now realize that the presence of water in soils is very important. Water very strongly affects the engineering behavior of most soils, especially fine-grained soils, and water is an important factor in most geotechnical engineering problems. A few examples include capillarity, swelling, and frost action in soils, discussed in this chapter, and seepage of water through dams, levees, etc., discussed in Chapter 7. Problems of settlement of structures constructed on clay soils and the stability of foundations and slopes also involve water to some extent. As an indication of the practical importance of water in soils, it has been estimated that more people have lost their lives as a result of failures of dams and levees due to seepage and "piping" (Chapter 7) than to all the other failures of civil engineering works combined. In the United States, damage from swelling soils annually causes a greater economic loss than floods, hurricanes, tornadoes, and earthquakes combined.

In general, water in soils can be thought of as either static or dynamic. The ground water table, even though it actually fluctuates throughout the year, is considered to be static for most engineering purposes. Adsorbed water (Chapter 4) is generally static. Similarly, capillary water is usually taken to be static, although it too can fluctuate, depending on climatic conditions and other factors. In this chapter we shall concentrate on static soil-water problems.

The following notation is introduced in this chapter.

Symbol	Dimension	Unit	Definition
d	L	m, mm	Diameter
F	MLT^{-2}	N	Force
h_c	L	m	Height of capillary rise
p_{atm}	$ML^{-1}T^{-2}$	kPa	Atmospheric pressure
r_m	L	m, mm	Radius of meniscus
T	MT^{-2}	N/m	Surface Tension
u	$ML^{-1}T^{-2}$	kPa	Pore water pressure
u_c	$ML^{-1}T^{-2}$	kPa	Capillary pressure
α	—	degree	Contact angle

6.2 CAPILLARITY

Capillarity arises from a fluid property known as *surface tension*, which is a phenomenon that occurs at the interface between different materials. For soils, it occurs between surfaces of water, mineral grains, and air. Fundamentally, surface tension results from differences in forces of attraction between the molecules of the materials at the interface.

The phenomenon of capillarity may be demonstrated in many ways. Placing the end of a dry towel in a tub of water will eventually result in a saturated towel. To illustrate the effects of capillarity in soils, we can use the analogy of small diameter glass tubes to represent the voids between the soil grains. Capillary tubes demonstrate that the adhesion forces between the glass walls and water causes the water to rise in the tubes and form a meniscus* between the glass and the tube walls. The height of rise is inversely proportional to the diameter of the tubing; the smaller the inside diameter of the tube, the greater the height of capillary rise. The meniscus formed is concave upward with the water "hanging," so to speak, on the walls of the glass tube (Fig. 6.1a). With some materials the internal cohesion forces are greater than the adhesion forces, and the substance will not "wet" the glass tube. Mercury, for example, has a depressed meniscus; its shape is convex (Fig. 6.1b).

If we look more closely at the meniscus geometry for water in a fine capillary tube (Fig. 6.2), we can write equations for the forces acting in the

*After Giacomo Meniscus (1449-1512), a Venetian physician and friend of Leonardo da Vinci. (We are indebted to Prof. M. E. Harr, Purdue University, for this little known fact.)

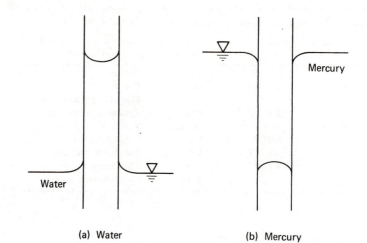

(a) Water   (b) Mercury

Fig. 6.1   Menisci in glass tubes in (a) water and (b) mercury.

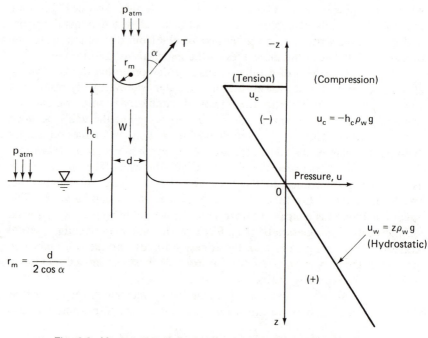

Fig. 6.2   Meniscus geometry of capillary rise of water in a glass tube.

water column. The force acting downward, considered positive, is the weight of the column of water, or

$$\sum F_{\text{down}} = \text{volume } (\rho_w)g = h_c \left(\frac{\pi}{4}d^2\right)\rho_w g \qquad (6\text{-}1)$$

The upward force is the vertical component of the reaction of the meniscus against the tube circumference, or

$$\sum F_{\text{up}} = \pi d T \cos \alpha \qquad (6\text{-}2)$$

where T is the *surface tension* of the water-air interface which acts around the circumference of the tube. The surface tension has dimensions of force/unit length. The other terms are functions of the geometry of the system and are defined in Fig. 6.2.

For equilibrium $\sum F_v = 0$, and

$$-(h_c)\frac{\pi}{4}d^2\rho_w g - \pi d T \cos \alpha = 0 \qquad (6\text{-}3)$$

Solving for the height of capillary rise, for clean glass tubes and pure water, $\alpha \to 0$ and $\cos \alpha \to 1$. Thus

$$h_c = \frac{-4T}{\rho_w g d} \qquad (6\text{-}4)$$

The capillary rise is upward, above the *free water surface*, but is has a negative value because of the sign convention shown in Fig. 6.2. Surface tension T is a physical property of water and, from the *Handbook of Chemistry and Physics* (1977), at 20°C, T is about 73 dynes/cm or 73 mN/m. Since $\rho_w = 1000$ kg/m^3 and $g = 9.81$ m/s^2, for pure water in clean glass tubes Eq. 6-4 reduces to

$$h_c = \frac{-3(10^{-5})\text{m}}{d(\text{in m})} = \frac{-0.03 \text{ m}}{d(\text{in mm})} \qquad (6\text{-}5)$$

This formula is easy to remember. For the height of capillary rise in metres, divide 0.03 by the diameter in millimetres.

All the preceding discussion is for clean glass tubes and pure water under laboratory conditions. In reality, the actual height of capillary rise is likely to be somewhat less due to the presence of impurities and imperfectly clean surfaces.

EXAMPLE 6.1

Given:

The diameter of a clean glass capillary tube is 0.1 mm.

Required:

Expected height of capillary rise of water.

Solution:

Use Eq. 6-5.

$$h_c = \frac{0.03}{0.1 \text{ mm}} (\text{m}) = 0.3 \text{ m}$$

Also shown in Fig. 6.2 is the pressure or stress distribution in the water. Below the surface of the water reservoir, the pressure increases linearly with depth (hydrostatic pressure). Above the reservoir surface, the water pressure in the capillary tube is negative or less than zero gage pressure (referenced to atmospheric pressure). From Eq. 6-4 and for $\alpha \simeq 0$, its magnitude is

$$u_c = h_c \rho_w g = -\frac{4T}{d} = \frac{-2T}{r_m} \tag{6-6}$$

The shape of the meniscus is actually spherical (a minimum energy condition) with radius r_m (Fig. 6.2). The radius is greater than or equal to the radius of the tube, depending on the contact angle α. When α is approximately zero, then $r_m = d/2$.

What is the maximum negative pressure that can be attained? In large tubes, the limitation is the vapor pressure of water. When the pressure goes negative, that is, less than atmospheric, water will cavitate, or "boil," when the ambient pressure reaches the vapor pressure. In absolute terms, the vapor pressure of water is 17.54 mm Hg or 2.34 kPa absolute at 20°C [from the *Handbook of Chemistry and Physics* (1977)]. The relationships between absolute, gage, and vapor pressure of water is shown in Fig. 6.3. The equivalent capillary tube diameter at the vapor pressure is about 3 μm. Now, if the tube is smaller than this diameter, then the water cannot cavitate because the surface tension is too high and a bubble cannot form. In this case, the height of capillary rise in smaller tubes depends only on the tube diameter, and the rise may be much greater than 10 m. Similarly, the capillary pressure (pore water tension) in this case may be much greater than -1 atm or -100 kPa. It should be noted that for *large tubes* the maximum allowable tension or suction in water depends only on the atmospheric pressure and has nothing to do with the diameter of the tube. Capillary rise in *small tubes*, on the other hand, has no relation to atmospheric pressure and is a function of the tube diameter only (Terzaghi and Peck, 1967).

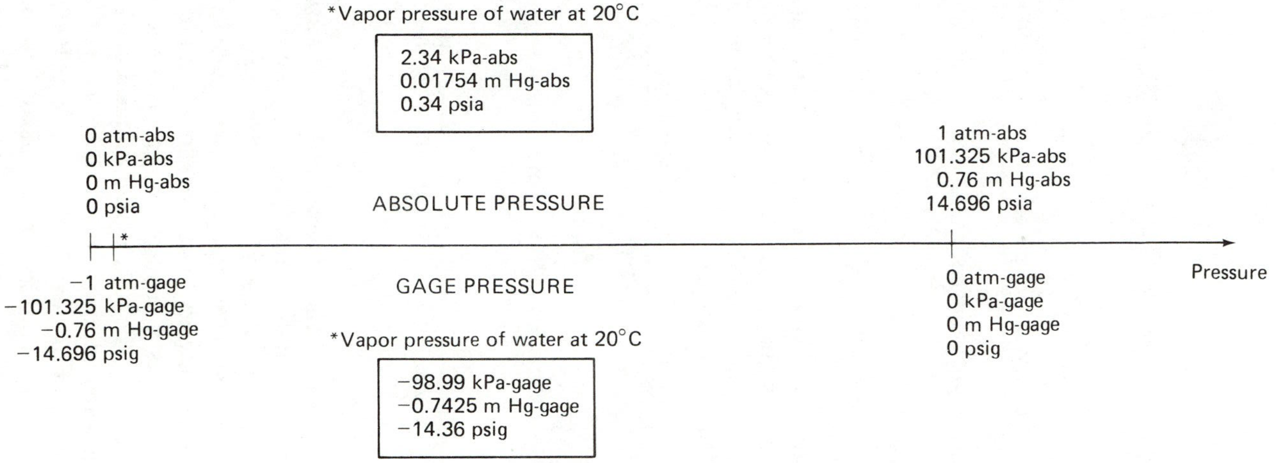

*Vapor pressure of water at 20°C

| 2.34 kPa-abs |
| 0.01754 m Hg-abs |
| 0.34 psia |

ABSOLUTE PRESSURE

1 atm-abs
101.325 kPa-abs
0.76 m Hg-abs
14.696 psia

GAGE PRESSURE

*Vapor pressure of water at 20°C

| −98.99 kPa-gage |
| −0.7425 m Hg-gage |
| −14.36 psig |

0 atm-gage
0 kPa-gage
0 m Hg-gage
0 psig

0 atm-abs
0 kPa-abs
0 m Hg-abs
0 psia

−1 atm-gage
−101.325 kPa-gage
−0.76 m Hg-gage
−14.696 psig

Pressure

Fig. 6.3 The relationship between atmospheric and vapor pressures (water) in terms of absolute and gage pressures.

EXAMPLE 6.2

Given:

The pressure relationships shown in Fig. 6.3.

Required:

a. Show that the maximum height of a water column in a large tube is about 10 m.

b. Show that the equivalent pore diameter at the vapor pressure is about 3 μm.

Solution:

a. In large tubes, the maximum height of a water column is governed by the vapor pressure or the maximum negative pressure in the water. From Fig. 6.3, at the vapor pressure, the pressure is -98.99 kPa. Using Eq. 6-6 we have,

$$h_c = \frac{u_c}{\rho_w g} = \frac{-98.99 \text{ kPa}}{(1000 \text{ kg/m}^3)(9.81 \text{ m/s}^2)}$$

$$= -10.1 \text{ m (rise)}$$

since 1 kPa = 10^3 kg·m/s^2/m^2 (Appendix A).

b. Use Eq. 6-5.

$$d_c = \frac{-3(10^{-5}) \text{ m}}{h_c(\text{in m})} = \frac{3(10^{-5}) \text{ m}}{-10.1 \text{ m}} = 3(10^{-6}) \text{ m}$$

Even though soils are random assemblages of particles and the resulting voids are similarly random and highly irregular, the capillary tube analogy, although imperfect, helps explain capillary phenomena observed in real soils. In principle, capillary or negative pressures and capillary rise will be similar in soils and in glass tubes. Let's look at a series of capillary tubes in Fig. 6.4. Tube 1 has a diameter d_c, and thus the corresponding height of capillary rise is h_c. The fully developed meniscus has a radius r_c. In tube 2, $h < h_c$; the water will try to rise to h_c but cannot. Consequently, the radius of the meniscus in tube 2 will be *greater* than r_c since it is physically impossible for the corresponding capillary pressure (and therefore the r_c) to develop. In tube 3, a large bubble or void exists, and there is no way for the water to be pulled above a void with diameter greater than

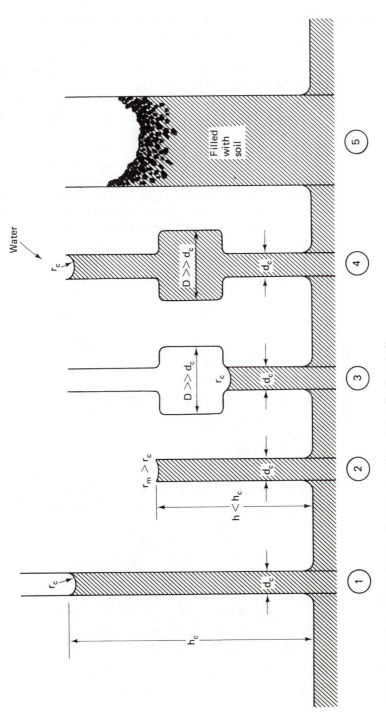

Fig. 6.4 Capillary rise in tubes of different shapes (after Taylor, 1948).

d_c. However, if as is shown in tube 4 water infiltrates down from the top, then it is possible for the meniscus at the top of the tube to support the entire column of water. The walls of the void support the water in the void outside of a column of water of diameter d_c. Tube 5 is filled with soil, and the water would rise to the surface of the soil since the average or *effective* pore diameter is much less than d_c. The capillary menisci hang on the particles, which increases the contact forces between the particles. A magnified picture of two sand particles connected by menisci of radius r_m is shown in Fig. 6.5a. The intergranular contact stress is σ'.

In soils, it is common to assume the effective pore diameter is about 20% of the effective grain size (D_{10}). Using this assumption, we can estimate a theoretical height of capillary rise and the corresponding capillary pressure for a fine-grained soil. This assumption points up the importance of *pore size*, not grain size, as the controlling factor in capillar-

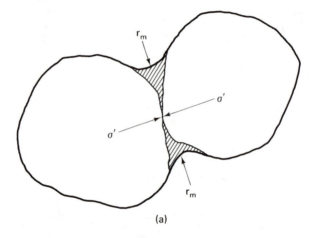

(a)

Fig. 6.5   (a) Two soil grains held together by a capillary film.

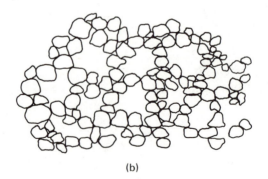

(b)

Fig. 6.5   (b) Bulking structure in sand.

ity. Research at Purdue University (Garcia-Bengochea, et al., 1979) has shown that, depending on the type of compaction and the molding water content, you can get very different pore size distributions in the same soil. The geologic history, soil structure, and fabric of natural soils also varies tremendously, and it is possible to have very different pore size distributions in soils with the same D_{10}.

EXAMPLE 6.3

Required:

Calculate (a) the theoretical height of capillary rise and (b) the capillary pressure in a clay soil with D_{10} of 1 μm.

Solution:

Effective pore diameter = 0.2 (D_{10}) = 0.2 μm = 0.2 $\times$ 10^{-3} mm

a. Capillary rise (Eq. 6-5):

$$h_c = \frac{-0.03 \text{ m}}{0.2 \times 10^{-3} \text{ mm}} = 150 \text{ m (about 500 ft)}$$

b. Capillary pressure (Eq. 6-6):

$$u_c = h_c \rho_w g = -150 \text{ m } (1000 \text{ kg/m}^3)(9.81 \text{ m/s}^2)$$

$$\simeq -1500 \text{ kPa} \simeq -15 \text{ atm} \simeq -225 \text{ psi}$$

These are very large pressures indeed and we can show that the intergranular stresses between the soil grains are of the same order of magnitude. The *intergranular* or *effective stress* σ' (Fig. 6.5a), which is discussed in more detail in the next chapter, is defined as the total stress σ minus the pore water pressure u,

$$\sigma' = \sigma - u \tag{6-7}$$

In Example 6.3, the clay sample in the laboratory is acted on by atmospheric pressure, or the total stress $\sigma = 0$ (zero gage pressure). Then $\sigma' = -(-u_c) = u_c$ or, for Example 6.3, $\sigma' \simeq +1500$ kPa. Rarely in nature do the capillary pressures actually reach such a magnitude. Some of the voids in natural soils are large enough that the water can vaporize and bubbles can form. Thus, menisci are destroyed and the actual height of capillary rise is reduced. But still heights of capillary rise can be significant

TABLE 6-1   Approximate Height of Capillary Rise in Different Soils*

	Loose		Dense
Coarse sand	0.03–0.12 m		0.04–0.15 m
Medium sand	0.12–0.50 m		0.35–1.10 m
Fine sand	0.30–2.0 m		0.40–3.5 m
Silt	1.5–10 m		2.5–12 m
Clay		≥ 10 m	

*After Hansbo (1975).

in especially fine-grained soils, and capillary pressures can be important. Table 6-1 lists some typical heights of capillary rise for several soil types.

At the top of the soil-water column, the tension in the water pulls the grains of soil together, as was shown in Fig. 6.5a. The greater the capillary tension, the greater the intergranular contact stress, and therefore a higher frictional resistance develops between the grains. The effect is similar to what happens when some sand is placed in a rubber membrane, sealed, and a vacuum drawn on the sample. The external air pressure holds the grains tightly together and thereby increases their strength considerably.

The moisture film surrounding the individual grains results in "apparent cohesion." It is not true cohesion in a physical sense. In some cases, for example, end dumping of moist sands, a very loose honeycombed structure (similar to Fig. 4.20) results, as is shown in Fig. 6.5b. The grains are all held together by capillary films, and the resulting structure, although it has a very "loose" relative density, is fairly stable as long as the capillary menisci are present. This condition is called *bulking*, and it occurs only in moist sands. It is possible to destroy the capillary menisci by flooding and thereby decrease the volume significantly. Still, flooding is not a very good way to increase the overall density of a sand fill; the relative density of flooded fills will still be very low and thus not a very good foundation material. Figure 6.5b also shows why it is not a good idea to purchase moist sand by the volume—one may end up buying a lot of air!

Another important consequence of the increase in effective stress that occurs due to capillarity is illustrated by the racetrack at Daytona Beach, Florida (Fig. 6.6). The sands are very fine and have been densified somewhat by wave action. The capillary zone, which is relatively wide due to the flat slope of the beach, provides excellent driving conditions due to high capillary pressures. As in tube 5 of Fig. 6.4, the confining pressure results from the columns of water hanging on the menisci at the surface of the beach. Where the ocean water destroys the menisci, the bearing capacity is very poor, as anyone who has ever tried to escape a rising tide at a beach in an automobile knows!

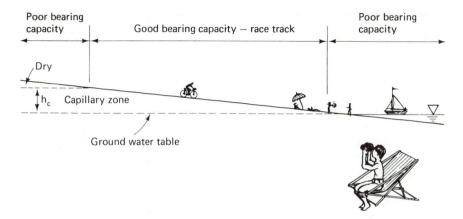

Fig. 6.6 Cross section of racetrack at Daytona Beach, Florida.

Similarly, above the zone of capillary rise, the sand is dry and also has relatively poor bearing capacity, especially for moving vehicles. The relative density throughout the beach zone is essentially the same, yet the bearing capacity is significantly different simply due to capillarity.

Capillarity also allows excavations to be constructed in silts and very fine sands—materials that, if dry, would readily fall to their natural *angle of repose* (Fig. 6.7 and Chapter 11). Below the ground water table, excavations in granular soils will collapse since the menisci obviously do not exist

Fig. 6.7 Illustration of the angle of repose (photograph by M. Surendra).

there! Above the ground water table and within the zone of capillarity, capillary menisci at the surface of the excavation provide the stability for the cut. However such excavations are extremely unstable. They have been known to collapse due to even very slight vibrations, such as from trucks on nearby streets or nearby construction operations like pile driving. Sometimes well points and other methods for lowering the ground water table are used to cause a tension in the pore water (Terzaghi and Peck, 1967). If pumping is stopped because of a power failure, for example, the excavation can fail. High humidity, rain, or even the sweat off workpersons' backs have been blamed for the collapse of unsupported excavations in the organic silts (called *bull's liver*) along the Hudson River in New York.

6.3 SHRINKAGE PHENOMENA IN SOILS

We can get an idea of how capillary stresses cause shrinkage in clay soils by studying the analogy of a horizontal tube with compressible elastic walls (Terzaghi, 1927). In Fig. 6.8a the tube at the beginning is completely filled with water and the radii of the menisci, which haven't reached their final shape yet, are very large. As evaporation occurs, pressure in the water decreases and the menisci start to form (Fig. 6.8b). As evaporation continues, the radii become smaller and smaller, the compression in the compressible walls of the tube increases, and the tube shrinks in length and diameter. The limiting case, shown in Fig. 6.8c, is when the radii of the menisci are at the minimum (equal to one-half the diameter of the tube) and are fully developed. The negative pressure in the capillary tube is then equal to the value computed from Eq. 6-6, and the walls of the tube have shrunk to an equilibrium condition between the rigidity of the walls and the capillary forces. If the tube is immersed in water, the menisci are destroyed and the tube can expand because the capillary forces acting on the tube walls are destroyed.

Unless the walls of the tube are perfectly elastic, the tube will not return completely to its original length and diameter.

Another analogy illustrating the development of menisci in soils is shown in Fig. 6.9. As with the previous case, the tube initially is completely filled with water. As evaporation takes place, the menisci begin to form and, at the beginning, the largest possible radius is the radius of the larger end, r_l. At the smaller end, the radius is also equal to r_l. It cannot be any smaller because then the pressure would have to be lower (more negative) and that cannot happen. By hydrostatics, the pressure in the water must be the same at both ends, otherwise flow would occur towards the end with

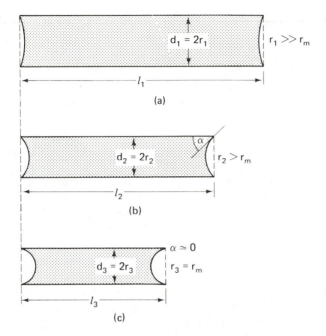

Fig. 6.8 Compressible elastic capillary tube shrinking due to evaporation and surface tension (after Terzaghi, 1927).

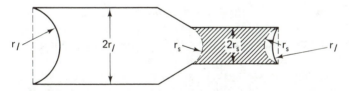

Fig. 6.9 Capillarity in a tube of unequal radii (after A. Casagrande).

the lower (more negative) pressure. As evaporation continues, the menisci retreat until the condition indicated by the cross-hatched section of the tube occurs. At this time, the menisci have radii equal to r_s, the radius of the smaller section of the capillary tube; the capillary pressure may go no further (no more negative) and corresponds to the pressure which can be supported by the smaller diameter radii. This pressure is given by Eq. 6-6. Eventually, the tube will become empty if evaporation continues.

Another simple analogy was used by Terzaghi to illustrate the effects of capillary pressures in a porous material (Casagrande, 1938). A loose ball of absorbent cotton is submerged in a beaker and allowed to become completely saturated. If the ball is compressed then released, the fibers will quickly swell again. However, if the compressed ball is removed from the water and released, it will essentially retain its compressed shape because of the capillary menisci that form around the fibers. In fact, the ball will be rather firm as long as it doesn't dry out too much. If the ball is again immersed in water, the menisci are destroyed and the fibers again become extremely loose and soft. Similar behavior results when dry cotton is compressed; it is quite elastic and becomes loose once the compression is released.

If the tube of Fig. 6.9 is assumed to be compressible, then the analogy with shrinking soils is very useful. A soil sample slowly drying (that is, undergoing desiccation) will form capillary menisci between the individual soil grains. As a result, the stresses between the grains (intergranular or effective stresses) will increase and the soil will decrease in volume. As shrinkage continues, the menisci become smaller and the capillary stresses increase, which further reduces the volume. A point is reached where no further volume decrease occurs, but with the degree of saturation still essentially 100%. The water content at which this occurs is defined as the *shrinkage limit* (SL or w_S), and it is one of the Atterberg limits mentioned in Chapter 2. At this point, the capillary menisci just begin to retreat below the soil surface, and the color of the surface changes from a shiny one to a dull appearance. (The same effect is observed when a dilatant soil [Chapter 3] is stressed—the menisci retreat below the surface, which becomes dull in appearance because the reflectivity of the surface changes.)

How is the shrinkage limit determined? Atterberg's (1911) original work was with small clay bars which he allowed to dry slowly. He observed the point at which the color changed and at the same time he noted that the length was essentially a minimum at that point. Terzaghi figured out that one could just as well measure the dry volume and dry mass and back calculate the water content at the point of minimum volume. Figure 6.10 illustrates this procedure. A small amount of soil of total mass M_i is placed in a small dish of known volume V_i and allowed to slowly dry. After the oven dry mass M_s is obtained, the volume of the dry soil V_{dry} is measured by weighing the amount of mercury the soil sample displaces. The shrinkage limit SL is calculated from

$$\text{(a)} \quad \text{SL} = \left(\frac{V_{dry}}{M_s} - \frac{1}{\rho_s} \right) \rho_w \times 100 \ (\%) \tag{6-8}$$

or

$$\text{(b)} \quad \text{SL} = w_i - \left(\frac{(V_i - V_{dry}) \rho_w}{M_s} \right) \times 100 \ (\%) \tag{6-9}$$

The two equations correspond to the two parts of Fig. 6.10 and may readily be derived from the figure and the fundamental phase relationships of Chapter 2.

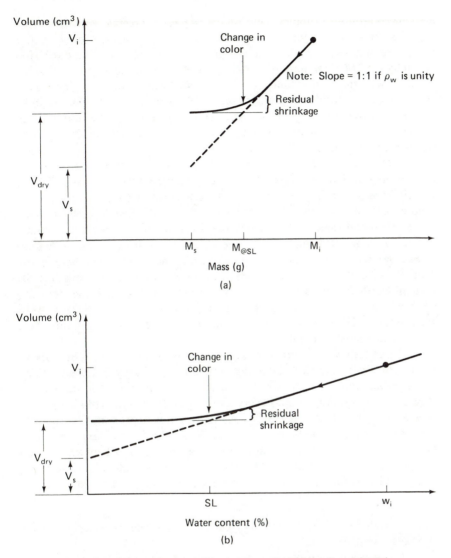

Fig. 6.10 Determination of the shrinkage limit, based on (a) total mass, and (b) water content.

Although the shrinkage limit was a popular classification test during the 1920's, it is subject to considerable uncertainty and thus is no longer commonly conducted. The test has some undesirable features. They include errors resulting from entrapped air bubbles in the dry soil specimen, cracking during drying, weighing and other measurement errors, and the danger of mercury poisoning to the operator. Casagrande suggested drying large samples and physically measuring their dimensions to avoid the mercury poisoning problem. One of the biggest problems with the shrinkage limit test is that the amount of shrinkage depends not only on the grain size but also on the initial fabric of the soil. The standard (for example, ASTM Designation D 427) procedure is to start with the water content near the liquid limit. However, especially with sandy and silty clays, this often results in a shrinkage limit greater than the plastic limit, which is meaningless (Fig. 2.6). Casagrande suggests that the initial water content be slightly greater than the PL, if possible, but admittedly it is difficult to avoid entrapping air bubbles.

If the soil is in a natural undisturbed state, then the shrinkage limit often is greater than the plastic limit due to the structure of the soil. This is especially true for highly sensitive clays according to Karlsson (1977). Figure 6.11 shows the results of tests on several Swedish clays, both undisturbed and remolded, in which the shrinkage limits are plotted versus the plastic limits. For highly sensitive clays, the shrinkage limit of undisturbed samples is considerably greater than the plastic limit, while for medium sensitive clays the SL is close to the PL. For organic soils, the SL is significantly below the PL for both types of samples. Although sensitivity has a precise engineering definition (Chapter 11), you already have some notion of it from our previous discussions of Atterberg limits (Sec. 2.7) and clay microstructure (Sec. 4.8).

If one follows Casagrande's advice and begins the test slightly above the plastic limit, then the following results are generally obtained. When the Atterberg limits for the soil plot near the A-line on the plasticity chart (Fig. 3.2), the shrinkage limit is very close to 20. If the limits plot above the A-line, then the SL is less than 20 by an amount approximately equal to how far the limits are above the A-line. Similarly, for ML and MH (and OL and OH) soils, the shrinkage limit is greater than 20 by an amount approximately equal to how far below the A-line the limits plot. Therefore, if the vertical distance above or below the A-line is Δp_i, then the

$$SL = 20 \pm \Delta p_i \qquad (6\text{-}10)$$

This procedure and equation have been found to be as accurate as the shrinkage limit test itself.

An even simpler procedure has been suggested by Prof. A. Casagrande in his lectures at Harvard University. If the U-line and A-line of the plasticity chart (Fig. 3.2) are extended, they converge at a point with coordinates $(-43.5, -46.4)$, as shown in Fig. 6.12. A line is extended from that point to the coordinates of the liquid limit and plasticity index on the plasticity chart; the shrinkage limit is where that line crosses the liquid limit axis. Although not an exact procedure, it is close enough for geotechnical work. However, the shrinkage limit test, as conducted according to ASTM, for example, is not particularly exact either. If you can obtain a

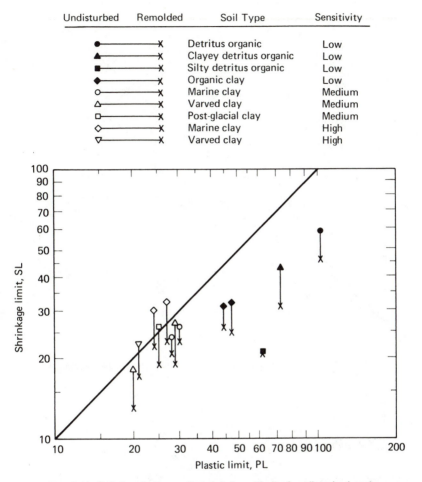

Undisturbed	Remolded	Soil Type	Sensitivity
●————X		Detritus organic	Low
▲————X		Clayey detritus organic	Low
■————X		Silty detritus organic	Low
◆————X		Organic clay	Low
○————X		Marine clay	Medium
△————X		Varved clay	Medium
□————X		Post-glacial clay	Medium
◇————X		Marine clay	High
▽————X		Varved clay	High

Fig. 6.11 Relation between the shrinkage limit of undisturbed and remolded samples and the plastic limit for several Swedish clays (after Karlsson, 1977).

reasonable estimate of the shrinkage limit from the plasticity chart (Fig. 6.12) then the shrinkage limit test need not be performed since it provides no additional information.

Note that the capillary pressures must be very large for very fine-grained clay soils with highly active clay minerals (near the U-line). These soils will have shrinkage limits around 8, according to the Casagrande procedure. In fact, Prof. Casagrande has observed shrinkage limits as low as 6 for montmorillonitic clays. Soils at the shrinkage limit will have a very low void ratio because the capillary pressures are so large, much greater than can be achieved by compaction, for example.

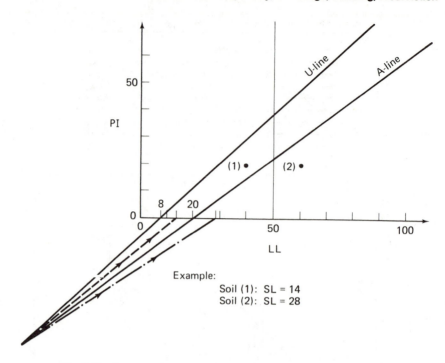

Fig. 6.12 Casagrande's procedure for estimating the shrinkage limit.

EXAMPLE 6.4

Required:

Calculate the void ratio and dry density of a soil with a shrinkage limit of 8. Assume $\rho_s = 2.70 \text{ Mg/m}^3$.

Solution:

Use Eqs. 2–12 and 2–15. Assume $S = 100\%$.

$$e = \frac{w\rho_s}{\rho_w} = \frac{0.08(2.70 \text{ Mg/m}^3)}{1 \text{ Mg/m}^3} = 0.22$$

$$\rho_d = \frac{\rho_s}{1+e} = \frac{2.70 \text{ Mg/m}^3}{1.22} = 2.21 \text{ Mg/m}^3 = 137.9 \text{ lbf/ft}^3$$

Since the density of concrete is about 2.4 Mg/m^3, you can see that the capillary pressures must be very large to cause soil to become so dense. It should not be surprising then that some clay soils have very high dry strengths.

One way to show that high capillary pressures can exist in soils is to allow a fat clay (CH) soil at a high water content to dry slowly on the skin. The high shrinkage pressures will actually cause pain; in fact, this process was used during ancient times as a torture system. A human body covered with clay drying slowly in the sun has ultimately very little resistance to pressures which can reach several atmospheres! (See Example 6.3).

Another phenomenon that depends on capillarity is *slaking*, which occurs when a clod of dry soil is immersed in a beaker of water. The clod immediately starts to disintegrate, and with some soils the disintegration is so rapid that the clod appears to almost explode. Slaking is a very simple way to distinguish between soil and a rock; rocks don't slake, whereas soils do. The soil clump has to be dry since, if it is above the shrinkage limit, it will gradually swell. Below the shrinkage limit, the capillary stresses draw water in and the air bubbles trapped in the voids are compressed by the menisci. Eventually, the internal air pressure gets high enough to exceed the tensile strength of the soil. In a rock fragment the internal cohesion is sufficiently strong to resist the capillary forces.

Terzaghi (1943) used the tube analogy similar to that shown in Fig. 6.8 to illustrate slaking. The difference is that the capillary tubes are submerged and the capillary menisci are trying to pull water into the voids, as shown in Fig. 6.13. By drawing a free-body diagram of the tube walls, one can see that the walls are in tension, and if the tensile strength is less than the tension applied by the menisci, the walls will fracture, which is exactly what occurs when a soil slakes.

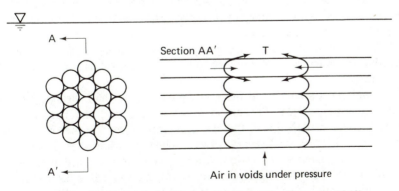

Fig. 6.13   Capillary tube analogy for slaking (after Terzaghi, 1943).

6.4 ENGINEERING SIGNIFICANCE OF SHRINKAGE AND SWELLING

The effects of shrinkage of fine-grained soils can be of considerable significance from an engineering point of view. For example, shrinkage cracks can occur locally when the capillary pressures exceed the cohesion or the tensile strength of the soil. These cracks, part of the clay macrostructure (Chapter 4), are zones of weakness which can significantly reduce the overall strength of a soil mass and affect the stability of clay slopes and the bearing capacity of foundations. The desiccated and cracked dry crust usually found over deposits of soft clay affects the stability of, for example, highway embankments constructed on these deposits. Shrinkage and shrinkage cracks are caused by evaporation from the surface in dry climates, lowering of the ground water table, and even desiccation of soil by trees during temporary dry spells in otherwise humid climates. When the climate changes and the soils have access to water again, they tend to increase in volume or swell. The volume changes resulting from both shrinkage and swelling of fine-grained soils are often large enough to seriously damage small buildings and highway pavements. Jones and Holtz (1973) have estimated that shrinking and swelling soils cause about $2.3 *billions* of damage annually in the United States alone which, to put things in perspective, is more than twice the annual cost of damage from floods, hurricanes, tornadoes, and earthquakes combined!

A common occurrence is that a pavement or building is constructed when the top soil layer is relatively dry. The structure covering the soil prevents further evaporation from occurring and the soils increase in water content due to capillarity; then the soils may swell. If the pressure exerted by the pavement or building is less than the swelling pressure, then heave will result. It is usually uneven and causes structural damage.

The process of shrinking and swelling is not completely reversible—the soil always has a memory of its stress history and will show the effects of previous shrinkage and drying cycles. Thus soft clays become what is called *overconsolidated* and less compressible because of the increase in effective stress caused by capillary action. Overconsolidation is discussed in Chapter 8.

Swelling is a somewhat more complex process than shrinkage (Yong and Warkentin, 1975). The amount of swelling and the magnitude of swelling pressure depend on the clay minerals present in the soil, the soil structure and fabric, and several physico-chemical aspects of the soil such as the cation valence, salt concentration, cementation, and presence of organic matter. Everything else being equal, montmorillonites swell more than illites, which swell more than kaolinites. Soils with random fabrics tend to swell more than soils with oriented fabrics. Disturbance of or remolding of old natural clays may increase the amount of swelling. Monovalent cations in a clay (for example, sodium montmorillonite) will swell more than divalent clays (for example, calcium montmorillonite). Cementation and organic substances tend to reduce swelling.

Swelling, like shrinkage, is generally confined to the upper portions of a soil deposit. Thus swelling damages lighter structures such as small buildings, highway pavements, and canal linings. Swelling pressures as high as 1000 kPa have been measured, which is equivalent to an embankment thickness of 40 to 50 m. Ordinarily, such high pressures do not occur, but even with more modest swelling pressures of 100 or 200 kPa an embankment of 5 or 6 m would be required to prevent all swelling of subgrade, for example. (For comparison, an ordinary building weighs something on the order of 10 kPa per story.) Practically speaking, the three ingredients generally necessary for potentially damaging swelling to occur are (1) presence of montmorillonite in the soil, (2) the natural water content must be around the PL, and (3) there must be a source of water for the potentially swelling clay (Gromko, 1974).

How is swelling predicted? Many methods and soil tests have been proposed. These include swelling tests and other simple laboratory tests, chemical and mineralogical analyses, and correlation with the classification and index properties of the soil. Table 6-2 is a summary of the experience of the U.S. Water and Power Resources Service (formerly U.S.B.R.) based on considerable research on swelling clays and expansive soils. Gromko (1974) provides some additional correlations with soil tests that have been successfully applied to predict swelling. Figure 6.14 relates swelling and collapse to the liquid limit and the in situ dry density of soils —again, based on the experience of the U.S. Water and Power Resources Service.

TABLE 6-2   Probable Expansion as Estimated from Classification Test Data*

Degree of Expansion	Probable Expansion as a % of the Total Volume Change (Dry to Saturated Condition)†	Colloidal Content (% −1 μm)	Plasticity Index, PI	Shrinkage Limit, SL
Very high	> 30	> 28	> 35	< 11
High	20–30	20–31	25–41	7–12
Medium	10–20	13–23	15–28	10–16
Low	< 10	< 15	< 18	> 15

*After Holtz (1959) and U.S.B.R. (1974).
†Under a surcharge of 6.9 kPa (1 psi).

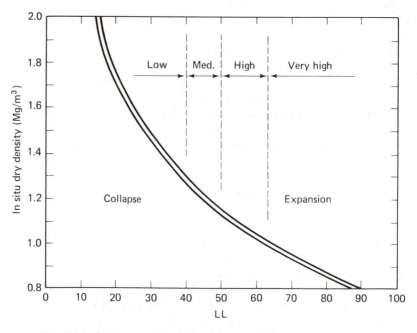

Fig. 6.14 Guide to collapsibility, compressibility, and expansion based on in situ dry densities and the liquid limit (adapted from Mitchell and Gardner, 1975, and Gibbs, 1969).

For compacted clays, Seed, et al. (1962) developed the relationships shown in Fig. 6.15 for artificial mixtures of sands and clays compacted to maximum unit weight by standard Proctor compaction and allowed to swell against a 6.9 kPa (1 psi) surcharge. These relationships between activity and percent clay sizes have also been shown to be applicable to natural soils if differences in the liquid limit devices used in the United States and Great Britain are considered. The term *activity* was originally defined by Skempton (Sec. 2.8) as the ratio of plasticity index to clay fraction (percent finer than 0.002 mm or 2 μm). Consequently, for natural soils, the following definition of activity has been proposed:

$$\text{activity, } A = \frac{\text{plasticity index}}{(\% - 2\,\mu\text{m}) - 5} \tag{6-11}$$

The purpose of Fig. 6.15 is to identify a potentially swelling soil, which may require further study and tests such as an expansion or swell test. Figure 6.15 should not be used for design purposes.

One of the simple swelling identification tests developed by the U.S. Water and Power Resources Service is called the *free-swell test* (Holtz and Gibbs, 1956). The test is performed by slowly pouring 10 cm³ of dry soil,

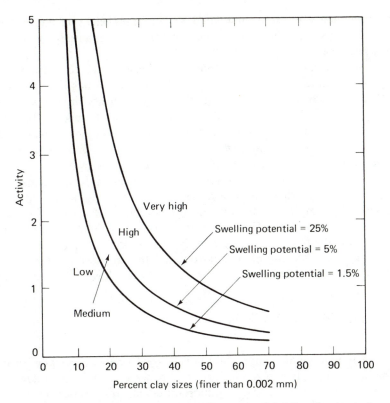

Fig. 6.15 Classification chart for swelling potential (after Seed, et al., 1962).

which has passed the No. 40 sieve, into a 100 cm^3 graduated cylinder filled with water, and observing the equilibrium swelled volume. Free swell is defined as

$$\text{free swell} = \frac{(\text{final volume}) - (\text{initial volume})}{\text{initial volume}} \times 100 \; (\%) \quad (6\text{-}12)$$

For comparison, highly swelling bentonites (mostly Na-montmorillonite) will have free-swell values of greater than 1200%. Even soils with free swells of 100% may cause damage to light structures when they become wet; soils with free swells less than 50% have been found to exhibit only small volume changes.

Other laboratory tests for swelling and swell pressures resemble the one-dimensional oedometer test described later. A specimen of soil is confined in a rigid brass ring, usually about 20 to 25 mm high and 50 to 100 mm in diameter. Sometimes the specimen is surcharged, flooded, and the percentage of swell is observed. Another variation is to keep loading the specimen after it is inundated so that the height of the specimen remains constant. The vertical stress necessary to maintain zero volume change is the swelling pressure.

What can engineers do to prevent damage to structures from shrinking and swelling soils? For compacted fill material, it has been found that soils compacted wet of optimum and at a lower density show less tendency to swell, probably because of a more oriented soil structure. Prewetting suspected soils will allow potentially damaging swelling or collapse to take place prior to construction. Moisture barriers and waterproof membranes have been used to prevent water from reaching the swelling soil. If the water content of the foundation soil is prevented from changing, no volume change will take place. Chemical stabilization has also been successfully employed to reduce the swelling of especially sodium montmorillonitic clays. The reason why it works is discussed in Chapter 4.

Since the potential damage to light structures and pavements due to shrinkage and swelling of soils is so great, the engineer must pay special attention to this problem if it is suspected that such soils exist at a site.

6.5 FROST ACTION

Anytime the air temperature falls below freezing, especially for more than a few days, it is possible for the pore water in soils to freeze. Frost action in soils can have several important engineering consequences. First, the volume of the soil can immediately increase about 10% just due to the volumetric expansion of water upon freezing. A second but significantly more important factor is the formation of ice crystals and lenses in the soil. These lenses can even grow to several centimetres in thickness and cause heaving and damage to light surface structures such as small buildings and highway pavements. If soils simply froze and expanded uniformly, structures would be evenly displaced since ice is significantly stronger than these light structures. However, just as with swelling and shrinking soils, the volume change is usually uneven, differential movement occurs, and that is what causes structural damage.

The problems do not end here. During the spring, the ice lenses melt and greatly increase the water content and decrease the strength of the soil. Highway pavements especially can suffer serious structural damage during the spring thaw (called, for obvious reasons, the "spring breakup").

An understanding of the real mechanism of ice lens formation as well as the conditions necessary for detrimental frost action only came about relatively recently. Prior to the 1920's and the rapid development of automobile traffic, roads were left snow covered for sleds during the winter. Since snow is a good insulator, depths of frost penetration were limited and rarely was frost heave a problem. Because the traffic loads were light, there were also few problems during the spring thaw.

The problems began when it became necessary to remove snow for car traffic. At first, frost heave was popularly attributed solely to the 10% volumetric expansion of water upon freezing. But some enterprising young engineers made some measurements, both of the magnitude of heave and of the water content of highway subgrades. Prof. Casagrande relates that, on one stretch of badly frost heaving road in New Hampshire, measurements during the winter of 1928–29 showed that the depth of frost penetration was about 45 cm, and the total surface heave was about 13 cm. The water content, normally between 8% and 12%, had increased and ranged between 60% and 110%. When a test pit was excavated, the subgrade was full of ice lenses with a total thickness of (you guessed it!) 13 cm! The water table had been located at some 2 m depth in the autumn, yet during the spring it was right below the pavement. When the soil began to thaw in the spring, the upper layers became water saturated and very soft—the water was trapped in the subgrade between the thawed surface layer and the top of the still frozen soil below.

Now the question was: how did the water get up there? It wasn't there before the winter season. Capillarity seemed to be involved in the cause. Also, the observation was made that there was very little ice in clean sands and gravels, but with silty soils, ice lenses were plentiful. Further investigations showed that the formation of ice lenses also depended on the rate of freezing of the soil. If the soil froze rapidly, as might occur during a cold snap early in the winter before there was significant snow, then less ice lenses tended to form. With a slower rate of freezing, it seemed that there were more ice lenses, and the thicker lenses tended to form nearer the bottom of the frozen layer. So one of the conditions for ice lens formation must be that there is a source of water nearby.

Research during the past 40 years has explained much of the observed phenomena associated with soil freezing and frost action. As might be expected, the process, especially with fine-grained soils, is a rather complicated heat-diffusion (thermodynamic) and pore water-chemistry problem and is related to the soil-water potential and water movement in frozen soils (Yong and Warkentin, 1975).

Basically, three conditions must exist for frost action and the formation of ice lenses in soils to occur:

1. Temperatures below freezing.
2. Source of water close enough to supply capillary water to the frost line.
3. Frost susceptible soil type and grain (pore) size distribution.

Freezing temperatures depend on the climatic conditions at the site. Ground cover, topography, presence of snow, and other factors locally affect the rate and depth of frost penetration. A ground water table within the height of capillary rise provides the water to feed growing ice lenses. The soil must be fine enough for relatively high capillary pressures to develop and yet not so fine that the flow of water (permeability) is restricted. As is discussed in the next chapter, the permeability of clay soils is very low. Even though the capillary pressures are very high, unless the clay is relatively sandy or silty, the amount of water that can flow during a freezing spell is so small that ice lenses have little chance to form. However, practically speaking, clay soils near the surface are often cracked and fissured, as described previously, which may allow some water movement to the frost line.

Figure 6.16 shows a sample of fissured clay which froze down from the top. Note how the water content increased within the frozen zone and how this compared with the value before freezing. Note too how the ice lenses developed in the frozen zone. They were continually supplied from the water table through the fissures and cracks in the clay.

What are frost susceptible soils? As suggested above, ice lenses will simply not form in coarse-grained soils. Casagrande (1932a) and other researchers like Beskow (1935) in Sweden in the early thirties found that ice lens formation in fine-grained soils depended on both a critical grain size and the grain size distribution of the soil. Beskow found 0.1 mm to be the maximum size that would permit ice lens formation under any conditions. Casagrande found 0.02 mm to be a critical grain size; even gravels with only 5% to 10% of 0.02 mm silt were frost susceptible. Casagrande also found that with well-graded soils, only 3% of the material finer than 0.02 mm was required to produce frost heaving, whereas fairly uniform soils must have at least 10% of that size to be dangerous. It seems that soils with less than 1% smaller than 0.02 mm also rarely frost heaved. Beskow's (1935) limiting grain size curves are shown in Fig. 6.17 for Swedish glacial tills and similar soils. Soils above the top curve were found to be frost heaving; those below the bottom curve never heaved. Current Swedish practice is given in Table 6-3.

As with other capillary phenomena, it is the pore sizes and not the grain sizes that really control frost action. Recent research at Purdue (Reed, et al., 1979) has shown that an intrinsically frost susceptible soil, as predicted by texture and/or gradation, can actually have many levels of susceptibility, depending on the details of its compaction. The explanation for these different responses lies in the distribution of porosity into various pore sizes, which was measured by mercury intrusion.

Just as with swelling and shrinking soils, frost action seriously affects structures such as small buildings and highway pavements which are found

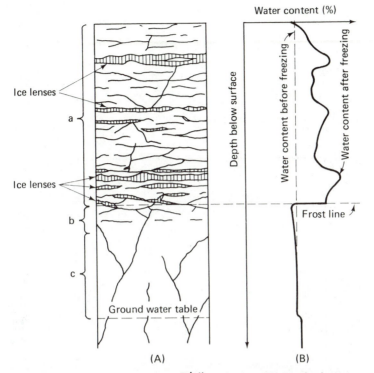

Fig. 6.16 Diagram showing the relation between different ice layers in frozen soil (A) and the water content distribution curve (B). The soil is assumed to be a medium clay with permanent cracks: a = frozen part, b = dried out zone below the frost line. Since the clay contains permanent cracks, the ground water surface is real, that is, the level where the cracks contain free water, and is therefore noticed as a kink in the water distribution curve (after Beskow, 1935).

directly on the ground surface. Damage to highways in the United States and Canada because of frost action is estimated to amount to millions of dollars annually. But because of the rather good fundamental understanding of the factors involved in frost action and heave, engineers have developed relatively successful methods for dealing with these problems. Load restrictions on secondary roads during the spring "breakup" are common in the northern United States and in Canada. More positive measures for dealing with the potential damage to structures and highways include lowering of the ground water table and, depending on the depth of frost penetration, removal of frost susceptible soils in the subgrade or foundation. Use of impervious membranes, chemical additives, and even foamed insulation (Styrofoam) under highways, buildings, and railroads have been successfully employed. Building foundations as well as water and sewer lines should be placed well below the maximum depth of frost penetration.

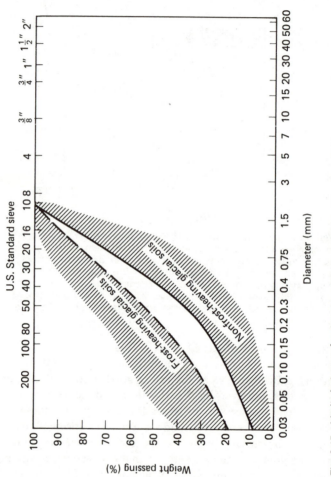

Fig. 6.17 Limits between frost susceptible and non-frost-susceptible mixtures of glacial tills or similar mixtures (after Beskow, 1935).

TABLE 6-3 Frost Susceptibility Soil Groups*

Group	Frost Susceptibility or Danger	Soils
I	None	Gravel, sand, gravelly tills
II	Moderate	Fine clay ($\geq 40\%$ clay[†] content); sandy tills, clayey tills with 16% fines[‡]
III	Strong	Silt, coarse clay (clay[†] content 15–25%); silty tills

*After Hansbo (1975).
[†]Defined as $-2\ \mu$m.
[‡]Defined as -0.06 mm.

PROBLEMS

6-1. The end of a clean glass tube is inserted in pure water. What is the height of capillary rise if the tube is (a) 0.1 mm, (b) 0.01 mm, and (c) 0.001 mm in diameter?

6-2. Calculate the maximum capillary tension for the tubes in Problem 6-1.

6-3. Calculate the theoretical height of capillary rise and the capillary tension of the three soils whose grain size distribution is shown in Fig. 2.4.

6-4. A tube, similar to Fig. 6.8, has a 0.002 mm inside diameter and is open at both ends. The tube is held vertically and water is added to the top end. What is the maximum height h of the column of water that will be supported? Hint: A meniscus will form at the top and at the bottom of the column of water, as shown in Fig. P6-4. (After Casagrande, 1938.)

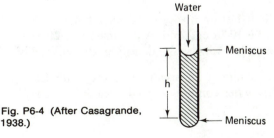

Fig. P6-4 (After Casagrande, 1938.)

6-5. The tube shown in Fig. 6.9 is filled with water. When evaporation takes place, the meniscus will first form at the larger end, as was explained in the text. Assuming this meniscus to be fully formed, derive an expression for the contact angle at the other end of the tube in terms of the two radii, r_l and r_s.

6-6. On a postcard from Daytona Beach, Florida, four answers were given to the question "Why can I drive on the beach here... but not other places?"

    (a) A gently sloping ocean bottom. The water 15 miles out is about 40 ft deep; at 50 miles, only 600 ft. This long gradual slope causes surf to have a downward pounding motion, thus packing the sand.

    (b) A silica compound in quantity in sand helps make a firm, hard base when mixed with salt water.

    (c) The presence of titanium (a very hard metal) contributes to firmness.

    (d) During the long tumbling journey from the Georgia and Carolina coasts, each grain becomes smooth and round, with sharp edges worn down, allowing it to pack closely.

    Are these reasons valid? Explain.

6-7. (a) Would the shrinkage limit of a clay be different if the water in the voids were replaced by some other liquid with a *smaller* surface tension? Why?
    (b) Would there be more or less shrinkage? Why?

6-8. Assume that equations developed for height of capillary rise in constant-diameter tubes can be applied. Calculate the *net* compressive stress on a soil pat at the shrinkage limit where the average diameter of the surface pores is 0.001 mm.

6-9. Estimate the shrinkage limits of the soils A–F in Problem 2-35.

6-10. During a shrinkage limit test on a silty clay, the volume of the dry soil pat was found to be 10.76 cm^3 and its dry mass was 22.68 g. If the shrinkage limit was 11.1, what is the density of the soil solids?

6-11. Estimate the volume change of an organic silty clay with LL = 62 and PL = 42, when its water content reduced from 50% to 20%.

6-12. Comment on the validity of Casagrande's procedure (Fig. 6.12) for estimating the shrinkage limit for undisturbed soils. Does it matter whether the soils are sensitive or not? Why?

6-13. A saturated sample of clay with an SL of 22 has a natural water content of 35%. What would its dry volume be as a percentage of its original volume if ρ_s is 2.70?

6-14. A sample of clayey silt is mixed at about its LL of 40. It is placed carefully in a small porcelain dish with a volume of 19.3 cm^3 and weighs 34.67 g. After oven drying, the soil pat displaced 216.8 g of mercury. (a) Determine the SL of the soil sample. (b) Estimate the ρ_s of the soil.

6-15. The LL of a bentonitic clay is 450 and the PL is 75. The SL was determined to be about 10. Calculate the expected volumetric decrease when a sample of this bentonite is dried, if its natural water content was 85%.

6-16. The shrinkage limit of a 0.1 m^3 sample of a clay is 15, and its natural water content is 34%. Assume the density of the soil solids is 2.68 Mg/m^3, and estimate the volume of the sample when the water content is 12.7%.

6-17. During the determination of the shrinkage limit of a sandy clay, the following laboratory data was obtained:

$$\text{Wet wt. of soil + dish} = 87.85 \text{ g}$$
$$\text{Dry wt. of soil + dish} = 76.91 \text{ g}$$
$$\text{Wt. of dish} \qquad\quad = 52.70 \text{ g}$$

Volumetric determination of soil pat:

$$\text{Wt. of dish + mercury} = 430.80 \text{ g}$$
$$\text{Wt. of dish} \qquad\quad\; = 244.62 \text{ g}$$

Calculate the shrinkage limit of the soil, assuming $\rho_s = 2.65$ Mg/m^3.

6-18. The LL of a medium sensitive Swedish post-glacial clay is 56 and the PI is 28. At its natural water content, the void ratio is 1.03 while after shrinkage the minimum void ratio is 0.72. Assuming the density of the soil solids is 2.72, calculate the shrinkage limit of the clay.

6-19. Derive Eqs. 6-8 and 6-9 for the shrinkage limit, using phase relationships. Show that they are identical.

6-20. Estimate the swelling potential of soils A–F, Problems 2-33 and 2-35. Use both Table 6-2 and Fig. 6.15.

6-21. Estimate the frost susceptibility of soils A–F, Problems 2-33 and 2-35, according to Beskow (Fig. 6.17) and current Swedish practice.

seven

Water in Soils, II: Permeability, Seepage, Effective Stress

7.1 INTRODUCTION

The importance in civil engineering of water in soils is mentioned at the beginning of Chapter 6. Most geotechnical engineering problems somehow have water associated with them in various ways, either because of the water flowing through the voids and pores in the soil mass or because of the state of stress or pressure in the water in the pores. Some of these effects are described in this chapter.

The following notation is introduced in this chapter.

Symbol	Dimension	Unit	Definition
A	L^2	m^2	Area
a, a'	L	m	Distance from impervious boundary to bottom of sheet pile (Table 7-2)
h	L	m	Energy or head, head loss, (also with subscripts f, L, m), layer thickness
h_p	L	m	Pressure head (Eq. 7-4)
i	—	—	Hydraulic gradient (Eq. 7-1)
i_c	—	—	Critical hydraulic gradient (Eq. 7-21)
i_E	—	—	Exit gradient
j	$ML^{-2}T^{-2}$	kN/m^3	Seepage force per unit volume (Eq. 7-23)
k	LT^{-1}	m/s	Darcy coefficient of permeability (Eqs. 7-2, 7-5)
K	—	—	Ratio of horizontal to vertical stress (Eq. 7-18)
K_o	—	—	Coefficient of lateral earth pressure at rest (Eq. 7-19)

Symbol	Dimension	Unit	Definition
K, K'	—	—	Parameters related to form factor Φ (Table 7-3)
L	L	m	Length of sample
l	L	m	Unit or characteristic length
m	—	—	Function of s and T (Eq. 7-31)
N_d	—	—	Number of equipotential "drops" in a flow net (Eq. 7-26)
N_f	—	—	Number of flow channels in a flow net (Eq. 7-26)
P	MLT^{-2}	N	Force or load (Eq. 7-16)
p	$ML^{-1}T^{-2}$	kPA	Pressure (Eq. 7-4)
q	L^3T^{-1}	m³/s	Flow rate (sometimes per unit width) (Eqs. 7-3, 7-5)
Q	L^3	m³	Volume of flow (Eq. 7-8)
R_{15}, R_{50}	—	—	Filter ratios (Table 7-5)
s, s', s''	L	m	Length of sheet pile (Table 7-2)
T	L	m	Thickness of layer
u	$ML^{-1}T^{-2}$	kPa	Pore water pressure (Eqs. 7-13, 7-15)
v	LT^{-1}	m/s	Velocity (Eq. 7-2)
v_s	LT^{-1}	m/s	Seepage velocity (Eq. 7-6)
z	L	m	Potential head; depth
z_w	L	m	Depth to the water table (Eq. 7-15)
σ	$ML^{-1}T^{-2}$	kPa	Total normal stress (Eq. 7-13)
σ'	$ML^{-1}T^{-2}$	kPa	Effective normal stress (Eq. 7-13)
σ_h	$ML^{-1}T^{-2}$	kPa	Total horizontal stress (Eq. 7-18)
σ_h'	$ML^{-1}T^{-2}$	kPa	Effective horizontal stress (Eq. 7-19)
σ_v	$ML^{-1}T^{-2}$	kPa	Total vertical stress (Eq. 7-14)
σ_v'	$ML^{-1}T^{-2}$	kPa	Effective vertical stress (Eq. 7-19)
Φ	—	—	Form factor

7.2 DYNAMICS OF FLUID FLOW

You may recall from your basic fluid mechanics courses that there are several different ways to describe or classify fluid flow. Flow can be *steady* or *unsteady*, which corresponds, respectively, to conditions that are constant or vary with time. Flow can also be classified as *one-*, *two-*, or *three-dimensional*. One-dimensional flow is flow in which all the fluid parameters such as pressure, velocity, temperature, etc., are constant in any cross section perpendicular to the direction of flow. Of course, these parameters can vary from section to section along the direction of flow. In two-dimensional flow, the fluid parameters are the same in parallel planes, whereas in three-dimensional flow, the fluid parameters vary in the three coordinate directions. For purposes of analysis, flow problems in geotechnical engineering are usually assumed to be either one- or two-dimensional, which is adequate for most practical cases.

Because density changes can be neglected at ordinary stress levels for most geotechnical engineering applications, flow of water in soils can be considered *incompressible*.

Flow can also be described as *laminar*, where the fluid flows in parallel layers without mixing, or *turbulent*, where random velocity fluctuations result in mixing of the fluid and internal energy dissipation. There can also be intermediate or *transition* states between laminar and turbulent flow. These states are illustrated in Fig. 7.1, which shows how the *hydraulic gradient* changes with increasing velocity of flow. Hydraulic gradient i, a very important concept, is defined as the energy or *head loss h* per unit length l, or

$$i = \frac{h}{l} \tag{7-1}$$

The energy or head loss increases linearly with increasing velocity as long as the flow is laminar. Once the transition zone is passed, because of internal eddy currents and mixing, energy is lost at a much greater rate (zone III, Fig. 7.1) and the relationship is nonlinear. Once in the turbulent zone, if the velocity is decreased, the flow remains turbulent well into transition zone II until the flow again becomes laminar.

For flow in most soils, the velocity is so small that the flow can be considered laminar. Thus, from Fig. 7.1, we could write that v is propor-

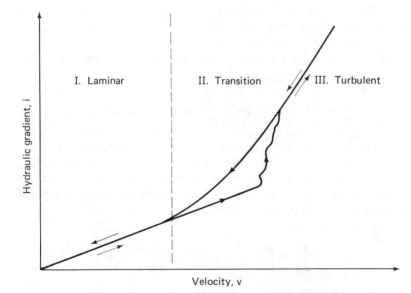

Fig. 7.1   Zones of laminar and turbulent flow (after Taylor, 1948).

tional to i, or

$$v = ki \qquad (7\text{-}2)$$

Equation 7-2 is an expression for *Darcy's law*, which is discussed later in this chapter.

Another important concept from fluid mechanics is the *law of conservation of mass*. For incompressible steady flow, this law reduces to the *equation of continuity*, or

$$q = v_1 A_1 = v_2 A_2 = \text{constant} \qquad (7\text{-}3)$$

where q = rate of discharge (units: volume/time, m^3/s),
v_1, v_2 = velocities at sections 1 and 2, and
A_1, A_2 = the cross-sectional areas at sections 1 and 2.

The other well-known equation from fluid mechanics that we shall use is the *Bernoulli energy equation* for incompressible steady flow of a fluid:

$$\frac{v_1^2}{2} + \frac{p_1}{\rho_w} + gz_1 = \frac{v_2^2}{2} + \frac{p_2}{\rho_w} + gz_2 = \text{constant energy} \qquad (7\text{-}4a)$$

where v_1, v_2 = velocities at sections 1 and 2,
g = acceleration of gravity,
ρ_w = density of the fluid (water),
p_1, p_2 = pressures at sections 1 and 2, and
z_1, z_2 = distance above some arbitrary datum plane at sections 1 and 2.

This equation is the steady-flow energy equation in terms of energy per unit of mass of fluid (units: J/kg). However in hydraulics it is more common to express the Bernoulli equation in terms of energy per unit weight by dividing each term in the equation by g, the acceleration of gravity, or

$$\frac{v_1^2}{2g} + \frac{p_1}{\rho_w g} + z_1 = \frac{v_2^2}{2g} + \frac{p_2}{\rho_w g} + z_2 = \text{constant total head} \qquad (7\text{-}4b)$$

Equation 7-4b states that the *total energy* or *head* in the system is the sum of the *velocity head* $v^2/2g$, the *pressure head* $p/\rho_w g\, (= p/\gamma_w)$, and the *potential (position) head* z. Whether the flow is in pipes, open channels, or through porous media, there are energy or head losses associated with the fluid flow. Usually an energy or head loss term h_f is added to the second part of Eq. 7-4b; thus

$$\frac{v_1^2}{2g} + \frac{p_1}{\rho_w g} + z_1 = \frac{v_2^2}{2g} + \frac{p_2}{\rho_w g} + z_2 + h_f \qquad (7\text{-}4c)$$

Why do we say head for each term in the Bernoulli equation? Because each term has units of length, and each is called the velocity head, pressure head, or potential head, as the case may be. For most soil flow problems, the velocity head is usually neglected since it is small in comparison to the other two heads.

7.3 DARCY'S LAW FOR FLOW THROUGH POROUS MEDIA

We have already mentioned that the flow of water through the pores or voids in a soil mass can in most cases be considered laminar. We also stated that for laminar flow the velocity is proportional to the hydraulic gradient, or $v = ki$ (Eq. 7-2). Over one-hundred years ago, a French waterworks engineer named Darcy (D'Arcy, 1856) showed experimentally that the rate of flow in *clean sands* was proportional to the hydraulic gradient (Eq. 7-2). Equation 7-2 is usually combined with the continuity equation (Eq. 7-3) and the definition of hydraulic gradient (Eq. 7-1). Using the notation as defined in Fig. 7.2, *Darcy's law* is usually written as

$$q = vA = kiA = k\frac{\Delta h}{L}A \qquad (7\text{-}5)$$

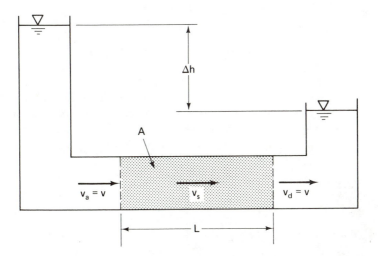

Fig. 7.2   Superficial and seepage velocities in uniform flow (after Taylor, 1948).

where q is the total rate of flow through the cross-sectional area A, and the proportionality constant k is called the *Darcy coefficient of permeability*. Commonly, in civil engineering, it is called simply the *coefficient of permeability* or, even more simply, the *permeability*. The permeability is a soil *property* which expresses or describes how water flows through soils. Knowledge of permeability is required for the design of engineering works where seepage of water is involved. Note that permeability has units of *velocity* because i is dimensionless. Units commonly used are m/s, cm/s for laboratory work, or ft/day in the British Engineering system of units.

Why do we use the total cross-sectional area in Eq. 7-5? Obviously, the water cannot be flowing through the solid particles but only through the voids or pores between the grains. So why don't we use that area and compute the velocity based on the area of the voids? It would be relatively easy to compute the area of the voids from the void ratio (Eq. 2-1), even though the void ratio is a volumetric ratio. For a unit width of sample in Fig. 7.2, we can write $e = V_v/V_s = A_v/A_s$. Now the approach velocity v_a and the discharge velocity v_d in Fig. 7.2 both equal $v = q/A$, the discharge q divided by the total cross-sectional area A. Thus the v in this relationship is really a *superficial* velocity, a fictitious but statistically convenient "engineering" velocity. The actual *seepage* velocity v_s, the actual velocity of the water flowing in the voids, is greater than the superficial velocity. We can show this by

$$q = v_a A = v_d A = vA = v_s A_v \qquad (7\text{-}6)$$

From Fig. 7.3 and Eq. 2-2, $A_v/A = V_v/V = n$; then,

$$v = nv_s \qquad (7\text{-}7)$$

Since $0\% \leqslant n \leqslant 100\%$, it follows that the seepage velocity is always greater than the superficial or discharge velocity.

From the preceding discussion you can see that the void ratio or porosity of a soil affects how water flows through it and thus the value of the permeability of a particular soil. From theoretical relationships for flow through capillary tubes developed by Hagen and Poiseuille about 1840 and from the more recent hydraulic radius models of Kozeny and Carman, we

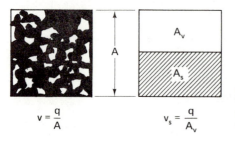

Fig. 7.3   Phase diagram for seepage and superficial velocities of flow (flow is perpendicular to page).

know that several other factors also affect permeability. (Leonards, 1962, Chapter 2, provides an excellent summary of these developments.) The effective grain size (or, better, the effective pore size) has an important influence, much as it does on the height of capillary rise (Sec. 6.2). The shapes of the voids and flow paths through the soil pores, called *tortuosity* also affect k. All of the previous discussion of permeability was for saturated soils only, so the degree of saturation S must influence the actual permeability. Finally the properties of the fluid have some effect; viscosity, which depends on the temperature, and the density immediately come to mind.

Since Darcy originally developed his relationship for clean filter sands, how valid is his law for other soils? Careful experiments have shown that Eq. 7-5 is valid for a wide range of soil types at engineering hydraulic gradients. In very clean gravels and open-graded rock fills, flow may be turbulent and Darcy's law would be invalid. At the other end of the spectrum, careful investigations by Hansbo (1960) found that in clays at very low hydraulic gradients the relationship between v and i is nonlinear (Fig. 7.4). Field measurements (Holtz and Broms, 1972) showed that the exponent n has an average value of about 1.5 in typical Swedish clays. However, there is by no means complete agreement with the concept shown in Fig. 7.4. Mitchell (1976, pp. 349–351) summarizes several investigations about this point, and he concludes that "... all other factors held constant, Darcy's law is valid."

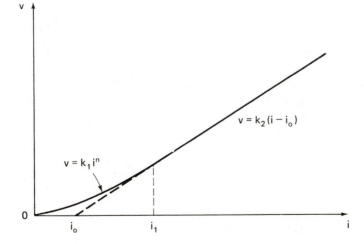

Fig. 7.4 Deviation from Darcy's law observed in Swedish clays (after Hansbo, 1960).

7.4 MEASUREMENT OF PERMEABILITY

How is the coefficient of permeability determined? A device called a *permeameter* is used in the laboratory, and either a *constant-head test* or a *falling-head test* is conducted (Figs. 7.5a and b). In the field, pumping tests are usually conducted, although it would be possible in principle to utilize either constant- or falling-head tests.

For the constant-head test, the volume of water Q collected in time t (Fig. 7.5a) is

$$Q = Avt$$

From Eq. 7-5,

$$v = ki = k\frac{h}{L}$$

so

$$k = \frac{QL}{hAt} \qquad (7\text{-}8)$$

where Q = total discharge volume, m^3, in time t, s, and
$\qquad A$ = cross-sectional area of soil sample, m^2.

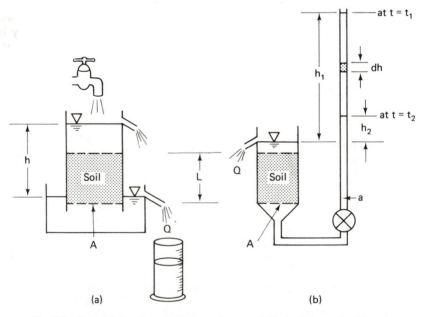

Fig. 7.5 Determining the coefficient of permeability by (a) constant-head test, (b) falling-head test.

EXAMPLE 7.1

Given:

A cylindrical soil sample, 7.3 cm in diameter and 16.8 cm long, is tested in a constant-head permeability apparatus. A constant head of 75 cm is maintained during the test. After 1 min of testing, a total of 945.7 g of water was collected. The temperature was 20°C. The void ratio of the soil was 0.43.

Required:

Compute the coefficient of permeability in centimetres per second and in furlongs per fortnight.

Solution:

First, calculate the cross-sectional area of the sample.

$$A = \frac{\pi D^2}{4} = \frac{\pi}{4}(7.3 \text{ cm})^2 = 41.9 \text{ cm}^2$$

From Eq. 7-8, solve for k:

$$k = \frac{QL}{hAt}$$

$$= \frac{945.7 \text{ cm}^3 \times 16.8 \text{ cm}}{75 \text{ cm} \times 41.9 \text{ cm}^2 \times 1 \text{ min} \times 60 \text{ s/min}}$$

$$= 0.08 \text{ cm/s}$$

To convert to furlongs per fortnight (see Appendix A):

$$k = \left(0.08 \frac{\text{cm}}{\text{s}}\right)\left(60 \frac{\text{s}}{\text{min}}\right)\left(60 \frac{\text{min}}{\text{h}}\right)\left(24 \frac{\text{h}}{\text{d}}\right)\left(14 \frac{\text{d}}{\text{fortnight}}\right)$$

$$\times \left(\frac{1 \text{ in.}}{2.54 \text{ cm}}\right)\left(\frac{1 \text{ ft}}{12 \text{ in.}}\right)\left(\frac{\text{mi}}{5280 \text{ ft}}\right)\left(\frac{8 \text{ furlongs}}{\text{mi}}\right)$$

$$= 4.8 \frac{\text{furlongs}}{\text{fortnight}}$$

For the falling-head test (Fig. 7.5b) the velocity of fall in the standpipe is

$$v = -\frac{dh}{dt}$$

and the flow into the sample is

$$q_{in} = -a\frac{dh}{dt}$$

From Darcy's law (Eq. 7-5), the flow out is

$$q_{out} = kiA = k\frac{h}{L}A$$

By Eq. 7-3 (continuity), $q_{in} = q_{out}$, or

$$-a\frac{dh}{dt} = k\frac{h}{L}A$$

Separating variables and integrating over the limits,

$$-a\int_{h_2}^{h_1}\frac{dh}{h} = k\frac{A}{L}\int_{t_1}^{t_2}dt$$

we obtain

$$k = \frac{aL}{A\Delta t}\ln\frac{h_1}{h_2} \tag{7-9a}$$

where $\Delta t = t_2 - t_1$. In terms of $\log_{10}$,

$$k = 2.3\frac{aL}{A\Delta t}\log_{10}\frac{h_1}{h_2} \tag{7-9b}$$

where a = area of standpipe,
    A, L = soil sample area and length,
    Δt = time for standpipe head to decrease from h_1 to h_2.

EXAMPLE 7.2

Given:

A laboratory falling-head permeability test is performed on a light-gray gravelly sand (SW), and the following data is obtained:

$$a = 6.25 \text{ cm}^2$$
$$A = 10.73 \text{ cm}^2$$
$$L = 16.28 \text{ cm}$$
$$h_1 = 160.2 \text{ cm}$$
$$h_2 = 80.1 \text{ cm, and}$$
$$\Delta t = 90 \text{ s for the head to fall from } h_1 \text{ to } h_2.$$

Water temperature = 20°C.

Required:

Compute the coefficient of permeability in cm/s.

Solution:

Use Eq. 7-9b and solve for k.

$$k = 2.3 \times \frac{6.25}{10.73} \times \frac{16.28}{90} \log \frac{160.2}{80.1}$$
$$= 0.07 \text{ cm/s at } 20°C.$$

Note: If the water temperature is different than 20°C, a correction for differences in the value of the viscosity is made.

Several factors may influence the reliability of the permeability test in the laboratory. Air bubbles may be trapped in the test specimen, or air may come out of solution from the water. The degree of saturation could thus be less than 100%, which would affect the test results significantly. Migration of fines in testing sands and silts also affects the measured values. Temperature variation, especially in tests of long duration, may affect the measurements, and if the ground temperature is significantly less than the laboratory test temperature, a viscosity correction should be made. Although the small samples used in the laboratory are assumed to be representative of field conditions, it is difficult to duplicate the in situ soil structure, especially of granular deposits and of stratified and other nonhomogeneous materials. To properly account for the natural variability and inhomogeneity of soil deposits and difficulties in laboratory tests, the engineer may resort to field pumping tests from wells to measure the overall average coefficient of permeability.

The coefficient of permeability may also be obtained by performing a laboratory one-dimensional compression (consolidation) test or by testing a soil sample in the triaxial cell. The use of these devices is discussed in Chapters 8 and 10.

Besides the direct determination of permeability in the laboratory, useful empirical formulas and tabulated values of k exist for various soil types. For example, Fig. 7.6, adapted from Casagrande (1938), is useful. In this figure, the coefficient of permeability is plotted on a log scale since the range of permeabilities in soils is so large. Note that certain values of k, 1.0, 10^{-4}, and 10^{-9} cm/s are emphasized. These are Casagrande's *bench-mark values* of permeability, and they are useful reference values for engineering behavior. For example, 1.0 cm/s is the approximate boundary between laminar and turbulent flow and separates clean gravels from clean

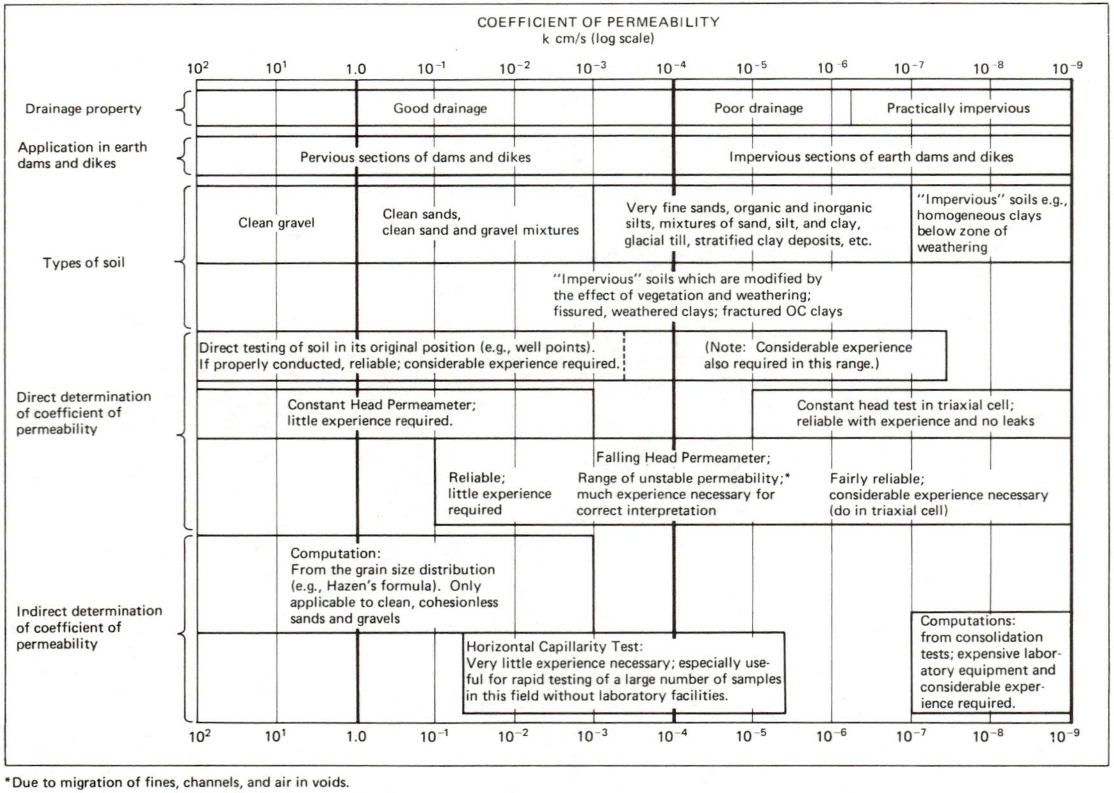

Fig. 7.6 Permeability, drainage, soil type, and methods to determine the coefficient of permeability (after A. Casagrande, 1938, with minor additions).

*Due to migration of fines, channels, and air in voids.

sands and sandy gravels. A k of 10^{-4} cm/s is the approximate boundary between pervious and poorly drained soils under low gradients. Soils around this value are also highly susceptible to migration of fines or *piping*. The next boundary, 10^{-9} cm/s, is approximately the lower limit of the permeability of soils and concrete, although some recent measurements have found permeabilities as low as 10^{-11} for highly plastic clays at the shrinkage limit. Prof. Casagrande recommends that k be related to the nearest benchmark value, for example, 0.01×10^{-4} cm/s rather than 10^{-6} cm/s. For various soil types, Fig. 7.6 also indicates their general drainage properties, applications to earth dams and dikes, and the means for direct and indirect determination of the coefficient of permeability.

An empirical equation relating the coefficient of permeability to D_{10}, the *effective grain size*, was proposed by A. Hazen (1911). For *clean* sands (with less than 5% passing the No. 200 sieve) with D_{10} sizes between 0.1 and 3.0 mm, the coefficient of permeability k is

$$k = CD_{10}^2 \qquad (7\text{-}10)$$

where the units of k are in cm/s and those of the effective grain size are in mm. The constant C varies from 0.4 to 1.2, with an average value of 1, and takes into account the conversion of units. The equation is valid for $k \geqslant 10^{-3}$ cm/s (Fig. 7.6).

To estimate k at void ratios other than the test void ratio, Taylor (1948) offers the relationship

$$k_1 : k_2 = \frac{C_1 e_1^3}{1 + e_1} : \frac{C_2 e_2^3}{1 + e_2} \qquad (7\text{-}11)$$

where the coefficients C_1 and C_2, which depend on the soil structure, must be determined empirically. *Very approximately for sands*, $C_1 \simeq C_2$. Another relationship which has been found to be useful for sands is

$$k_1 : k_2 = C_1' e_1^2 : C_2' e_2^2 \qquad (7\text{-}12)$$

As before, *approximately for sands*, $C_1' \simeq C_2'$.

For silts and clays, neither of these relationships works very well. For kaolinites over a rather *narrow* range of permeabilities (say one order of magnitude), e versus $\log_{10} k$ has been found to be approximately linear, all other factors being equal. However for compacted silts and silty clays Garcia-Bengochea, et al. (1979) found that the relationship between void ratio e and the logarithm of permeability k is far from linear (Fig. 7.7). They showed that pore size distribution parameters provide a better relationship with the ratio for some compacted soils.

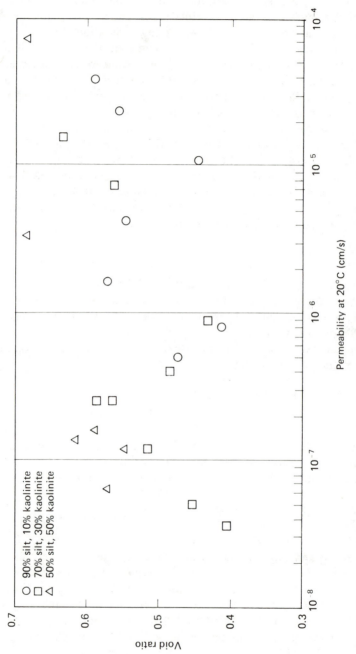

Fig. 7.7 Void ratio *e* versus permeability *k* for several compacted soils (after Garcia-Bengochea, et al., 1979).

7.5 INTERGRANULAR OR EFFECTIVE STRESS

The concept of intergranular or *effective stress* was introduced in Sec. 6.2. By definition,

$$\sigma = \sigma' + u \qquad (7\text{-}13)$$

where σ = total normal stress,
σ' = intergranular or effective normal stress, and
u = pore water or neutral pressure.

Both the total stress and pore water pressure may readily be estimated or calculated with knowledge of the densities and thicknesses of the soil layers and location of the ground water table. The effective stress cannot be measured; it can only be calculated!

The total vertical stress is called the *body stress* because it is generated by the mass (acted upon by gravity) in the body. To calculate the total vertical stress σ_v at a point in a soil mass, you simply sum up the densities of all the material (soil solids + water) above that point multiplied by the gravitational constant g, or

$$\sigma_v = \int_0^h \rho g\, dz \qquad (7\text{-}14a)$$

If ρg is a constant throughout the depth, then

$$\sigma_v = \rho g h \qquad (7\text{-}14b)$$

Typically, we divide the soil mass into n layers and evaluate the total stress incrementally for each layer or

$$\sigma_v = \sum_{i=1}^n \rho_i g z_i \qquad (7\text{-}14c)$$

As an example, if a soil could have zero voids, then the total stress exerted on a particular plane would be the depth to the given point times the density of the material or, in this case, ρ_s times the gravitational constant g. If the soil were dry, then you would use ρ_d instead of ρ_s.

The neutral stress or pore water pressure is similarly calculated for static water conditions. It is simply the depth below the ground water table to the point in question, z_w, times the product of the density of water ρ_w and g, or

$$u = \rho_w g z_w \qquad (7\text{-}15)$$

It is called the *neutral stress* because it has no shear component. Recall from fluid mechanics that by definition a liquid cannot support static shear stress. It has only normal stresses which act equally in all directions. On the other hand, total and effective stresses can have both normal and shear components. By Eq. 7-13, the effective stress σ' is simply the difference between the total and neutral stresses.

What is the physical meaning of effective stress? First, let's discuss the concept of stress itself. You may recall from basic mechanics that stress is really a fictitious quantity. It is defined as a differential force divided by a differential area, as the area shrinks to a point in the limit. This concept is useful even though in reality, on the micro scale, it has no meaning physically. For example, what would happen in a sand or gravel when the particular differential area you chose ended up in a void? Of course the stress would have to be zero. Yet right next door, where two gravel particles might be in point-to-point contact, the contact stress might be extremely high; it could even exceed the crushing strength of the mineral grains. Stress then really calls for a continuous material and, depending on the scale, real materials are not really continuous. Soils, especially, are not continuous, as we have seen in Chapter 4. Even fine-grained clay soils are collections of discrete mineral particles held together by gravitational, chemical, ionic, van der Waals, and many other kinds of forces. Still the concept of stress is useful in engineering practice, and that is why we use it.

So now what does effective stress mean physically? In a granular material such as a sand or gravel, it is sometimes called the *intergranular stress*. However it is not really the same as the grain-to-grain contact stress since the contact area between granular particles can be very small. In fact, with rounded or spherical grains the contact area can approach a point. Therefore the actual contact stress can be very large. Rather, the intergranular stress is the sum of the contact forces divided by the total or gross (engineering) area, as shown in Fig. 7.8. If we look at forces, the total vertical force or load P can be considered to be the sum of the intergranular contact forces P' plus the hydrostatic force $(A - A_c)u$ in the pore water. Since the neutral stress can obviously act only over the void or pore area, to get *force* the neutral stress u must be multiplied by the area of the voids $A - A_c$, or

$$P = P' + (A - A_c)u \qquad (7\text{-}16a)$$

where A = total or gross (engineering) area, and
  A_c = contact area between grains.

Dividing by the gross area A to obtain stresses, we have

$$\frac{P}{A} = \frac{P'}{A} + \left(\frac{A - A_c}{A}\right)u \qquad (7\text{-}16b)$$

or

$$\sigma = \sigma' + \left(1 - \frac{A_c}{A}\right)u \qquad (7\text{-}16c)$$

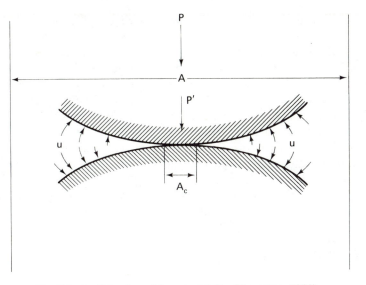

Fig. 7.8 Particles in solid contact (after Skempton, 1960).

or

$$\sigma = \sigma' + (1 - a)u \qquad (7\text{-}16d)$$

where a = contact area between particles per unit gross area of the soil (Skempton, 1960).

In granular materials, since the contact areas approach point areas, a approaches zero. Thus Eq. 7-16d reduces to Eq. 7-13, or $\sigma = \sigma' + u$. This equation, which defines effective stress, was first proposed in the 1920's by Terzaghi, who is considered to be the father of soil mechanics. Equation 7-13 is an extremely useful and important equation. We believe that the effective stresses in a soil mass actually control or govern the engineering behavior of that mass. The response of a soil mass to changes in applied stresses (compressibility and shearing resistance) depend almost exclusively on the effective stresses in that soil mass. The principle of effective stress is probably the single most important concept in geotechnical engineering.

We have discussed effective stresses for granular particulate materials. What does the concept mean for fine-grained cohesive soils? From the discussion in Chapter 4, it is doubtful that the mineral crystals are in actual physical contact since they are surrounded by a tightly bound water film. On the micro scale, the interparticle force fields which would contribute to effective stress are extremely difficult to interpret and philosophically impossible to measure. Any inference about these force fields comes from a study of the fabric of the soil. So, in view of this complexity, what place does such a simple equation as 7-13 have in engineering practice? Experimental evidence as well as a careful analysis by Skempton (1960) has shown that for saturated sands and clays the principle of effective stress is an excellent approximation to reality. However it is not so good for partially saturated soils or saturated rocks and concrete. Whatever it is physically, effective stress is defined as the difference between an engineering total stress and a measurable neutral stress (pore water pressure). The concept of effective stress is extremely useful, as we shall see in later chapters, for understanding soil behavior, interpreting laboratory test results, and making engineering design calculations. The concept works, and that is why we use it.

Now we shall work through some examples to show you how to calculate the total, neutral, and effective stresses in soil masses.

EXAMPLE 7.3

Given:

The container of soil shown in Fig. Ex. 7.3. The saturated density is 2.0 Mg/m^3.

Required:

Calculate the total, neutral, and effective stresses at elevation A when (a) the water table is at elevation A and (b) when the water table rises to elevation B.

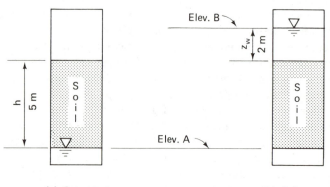

(a) Saturated   (b) Submerged

Fig. Ex. 7.3

Solution:

a. Assume the soil in the container to be initially *saturated* (but not submerged). The water table is located at elevation A. Use Eqs. 7-14b, 7-15, and 7-13 to calculate the stresses at elevation A.

Total stress (Eq. 7-14b):

$$\sigma = \rho_{sat}gh = 2.0 \text{ Mg/m}^3 \times 9.81 \text{ m/s}^2 \times 5 \text{ m}$$

$$= 98\ 100 \text{ N/m}^2 = 98.1 \text{ kPa}$$

Neutral stress (Eq. 7-15):

$$u = \rho_w g z_w = 1 \text{ Mg/m}^3 \times 9.81 \text{ m/s}^2 \times 0 = 0$$

From Eq. 7-13:

$$\sigma' = \sigma = 98.1 \text{ kPa}$$

Recall that $1 \text{ N} = 1 \text{ kg·m/s}^2$ and that $1 \text{ N/m}^2 = 1$ Pa (Appendix A).

b. If we raise the water table to elevation B, a change in effective stresses at elevation A occurs since the saturated soil becomes submerged or buoyant. The stresses at elevation A due to the soil and water above are as follows:

Total stress:

$$\sigma = \rho_{sat}gh + \rho_w g z_w$$

$$= (2.0 \times 9.81 \times 5) + (1 \times 9.81 \times 2)$$

$$= 117.7 \text{ kPa}$$

Neutral stress:

$$u = \rho_w g(z_w + h)$$

$$= 1 \times 9.81 \times (2 + 5)$$

$$= 68.7 \text{ kPa}$$

Effective stress at elevation A:

$$\sigma' = \sigma - u = (\rho_{sat}gh + \rho_w g z_w) - \rho_w g(z_w + h)$$

$$= 117.7 - 68.7 = 49.0 \text{ kPa}$$

Thus by raising the elevation of the ground water table the intergranular pressure or effective stress in Example 7.3 drops from 98 kPa to 49 kPa, or a reduction of 50%! When the ground water table is *lowered*, the reverse occurs and the soil is subjected to an *increase* in effective stress. This overall increase in vertical stress may lead to substantial areal subsidence as is occurring, for example, in Mexico City and Las Vegas.

Ground water is being pumped for municipal water supply, and the resulting settlements have caused substantial damage to streets, buildings, and underground utilities.

Another way to calculate the effective stress in part (b) of Example 7.3 is to use the submerged or buoyant density (Eq. 2-11). Note that

$$\sigma' = (\rho_{sat}gh + \rho_w gz_w) - \rho_w g(z_w + h)$$

$$= (\rho_{sat} - \rho_w)gh$$

$$= \rho'gh .\tag{7-17}$$

EXAMPLE 7.4

Given:

The data of Example 7.3.

Required:

Use Eq. 7-17 to compute the effective stress at elevation A when the water table is at elevation B.

Solution:

$$\rho' = \rho_{sat} - \rho_w = 2.0 - 1.0 = 1.0 \text{ Mg/m}^3$$

$$\sigma' = \rho'gh = 1.0 \times 9.81 \times 5 = 49.0 \text{ kPa}$$

EXAMPLE 7.5

Given:

The soil profile as shown in Fig. Ex. 7.5

Required:

What are the total and effective stresses at point A?

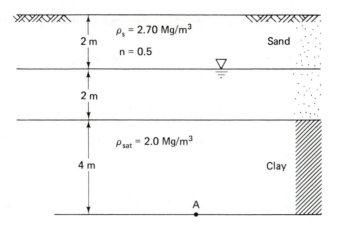

Fig. Ex. 7.5

Solution:

First find ρ_d and ρ_{sat} of the sand. This will be a review of phase relations. Let $V_t = 1 \text{ m}^3$; therefore $n = V_v$ and,

$$V_s = 1 - V_v = 1 - n. \text{ From Eq. 2-7,}$$

$$M_s = \rho_s(1 - n)$$

$$M_s = 2.70 \text{ Mg/m}^3 \, (1 - 0.5) \text{ m}^3 = 1.35 \text{ Mg (or 1350 kg)}$$

$$\rho_d = \frac{M_s}{V_t} = \frac{1.35 \text{ Mg}}{1 \text{ m}^3} = 1.35 \text{ Mg/m}^3 \text{ (or 1350 kg/m}^3)$$

$$\rho_{sat} = \frac{M_s + M_w}{V_t} = \frac{M_s + \rho_w V_v}{V_t}$$

$$\rho_{sat} = \frac{1.35 \text{ Mg} + 1 \text{ Mg/m}^3(0.5\text{m}^3)}{1 \text{ m}^3} = 1.85 \text{ Mg/m}^3$$

The total stress at A is $\Sigma \rho_i g h_i$:

$$1.35 \text{ Mg/m}^3 \times 9.81 \text{ m/s}^2 \times 2 \text{ m} = 26.49 \text{ kN/m}^2$$
$$+ \ 1.85 \text{ Mg/m}^3 \times 9.81 \text{ m/s}^2 \times 2 \text{ m} = 36.30 \text{ kN/m}^2$$
$$+ \ 2.0 \text{ Mg/m}^3 \times 9.81 \text{ m/s}^2 \times 4 \text{ m} = \underline{78.48 \text{ kN/m}^2}$$
$$141.27 \text{ kN/m}^2, \text{ or } 141.3 \text{ kPa.}$$

The effective stress at A is

$$\sigma' = \sigma - \rho_w g h$$

$$= 141.3 - (1 \text{ Mg/m}^3 \times 9.81 \text{ m/s}^2 \times 6 \text{ m}) = 82.4 \text{ kPa}$$

The effective stress may also be computed by the Σ ρgh above the water table and the Σ $\rho' gh$ below the water table, or

$$1.35 \text{ Mg/m}^3 \times 9.81 \text{ m/s}^2 \times 2 \text{ m} = 26.49 \text{ kPa}$$
$$+ (1.85 - 1.0) \times 9.81 \times 2 \text{ m} = 16.68 \text{ kPa}$$
$$+ (2.0 - 1.0) \times 9.81 \times 4 \text{ m} = \underline{39.24 \text{ kPa}}$$
$$82.41 \text{ kPa} \text{ (checks)}$$

Note: In practice, computations would probably only be carried out to the nearest whole kPa.

EXAMPLE 7.6

Given:

The soil profile of Example 7.5.

Required:

Plot the total, neutral, and effective stresses with depth for the entire soil profile.

Solution:

Figure Ex. 7.6. You should verify that the numerical values shown on the figure are correct. As in the previous example, computations to the nearest whole kPa are generally accurate enough.

Note how the slopes of the stress profiles change as the density changes. Profiles such as those shown in Fig. Ex. 7.6 are useful in foundation engineering, so you should become proficient in computing them. In engineering practice, the basic soils information comes from site investigations and borings which determine the thicknesses of the significant soil layers, the depth to the water table, and the water contents and densities of the various materials. Stress profiles are also useful for illustrating and understanding what happens to the stresses in the ground when conditions change, for example, when the ground water table is raised or lowered as a result of some construction operation, pumping, or flooding. Some of these effects are illustrated in the following examples.

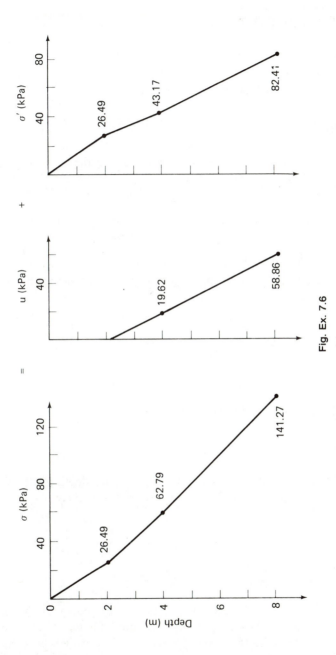

Fig. Ex. 7.6

EXAMPLE 7.7

Given:

The soil profile of Example 7.5.

Required:

Plot the total, neutral, and effective stresses with depth if the ground water table rises to the ground surface.

Solution:

Figure Ex. 7.7.
   Note that the effective stress at point A (at $z = 8$ m) is *reduced*! Had the ground water table dropped below its original elevation, the effective stress at point A would have increased.

EXAMPLE 7.8

Given:

The soil profile of Example 7.5.

Required:

Plot the total, neutral, and effective stresses with depth for the case where the ground water table is 2 m *above* the ground surface.

Solution:

Figure Ex. 7.8. (See page 224.)

Consider carefully how the stress profiles change as the water table elevation changes. Note especially how the effective stresses decrease as the water table rises (Example 7.6 versus 7.7) and then how the effective stress is *not* changed even when the ground water table is *above* the ground

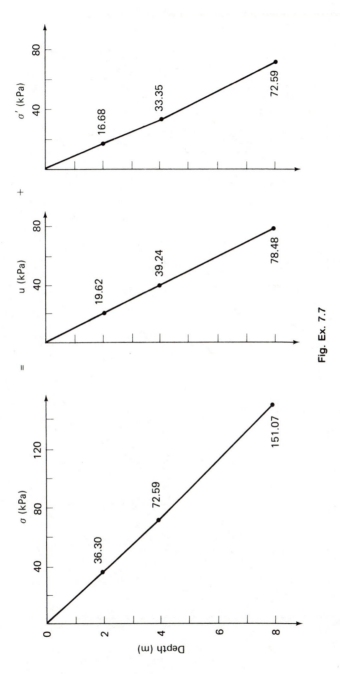

Fig. Ex. 7.7

223

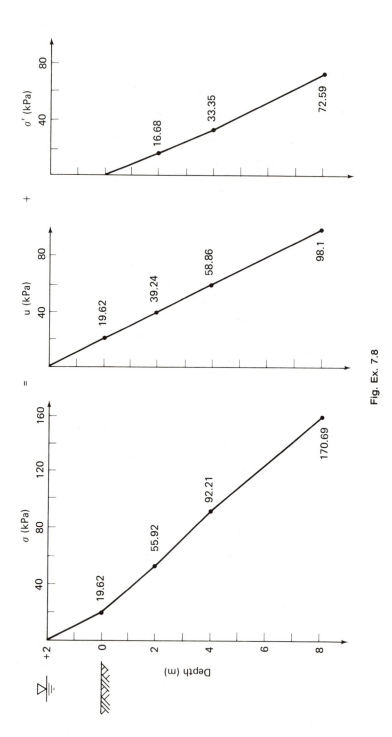

Fig. Ex. 7.8

surface (Example 7.8). Of course, for that case both the total and neutral stresses increase as the water table rises above the ground surface, but the effective stresses remain *unchanged*. The reasoning behind why the effective stresses remain unchanged is a very important concept, and you should be sure you understand why it happens.

Similar but opposite changes in effective stresses occur when the ground water table is lowered. For example, this might be caused by pumping from a deeper pervious layer. If this happens, as you might suspect by analogy, the effective stresses in the clay layer actually *increase*, causing compression of the clay and consequent surface settlements, as we shall see later. In a clay this process doesn't occur overnight; on the contrary it may take several decades for the compression to occur. These processes are discussed in detail in Chapters 8 and 9.

7.6 RELATIONSHIP BETWEEN HORIZONTAL AND VERTICAL STRESSES

You may recall from hydrostatics that the pressure in a liquid is the same in any direction—up, down, sideways, or at any inclination, it doesn't matter. However this is not true in soils. Rarely in natural soil deposits is the horizontal stress in the ground equal exactly to the vertical stress. In other words, the stresses in situ are not necessarily hydrostatic. We can express the ratio of the horizontal to vertical stress in the ground as

$$\sigma_h = K\sigma_v \qquad (7\text{-}18)$$

where K is an *earth pressure coefficient*. Since the ground water table can fluctuate and the total stresses can change, the coefficient K is *not* a constant for a particular soil deposit. However, if we express this ratio in terms of effective stresses, we take care of the problem of a variable water table, or

$$\sigma_h' = K_o\sigma_v' \qquad (7\text{-}19)$$

K_o is a very important coefficient in geotechnical engineering. It is called the *coefficient of lateral earth pressure at rest*. It expresses the stress conditions in the ground in terms of *effective stresses*, and it is independent of the location of the ground water table. Even if the depth changes, K_o will be a constant as long as we are in the same soil layer and the density remains the same. However this coefficient is very sensitive to the geologic and engineering stress history, as well as to the densities of the overlying soil layers (see for example, Massarsch, et al., 1975). The value of K_o is important in stress and analyses, in assessing the shearing resistance of particular soil layers, and in such geotechnical problems as the design of earth-retaining structures, earth dams and slopes, and many foundation engineering problems.

The K_o in natural soil deposits can be as low as 0.4 or 0.5 for sedimentary soils that have never been preloaded or up to 3.0 or greater for some very heavily preloaded deposits. Typical values of K_o for different geologic conditions are given in Chapter 11.

EXAMPLE 7.9

Given:

The stress conditions of Example 7.5. Assume K_o for this soil deposit is 0.6.

Required:

Calculate both the horizontal total and effective stresses at depths of 4 m and 8 m in the deposit. Also, determine the value of K at these depths.

Solution:

From Fig. Ex. 7.6, at 4 m, σ_v' is 43 kPa. From Eq. 7-19, $\sigma_h' = 0.6 \times 43$ kPa = 26 kPa. At 8 m, $\sigma_h' = 0.6 \times 82 = 49$ kPa. For the total horizontal stresses, we cannot use Eq. 7-18 directly because we do not know K. So we use Eq. 7-13 to get σ_h, or $\sigma_h = \sigma_h' + u$. At 4 m, $\sigma_h = 26 + 20 = 46$ kPa. At 8 m, $\sigma_h = 49 + 59 = 108$. Using Eq. 7-18, we can determine the value of the total stress coefficient K.

At 4 m,

$$K = \frac{\sigma_h}{\sigma_v} = \frac{46}{63} = 0.73$$

At 8 m,

$$K = \frac{\sigma_h}{\sigma_v} = \frac{108}{141} = 0.77$$

Note that K is *not* necessarily equal to K_o. To get K, we have to go through K_o and add the pore water pressure to the effective stress for the depth in question.

7.7 HEADS AND ONE-DIMENSIONAL FLOW

Very early in this chapter we mentioned the three types of heads associated with the Bernoulli energy equation (Eq. 7-4). They were the velocity head $v^2/2g$, the pressure head $h_p = p/\rho_w g$, and the position or elevation head z. We discussed why the energy per unit mass (or weight) was called *head* and had units of length. And we also stated that, for most seepage problems in soils, the velocity head was small enough to be neglected. Thus *total head h* becomes the sum of the *pressure head* and the *elevation head*, or $h = h_p + z$.

The elevation head at any point is the vertical distance above or below some *reference elevation* or *datum plane*. It is most often convenient to establish the datum plane for seepage problems at the tailwater elevation, but you could just as well use the bedrock or some other convenient plane as the datum. Pressure head is simply the water pressure divided by $\rho_w g$ (Eq. 7-4). The quantity $(h_p + z)$ is also called the *piezometric head* since this is the head that would be measured by an open standpipe or *piezometer* referenced to some datum plane. The elevation of the water level in the standpipe is the total head whereas the actual height of rise of the water column in the standpipe is the pressure head h_p. These concepts are illustrated in Fig. 7.9. Here we have an open-ended cylinder of soil similar to the permeameter of Fig. 7.5a. The flow into the cylinder is sufficient to maintain the water elevation at A, and the tail water is constant at elevation E. All energy or head is lost in the soil.

Note that for piezometer c in the figure the pressure head h_p is the distance AC and the elevation head z is the distance CE. Thus the total head at point C is the sum of these two distances, or AE. Determinations of the piezometric heads at the other points in Fig. 7.9 are made in a similar manner, and these are shown in the table below the figure. Be sure you understand how each of the heads, including the head loss through the soil, is obtained in Fig. 7.9. Note that it is possible for the elevation head (as well as the pressure head) to be negative, depending on the geometry of the problem. The important thing is that the total head must equal the sum of the pressure and elevation head at all times.

As mentioned, we assume that all the energy or head lost in the system is lost in flowing through the soil sample of Fig. 7.9. Thus at elevation C no head loss has yet occurred; at D, the midpoint of the sample, half the head is lost ($\frac{1}{2} AE$); and at F, all of the head has been lost (AE).

The following examples illustrate how you determine the various types of heads and head loss in some simple one-dimensional flow systems.

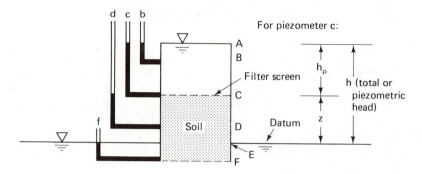

Point	Pressure Head	Elevation Head	Total Head	Head Loss through Soil
B	AB	BE	AE	0
C	AC	CE	AE	0
D	CD	DE	CE	$\frac{1}{2}$ AE
F	EF	−EF	0	AE

Fig. 7.9   Illustration of types of head (after Taylor, 1948).

EXAMPLE 7.10

Given:

The test setup of Fig. 7.9 has the dimensions shown in Fig. Ex. 7.10a.

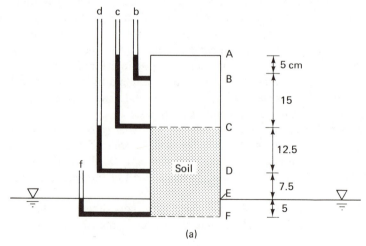

(a)

Fig. Ex. 7.10a

Required:

(a) Calculate the actual magnitude of pressure head, elevation head, total head, and head loss at points B, C, D, and F, in centimetres of water. (b) Plot the heads versus the elevation.

Solution:

    a. List dimensions and heads in a table as in Fig. 7.9, as shown below; the heads are in units of centimetres of water.

Point	Pressure Head	Elevation Head	Total Head	Head Loss
B	5	35	40	0
C	20	20	40	0
D	12.5	7.5	20	20
F	5	− 5	0	40

    b. See Fig. Ex. 7.10b.

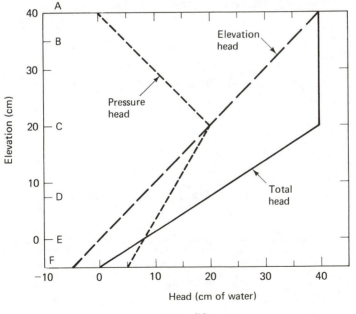

(b)

Fig. Ex. 7.10b

EXAMPLE 7.11

Given:

The cylinder of soil and standpipe arrangement shown in Fig. Ex. 7.11.

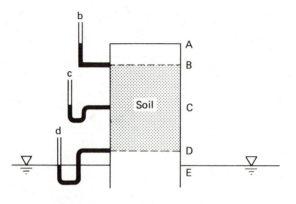

Fig. Ex. 7.11   (After Taylor, 1948.)

Required:

Determine the pressure head, elevation head, total head, and head loss at points B, C, and D.

Solution:

Set up a table similar to that of Fig. 7.9 and Example 7.10.

Point	Pressure Head	Elevation Head	Total Head	Head Loss
B	AB	BE	AE	0
C	0	CE	CE	$\frac{1}{2}AE$
D	$-DE$	DE	0	AE

EXAMPLE 7.12

Given:

The horizontal cylinder of soil as shown in Fig. 7.2. Assume $L = 10$ cm, $A = 10$ cm², and $\Delta h = 5$ cm. Tailwater elevation is 5 cm above the centerline of the cylinder. The soil is a medium sand with $e = 0.68$.

Required:

Determine the pressure, elevation, and total head at sufficient points to be able to plot them versus horizontal distance.

Solution:

Redraw Fig. 7.2 in Fig. Ex. 7.12a with the key dimensions. Estimate the other required dimensions. Label the key points as shown. Assume the datum is the elevation of the tailwater. Set up a table as in Fig. 7.9, and fill in the blanks. The units are in centimetres of water.

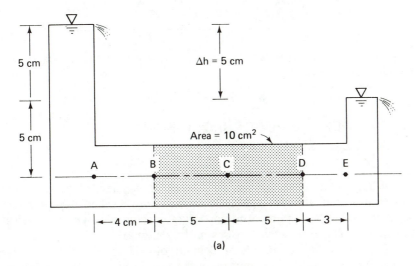

(a)

Fig. Ex. 7.12a

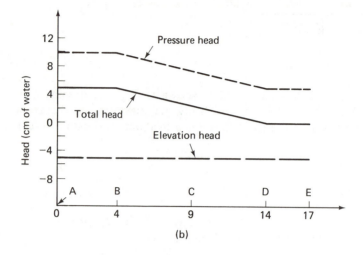

(b)

Fig. Ex. 7.12b

Point	Pressure Head	Elevation Head	Total Head	Head Loss
A	10	− 5	5	0
B	10	− 5	5	0
C	$7\frac{1}{2}$	− 5	$2\frac{1}{2}$	$2\frac{1}{2}$
D	5	− 5	0	5
E	5	− 5	0	5

The plot of heads versus horizontal distance is in Fig. 7.12b for the centerline of the cylinder.

7.8 SEEPAGE FORCES, QUICKSAND, AND LIQUEFACTION

When water flows through soils (such as the flow of water in the permeability tests already discussed) it exerts forces called *seepage forces* on the individual soil grains. And you might imagine, seepage forces affect the intergranular or effective stresses in the soil mass.

Let us reconsider the 5 m column of soil of Example 7.3. By connecting a riser tube to the bottom of the sample, we can flow water into the column of soil, as shown in Fig. 7.10. When the water level in the riser tube is at elevation B, we again have the static case and all the standpipes

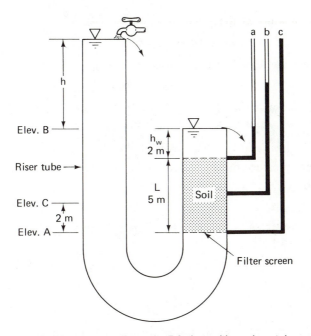

Fig. 7.10 Sample of soil from Example 7.3, but with a riser tube connected to the bottom of the sample. Standpipes are shown for the case where the water level in the riser pipe is at a distance *h* above elevation *B*.

would be at elevation *B*. If the water in the riser tube is below elevation *B*, water will flow *downward* through the soil; the reverse is true when the water elevation in the riser tube is above elevation *B*. This is the same case as the falling-head permeameter test setup of Fig. 7.5b in which water flows *upward* through the soil: when this happens, the water loses some of its energy through friction. The greater the head *h* above elevation *B* in Fig. 7.10, the larger the energy or head loss and the larger the seepage forces transmitted to the soil. As the seepage forces increase, they gradually overcome the gravitational forces acting on the soil column, and eventually a *quick* condition (quick = alive) or *boiling* will occur. Another name for this phenomenon is *quicksand*. To have a sand mass in a quick condition, the effective stresses throughout the sample must be zero.

What is the height *h* above elevation *B* that causes the soil to become quick? First from Fig. 7.10 we can calculate the total, neutral, and effective stress at elevation *A* when the water level in the riser tube is at elevation *B*. We will neglect any friction losses in the riser tube. Total stress at the bottom of the sample (elevation *A*) is

$$\sigma = \rho_{\text{sat}}gL + \rho_w gh_w = \rho'gL + \rho_w g(L + h_w) \qquad \text{(a)}$$

The pore pressure at that point is

$$u = \rho_w g(L + h_w)$$ (b)

Therefore the effective stress is [Eqs. (a) − (b)]:

$$\sigma' = \sigma - u = \rho' g L$$ (c)

Let the water level rise a distance h *above* elevation B (Fig. 7.10). Now the pore water pressure at the bottom of the sample is

$$u = \rho_w g(L + h_w + h)$$ (d)

or the pore pressure difference acting at the bottom of the sample is [Eqs. (d) − (b)]:

$$\Delta u = \rho_w g h$$ (e)

The effective stress at the bottom of the soil column (elevation A) is now [Eqs. (a) − (d)]:

$$\sigma' = \left[\rho' g L + \rho_w g(L + h_w) \right] - \left[\rho_w g(L + h_w + h) \right]$$

or

$$\sigma' = \rho' g L - \rho_w g h$$ (f)

Thus the effective stress decreased by exactly the increase in pore water pressure Δu at the base of the sample [Eqs. (f) − (c) = (e)].

What happens when the effective stress at the bottom of the soil column is zero? (Note that σ' cannot be less than zero.) Set Eq. (f) equal to zero and solve for Eq. (h), which is the head above elevation B to cause a *quick* condition, or

$$h = \frac{L\rho'}{\rho_w}$$

Rearranging,

$$\frac{h}{L} = i = \frac{\rho'}{\rho_w} = i_c$$ (7-20)

By Eq. 7-1, the head h divided by the sample length L equals the hydraulic gradient i. The value i when a quick condition occurs is called the *critical hydraulic gradient* i_c.

In Example 2.6 we obtained the following relationship for the submerged density ρ':

$$\rho' = \frac{\rho_s - \rho_w}{1 + e}$$ (2-18)

Combining Eqs. 7-20 and 2-18 we obtain an expression for the critical hydraulic gradient necessary for a quick condition to develop,

$$i_c = \frac{\rho_s - \rho_w}{(1 + e)\rho_w}$$ (7-21)

or

$$i_c = \frac{1}{1 + e}\left(\frac{\rho_s}{\rho_w} - 1\right)$$ (7-22)

The approach just used to obtain i_c is based on the premise that quick conditions occur when the effective stress at the bottom of the soil column is zero.

Another way to obtain the formula for the critical gradient is to consider the *total boundary pore water pressure* and the *total weight* of all the material above that boundary. Quick conditions then occur if these forces are just equal. From Fig. 7.10, the upward force equals the pore water pressure acting on the filter screen at elevation A on the bottom of the soil column, or

$$F_{\text{water}}\uparrow = (h + h_w + L)\rho_w gA$$

where A is the cross-sectional area of the sample.

The total weight of soil and water acting downward at the bottom of the sample (elevation A) is

$$F_{\text{soil}+\text{water}}\downarrow = \rho_{\text{sat}} gLA + \rho_w gh_w A$$

Equating these two forces, we obtain

$$(h + h_w + L)\rho_w gA = \rho_{\text{sat}} gLA + \rho_w gh_w A$$ (g)

Use Eq. 2-17 for ρ_{sat} and do the algebra to satisfy yourself that Eq. (g) reduces to Eq. 7-21. Therefore both approaches, total and effective, will give the same results.

We can compute typical values of the critical hydraulic gradient, assuming a value of $\rho_s = 2.68$ Mg/m³ and void ratios representative of loose, medium, and dense conditions. The values of i_c are presented in Table 7-1. Thus, for estimation purposes, i_c is often taken to be about unity, which is a relatively easy number to remember.

TABLE 7-1 Typical Values of i_c for $\rho_s = 2.68$ Mg/m³

Void Ratio	Approximate Relative Density	i_c
0.5	Dense	1.12
0.75	Medium	0.96
1.0	Loose	0.84

EXAMPLE 7.13

Given:

The soil sample and flow conditions of Fig. 7.10 and Example 7.3.

Required:

   a. Find the head required to cause quick conditions.
   b. Find the critical hydraulic gradient.

Solution:

   a. From Eq. 7-20,

$$h = \frac{\rho' L}{\rho_w} = \frac{\rho_{sat} - \rho_w}{\rho_w} L$$

$$= \left(\frac{2.0 - 1.0}{1.0}\right) 5 \text{ m} = 5.0 \text{ m}$$

   b. The critical hydraulic gradient (Eq. 7-20) is

$$i_c = \frac{\rho'}{\rho_w} = \frac{(2.0 - 1.0)}{1.0} = 1.0$$

We could also use Eq. 7-21 if we knew the value of ρ_s and e. Assume $\rho_s = 2.65$ Mg/m^3. From Eq. 2-17, solve for $e = 0.65$. Therefore,

$$i_c = \frac{(2.65 - 1.0)}{(1 + 0.65)(1.0)} = 1.0$$

Seepage forces, which may but not necessarily cause quicksand to develop, are always present in soils where there is a gradient causing the flow of water. Seepage forces affect sands more than clays because sands are cohesionless whereas clay soils have some inherent cohesion which holds the particles together. To evaluate the seepage forces, let's look again at Fig. 7.10. For quick conditions to develop, the upward force of water due to the head h on the left side of the figure must just equal the *effective* downward force exerted by the submerged soil column on the right side of the figure, or

$$\text{upward force} = \text{downward force}$$

$$\rho_w ghA = \rho' gLA \tag{7-23a}$$

Substituting Eq. 2-18 into this equation we get

$$\rho_w ghA = \frac{\rho_s - \rho_w}{1 + e} gLA \tag{7-23b}$$

After algebraic manipulation, this equation is identical to Eq. 7-22. In uniform flow the upward force $\rho_w ghA$, the left-hand side of Eq. 7-23a, is distributed (and dissipated) uniformly throughout the volume LA of the soil column. Thus

$$\frac{\rho_w g h A}{LA} = \rho_w g i = j \qquad (7\text{-}23c)$$

The term $i\rho_w g$ is the *seepage force per unit volume*, and it is commonly represented by the symbol j. The value of this force at quick conditions equals $i_c\rho_w g$, and it acts in the direction of fluid flow in an isotropic soil. If the right-hand side of Eq. 7-23a is divided by LA, the unit volume, then we have

$$j = \rho'g \qquad (7\text{-}23d)$$

These expressions, Eqs. 7-23c and 7-23d, can be shown to be identical when quick conditions occur (see Eq. 7-21).

EXAMPLE 7.14

Given:

The soil sample and flow conditions of Fig. 7.10 and Example 7.3.

Required:

    a. Find the head required to cause a quick condition.
    b. Compute the seepage force per unit volume at quick conditions.
    c. Show, using seepage forces, that quick conditions really develop under the head of part (a).
    d. Compute the total seepage force at elevation A.

Solution:

    a. From Example 7.13, h above elevation B to cause a quick condition is 5.0 m.
    b. The seepage force per unit volume is computed from Eq. 7-23c.

$$j = i\rho_w g = \frac{5\ m}{5\ m} \times 1\frac{Mg}{m^3} \times 9.81\frac{m}{s^2} = 9.81\frac{kN}{m^3}$$

We also could use Eq. 7-23d if we knew the value of ρ_s or e. Assume, as in Example 7.13, $\rho_s = 2.65$ Mg/m³. Then $e = 0.65$. Therefore

$$j = \frac{2.65 - 1.0}{1.65}g = 9.81\frac{kN}{m^3}$$

Note that the units check $(F/L^3 = ML^{-2}T^{-2})$.

c. Quick conditions develop when the upward seepage force just equals the downward buoyant force of the soil. Or, from Eq. 7-23c and d:

$$\frac{j}{vol}(vol) \uparrow = \rho'g(vol)\downarrow$$

$$9.81\frac{kN}{m^3} \times 5\ m \times 1\ m^2 = (2.0 - 1.0)\ \frac{Mg}{m^3} \times 9.81\frac{m}{s^2} \times 5\ m \times 1\ m^2$$

$$49.05\ kN\uparrow = 49.05\ kN\downarrow$$

d. The total seepage force at elevation A is

$$j\,(vol) = 9.81\frac{kN}{m^3} \times 5\ m \times 1\ m^2 = 49.05\ kN$$

This force is distributed uniformly through the volume of the soil column.

The seepage force is a real force and it is added vectorially to the body or gravitational forces to give the net force acting on the soil particles. We can represent these forces in two different ways but each way gives identical results. In Example 7.14 we treat the problem considering seepage forces and submerged densities. A quick condition resulted because the effective or buoyant density of the soil volume (acting downward) just equalled the seepage force (acting upward).

An alternative approach is to consider the *total* saturated weight of soil and the boundary water forces acting on the soil, top and bottom, as shown in Example 7.15.

EXAMPLE 7.15

Given:

The soil sample and conditions of Fig. 7.10 and Examples 7.3 and 7.14.

Required:

Show, using total (saturated) weight of the soil above elevation A and the boundary water forces, that quick conditions develop when the head h is 5 m.

Solution:

For a quick condition, $\sum F_v = 0$.

$$F_{soil}\downarrow = \rho_{sat}gLA$$

$$= 2.0\frac{Mg}{m^3} \times 9.81\frac{m}{s^2} \times 5\text{ m} \times 1\text{ m}^2 = 98.1\text{ kN}$$

$$F_{water\ top}\downarrow = \rho_w gh_w A$$

$$= 1\frac{Mg}{m^3} \times 9.81\frac{m}{s^2} \times 2\text{ m} \times 1\text{ m}^2 = 19.6\text{ kN}$$

$$F_{water\ bottom}\uparrow = \rho_w g(L + h_w + h)A$$

$$= 1\frac{Mg}{m^3} \times 9.81\frac{m}{s^2} \times (5 + 2 + 5)\text{ m}$$

$$= 117.7\text{ kN}$$

Therefore $\sum F_{down} = \sum F_{up}$ for a quick condition (117.7 kN = 117.7 kN).

EXAMPLE 7.16

Given:

The soil and flow condition of Fig. 7.10, except that the left-hand riser tube is at elevation C, or 2 m above elevation A. Assume the water level is maintained constant at elevation C.

Required:

Compute (a) the hydraulic gradient, (b) effective stress, and (c) seepage force at elevation A.

Solution:

In this case the flow of water is downward through the soil. Assume the datum plane is at the tailwater elevation, or at elevation B.

    a. Use Eq. 7-1; since the head loss is -5 m (below elevation B),

$$i = \frac{-H}{L} = \frac{-5}{5} = -1$$

    b. The effective stress at elevation A may be computed in the two ways just described.

1. Using *boundary* water forces and saturated densities we get (units are the same as Example 7.15)

$$F_{soil}\downarrow = \rho_{sat}gLA$$
$$= 2.0(9.81)(5)(1) = 98.1 \text{ kN}\downarrow$$
$$F_{water\ top}\downarrow = \rho_w gh_w A$$
$$= 1(9.81)(2)(1) = 19.6 \text{ kN}\downarrow$$
$$F_{water\ bottom}\uparrow = \rho_w ghA$$
$$= 1(9.81)(2)(1) = 19.6 \text{ kN}\uparrow$$
$$\sum F_{v_A} = 19.6 + 98.1 - 19.6$$
$$= 98.1 \text{ kN}\downarrow(\text{net or effective force})$$

$$\text{the effective stress} = \frac{F}{A} = 98.1 \text{ kN/m}^2$$

Thus the filter screen at elevation A must support a force of 98.1 kN per unit area or a stress of 98.1 kN/m^2 in this case.

2. The other way to compute the effective stress at elevation A is to use seepage forces, buoyant densities, and Eq. 7-23. Note that $h = -5$ m referenced to elevation B.

$$j = \rho_w gi(\text{vol}) = 1(9.81)\left(\frac{-5}{5}\right)(5)(1)$$
$$= 49.05 \text{ kN acting down in the direction of flow}$$

To this we add the effective or buoyant weight:

$$F_{down}\downarrow = \rho'gLA = (\rho_{sat} - \rho_w)gLA$$
$$= (2.0 - 1)(9.81)(5)(1) = 49.05 \text{ kN}\downarrow$$

Therefore adding vectorially these two forces, we get the seepage force plus the effective soil force acting on area A, or 49.05 + 49.05 = 98.1 kN per unit area as before. Or the effective stress at $A = 98.1$ kN/m^2. Note that this second approach also automatically gives the solution to part (c), the seepage force at A. Note that the seepage force at the top of the soil is zero and increases linearly to 49 kN at elevation A.

An apparatus that is commonly used in soil mechanics teaching laboratories to demonstrate the phenomenon of quicksand is shown in Fig. 7.11. Instead of a standpipe as in Fig. 7.10, a pump is used to create the upward flow in the quicksand tank. The water flows through a porous

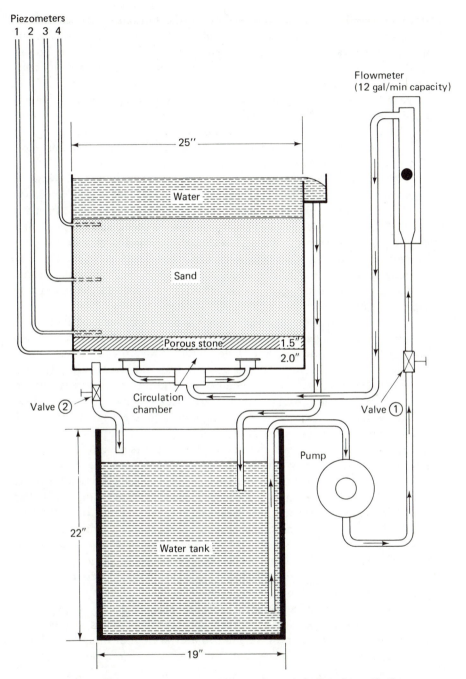

Fig. 7.11 Diagram of a quicksand tank (courtesy J. O. Osterberg, North-western University).

241

stone to distribute the pressure evenly on the bottom of the sand mass. Piezometers at various levels on the tank enable heads to be observed and measured. As valve 1 is gradually opened, the head applied to the bottom of the sand mass increases, and eventually it becomes sufficient to cause the entire sand mass to boil or *liquefy*. As in Examples 7.14 and 7.15, the seepage forces are acting upward and just balance the downward-acting gravitational forces. The effective stresses between the sand grains are zero, and the soil has no shear resistance. As long as the pump is running, the soil mass can easily be stirred with a rod or metre stick, and it acts like a dense liquid (Fig. 7.12a).

Next we shut off the pump, close valve 1, and open valve 2. Now the direction of the water flow is reversed, and the seepage forces act downward along with gravity and increase the effective stresses. A rod or metre stick left buried in the sand has resistance to movement, and the sand mass can no longer be stirred easily. Even though the sand is very loose, it can support some static loads at the surface, as shown in Fig. 7.12b. This case is similar to Example 7.16. Therefore, depending on their direction, seepage forces can significantly increase the effective stresses and the strength of the soil mass.

Some practical examples of quick conditions include excavations in granular materials behind cofferdams alongside rivers. To excavate and proceed with construction, the water table at the site is lowered by a system of wells and pumps. Of course water from the river invariably seeps into the excavation and must be pumped out to keep the excavation dry. If upward gradients approach unity, the sand can become quick and failure of the cofferdam can occur. As is explained in the next section, such failures are usually catastrophic so high safety factors must be used in design. Another place quick conditions often occur is behind levees during floods. The water seeps under the levee and, as in the case of the cofferdam, if the gradient is high enough localized quick conditions can occur. This phenomenon is known as a *sand boil* and must be halted quickly (usually by stacking sand bags in a ring around the boil), otherwise the erosion can spread and undermine the levee. Quick conditions are also possible almost any place where *artesian* pressures exist, that is, where the head is greater than the usual static water pressure. Such pressures occur where a pervious underground stratum is continuous and connected to a place where the head is higher.

Contrary to popular belief, it is not possible to drown in quicksand, unless you really work at it, because the density of quicksand is much greater than that of water. Since you can almost float in water, you should easily be able to float in quicksand.

(a)

(b)

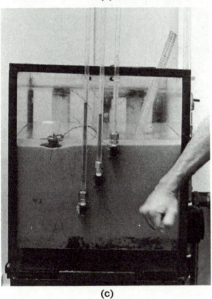

(c)

Fig. 7.12 Quicksand tank: (a) under an upward gradient, the sand mass is easily stirred with a metre stick; (b) gradient is downward; sand is able to support a static load; (c) after a shock load is applied to the side of the tank, the sand mass liquefies and momentarily loses all bearing capacity. Notice water level in piezometers (photograph by M. Surendra; hand by R. D. Holtz).

EXAMPLE 7.17

Given:

The conditions as shown in Fig. Ex. 7.17. The silty clay acts as an impervious layer and prevents flow of water up from the fine sand layer below it. Because of a river nearby, the fine sand layer is under a head of water greater than the existing ground surface (artesian conditions). If a standpipe or piezometer were installed through the silty clay layer, it would rise a distance h above the top of the sand layer. If H_s is not sufficiently large, the uplift pressure in the middle of the excavation could cause the bottom of the excavation to "blow up."

Required:

Calculate how deep an excavation can be made, based on the assumption that you neglect the shear force on the sides of the soil plug.

Solution:

At equilibrium, $\sum F_v = 0$.

$$H_s \rho g = \rho_w g h$$

or

$$H_s = \frac{\rho_w g h}{\rho g}$$

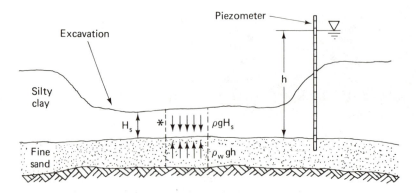

*Neglect shear along sides.

Fig. Ex. 7.17

Failure will occur if $H_s < \rho_w gh/\rho g$. If $H_s > \rho_w gh/\rho g$, then failure cannot happen. In practice, the factor of safety against blow up should be rather high since, if it occurs, it is catastrophic. You must be very conservative in designing such situations because of the possible consequences.

Another phenomenon related to quicksand is *liquefaction*, which can be demonstrated in a quicksand tank. After the sand has been made quick and in a very loose state, the flow is reversed and the water level is allowed to drop just slightly below the sand surface. A sharp blow is applied to the side of the tank, and instantly the entire soil mass liquefies and the sand loses all bearing capacity (Fig. 7.12c). This reaction is exactly what happens when a loose saturated deposit of sand is subjected to loads of very short duration, such as occur during earthquakes, pile driving, and blasting. The loose sand tried to densify during shear and this tends to squeeze the water out of the pores. Normally, under static loading, the sand has sufficient permeability so the water can escape and any induced pore water pressures can dissipate. But in this situation because the loading occurs in such a short time, the water doesn't have time to escape and the pore water pressure increases. Since the total stresses have not increased during loading, the effective stresses then tend toward zero (by Eq. 7-13), and the soil loses all strength. Note the position of the water level in the standpipes of Fig. 7.12c. The photograph was taken just after a sharp blow against the side of the tank.

Casagrande (1936a) was the first to explain liquefaction in terms of soil mechanics, and he also describes (1950, 1975) some situations in practice where liquefaction has occurred. Among these are the failure of Ft. Peck Dam in Montana in 1938 and *flow slides* along the lower Mississippi River. Here sands are deposited during floods in a very loose state. Somehow strains are induced in these deposits, and it seems that they almost spontaneously liquefy and flow out into the river. The problem is that they often take levees and other flood protection works along with them, and repairs of these features are expensive. Bank erosion leading to progressive liquefaction, seepage pressures from high water tables, and even traffic vibrations have been blamed for flow slides. Flow slides also occur in mine tailings dams. These structures are often very large and constructed hydraulically of very loose sands and silts. Since they are a waste dump, very little engineering and construction inspection goes into them. Failures are relatively common.

Since the Niigata, Japan, and the Anchorage, Alaska, earthquakes of 1964, where severe damage occurred due to liquefaction, there has been increasing interest in liquefaction. It has been found that liquefaction can occur in the laboratory in even moderately dense sands due to a repeated or cyclic application of shear stress, which means that if an earthquake lasted long enough, then even moderately dense saturated sands could possibly liquefy. This important, though controversial point, is discussed by Casagrande (1975) and Seed (1979). (See Sec. 11.8.)

7.9 SEEPAGE AND FLOW NETS: TWO-DIMENSIONAL FLOW

The concept of head and energy loss as water flows through soils has been mentioned several times in this chapter. When water flows through a porous medium such as soil, energy or head is lost through friction similar to what happens in flow through pipes and in open channels. As in the laboratory permeability test described earlier, for example, similar energy or head losses occur when water seeps through an earth dam or under a sheet pile cofferdam (Fig. 7.13).

Different kinds of heads and head losses were described in Sec. 7.7, and it might be a good idea to review that material before proceeding further in this section. Figure 7.14 shows how the piezometric heads $(h_p + z)$ might be determined from the positions and elevations of the

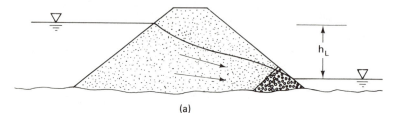

(a)

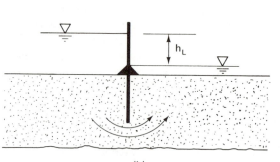

(b)

Fig. 7.13 Engineering examples of head loss because of seepage through soils.

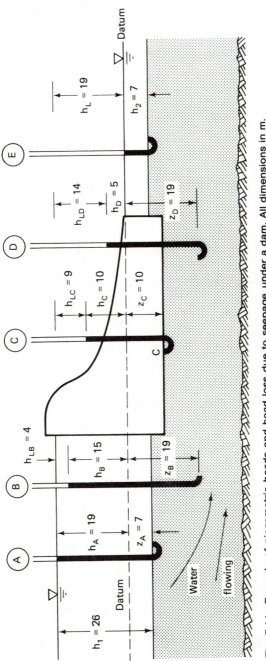

Fig. 7.14 Example of piezometric heads and head loss due to seepage under a dam. All dimensions in m.

water levels in the standpipes. Also shown in this figure is how energy or head is lost in flowing under a dam. Note how the water levels in each successive piezometer decrease as water flows from the heel to the toe of the dam. Example 7.18 explains in detail how head computations are made.

EXAMPLE 7.18

Given:

The dam with piezometers shown in Fig. 7.14.

Required:

    a. Calculate the pressure heads h_p and the total (piezometric) heads h for piezometers A through E.
    b. Determine the uplift pressure acting on the base of the dam at point C.

Solution:

    a. Pressure and total (piezometric) heads.
        Piezometer A: The pressure head is the length of the column of water in the standpipe, or

$$h_p = h_A + z_A = h_1 = 19 + 7 = 26 \text{ m}$$

Note that this dimension is also numerically equal to

$$h_L + h_2 = 19 + 7 = 26 \text{ m}$$

The total or piezometric head is

$$h = (h_p + z) = h_A + z_A - z_A = 19 \text{ m}$$

which is the height of rise above the datum.
        Piezometer B:

$$h_p = h_B + z_B = 15 + 19 = 34 \text{ m}$$

$$h = (h_p + z) = h_B + z_B - z_B = 15 \text{ m}$$

Note that h is also numerically the same as

$$h_L - h_{LB} \quad \text{or} \quad h = 19 - 4 = 15 \text{ m}$$

        Piezometer C:

$$h_p = h_C + z_C = 10 + 10 = 20 \text{ m}$$

$$h = h_C + z_C - z_C = 10 \text{ m}$$

(Check: $h = h_L - h_{LC} = 19 - 9 = 10$ m.)

Piezometer D:

$$h_p = h_D + z_D = 5 + 19 = 24 \text{ m}$$

$$h = h_D + z_D - z_D = 5 \text{ m}$$

(Check: $h = h_L - h_{LD} = 19 - 14 = 5$ m.)

Piezometer E:

$$h_p = h_2 = 7 \text{ m}$$

$$h = h_p - z_E = 7 - 7 = 0$$

$$h_L = 19 \text{ m}$$

Note that at the tailwater all of the head has been lost. Thus the total head at this point is zero.

b. Uplift pressure at point C:

$$p_C = h_p \rho_w g = (h_C + z_C)\rho_w g = (h_L - h_{LC} + z_C)\rho_w g$$

$$= 20 \text{ m } (1000 \text{ kg/m}^3)(9.81 \text{ m/s}^2) = 196.2 \text{ kPa}$$

We could represent the flow of water through the foundation under the dam in Fig. 7.14 by *flow lines*, which would be the average flow path of a particle of water flowing from the upstream reservoir down to the tailwater. Similarly, we could represent the energy of flow by lines of equal potential, called, naturally, *equipotential lines*. Along any equipotential line, the energy available to cause flow is the same; conversely, the energy lost by the water getting to that line is the same all along that line. The network of flow lines and equipotential lines is called a *flow net*, a concept that illustrates graphically how the head or energy is lost as water flows through a porous medium, as shown in Fig. 7.15.

You probably can see that we could, if we wanted to, draw an infinite number of flow lines and equipotential lines to represent the seepage shown in Fig. 7.15, but it is more convenient to select only a few representative lines of each type. The hydraulic gradient between any two adjacent equipotential lines is the drop in potential (head) between those lines divided by the distance traversed. Or, in Fig. 7.15 along flow line 2, the gradient between equipotential lines a and b is the head drop between those lines divided by l. Because in an *isotropic* soil the flow must follow paths of the largest gradient, the flow lines have to cross the equipotential lines at right angles, as shown in Fig. 7.15. Note that, as the equipotential lines become closer together, l decreases and the gradient increases.

Figure 7.15 represents a typical cross section of the dam and foundation. Thus the flow condition is *two dimensional*, like all seepage problems

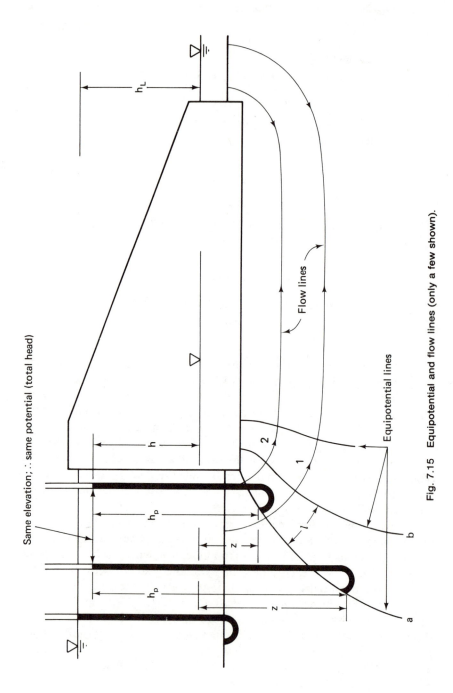

Fig. 7.15 Equipotential and flow lines (only a few shown).

considered in this text. Three-dimensional flow is the more general situation in many geotechnical problems, but seepage analyses of these problems are just too complicated to be practical, so we usually simplify the problem to two dimensions. Also in this text we will only consider the simple case of *confined* flow, that is, where the seepage is confined between two impervious surfaces. For a discussion of unconfined flow problems (such as for earth dams and seepage toward wells), see Casagrande (1937), Taylor (1948), Leonards (1962), and Cedergren (1977).

Flow nets are very useful in solving seepage problems in engineering practice, for example to estimate seepage losses from reservoirs, uplift pressures under dams, and check points of potential detrimental erosion where $i \rightarrow i_{cr}$. We shall explain the techniques in this section.

A flow net is actually a graphical solution of *Laplace's equation* in two dimensions,

$$\frac{\partial^2 h}{\partial x^2} + \frac{\partial^2 h}{\partial y^2} = 0 \qquad (7\text{-}24)$$

where x and y are the two coordinate directions. Laplace's equation, derived in Appendix B, is a very important equation in mathematical physics; it represents the energy loss through any resistive medium. For example, besides the flow of water through soils, it describes electron flow, the flow of people to hospitals, etc. If the *boundary conditions* (geometry, flow conditions, and head conditions at the boundaries) are simple, then it is even possible to solve the equation in closed form, that is, exactly. But for most practical engineering problems, it is simpler to solve such problems graphically albeit somewhat inexactly. Flow nets are such graphical solutions to Laplace's equation for a given set of boundary conditions.

How do you make a flow net? By sketching. For two-dimensional steady-state problems, you simply draw the medium with its boundaries to some convenient scale. By trial and error (mostly error, until you get some practice!) sketch a network of flow lines and equipotential lines spaced so that the enclosed figures resemble "squares." Their sides intersect at right angles. Look again at Fig. 7.15 and the "square" enclosed by flow lines 1 and 2 and equipotential lines *a* and *b*. Not all the "squares" in a flow net have to be the same size either. Since the squares are made of curved lines, they are only squares in the strictest sense when they can be subdivided down the truly equilateral figures. Note that a flow line cannot intersect an impervious boundary; in fact, an impervious boundary is a flow line. Note too that all equipotential lines must meet impervious boundaries at right angles. Neither the number of *flow channels* (channels between flow lines) nor the number of *equipotential drops* (a *drop* is the decrease in head Δh from one equipotential line to the next) need to be a whole number; fractional squares are allowed. Figure 7-16 defines some of the terms associated with flow nets. Look at the "square" with dimensions $a \times b$. Note that the gradient is

$$i = \frac{\Delta h}{\Delta l} = \frac{\Delta h}{b} = \frac{h_L/N_d}{b} \qquad (7\text{-}25)$$

where the length of the flow path in one square is $b = \Delta l$. The equipotential drop between two flow lines is $\Delta h = h_L/N_d$, where N_d is the total number of potential drops, and h_L is the total head lost in the system. From Darcy's law we know that the flow in each flow channel is

$$\Delta q = k\frac{\Delta h}{\Delta l}A = k\left(\frac{h_L/N_d}{b}\right)a$$

and the total discharge q per unit depth (perpendicular to the paper) is

$$q = \Delta q N_f = kh_L\left(\frac{a}{b}\right)\left(\frac{N_f}{N_d}\right) \qquad (7\text{-}26)$$

where N_f is the total number of flow channels in the flow net. If we sketched squares in our flow net, then $a = b$. Thus we can readily estimate the quantity of flow q by simply counting the number of potential drops N_d and the number of flow channels N_f, if we know the k of the material and the total head loss h_L. Even a crude flow net provides a fairly accurate estimate of the flow quantities.

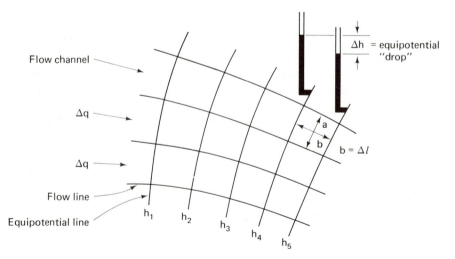

Fig. 7.16   Flow net illustrating some definitions.

With confined flow problems, where there is no phreatic (free) surface, sketching a flow net is not so difficult. Start with a sketch, to scale, of the soil mass, boundaries, etc. Keep the sketch small so you can observe the entire picture as it develops. Use good quality paper, a soft pencil, and have a good eraser handy—you'll need it! Draw the boundaries in ink on the reverse side of the sheet. Start with, at most, only three or four lines at first; by trial and error, sketch the net (lightly) until you get "squares" throughout the region of flow. It's easier if you can manage to keep the number of flow channels to a whole number. The flow lines and equipotential lines should be smooth, gradual curves, all intersecting at right angles. As mentioned, you should be able to subdivide each square to make additional small squares. The flow net shown in Fig. 7.17 is an example of a fairly well-drawn flow net for confined flow.

We mentioned earlier that flow nets were valid for isotropic soils only, a condition that is unlikely in natural soil deposits or even in earth dams. However it is easy to take the directional difference in permeability into account by transforming the scale to which you draw the flow net. For example, if the horizontal permeability is much greater than the vertical permeability, then you shorten the horizontal dimensions of the problem by the ratio $\sqrt{k_h/k_v}$. The proof of this transformation as well as examples for its use are shown in Taylor (1948). Equation 7-28 for the quantity of seepage then becomes

$$q = \frac{N_f}{N_d} \sqrt{k_v k_h} \, h_L$$

For unconfined flow, where there is a free surface at atmospheric pressure (for example, seepage through earth dams, levees, and toward wells), the major problem is to establish the shape of the top line of seepage. This is not so easy, and if you have such a problem it is best to consult one of the references cited earlier in this section. One of the serious problems, practically speaking, is when the top line of seepage intersects the top downstream surface of an earth dam or levee. This condition leads to surface erosion, piping, and eventually possible failure of the structure. Therefore, in design, keep the top line of seepage well below the downstream surface. In the dam of Fig. 7.13a, the toe drain tends to keep the top seepage line from intersecting the downstream surface.

Other methods besides sketching for obtaining flow nets include mathematical solutions (for example, Harr, 1962), electrical analogs, viscous flow (Hele-Shaw) models, small-scale laboratory flow models, and the method of fragments (Harr, 1962). This last method is so simple and practical that we explain it in the next section.

Example 7.18 indicates how the uplift pressure under a dam can be calculated. From the flow net, it is not difficult to determine what the h_p is

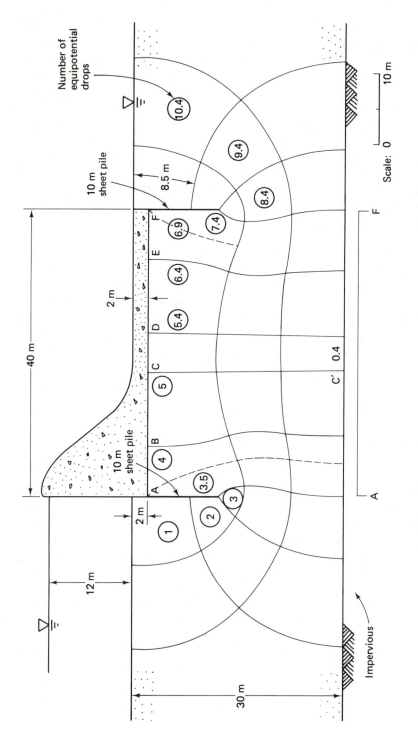

Fig. 7.17 Example of a reasonably well-drawn flow net for confined flow.

at the bottom of the dam. Then the distribution of uplift pressures can be drawn. This distribution is important for the analysis of the stability of concrete gravity dams.

Another important use of flow nets is to determine gradients, especially at certain critical points, for example, at the toe of a dam or any place where seepage water exits. From Sec. 7.8 you know that when the gradient approaches unity, critical conditions can occur, which leads to *piping* and *erosion* and may lead to complete failure of the structure. Piping is a phenomenon where seeping water progressively erodes or washes away soil particles, leaving large voids (pipes) in the soil. These voids simply continue to erode and work their way backwards under the structure, or they may collapse. Either way, if piping is not stopped promptly, failure is imminent. The critical place for piping is usually right at the corner of the toe of a dam, and we can see why if we study an enlargement of the flow net at the toe (Fig. 7.18).

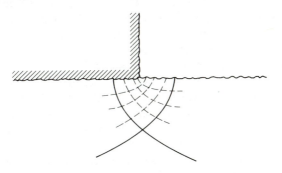

(a) Dam constructed directly on ground surface.

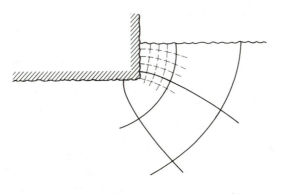

(b) Dam placed somewhat below ground surface

Fig. 7.18   Exit gradients at toe of dams.

For the case of the dam placed (foolishly) right on the ground surface (Fig. 7.18a), if we keep subdividing the squares, l rapidly approaches zero and, since Δh is still finite, the gradient rapidly increases. If this actually happened in a real structure, piping and probably failure of the structure (by undermining) would occur.

For the example shown in Fig. 7.18b, the dam is somewhat safer than in Fig. 7.18a since, for typical cases, the exit gradient is much less than critical. From Eq. 7-25, the exit gradient i_E equals $\Delta h_L / \Delta l$, where Δh_L equals the head loss h_L divided by the number of equipotential drops N_d. Thus if all other things are the same, an embedded foundation will have more equipotential drops and a lower exit gradient. Remember that the flow net in its enlarged condition in Fig. 7.18 merely shows the *concentration* of flow. As the squares get smaller and smaller, the tendency is to think that the exit gradient is steadily increasing! This is not so. As the number of equipotential drops increases, Δh_L also decreases per drop, and the ratio of $\Delta h_L / \Delta l$ remains about the same. For this example, too, you can see why the critical place is right next to the downstream toe. There, the Δl is the least for a given Δh_L. The next flow channel over, for example, is safer since the same head (Δh_L) is lost over a greater length (greater distance between equipotential lines).

For practical problems, where there is a danger that i could approach i_c, be very conservative in your design. Use a factor of safety of at least 5 or 6 for such cases. For one thing, failure is usually catastrophic and occurs rapidly and with little warning. For another, it is extremely difficult to know exactly what is going on underground, especially locally. Local defects, gravel pockets, etc., can significantly alter the flow regime and concentrate flow, for example, where you might not want it and not be prepared for it. Concentration of flow occurs, too, at corners of temporary structures like cofferdams. As Taylor (1948) points out, the entire flow regime may be widely different from that assumed in our (idealized) flow net. Great variation in horizontal and vertical permeability may exist from point to point under a foundation; the flow may not be entirely two-dimensional; geologic defects in the underlying subsoils may provide express routes for the water to concentrate and seep under and out of a foundation. If sheet piling is used, cutoff is often uncertain (for example, piling unknowingly driven into boulders), and you would be wise to assume that the worst possible conditions could happen—then prepare for such eventualities. Since failure of cofferdams is often catastrophic, it is extremely important that large factors of safety be used, especially where people's lives are at stake. Failures of earth structures resulting from piping have caused more deaths than all other failures of civil engineering structures combined. Therefore your responsibility is clear—be careful and conservative, and be sure of your ground conditions and design.

EXAMPLE 7.19

Given:

The dam and flow net shown in Fig. 7.17. The dam is 120 m long and has two 10 m sheet piles driven partially into the granular soil layer.

Required:

 a. The quantity of seepage loss under the dam when $k = 20 \times 10^{-4}$ cm/s per metre of dam.
 b. The exit gradient (at point X).
 c. The pressure distribution on the base of the dam.

Solution:

 a. From Eq. 7-26, the quantity of seepage is

$$q = kh_L \left(\frac{N_f}{N_d} \right) \times width$$

$$= \left(20 \times 10^{-4} \frac{cm}{s} \right) \left(\frac{m}{100 \ cm} \right) 12 \ m \ \frac{3}{10.4} \ 120 \ m$$

$$= 8.31 \times 10^{-3} \ m^3/s$$

 b. At point X, the exit gradient is

$$i_E = \frac{\Delta h_L}{L} = \frac{1.15}{8.5} = 0.14, \ \text{which is not critical}$$

 Note: $\Delta h_L = h_L/N_d = 12 \ m/10.4 = 1.15 \ m. \ L = 8.5 \ m,$ scaled from Fig. 7.17, is the length of square X.

 c. Pressure heads are evaluated for point A through F along the base of the dam in Fig. Ex. 7.19.

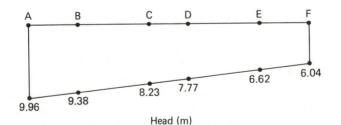

Fig. Ex. 7.19   Pressure head for locations A through F.

The pressure head at point A, at the base of the dam and just to the right of the left sheet pile is found this way: the percentage of the head loss is proportional to the number of equipotential drops. Of the total of 10.4 drops for the entire flow net, only 3.5 have occurred by point A. Thus the pressure head at point A is

$$h_A = 12 \text{ m} - 12 \text{ m} \times \frac{3.5}{10.4} + 2 \text{ m}$$

$$= 12 - 4.04 + 2 = 9.96 \text{ m}$$

The extra 2 m brings the head from the water-soil interface down to the base of the dam.

In a similar manner, we can calculate the head at point D:

$$h_D = 12 - 12 \times \frac{5.4}{10.4} + 2 = 7.77 \text{ m}$$

The heads at all the points under the dam are as follows:

Location	Head (m)	Pressure (kPa)
A	9.96	97.7
B	9.38	92.0
C	8.23	80.7
D	7.77	76.2
E	6.62	64.9
F	6.04	59.2

These values of head are plotted in Fig. Ex. 7.19. To compute the *uplift pressures* on the base of the dam, we multiply the head times the product $\rho_w g$. The pressures are given above. If the density of concrete is 2.4 Mg/m³, then the pressure exerted by 2 m of concrete is

$$2.4 \text{ Mg/m}^3 \times 9.81 \text{ m/s}^2 \times 2 \text{ m} = 47 \text{ kPa}$$

Thus at any point along the base of the dam from point C through F the uplift force exceeds the weight of the dam so the dam is *unstable* with this design.

7.10 THE METHOD OF FRAGMENTS

The *method of fragments* presents a useful, rapid, although approximate, analytical design method for the solution of confined flow problems. After you learn the procedure, many cases may be investigated in little more than the time it usually takes to assemble paper, pencils and erasers for drawing flow nets. The method originated with Pavlovsky (1956) and was brought to the attention of the western world by Harr (1962). The

TABLE 7-2 Summary of Fragment Types and Form Factors*

Fragment Type	Illustration	Form Factor, Φ (h is head loss through fragment)
I	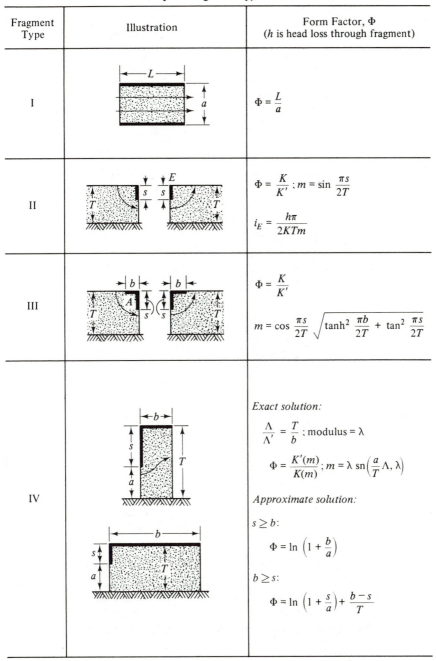	$\Phi = \dfrac{L}{a}$
II		$\Phi = \dfrac{K}{K'}\,;\ m = \sin\dfrac{\pi s}{2T}$ $i_E = \dfrac{h\pi}{2KTm}$
III		$\Phi = \dfrac{K}{K'}$ $m = \cos\dfrac{\pi s}{2T}\ \sqrt{\tanh^2\dfrac{\pi b}{2T} + \tan^2\dfrac{\pi s}{2T}}$
IV		*Exact solution:* $\dfrac{\Lambda}{\Lambda'} = \dfrac{T}{b}\,;$ modulus $= \lambda$ $\Phi = \dfrac{K'(m)}{K(m)}\,;\ m = \lambda\ \text{sn}\!\left(\dfrac{a}{T}\Lambda, \lambda\right)$ *Approximate solution:* $s \ge b:$ $\Phi = \ln\left(1 + \dfrac{b}{a}\right)$ $b \ge s:$ $\Phi = \ln\left(1 + \dfrac{s}{a}\right) + \dfrac{b-s}{T}$

TABLE 7-2 Continued

Fragment Type	Illustration	Form Factor, Φ (h is head loss through fragment)
V		$L \leq 2s$: $$\Phi = 2 \ln \left(1 + \frac{L}{2a} \right)$$ $L \geq 2s$: $$\Phi = 2 \ln \left(1 + \frac{s}{a} + \frac{L - 2s}{T} \right)$$
VI		$L > s' + s''$: $$\Phi = \ln \left[\left(1 + \frac{s'}{a'} \right) \left(1 + \frac{s''}{a''} \right) \right]$$ $$+ \frac{L - (s' + s'')}{T}$$ $L = s' + s''$: $$\Phi = \ln \left[\left(1 + \frac{s'}{a'} \right) \left(1 + \frac{s''}{a''} \right) \right]$$ $L < s' + s''$ $$\Phi = \ln \left[\left(1 + \frac{b'}{a'} \right) \left(1 + \frac{b''}{a''} \right) \right]$$ where $$b' = \frac{L + (s' - s'')}{2}$$ $$b'' = \frac{L - (s' - s'')}{2}$$

basic assumption in this approach is that the equipotential lines at selected critical points in a flow net are *vertical* and that they divide the flow net into *fragments*. Table 7-2 summarizes fragment types and form factors.

Perhaps the best way to illustrate the procedure is by an example. If you are interested in the theory behind the method, consult Harr (1962 and 1977).

EXAMPLE 7.20

Given:

The dam with the same data as in Example 7.19 (shown in Fig. Ex. 7.20a).

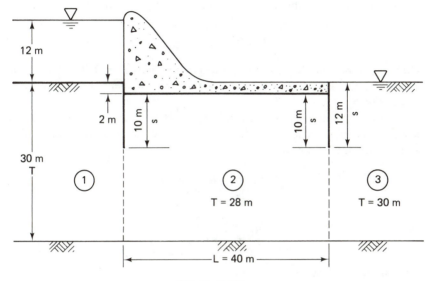

Fig. Ex. 7.20a

Required:

By means of the method of fragments, compute:

    a. The quantity of seepage loss under the dam when $k = 20 \times 10^{-4}$ cm/s, per metre of dam.
    b. The exit gradient (at point E).
    c. The pressure distribution on the base of the dam.
    d. Compare these values with those obtained from the flow net in Example 7.19.

Solution:

    a. Divide the flow system into fragments. The critical points chosen are the bottoms of the sheet piles. Refer to Table 7-2 and review the fragments. The heavy lines represent impervious boundaries which can be in vertical or horizontal directions. You choose the fragments to match the particular boundary conditions of your problem. Notice how the definitions of s and T are used. Their values are shown in Fig. Ex. 7.20a. The vertical, dashed (equipotential) lines separate the flow regime into three fragments, as shown circled in the figure. Clearly, the flow q through each fragment must be the same and is stated by Eq. 7-26. However in the method of fragments this equation is changed slightly to

$$q = kh_L \frac{N_f}{N_d} = \frac{kh_m}{\Phi_m} \tag{7-27}$$

where h_m is the head loss in the mth fragment where $m = 1, 2, 3, \ldots, n$, and Φ_m is the dimensionless *form factor* for the mth fragment. The form factor is equal to N_d/N_f.

In this example $h_L = 12$ m $= \Sigma h_m$, that is, the sum of the head loss in each fragment. Also, since the flow is equal in each fragment and is equal to the total flow, we have

$$\frac{q}{k} = h_1 \frac{N_f}{N_d} = h_2 \frac{N_f}{N_d} = h_3 \frac{N_f}{N_d} \tag{7-28}$$

or

$$\frac{q}{k} = \frac{h_1}{\Phi_1} = \frac{h_2}{\Phi_2} = \frac{h_3}{\Phi_3} = \frac{h}{\Sigma\Phi} \tag{7-29}$$

and, finally, the flow is

$$q = \frac{kh}{\displaystyle\sum_{m=1}^{n} \Phi} \tag{7-30}$$

The next step is to define the types of fragments for our problem and to determine the value of the form factors Φ for each fragment. Six general types of fragments are shown in Table 7-2, where the heavy lines represent impervious boundaries. Other fragments are also available in the literature. Also given are the values of Φ in terms of the geometry of each problem. If you study Fig. Ex. 7.20a, you can see that fragments 1 and 3 are type II but fragment 2 is a type V fragment. Had the sheet piles been of different lengths, then fragment 2 would be a type VI instead of type V fragment.

Next, we have to determine the form factors for our two types of fragments. For type II fragments, we see from Table 7-2 that $\Phi = K/K'$. Both K and K' are functions of m, which is defined as

$$m = \sin\frac{\pi s}{2T} \tag{7-31}$$

where $s =$ depth of the sheet pile, and
$T =$ thickness of the soil layer.

Substituting the values for s and T of our example into Eq. 7-31, we find that

$$m = \sin\frac{\pi s}{2T} = \sin\frac{\pi}{2}\frac{12}{30} = 0.588$$

The value of K/K' can be found from Table 7-3. For $m = 0.588$, $m^2 = 0.345$, so K/K' is equal to about 0.865 (by interpolation), which equals Φ_1. By inspection, Φ_1 is also equal to Φ_3. These values are tabulated in Table Ex. 7.20a.

TABLE EX. 7.20a

Fragment	Type	Φ
1	II	$K/K' = 0.865$
2	V	1.598
3	II	$K/K' = 0.865$

For fragment 2, which is a type V, we need to compare L and $2s$ to obtain Φ. For our example, $L = 40$ m and $2s = 20$ m. Since $L > 2s$, Φ is given by

$$\Phi_2 = 2\ln\left(1 + \frac{s}{a}\right) + \frac{L - 2s}{T}$$

$$= 2\ln\left(1 + \frac{10}{18}\right) + \frac{40 - 2 \times 10}{28}$$

$$= 0.884 + 0.714 = 1.598$$

Note that the distance $a = 18$ m is the distance from the bottom impervious boundary to the bottom end of the sheet pile.

The quantity of flow is found from Eq. 7-30,

$$q = \frac{kh}{\displaystyle\sum_{m=1}^{n}\Phi} = \frac{20 \times 10^{-4}\,\frac{\text{cm}}{\text{s}}\left(\frac{\text{m}}{100\ \text{cm}}\right)(12\ \text{m})}{0.865 + 1.598 + 0.865} \quad \text{per metre}$$

$$= 7.21 \times 10^{-5}\ \text{m}^2/\text{s per metre of dam}$$

$$= 8.65 \times 10^{-3}\ \text{m}^3/\text{s for a 120 m long dam}$$

This compares satisfactorily with the value of $8.31 \times 10^{-3}\ \text{m}^3/\text{s}$ obtained in Example 7.19.

An alternative way to determine the form factor is to use Fig. 7.19. Since $s/T = 10/28 = 0.36$, find $1/(2\Phi)$ equal to 0.60 for $b/T = 0$. Solving for Φ, we obtain 0.83, which is close to our previously determined value of 0.865.

b. Computation of the exit gradient i_E at point E is easy. From

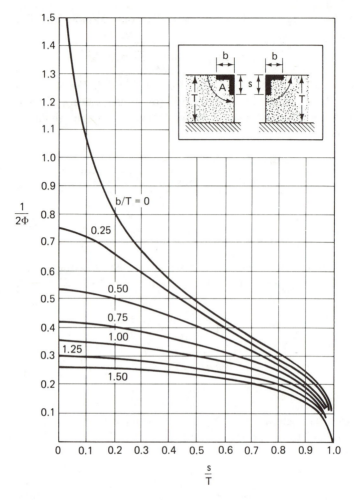

Fig. 7.19 Relationship between form factor Φ and s/T ratio for type II and type III fragments. (After Harr, 1977, © copyright McGraw-Hill Book Company. Used with permission.)

Table 7-2, fragment type II, we find the formula for the exit gradient is

$$i_E = \frac{h\pi}{2KTm} \tag{7-32}$$

where the value of m is from Eq. 7-31 and equals 0.588; the value of h is the head loss in the third (exit) fragment. The value of K is found in Table 7-3 for $m^2 = 0.345$; interpolating, $K = 1.741$. The value of h to use in Eq. 7-32 is the head loss in the *third fragment*, where the water exits, and it is

TABLE 7-3 Values of Parameters Used for Types II and III Fragments Described in Table 7-2*

m'^2	K	K'	K'/K	K/K'	m^2
0.000	1.571	∞	0.000	∞	1.000
0.001	1.571	4.841	0.325	3.08	0.999
0.002	1.572	4.495	0.349	2.86	0.998
0.003	1.572	4.293	0.366	2.73	0.997
0.004	1.572	4.150	0.379	2.64	0.996
0.005	1.573	4.039	0.389	2.57	0.995
0.006	1.573	3.949	0.398	2.51	0.994
0.007	1.574	3.872	0.406	2.46	0.993
0.008	1.574	3.806	0.413	2.42	0.992
0.009	1.574	3.748	0.420	2.38	0.991
0.01	1.575	3.696	0.426	2.35	0.99
0.02	1.579	3.354	0.471	2.12	0.98
0.03	1.583	3.156	0.502	1.99	0.97
0.04	1.587	3.016	0.526	1.90	0.96
0.05	1.591	2.908	0.547	1.83	0.95
0.06	1.595	2.821	0.565	1.77	0.94
0.07	1.599	2.747	0.582	1.72	0.93
0.08	1.604	2.684	0.598	1.67	0.92
0.09	1.608	2.628	0.612	1.63	0.91
0.10	1.612	2.578	0.625	1.60	0.90
0.11	1.617	2.533	0.638	1.57	0.89
0.12	1.621	2.493	0.650	1.54	0.88
0.13	1.626	2.455	0.662	1.51	0.87
0.14	1.631	2.421	0.674	1.48	0.86
0.15	1.635	2.389	0.684	1.46	0.85
0.16	1.640	2.359	0.695	1.44	0.84
0.17	1.645	2.331	0.706	1.42	0.83
0.18	1.650	2.305	0.716	1.40	0.82
0.19	1.655	2.281	0.726	1.38	0.81
0.20	1.660	2.257	0.735	1.36	0.80
m'^2	K'	K	K'/K	K/K'	m^2

m^2	K	K'	K/K'	K'/K	m'^2
0.21	1.665	2.235	0.745	1.34	0.79
0.22	1.670	2.214	0.754	1.33	0.78
0.23	1.675	2.194	0.763	1.31	0.77
0.24	1.680	2.175	0.773	1.29	0.76
0.25	1.686	2.157	0.782	1.28	0.75
0.26	1.691	2.139	0.791	1.26	0.74
0.27	1.697	2.122	0.800	1.25	0.73
0.28	1.702	2.106	0.808	1.24	0.72
0.29	1.708	2.090	0.817	1.22	0.71
0.30	1.714	2.075	0.826	1.21	0.70
0.31	1.720	2.061	0.834	1.20	0.69
0.32	1.726	2.047	0.843	1.19	0.68
0.33	1.732	2.033	0.852	1.17	0.67
0.34	1.738	2.020	0.860	1.16	0.66
0.35	1.744	2.008	0.869	1.15	0.65
0.36	1.751	1.995	0.877	1.14	0.64
0.37	1.757	1.983	0.886	1.13	0.63
0.38	1.764	1.972	0.895	1.12	0.62
0.39	1.771	1.961	0.903	1.11	0.61
0.40	1.778	1.950	0.911	1.10	0.60
0.41	1.785	1.939	0.920	1.09	0.59
0.42	1.792	1.929	0.929	1.08	0.58
0.43	1.799	1.918	0.938	1.07	0.57
0.44	1.806	1.909	0.946	1.06	0.56
0.45	1.814	1.899	0.955	1.05	0.55
0.46	1.822	1.890	0.964	1.04	0.54
0.47	1.829	1.880	0.973	1.03	0.53
0.48	1.837	1.871	0.982	1.02	0.52
0.49	1.846	1.863	0.991	1.01	0.51
0.50	1.854	1.854	1.000	1.00	0.50
m^2	K'	K	K'/K	K/K'	m'^2

*After Aravin and Numerov (1955).

from Eq. 7-29

$$h_3 = \frac{\Phi_3 h}{\sum\limits_{m=1}^{n} \Phi} = \frac{0.865 \times 12 \text{ m}}{3.328} = 3.12 \text{ m} \qquad (7\text{-}33)$$

Substituting these values and $m_3 = 0.588$ into Eq. 7-32, we obtain

$$i_E = \frac{3.12 \text{ m} (\pi)}{2 \times 1.741 \times 30 \text{ m} \times 0.588} = 0.16$$

This result compares well with the value of 0.14 found in Example 7.19.

An alternative procedure to find the exit gradient at point E is to use Fig. 7.20. For this example, $s/T = 12/30 = 0.40$; enter the graph and find $(i_E s)/h_m = 0.6$. Solving for i_E, we find that

$$i_E = \frac{0.6 h_m}{s} = \frac{0.6 \times 3.12 \text{ m}}{12 \text{ m}} = 0.16$$

c. To compute the pressure distribution under the dam, the assumption is made that the head loss varies *linearly* from fragment 1 to fragment 3.

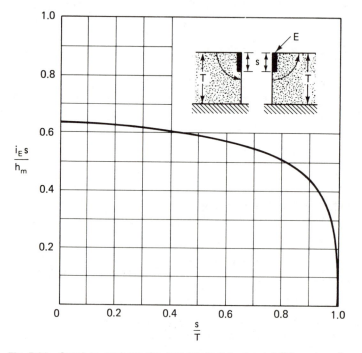

Fig. 7.20 Graph to evaluate the exit gradient i_E at point E for a type II fragment for a given s/T ratio. (After Harr, 1977, © copyright McGraw-Hill Book Company. Used with permission.)

Compute the head loss per fragment:

$$\sum \Phi = \Phi_1 + \Phi_2 + \Phi_3$$
$$= 0.865 + 1.598 + 0.865 = 3.328$$

The head loss per fragment is given by Eq. 7-33,

$$h_1 = \frac{\Phi_1 h}{\sum \Phi} = \frac{0.865 \times 12 \text{ m}}{3.328} = 3.12 \text{ m}$$

$$h_2 = \frac{\Phi_2 h}{\sum \Phi} = \frac{1.598 \times 12 \text{ m}}{3.328} = 5.76 \text{ m}$$

$h_3 = h_1$ due to symmetry

Redraw the dam to scale and place the values of head at selected points (Fig. Ex. 7.20b). At equipotential line A', the head loss is $h_1 = 3.12$ m; the head loss is therefore $h - h_1 = 12$ m $- 3.12$ m $= 8.88$ m. Similarly, at equipotential line F', $h_2 = 5.76$ m. Therefore the head at F' is

$$h - h_1 - h_2 = 12 - 3.12 - 5.76 = 3.12 \text{ m}$$

Assuming that the head loss varies linearly from points $A'-A-F-F'$, which is equal to the total distance of 10 m + 40 m + 10 m = 60 m, then the head loss per metre is $h_2/60$ m, or 5.76 m/60 m = 0.096 m/m. Thus the head at point $A =$ the head at $A' - 10$ m $\times$ 0.096 m per metre, or $8.88 - 10 \times 0.096 = 7.92$ m. Likewise, the head at F is 4.08 m.

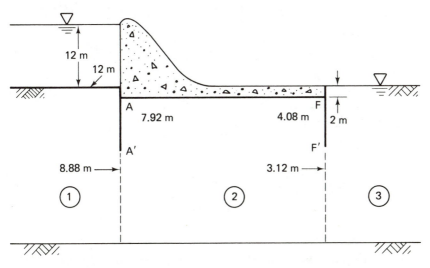

Fig. Ex. 7.20b

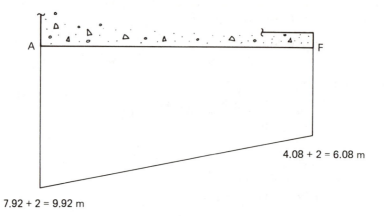

7.92 + 2 = 9.92 m

Fig. Ex. 7.20c

To this head we add the tailwater head of 2 m. The vertical uplift pressures may now be computed, as shown in Fig. Ex. 7.20c. These values compare almost exactly with those shown in Fig. Ex. 7.19.

d. Comparison of values of flow, exit gradient, and uplift pressures as determined by the two different procedures are summarized in Table Ex. 7.20b.

TABLE EX. 7.20b

Parameter	From Flow Net	From Method of Fragments
q	8.31×10^{-3} m³/s	8.65×10^{-3} m³/s
i_E	0.14	0.16
Uplift pressures at A*	9.96 m	9.92 m
Uplift pressures at F*	6.04 m	6.08 m

*In metres of water.

EXAMPLE 7.21

Given:

Several cases of confined flow.

Required:

Without computing the quantities as we did in Example 7.20, just identify the fragment types shown in Fig. Ex. 7.21.

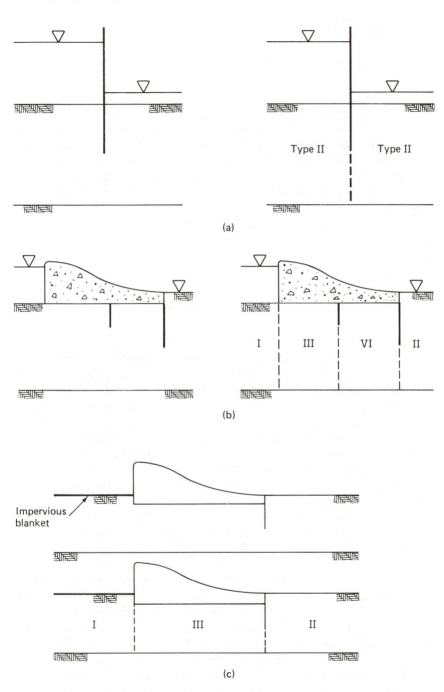

(a)

(b)

Impervious blanket

(c)

Fig. Ex. 7.21

269

Solution:

Fig. Ex. 7.21.

It should be obvious now that the method of fragments is a powerful analytical and design technique. Solutions to many complex problems can be found quickly, whereas to solve a series of complex problems by means of flow nets would require a large amount of time. Take, for example, the problem illustrated by Example 7.21c. A practical design question is: How long should the drainage blanket be constructed to reduce the quantity of flow by one-third or to reduce the uplift pressures by one-half? Many trial and error flow net solutions would be required to solve the problem, whereas with the method of fragments the solution could be obtained directly!

7.11 CONTROL OF SEEPAGE AND FILTERS

In the discussion of seepage forces and flow nets, piping and erosion were mentioned as a possibility if, somewhere in the porous medium, the gradient exceeded the critical gradient. Piping can occur any place in the system, but usually it occurs where the flow is concentrated, as shown in Fig. 7.18. Once the seepage forces are large enough to move particles, piping and erosion can start, and it usually continues until either all the soils in the vicinity are carried away or the structure collapses. Cohesionless soils, especially silty soils, are highly susceptible to piping, and if you must use such soils in an embankment dam, for example, then you must be very careful to see that the seepage is controlled and that the chance for piping to occur is very small.

How is seepage controlled? Methods used depend on the situation, but sometimes a cutoff wall or trench is constructed to completely block the seeping water. Sometimes the drainage path is lengthened by an impervious blanket, so that more of the head is lost and thus the gradient in the critical region is reduced. If properly designed and constructed, relief wells and other kinds of drains can be used to positively relieve high uplift pressures at the base of hydraulic structures (Cedergren, 1977).

Another way to prevent erosion and piping and to reduce potentially damaging uplift pressures is to use a *protective filter*. A filter consists of one or more layers of free-draining granular materials placed in less pervious foundation or base materials to prevent the movement of soil particles that are susceptible to piping while at the same time allowing the seepage water

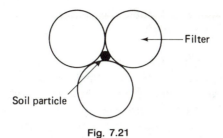

Fig. 7.21

to escape with relatively little head loss. Thus the seepage forces within the filter itself are reduced. Figure 7.21 illustrates the principle of the first requirement, that of preventing the movement of particles from the foundation soil into the filter. If the particles are perfect spheres, it can be shown that, for loose packing, the grain size of the filter material should be about 6.5 times larger than the foundation material. However, laboratory tests have shown that the grain size of a uniform filter material can be as large as 10 times the grain size of a uniform foundation soil and still prevent particle movement. It is clear that particle shape and the porosity of the filter material also affect this limiting size.

Hazen (1911) was working with water treatment filters around the turn of the century, and he found that the *effective size* of a filter was the D_{10} (for example, Eq. 7-10); that is, this size affected the performance of a filter sand as much as the remaining 90% of the sizes.

In 1922, Terzaghi delineated some requirements for protective filters based on the grain size distributions of both the filter and the material to be protected. These requirements have since been modified slightly after laboratory tests by the U.S. Army Corps of Engineers and the U.S. Water and Power Resources Service. The four main requirements (U.S.B.R.,1974) for a protective filter are:

1. The filter material should be more pervious than the base material in order that no hydraulic pressure will build up to disrupt the filter and adjacent structures.
2. The voids of the inplace filter material must be small enough to prevent base material particles from penetrating the filter and causing clogging and failure of the protective filter system.
3. The layer of the protective filter must be sufficiently thick to provide a good distribution of all particle sizes throughout the filter and also to provide adequate insulation for the base material where frost action is involved.
4. Filter material particles must be prevented from movement into the drainage pipes by providing sufficiently small slot openings or perforations, or additional coarser filter zones if necessary.

The last requirement could be fulfilled by some of the modern nonwoven and woven fabric filter materials that have been developed in recent years.

Gradation requirements for protective filters are given in Table 7-4. The first ratio, R_{15}, ensures that the small particles of the material to be protected are prevented from passing through the pores of the filter; the second ratio, R_{50}, ensures that seepage forces within the filter are reasonably small. If the criteria in this table cannot be met by one layer of filter material, then a zoned or multilayered filter can be designed and specified.

Some additional practical requirements for the design of filters are also shown in Table 7-4.

A special comment concerning the use of modern woven and nonwoven fabric materials for filters: these materials have recently been introduced into civil engineering practice and with considerable success. However, because of the rather limited research and experience with them, fabric filters are mostly built by trial and error. As natural deposits of sand and gravel become depleted, fabric filters will become increasingly important in drainage and seepage control. An additional advantage is that they are easy to use in the field, and thus construction costs are often less than with conventional granular filters. For additional information about filter fabrics, see Barrett (1966), Cedergren (1977), Dallaire (1976), Seemel (1976), and Steward, et al. (1977).

TABLE 7-4   Gradation Requirements for Filter Materials*

Filter Material Characterisitics	R_{15}	R_{50}
Uniform grain size filters, C_u = 3 to 4	—	5 to 10
Graded filters, subrounded particles	12 to 40	12 to 58
Graded filters, angular particles	6 to 18	9 to 30

$$R_{15} = \frac{D_{15} \text{ of filter material}}{D_{15} \text{ of material to be protected}}$$

$$R_{50} = \frac{D_{50} \text{ of filter material}}{D_{50} \text{ of material to be protected}}$$

Notes: Maximum size of the filter material should be less than 76 mm (3 in.). Use the minus No. 4 fraction of the base material for setting filter limits when the gravel content (plus No. 4) is more than 10%, and the fines (minus No. 200) are more than 10%. Filters must not have more than 5% minus No. 200 particles to prevent excessive movement of fines in the filter and into drainage pipes. The grain size distribution curves of the filter and the base material should approximately parallel in the range of finer sizes.

*After U.S.B.R. (1974).

PROBLEMS

7-1. A clean sand having a permeability of 5×10^{-2} cm/s and a void ratio of 0.5 is placed in a horizontal permeability apparatus, as shown in Fig. 7.2. Compute the discharge velocity and the seepage velocity as the head Δh goes from 0 to 100 cm. The cross-sectional area of the horizontal pipe is 100 cm², and the soil sample is 0.5 m long.

7-2. A sample of medium quartz sand is tested in a constant head permeameter. The sample diameter is 50 mm and its length is 120 mm. Under an applied head of 50 cm, 113 cm³ flows through the sample in 5 min. The M_s of the sample is 985 g. Calculate (a) the Darcy coefficient of permeability, (b) the discharge velocity, and (c) the seepage velocity. (After A. Casagrande.)

7-3. A permeability test was run on a compacted sample of dirty sandy gravel. The sample was 150 mm long and the diameter of the mold 150 mm. In 83 s the discharge under a constant head of 40 cm was 392 cm³. The sample had a dry mass of 5300 g and its ρ_s was 2680 kg/m³. Calculate (a) the coefficient of permeability, (b) the seepage velocity, and (c) the discharge velocity during the test.

7-4. During a falling-head permeability test, the head fell from 50 to 30 cm in 4.5 min. The specimen was 5 cm in diameter and had a length of 90 mm. The area of the standpipe was 0.5 cm². Compute the coefficient of permeability of the soil in cm/s, m/s, and ft/d. What was the probable classification of the soil tested? (After A. Casagrande.)

7-5. A falling-head permeability test is to be performed on a soil whose permeability is estimated to be 3×10^{-7} m/s. What diameter standpipe should you use if you want the head to drop from 27.5 cm to 20.0 cm in about 5 min? The sample's cross section is 15 cm² and its length is 8.5 cm. (After Taylor, 1948.)

7-6. Show that the units of Eq. 7-4a are in fact energy/mass. Show that Eq. 7-4b has units of energy/weight, and that this comes out as a length (head).

7-7. You have just been placed in charge of a small soils laboratory and there is a tight budget for permeability equipment. Make a decision as to what *type* of equipment you would order so as to perform laboratory tests on the widest range of soils.

7-8. Briefly describe exactly how you make a correction for temperature in a permeability test if the water is not exactly 20°C. State your reference.

7-9. In Example 7.1, the void ratio is specified as 0.43. If the void ratio of the same soil were 0.38, evaluate its coefficient of permeability. First estimate in which direction k will go, higher or lower; then proceed.

7-10. Estimate the coefficient of permeability for soil 2 of Fig. Ex. 3.1. Will the same approach work for soil 1 of the same figure?

7-11. A falling-head permeability test on a specimen of fine sand 16 cm² in area and 9 cm long gave a k of 7×10^{-4} cm/s. The dry mass of the sand specimen was 210 g and its ρ_s was 2.68 Mg/m³. The test temperature was 26°C. Compute the coefficient of permeability of the sand for a void ratio of 0.70 and the standard temperature of 20°C. (After A. Casagrande.)

7-12. The coefficient of permeability of a clean sand was 400×10^{-4} cm/s at a void ratio of 0.42. Estimate the permeability of this soil when the void ratio is 0.58.

7-13. Permeability tests on a soil supplied the following data:

Run No.	e	Temp. (°C)	$k\ (cm/s)$
1	0.70	25	0.32×10^{-4}
2	1.10	40	1.80×10^{-4}

Estimate the coefficient of permeability at 20°C and a void ratio of 0.85. (After Taylor, 1948.)

7-14. For the soil profile of Example 7.4 plot the total, neutral, and effective stresses with depth if the ground water table is *lowered* 2 m below the ground surface.

7-15. Soil borings made at a site near Chicago indicate that the top 6 m is a loose sand and miscellaneous fill, with the ground water table at 3 m below the ground surface. Below this is a fairly soft blue-gray silty clay with an average water content of 30%. The boring was terminated at 16 m below the ground surface when a fairly stiff silty clay was encountered. Make reasonable assumptions as to soil properties and calculate the total, neutral, and effective stresses at 3, 6, 11, and 16 m below the ground surface.

7-16. Plot the soil profile of Problem 7-15 and the total, neutral, and effective stresses with depth.

7-17. A soil profile consists of 5 m of compacted sandy clay followed by 5 m of medium dense sand. Below the sand is a layer of compressible silty clay 20 m thick. The initial ground water table is located at the bottom of the first layer (at 5 m below the ground surface). The densities are 2.05 Mg/m³ (ρ), 1.94 Mg/m³ (ρ_{sat}), and 1.22 Mg/m³ (ρ') for the three layers, respectively. Compute the effective stress at a point at middepth in the compressible clay layer. Then, assuming that the medium dense sand remains *saturated*, compute the effective stress in the clay layer at midpoint again, when the ground water table drops 5 m to the top of the silty clay layer. Comment on the difference in effective stress.

7-18. For the initial case in Problem 7-17, compute the head of water required at the top of the silty clay layer to cause a quick condition.

7-19. Sand is supported on a porous disc and screen in vertical cylinder, as shown in Fig. P7-19.

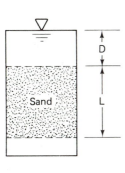

Case I

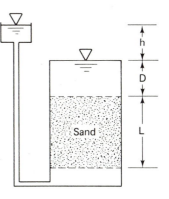

Case II

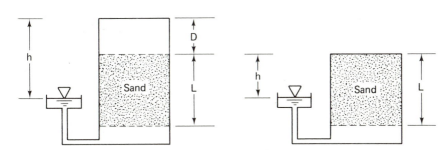

Case III

Case IV. Assume that the sand is fine enough to remain 100% saturated up to its top surface by capillarity.

Fig. P7-19

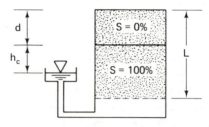

Case V. Assume an idealized case in
which the height of capillary rise is
h_c. All soil below that height is
100% saturated, and all soil above
that height is 0% saturated.

Fig. P7-19   Continued

(a) For each of the five cases, plot the total, neutral, and effective
stresses versus height. These plots should be approximately to
scale.

(b) Derive formulas for those three stresses in terms of the dimensions shown and e, ρ_s, and ρ_w for each case at both the top and
bottom of the sand layer. For case IV, assume the sand is 100%
saturated to the upper surface by capillarity. For case V, assume
the sand above level h_1 is completely dry and below h_1 is
completely saturated. (After A. Casagrande.)

7-20. Specify the conditions under which it is possible for K_o to equal K.

7-21. For the soil profile of Problem 7-15, calculate both the horizontal,
total, and effective stresses at 3, 6, 11 and 16 m depths, assuming (a)
K_o is 0.5 and (b) K_o is 1.5.

7-22. The value of K_o for the compressible silty clay layer of Problem 7-17
is 0.75. What are the total and effective horizontal stresses at
middepth of the layer?

7-23. Given, the soil cylinder and test setup of Example 7.11, with actual
dimensions as follows: $AB = 5$ cm, $BC = 10$ cm, $CD = 10$ cm, and
$DE = 5$ cm. Calculate the pressure, elevation, and total heads at
points A through E in centimetres of water, and plot these values
versus elevation.

7-24. For each of the cases I, II, and III of Fig. P7-24, determine the
pressure, elevation, and total head at the entering end, exit end, and
point A of the sample. (After Taylor, 1948.)

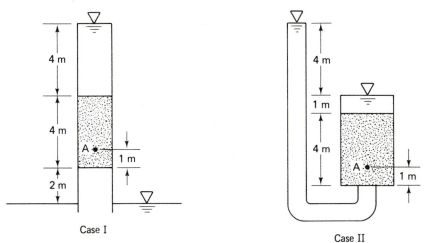

Case I

Case II

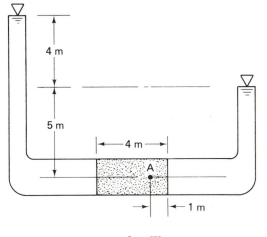

Case III

Fig. P7-24

7-25. For each of the cases shown in Fig. P7-24, determine the discharge velocity, the seepage velocity, and the seepage force per unit volume for (a) a permeability of 0.1 cm/s and a porosity of 50% and (b) a permeability of 0.001 cm/s and a void ratio of 0.67. (After Taylor, 1948.)

7-26. An inclined permeameter tube is filled with three layers of soil of

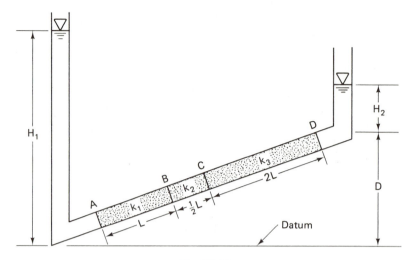

Fig. P7-26

different permeabilities as in Fig. P7-26. Express the head at points A, B, C, and D (with respect to the datum indicated) in terms of the different dimensions and permeabilities. (a) Work the problem first assuming $k_1 = k_2 = k_3$. (b) Then work it assuming $3k_1 = k_2 = 2k_3$. Plot the various heads versus horizontal distance for both parts (a) and (b). (After A. Casagrande.)

7-27. Assume the soil of Fig. 7.10 has a saturated unit weight of 1.94 Mg/m^3. If the head of water h above elevation B is 2.34 m, compute the effective stress at elevation A at the bottom of the soil sample during flow. What is the effective stress under these conditions at midheight in the soil column during steady-state flow?

7-28. (a) Show that Eq. 7-20 is identical to Eq. 7-21. (b) Show that Eq. (g) in Sec. 7.8 also reduces to Eq. 7-21. (c) Show that Eq. 7-23c is identical to Eq. 7-23d at a critical condition.

7-29. The foundation soil at the toe of a masonry dam has a porosity of 41% and a ρ_s of 2.68 Mg/m^3. To assure safety against piping, the specifications state that the upward gradient must not exceed 25% of the gradient at which a quick condition occurs. What is the maximum permissible upward gradient? (After Taylor, 1948.)

7-30. Show that it is impossible to drown in quicksand. Hint: Calculate the density of the quicksand.

7-31. A contractor plans to dig an excavation as shown in Fig. P7-31. If the river is at level A, what is the factor of safety against quick

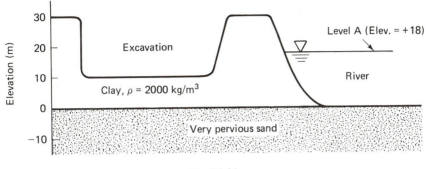

Fig. P7-31

conditions? Neglect any vertical shear. To what elevation can the water rise before a quick condition will develop? (After D. N. Humphrey.)

7-32. Given, the excavation as shown in Example 7.17, with $h = 20$ m and $\rho = 1850$ kg/m³. Calculate the minimum allowable H_s.

7-33. A sheet pile wall has been installed partially through a silty sand layer, similar to the one shown in Fig. 7.13b. Assume a 14 m long sheet pile penetrates 7 m (half way) into the silty sand layer of thickness 14 m. For this condition:

(a) Draw a flow net using three (or four at most) flow channels. Note that the flow net is completely symmetrical about the bottom of the sheet pile. (This part is needed for the solution of Problem 7-35.)

(b) If the water height on the upstream side is 5 m and 1 m on the downstream side, compute the amount of water flowing under the sheet pile per metre of wall if the coefficient of permeability is 2×10^{-4} cm/s.

(c) Compute the theoretical hydraulic gradient to cause a quick condition at the downstream side of the sheet pile.

7-34. Using the data of Fig. 7.17, compute the total head, piezometric head, pressure head, and elevation head for points C and C'. Assume any convenient datum.

7-35. Assuming that you have completed the flow net of Problem 7-33, compute the total head, piezometric head, pressure head, and elevation head for a point half way up the sheet pile from its base, on either side of the sheet pile. Assume the datum is at the bottom of the silty sand layer.

7-36. Draw a flow net for the case shown in Fig. P7-36. Assume three or four flow channels. Use any convenient scale, like 1 cm = 5 m.

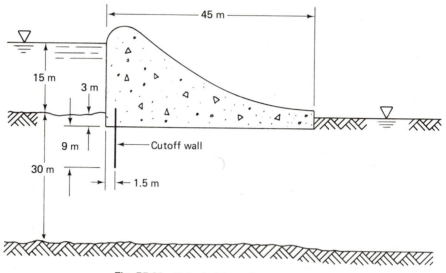

Fig. P7-36   (Adapted from Taylor, 1948.)

7-37. Draw a flow net for the case shown in Fig. P7-37. Use any convenient scale.

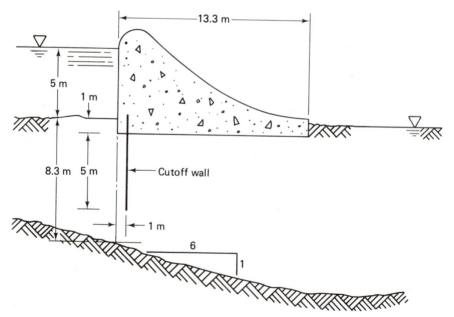

Fig. P7-37   (Adapted from Taylor, 1948.)

7-38. For the completed flow net of Fig. P7-38, compute the flow under the dam per metre of dam if the coefficient of permeability is 3.5×10^{-4} cm/s.

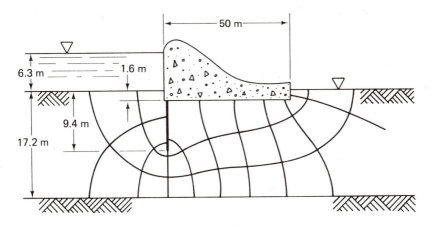

Fig. P7-38

7-39. For the completed flow net of Fig. P7-38, compute the uplift pressure all along the base of the dam.

7-40. Given, the data of Problem 7-33. Using the method of fragments determine:
    (a) The amount of water flowing under the sheet pile per metre of wall.
    (b) The exit gradient.

7-41. For the dam of Fig. 7.14 set up the problem and solve as far as you can by the method of fragments.

7-42. Using the method of fragments, show that one case of Fig. 7.18 is more critical than the other.

7-43. If one of the rows of sheet piles had to be removed for the problem given in Fig. Ex. 7.20a, which one when removed would cause the least reduction in flow?

7-44. Same as Problem 7-43, but do for the least amount of uplift pressure. Give your answer in terms of head.

7-45. Assume a row of sheet piles as shown in Fig. Ex. 7.21a. The total thickness of the soil layer is 20 m, while the difference between the head and tailwater is 10 m. Plot a graph showing how the flow under the sheet pile varies when the depth of the sheet pile goes for 8 m and approaches 20 m. Ignore problems associated with the exit gradient.

7-46. Suppose there is a problem with the exit gradients, as in Problem 7-45. One solution would be to place a horizontal filter over the soil where the water exits. How does this help? Is it better for the filter to have a similar coefficient of permeability? One that is much smaller? Or one that is much larger? Explain which one is most desirable. (We saved the best one for last!)

7-47. Soil 4 of Fig. Ex. 3.1 is used as the core material of a compacted earth dam. Design a filter to protect this soil. Present your results on a grain size distribution curve (or semilog plot) as in Fig. Ex. 3.1.

7-48. A protective three-layer filter is proposed between the foundation and rock drain located near the toe of a compacted earth-fill dam. Samples were taken and the grain sizes of the materials were determined to be as follows:

	D_{15} (mm)	D_{85} (mm)
Foundation, finest samples	0.024	0.1
Foundation, coarsest samples	0.12	0.9
Filter layer No. 1	0.3	1.0
Filter layer No. 2	2.0	3.5
Filter layer No. 3	5.0	10.0
Rock drain	15.0	40.0

Analyze this filter with the U.S.B.R. (1974) criteria. Does it meet the criteria? If not, comment briefly on any practical consequences. (After Taylor, 1948.)

eight

Consolidation and Consolidation Settlements

8.1 INTRODUCTION

You are undoubtedly aware that when materials are loaded or stressed they deform or strain. Sometimes, such as with elastic materials, the response under load is instantaneous. Other materials (some soils are very good examples) require a relatively long time for deformations to occur; this is especially true for clay soils. Most of this chapter is devoted to the compressibility of these kinds of soils.

The simplest type of stress-strain-time relationship applies to *elastic* materials—where, as mentioned, the stresses and strains occur simultaneously. In fact, elastic stress-strain relationships can either be *linear* or *nonlinear*. Materials which have time as a factor in their stress-strain response are called *visco-elastic*. Soils, then, are visco-elastic materials from the viewpoint of their mechanical behavior. The problem with using the theory of visco-elasticity is that, in its present state of development, the theory is only applicable to materials that are linear and, as we shall soon see, soils are highly nonlinear materials. In other words, the interrelationship between stress, strain, and time for soils is not simple and cannot be treated mathematically with present theory. Soils have another property that complicates matters: they have a "memory." Thus the material is *nonconservative*. When soils are stressed, they deform, and even when the stress is released some permanent deformation remains. Deformations in general can be either a change of shape (*distortion*), or a change of volume (*compression*), or both.

The following notation is introduced in this chapter.

Symbol	Dimension	Unit	Definition
a_v	$M^{-1}LT^2$	$(kPa)^{-1}$	Coefficient of compressibility (Eq. 8-5)
B	L	m	Width of footing (Eq. 8-22)
C_c	—	—	Compression index (Eq. 8-7)

Symbol	Dimension	Unit	Definition
$C_{c\epsilon}$	—	—	Modified compression index (Eq. 8-8)
C_r	—	—	Recompression index (Eq. 8-15); C_E and C_s are sometimes used
$C_{r\epsilon}$	—	—	Modified recompression index
D, E_{oed}	$ML^{-1}T^{-2}$	kPa	Constrained or oedometric modulus (Eq. 8-6)
e_o	—	(decimal)	Initial or in situ void ratio
Δe	—	(decimal)	Change in void ratio
H_o	L	m	Original thickness of a soil layer (Eq. 8-3)
ΔH	L	m	Change in thickness of a soil layer (Eq. 8-3)
I	—	—	Influence factor (Eq. 8-30)
L	L	m	Length of foundation (Eq. 8-23)
LIR	—	—	Load increment ratio (Eq. 8-20)
m, n	—	—	Ratios of foundation width to depth (Eqs. 8-28 and 8-29)
m_v	$M^{-1}LT^2$	$(kPa)^{-1}$	Coefficient of volume change (Eq. 8-6)
N_B	—	—	Influence factor (Eq. 8-25)
N_W	—	—	Influence factor (Eq. 8-34)
OCR	—	—	Overconsolidation ratio (Eq. 8-2)
P	MT^{-2}	kN/m	Line load (Eq. 8-26)
Q	MLT^{-2}	kN	Force or load (Eq. 8-24)
q_o	$ML^{-1}T^{-2}$	kPa	Surface or contact stress (Eq. 8-27)
r	L	m	Horizontal distance from load to a point (Eq.8-24)
s	L	m	Settlement (Eq. 8-4)
s_c	L	m	Consolidation settlement (Eq. 8-1)
s_i	L	m	Immediate or distortion settlement (Eq. 8-1)
s_s	L	m	Secondary compression (Eq. 8-1)
s_t	L	m	Total settlement (Eq. 8-1)
u	$ML^{-1}T^{-2}$	kPa	Pore water pressure
u_o	$ML^{-1}T^{-2}$	kPa	Initial or hydrostatic pore water pressure
z	L	m	Depth
ϵ_h	—	(%)	Horizontal strain (Eq. 8-33)
ϵ_v	—	(%)	Vertical strain (Eq. 8-3)
ν	—	—	Poisson's ratio (Eq. 8-33)
σ_o	$ML^{-1}T^{-2}$	kPa	Surface or contact stress (Eq. 8-22)
σ'_{vc}	$ML^{-1}T^{-2}$	kPa	Vertical effective consolidation stress
σ'_p	$ML^{-1}T^{-2}$	kPa	The preconsolidation stress or maximum past vertical effective stress (Eq. 8-2); p'_c and σ'_{vm} are sometimes used
σ'_{vo}	$ML^{-1}T^{-2}$	kPa	Vertical (effective) overburden stress (Eq. 8-2)
σ_z	$ML^{-1}T^{-2}$	kPa	Vertical stress at depth z (Eq. 8-22)

8.2 COMPONENTS OF SETTLEMENT

When a soil deposit is loaded, for example by a structure or a man-made fill, deformations will occur. The total vertical deformation at the surface resulting from the load is called *settlement*. The movement may be downward with an increase in load or upward (called *swelling*) with a decrease in load. Temporary construction excavations and permanent excavations such as highway cuts will cause a reduction in the stress, and swelling may result. As shown in Chapter 7, a lowering of the water table will also cause an increase in the effective stresses within the soil, which will lead to settlements. Another important aspect about settlements of especially fine-grained soils is that they are often time-dependent.

In the design of foundations for engineering structures, we are interested in how much settlement will occur and how fast it will occur. Excessive settlement may cause structural as well as other damage, especially if such settlement occurs rapidly. The total settlement, s_t, of a loaded soil has three components, or

$$s_t = s_i + s_c + s_s \qquad (8\text{-}1)$$

where s_i = the *immediate*, or *distortion*, settlement,

$\quad s_c$ = the *consolidation* (time-dependent) settlement, and

$\quad s_s$ = the *secondary* compression (also time-dependent).

The immediate, or distortion, settlement although not actually elastic is usually estimated by using elastic theory. The equations for this component of settlement are in principle similar to the deformation of a column under an axial load P, where the deformation is equal to PL/AE. In most foundations, however, the loading is usually three dimensional, which causes some distortion of the foundation soils. Problems arise concerning the proper evaluation of a compression modulus and the volume of soil that is stressed. Immediate settlements must be considered in the design of shallow foundations, and procedures for dealing with this problem can be found in textbooks on foundation engineering.

The consolidation settlement is a time-dependent process that occurs in saturated fine-grained soils which have a low coefficient of permeability. The rate of settlement depends on the rate of pore water drainage. Secondary compression, which is also time-dependent, occurs at constant effective stress and with no subsequent changes in pore water pressure. Settlement computations are discussed in this chapter; the time rate of consolidation and secondary compression are discussed in Chapter 9.

8.3 COMPRESSIBILITY OF SOILS

Assume for the time being that the deformations of our compressible soil layer will occur in only one dimension. An example of one-dimensional compression would be the deformation caused by a fill covering a very large area. Later on we shall discuss what happens when a structure of finite size loads the soil and produces deformation.

When a soil is loaded, it will compress because of:

1. deformation of soil grains,
2. compression of air and water in the voids, and/or
3. squeezing out of water and air from the voids.

At typical engineering loads, the amount of compression of the soil mineral grains themselves is small and usually can be neglected. Often, compressible soils are found below the water table, and they can be considered fully saturated. At least we usually assume 100% saturation for most settlement problems. Thus the compression of the pore fluid can be neglected. Therefore the last item contributes the most to the volume change of loaded soil deposits. As the pore fluid is squeezed out, the soil grains rearrange themselves into a more stable and denser configuration, and a decrease in volume and surface settlement results. How fast this process occurs depends on the permeability of the soil. How much re-arrangement and compression takes place depends on the rigidity of the soil skeleton, which is a function of the structure of the soil. Soil structure, as discussed in Chapter 4, depends on the geologic and engineering history of the deposit.

Consider the case where granular materials are one-dimensionally compressed. The curve shown in Fig. 8.1a is typical for sands in compression in terms of stress-strain; Fig. 8.1b shows the same data as a void ratio versus pressure curve. Note that it is common to rotate the coordinate axes 90° when plotting e versus σ_v data. Fig. 8.1c shows the compression versus time; note how rapidly compression occurs. The deformations take place in a very short time due to the relatively high permeability of granular soils. It is very easy for the water (and air) in the voids to squeeze out. Many times, for all practical purposes, the compression of sands occurs during construction and, as a result, most of the settlements have taken place by the time the structure is completed. However, because they occur so fast, even the relatively small total settlements of granular layers may be detrimental to a structure which is particularly sensitive to rapid settlements. The settlement of granular soils is estimated by using Eq. 8-1 with s_c and s_s neglected. Details of these analyses can be found in books on foundation engineering.

When clays undergo loading, because of their relatively low permeability their compression is controlled by the rate at which water is squeezed out of the pores. This process is called *consolidation*, which is a stress-strain-time phenomenon. Deformation may continue for months, years, or even decades. This is the fundamental and only difference between the compression of granular materials and the consolidation of cohesive soil: compression of sands occurs almost instantly, whereas consolidation is a very time-dependent process. The difference in settlement rates depends on the difference in permeabilities.

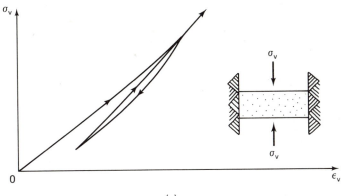

(a)

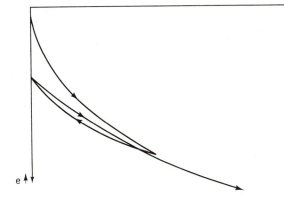

(b)

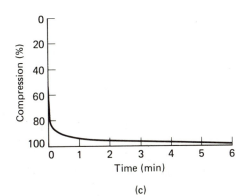

(c)

Fig. 8.1 Stress-strain and stress-time curves for a typical sand: (a) stress versus strain; (b) void ratio versus pressure; (c) compression versus time (after Taylor, 1948).

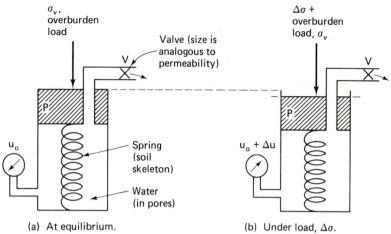

(a) At equilibrium.

(b) Under load, $\Delta\sigma$.
Note increased pore water
pressure and water flow.

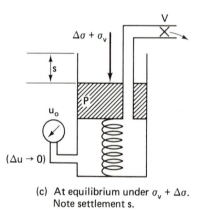

(c) At equilibrium under $\sigma_v + \Delta\sigma$.
Note settlement s.

Fig. 8.2   Spring analogy as applied to consolidation.

    The consolidation of clay is readily explained by the *spring analogy* shown in Fig. 8.2. A piston P is loaded vertically and compresses a spring inside the chamber, which is filled with water. The spring is analogous to the soil mineral skeleton, while the water in the cylinder represents the water in the soil voids. The valve V at the top of the piston represents the pore sizes in the soil, and at equilibrium when the valve is open no water flows out. This situation is analogous to one where a soil layer is at equilibrium with the weight of all soil layers (called *overburden*) above it. A pressure gage is connected to the cylinder and it shows the hydrostatic pressure u_o at this particular location in the soil. Now the soil layer is loaded by an additional load increment $\Delta\sigma$, Fig. 8.2b. At the start of the consolidation process, let us assume that the valve V is initially closed. Upon application of the load, the pressure is immediately transferred to the water inside the cylinder. Since the water is relatively incompressible and the valve is shut so that no water can get out, there is no deformation of the piston, and the pressure gage reads $\Delta u = \Delta\sigma$ (Fig. 8.2b). The pore water pressure Δu is called the *excess hydrostatic pressure* since it is in excess of the original hydrostatic pressure u_o.

To simulate a fine-grained cohesive soil with its low permeability, we can open the valve and allow water to slowly leave the cylinder under the initial excess pressure Δu. With time, as the water flows out, the water pressure decreases and gradually the load $\Delta\sigma$ is transferred to the spring, which compresses under the load. Finally, at equilibrium (Fig. 8.2c) no further water is squeezed out of the cylinder, the pore water pressure is again hydrostatic, and the spring is in equilibrium with the load, $\sigma_v + \Delta\sigma$.

Although the model is rather crude, the process is analogous to what happens when cohesive soils in the field and laboratory are loaded. Initially, all the external load is transferred into excess pore water or excess hydrostatic pressure. Thus at first there is no change in the effective stress in the soil. Gradually, as the water is squeezed out under a pressure gradient, the soil skeleton compresses, takes up the load, and the effective stresses increase. The compressibility of the spring is analogous to the compressibility of the soil skeleton. Eventually the excess hydrostatic pressure becomes zero and the pore water pressure is the same as the hydrostatic pressure prior to loading.

8.4 THE OEDOMETER AND CONSOLIDATION TESTING

When soil layers covering a large area are loaded vertically, the compression can be assumed to be one dimensional. To simulate one-dimensional compression in the laboratory, we compress the soil in a special device called an *oedometer* or *consolidometer*. Principal components of two types of oedometers are shown in Fig. 8.3.

An undisturbed soil specimen, which represents an element of the compressible soil layer under investigation, is carefully trimmed and placed into the confining ring. The ring is relatively rigid so that no lateral deformation takes place. On the top and bottom of the sample are porous "stones" which allow drainage during the consolidation process. Porous stones are made of sintered corundum or porous brass. Ordinarily, the top

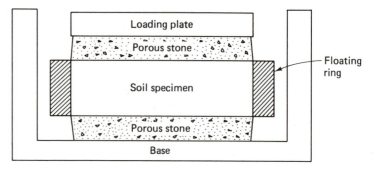

(a)

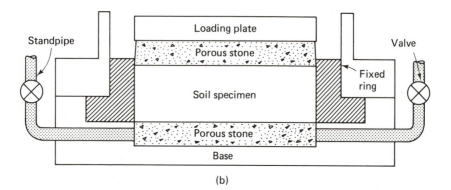

(b)

Fig. 8.3 Schematic of an oedometer or consolidation test apparatus: (a) floating-ring oedometer; (b) fixed-ring oedometer (after U.S. Army Corps of Engineers, 1970).

porous stone has a diameter approximately 0.5 mm smaller than the ring, so that it does not drag along the side of the ring when the specimen is being loaded. Usually the ratio of the diameter to height of the specimen is between 2.5 and 5, and the diameter depends on the diameter of the undisturbed soil samples tested. There is more trimming disturbance with larger ratios; on the other hand, taller specimens have greater side friction. Side friction can be reduced to some extent by the use of ceramic or Teflon-lined rings or by application of a lubricant like molybdenum disulphide.

In the *floating-ring test* (Fig. 8.3a) the compression takes place from both faces of the soil sample. It can be shown (Lambe, 1951) that the ring friction is somewhat less in this test than in a *fixed-ring test* (Fig. 8.3b), in which all movement is downward relative to the ring. The primary advantage of the fixed-ring test is that drainage from the bottom porous stone may be measured or otherwise controlled. In this manner, for example, permeability tests may be conducted in the oedometer.

To establish the relationship between load and deformation in the laboratory sample, during the consolidation test, the applied load as well as the deformation of the sample are carefully measured. Stress is of course computed by dividing the applied load by the area of the specimen. It is common practice to load the specimen incrementally, either through a mechanical lever-arm system or by an air or air-hydraulic pressure cylinder. Each stress increment is applied, and the sample is allowed to consolidate and come to equilibrium with little or no further deformation and with the *excess* pore water pressure within the sample approximately equal to zero. Thus the final or equilibrium stress is an *effective stress*. The process is repeated until sufficient points are obtained to adequately define the stress-deformation curve.

The object of the consolidation test, then, is to simulate the compression of the soil under given external loads. What we are in fact measuring is the modulus of the soil in confined compression (Fig. 8.1a). By evaluating the compression characteristics of an *undisturbed representative* sample, we can predict the settlement of the soil layer in the field.

Engineers use several methods to present load-deformation data. Two methods are shown in Fig. 8.4. In one, *percent consolidation* or *vertical strain* is plotted versus the equilibrium or *effective consolidation stress* σ'_{vc}. (The subscripts *vc* refer to vertical consolidation, and the prime mark indicates effective stress.) A second way is to relate the *void ratio* to the *effective consolidation stress*. Both of these graphs show that soil is a strain hardening material; that is, the (instantaneous) modulus increases as the stresses increase.

Since the stress-strain relationships shown in Fig. 8.4 are highly nonlinear, more common ways to present the results of a consolidation test are shown in Fig. 8.5. The data shown in Fig. 8.4 are now presented as percent consolidation (or vertical strain) and void ratio versus the *logarithm* of effective consolidation stress. It can be seen that both plots have two approximately straight line portions connected by a smooth transitional curve. The stress at which the transition or "break" occurs in the curves shown in Fig. 8.5 is an indication of the *maximum* vertical overburden stress that this particular sample has sustained in the past; this stress, which is very important in geotechnical engineering, is known as the *preconsolidation pressure*, σ'_p. Sometimes the symbol p'_c or σ'_{vm} is used, where the subscript *m* indicates maximum past pressure.

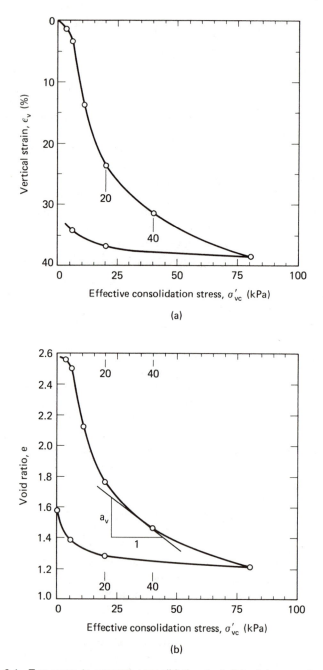

Fig. 8.4 Two ways to present consolidation test data: (a) percent consolidation (or strain) versus effective stress; (b) void ratio versus effective
stress. Test on a sample of San Francisco Bay Mud from −7.3 m.

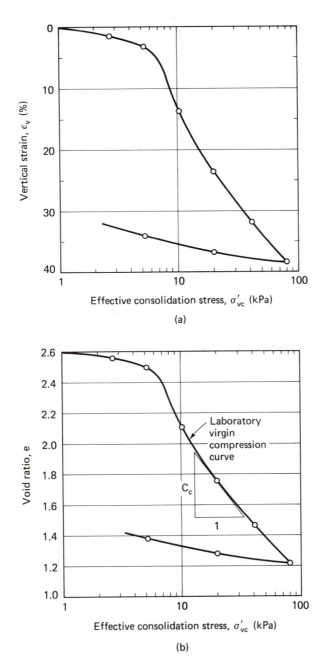

Fig. 8.5 Consolidation test data presented as: (a) percent consolidation (or strain) versus log effective stress and (b) void ratio versus log effective stress (same data as in Fig. 8.4).

8.5 PRECONSOLIDATION PRESSURE; NORMAL CONSOLIDATION, OVERCONSOLIDATION, UNDERCONSOLIDATION

Soils have a "memory," so to speak, of the stress and other changes that have occurred during their history, and these changes are preserved in the soil structure (Casagrande, 1932c). When the specimen or soil deposit in the field is loaded to a stress level greater than it ever "experienced" in the past, the soil structure is no longer able to sustain the increased load and the structure starts to break down. Depending on the type of soil and its geologic history, this breakdown may result in a quite drastic difference in the slopes of the two portions of the consolidation curve. In other words, the transition region may be small, and such soils are often very sensitive to even small changes in the applied stresses. With other less sensitive soils, for example silty soils, there never really is a "break" in the curve because the fabric gradually alters and adjusts as the applied stress increases. The initial flatter portion of the void ratio-log pressure consolidation curve is termed the *reconsolidation* curve, and the part after the "break" in the curve is called the *virgin compression curve* (Fig. 8.5b). As the latter name implies, the soil has never before "experienced" a stress greater than the preconsolidation stress.

We say that the soil is *normally consolidated* when the preconsolidation pressure σ_p' just equals the existing effective vertical overburden pressure σ_{vo}' (that is, $\sigma_p' = \sigma_{vo}'$). If we have a soil whose preconsolidation pressure is *greater* than the existing overburden pressure (that is, $\sigma_p' > \sigma_{vo}'$), then we say the soil is *overconsolidated* (or *preconsolidated*). We can define the *overconsolidation ratio*, OCR, as the ratio of the preconsolidation stress to the existing vertical effective overburden stress, or

$$\text{OCR} = \frac{\sigma_p'}{\sigma_{vo}'} \qquad (8\text{-}2)$$

Soils that are normally consolidated have an OCR = 1, and soils with an OCR > 1 are overconsolidated. Also, it is not impossible to find a soil that has an OCR < 1, in which case the soil would be *underconsolidated*. Underconsolidation can occur, for example, in soils that have only recently been deposited, either geologically or by man. Under these conditions, the clay layer has not yet come to equilibrium under the weight of the overburden load. If the pore water pressure were measured under conditions of underconsolidation, the pressure would be in excess of hydrostatic.

There are many reasons why a soil may be overconsolidated. It could be due to either a change in the total stress or a change in pore water pressure; both changes would alter the effective stress. Geologic deposition followed by subsequent erosion is an example of a change in the total stress that will preconsolidate the underlying soils. Desiccation of the upper layers due to surface drying will also produce overconsolidation. Sometimes an increase in σ_p' occurs due to changes in the soil structure and alterations of the chemical environment of the soil deposit. Table 8-1 lists some of the mechanisms leading to preconsolidation of soils.

How is the preconsolidation pressure determined? Several procedures have been proposed to determine the value of σ_p'. The most popular method is the Casagrande (1936b) construction, which is illustrated in Fig. 8.6, where a typical void ratio versus log pressure curve is plotted for a clay soil. The procedure is also applicable to ϵ_v versus log σ_{vc}' curves. The

TABLE 8-1 Mechanisms Causing Preconsolidation*

Mechanism	Remarks and References

Change in total stress due to:
   Removal of overburden
   Past structures
   Glaciation
	Geologic erosion or excavation
	by man

Change in pore water pressure due to:
   Change in water table elevation
   Artesian pressures
   Deep pumping; flow into tunnels
   Desiccation due to surface drying
   Desiccation due to plant life
	
	Kenney (1964) gives sea level changes
	Common in glaciated areas
	Common in many cities
	May have occurred during deposition
	May have occurred during deposition

Change in soil structure due to: Secondary compression (aging)†	 Raju (1956) Leonards and Ramiah (1959) Leonards and Altschaeffl (1964) Bjerrum (1967, 1972)
Environmental changes such as pH, temperature, and salt concentration	Lambe (1958a and b)
Chemical alterations due to "weathering," precipitation, cementing agents, ion exchange	Bjerrum (1967)
Change of strain rate on loading‡	Lowe (1974)

* After Brumund, Jonas, and Ladd (1976).

† The magnitude of σ_p'/σ_{vo}' related to secondary compression for mature natural deposits of highly plastic clays may reach values of 1.9 or higher.

‡ Further research is needed to determine whether this mechanism should take the place of secondary compression.

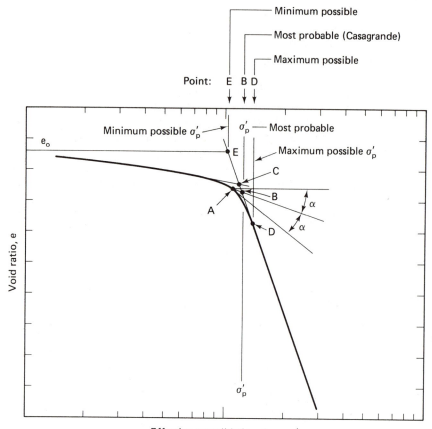

Fig. 8.6　The Casagrande (1936b) construction for determining the pre-consolidation stress. Also shown are the minimum possible, the most probable, and the maximum possible preconsolidation stresses.

Casagrande procedure is as follows:

1. Choose by eye the point of minimum radius (or maximum curvature) on the consolidation curve (point A in Fig. 8.6).
2. Draw a horizontal line from point A.
3. Draw a line tangent to the curve at point A.
4. Bisect the angle made by steps 2 and 3.
5. Extend the straight line portion of the *virgin* compression curve up to where it meets the bisector line obtained in step 4. The point of intersection of these two lines is the preconsolidation stress (point B of Fig. 8.6).

An even simpler method for estimating the preconsolidation stress is used by some engineers. The two straight line portions of the consolidation curve are extended; their intersection defines another "most probable" preconsolidation pressure (point C of Fig. 8.6). If you think about it, the maximum possible σ_p' is at point D, the minimum possible σ_p' is at point E, the intersection of the virgin compression curve with a horizontal line drawn from e_o.

How is it possible that these graphical procedures predict the preconsolidation pressure? To understand the reason, let us follow the complete stress-strain history of a sedimentary clay soil during deposition, sampling, and finally reloading in the laboratory by the consolidation test. This history is shown in Fig. 8.7. The line OA represents the relationship between the void ratio and the logarithm of effective stress of a particular element in the ground during deposition. In this case, additional material is deposited above our element, and the process consolidates the element to point A. This point represents the in situ e versus log σ_{vc}' coordinates of the normally consolidated clay element. When a boring is made and the soil is sampled, the overburden stresses are removed by the sampling operation and the sample rebounds or swells along the (dashed) curve AB. When the sample is transferred from the sampling tube into a consolidometer ring and then reloaded in the consolidation test, the (solid) reloading curve BC is obtained. About point C, the soil structure starts to break down, and if loading continues the laboratory virgin compression curve CD is obtained. Eventually the field and laboratory curves OAD and BCD will converge beyond point D. If you perform the Casagrande construction on the curve in Fig. 8.7, you will find that the most probable preconsolidation pressure is very close to point A on the graph, which is the actual maximum past pressure. Observations of this sort enabled Casagrande to develop his graphical procedure to find the preconsolidation stress. If the sampling operation was of poor quality and mechanical disturbance to the soil structure occurred, a different curve $BC'D$ (long dashes) would result upon reloading of the sample in the consolidometer. Note that with the "disturbed" curve the preconsolidation stress has all but disappeared with increasing mechanical disturbance; the reloading curve will move away from point A in the direction of the arrow. The preconsolidation pressure is much more difficult to define when sample disturbance has occurred.

In the consolidation test, after the maximum stress is reached, the specimen is rebounded incrementally to essentially zero stress (points D to E of Fig. 8.7). This process allows you to determine the final void ratio, which you need in order to plot the entire e versus log σ_{vc}' curve. Sometimes another reload cycle is applied, like curve E to F of Fig. 8.7. Just as with the initial reconsolidation curve (BCD), this loading curve eventually rejoins the virgin compression curve.

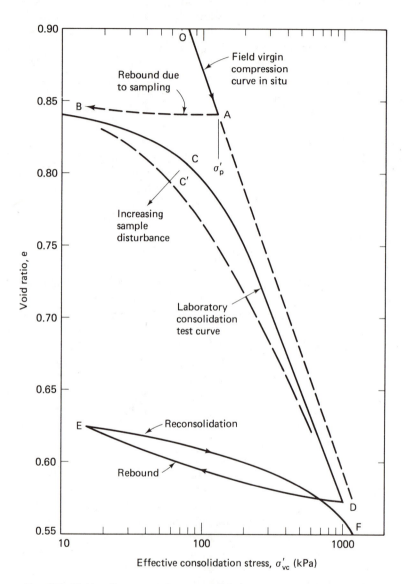

Fig. 8.7 Void ratio versus log pressure curve illustrating deposition, sampling (unloading) and reconsolidation in the consolidation test apparatus.

EXAMPLE 8.1

Given:

The results of the laboratory consolidation test of Fig. 8.7.

Required:

For the laboratory compression curve (BCD), determine (a) the precon-solidation stress using the Casagrande procedure; (b) find both the minimum and maximum possible values of this stress; and (c) determine the OCR if the in situ effective overburden stress is 80 kPa.

Solution:

a. Go through the steps of the Casagrande construction as shown on Fig. 8.6. The σ_p' is about 130 kPa.
b. Assume $e_o = 0.84$. Minimum possible σ_p' is about 90 kPa, and the maximum possible σ_p' is about 200 kPa.
c. Use Eq. 8-2.

$$\text{OCR} = \frac{\sigma_p'}{\sigma_{vo}'} = \frac{130}{80} = 1.6$$

Because of the uncertainties in determining both σ_p' and σ_{vo}', OCR's are usually given to only one decimal place.

8.6 CONSOLIDATION BEHAVIOR OF NATURAL SOILS

Typical consolidation curves for a wide variety of soils are presented in Figs. 8.8a through 8.8j. You should become familiar with the general shapes of these curves, especially around the preconsolidation stress, for the different soil types. Also study the amount of compression Δe as well as the slopes of the various curves.

The test results in Fig. 8.8a are typical of soils from the lower Mississippi River Valley near Baton Rouge, Louisiana. These soils, primarily silts and sand silts with clay strata, are slightly overconsolidated due to wetting and drying cycles during deposition (Kaufman and Sherman, 1964). Figs. 8.8b and 8.8c show test results from heavily overconsolidated

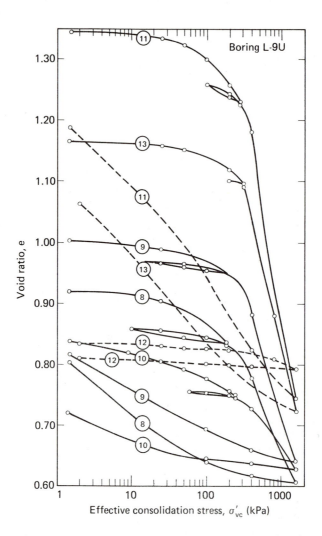

Test No.	Elev. (m)	Classification	Atterberg Limits			w_n (%)	e_o	σ'_{vo} (kPa)	σ'_p (kPa)	C_c
			LL	PL	PI					
8	−8.8	CL-clay, soft	41	24	17	34.0	0.94	160	200	0.34
9	−9.8	CL-clay, firm	50	23	27	36.4	1.00	170	250	0.44
10	−17.1	ML-sandy silt	31	25	6	29.8	0.83	230	350	0.16
11	−20.1	CH-clay, soft	81	25	56	50.6	1.35	280	350	0.84
12	−23.2	SP-sand	Nonplastic			27.8	0.83	320	—	—
13	−26.2	CH-clay w/silt strata	71	28	43	43.3	1.17	340	290	0.52

(a)

Fig. 8.8 (a) Nearly normally consolidated clays and silts (after Kaufman and Sherman, 1964).

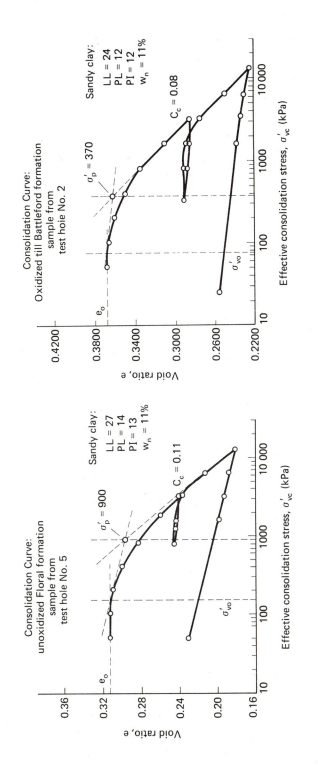

Fig. 8.8 (b) Overconsolidated clay tills (after MacDonald and Sauer, 1970).

(b)

301

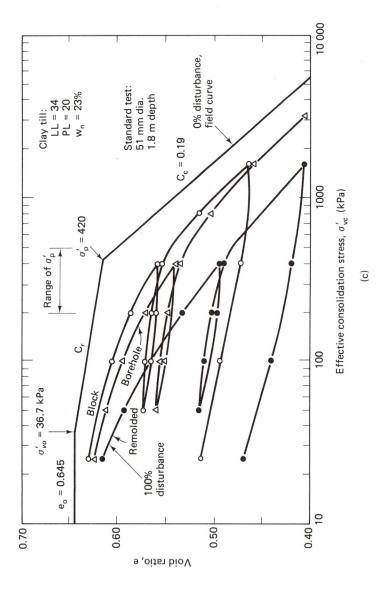

Fig. 8.8 (c) Overconsolidated clay tills, showing effects of different types of sampling (after Soderman and Kim, 1970).

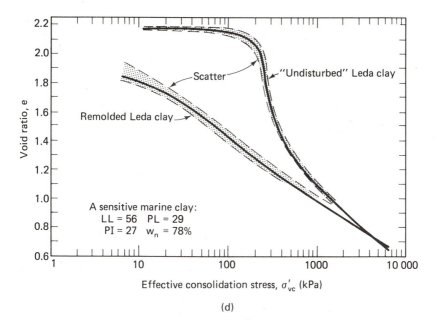

(d)

Fig. 8.8 (d)   Leda clay (after Quigley and Thompson, 1966).

clays. Note the very low void ratios for the precompressed glacial till soils from Canada in Fig. 8.8b (MacDonald and Sauer, 1970). The effects of sample disturbance on clay tills are shown in Fig. 8.8c. Note how the consolidation curves move downward and to the left (see Fig. 8.7) as disturbance increases (Soderman and Kim, 1970).

Compression curves for another Canadian clay, a sensitive marine clay called *Laurentian* or *Leda clay*, are shown in Fig. 8.8d (Quigley and Thompson, 1966). Both the undisturbed and remolded curves are shown. The very sharp "break" or drop-off in the undisturbed curve when the preconsolidation stress is reached is typical of highly sensitive clays. Until then the compression curve is very flat, but once a "critical" or precon-solidation stress is reached, the soil structure breaks down quickly and dramatically.

Figure 8.8e shows the consolidation characteristics of Mexico City clay (Rutledge, 1944). This sediment is not really a clay but is composed primarily of microfossils and diatoms. The porous structure of the fossils gives the soil a very high void ratio, natural water content, and compressi-bility. Mexico City clay was previously thought to be composed primarily of volcanic ash which weathered to allophane (Chapter 4) since it is amorphous to X-rays. Note the extremely high void ratios of Mexico City

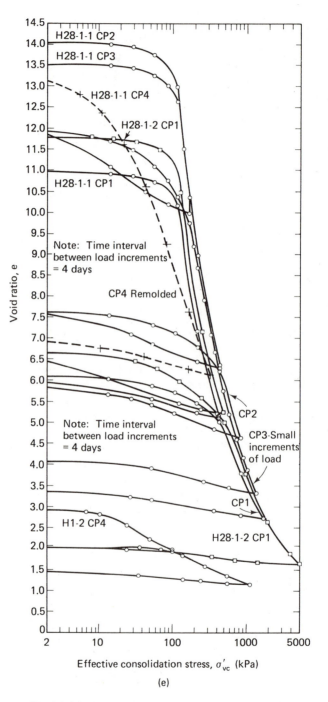

Fig. 8.8 (e)   Mexico City clay (after Rutledge, 1944).

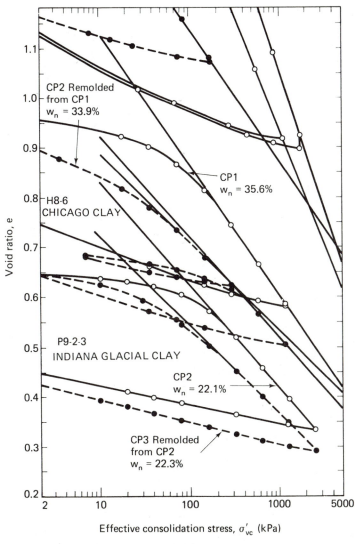

Effective consolidation stress, σ'_{vc} (kPa)

(f)

Fig. 8.8 (f)   Chicago and Indiana glacial clay (after Rutledge, 1944).

305

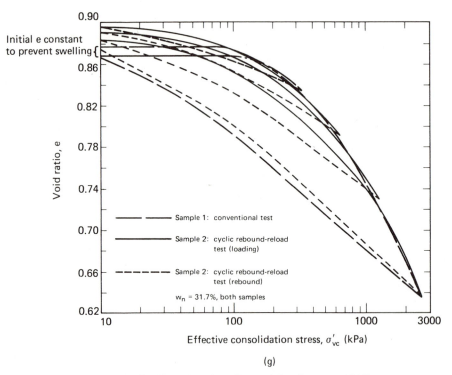

Fig. 8.8 (g)   Swelling clays from Texas (after Dawson, 1944).

clay. Also, see how the compression increases markedly once the preconsolidation stress is reached. As expected, remolding almost completely destroys the preconsolidation effect (see the dashed curve).

Fig. 8.8f shows the consolidation curves for two typical glacial lake clays (Rutledge, 1944). Both of these clays are rather silty and have much lower in situ void ratios and natural water contents than either the Leda or Mexico City clays.

Highly expansive or swelling clays from the southwest United States have compression curves like those shown in Fig. 8.8g. Both tests started out at about the same void ratio and water content. Both were initially loaded so that the void ratios remained constant. Then one sample (1) was loaded incrementally and continuously in the conventional manner; the other was repeatedly rebounded and reloaded. Notice how much rebound (swell) occurred and also that the cyclic test (2) had essentially the same compression characteristics as the conventional test. These variations probably occurred because the samples had a long history of alternate wetting and drying (desiccation), which caused the soils to be heavily overconsolidated (Chapter 6 and Table 8-1).

Consolidation curves for windblown silts (loess) are shown in Fig. 8.8h. The first figure from Clevenger (1958) shows dry density versus

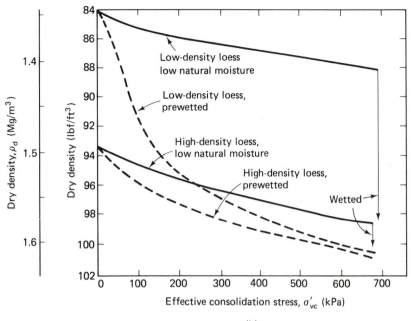

(h)

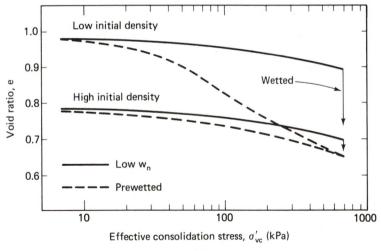

(h)

Fig. 8.8 (h) Loessial soils from the Missouri River Basin, showing effect of prewetting on consolidation. Note the drastic reduction in the void ratio when the low natural water content soil is wetted (after Clevenger, 1958).

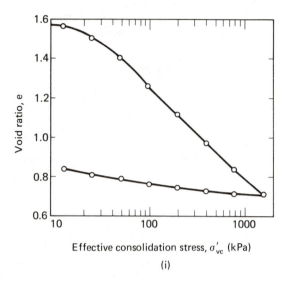

Fig. 8.8 (i)   Newfoundland silt (after Taylor, 1948).

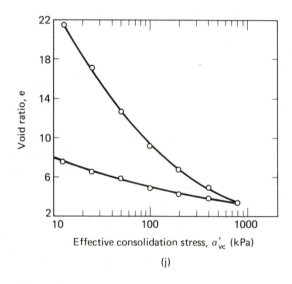

Fig. 8.8 (j)   Newfoundland peat (after Taylor, 1948).

applied load (arithmetic) for an initially low and an initially high density sample. The second figure is the same data conventionally plotted as an e versus $\log \sigma'_{vc}$ curve. Notice what happens when the samples are prewetted. In its natural state, loess is typically partially saturated, and when it is submerged or innundated collapse of the soil structure occurs. This condition is shown by the prewetted (dashed) curves of Fig. 8.8h. The amount of collapse upon wetting depends, as you might expect, on the initial density. Had the water not been added, the consolidation would have followed the upper curve. Sometimes prewetting loessial soils may be desirable to reduce settlements after construction.

Consolidation characteristics of another undisturbed silt are shown in Fig. 8.8i. Notice the lack of a "break" in the curve; this is typical of silty soils, and it makes determination of the preconsolidation stress difficult in practice.

Besides Mexico City clay, peats and other highly organic soils also have high void ratios and high natural water contents. The very high void ratio and concave upward shape of the compression curve is typical for peat, as shown in Fig. 8.8j. Just as with silts, determination of the preconsolidation stress is often difficult for such soils.

8.7 SETTLEMENT CALCULATIONS

How are settlements calculated? Figure 8.9 shows a soil layer of height H that is composed of both solids and voids, as shown in the middle of the figure. From the phase relationships described in Chapter 2, we can assume that the volume of solids V_s is equal to unity, and therefore the volume of voids is equal to e_o, the initial or original void ratio. Finally, upon completion of consolidation, the column of soil would look like that shown at the right side of Fig. 8.9. The volume of solids remains the same, of course, but the void ratio has decreased by the amount Δe. As you know, linear strain is defined as a change in length divided by the original length. Likewise, we may define the vertical strain in a soil layer as the ratio of the change in height to the original height of our soil column. Strain may be related to void ratio by using Fig. 8.9, or

$$\epsilon_v = \frac{\Delta L}{L_o} \quad \text{or} \quad \frac{\Delta H}{H_o} = \frac{s}{H_o} = \frac{\Delta e}{1 + e_o} \tag{8-3}$$

Solving for the settlement s in terms of the void ratio, we obtain:

$$s = \frac{\Delta e}{1 + e_o} H_o = \epsilon_v H_o \tag{8-4}$$

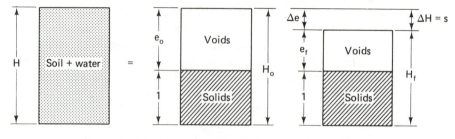

Fig. 8.9   Calculation of settlement from the phase diagram.

Note that Eq. 8-4 is based only on phase relationships and applies for both sands and clays.

EXAMPLE 8.2

Given:

Prior to placement of a fill covering a large area at a site, the thickness of a compressible soil layer was 10 m. Its original in situ void ratio was 1.0. Some time after the fill was constructed, measurements indicated that the average void ratio was 0.8.

Required:

Estimate the settlement of the soil layer.

Solution:

Use Eq. 8-4.

$$s = \frac{\Delta e}{1 + e_o} H_o = \frac{1.0 - 0.8}{1 + 1.0} 10 \, \text{m} = 1.0 \, \text{m}$$

By knowing the relationship between void ratio and effective stress it is possible to compute the settlement of a compressible layer due to the applied load. This relationship is, of course, determined from a one-dimensional compression or consolidation test, and we have already shown several ways to display the test results. The slope of the compression curve, when the results are plotted arithmetically, is called the *coefficient of compressibility*, a_v, or

$$a_v = \frac{-de}{d\sigma_v'} \qquad (8\text{-}5a)$$

Since the curve is not linear (see Figs. 8.1b and 8.4b), a_v is approximately constant over only a small pressure range, σ_1' to σ_2'; or

$$a_v = \frac{-\Delta e}{\Delta\sigma_v'} = \frac{e_1 - e_2}{\sigma_2' - \sigma_1'} \qquad (8\text{-}5b)$$

where the void ratios e_1 and e_2 correspond to the respective pressures σ_1' and σ_2'.

EXAMPLE 8.3

Given:

The compression curve shown in Fig. 8.4b.

Required:

Compute the coefficient of compressibility a_v for the stress increment from 20 to 40 kPa.

Solution:

From Fig. 8.4b, we find the void ratios corresponding to these stresses are $e_1 = 1.76$ and $e_2 = 1.47$. Using Eq. 8-5b we have,

$$a_v = \frac{1.47 - 1.76}{40 - 20} = -0.0145 \text{ per kPa}$$

Note that the units of a_v are the *reciprocal* of stress, or $1/\text{kPa}$ or m^2/kN; a_v could be reported as 15 m^2/N.

When the test results are plotted in terms of the percent consolidation or strain as in Fig. 8.4a, then the slope of the compression curve is the *coefficient of volume change*, m_v, or

$$m_v = \frac{d\epsilon_v}{d\sigma_v'} = \frac{\Delta\epsilon_v}{\Delta\sigma_v'} = \frac{a_v}{1 + e_0} = \frac{1}{D} \qquad (8\text{-}6)$$

where ϵ_v is the vertical compression or strain (Eq. 8-3), and D is the *constrained* or *oedometric modulus*. Sometimes the symbol E_{oed} is used for D. In one-dimensional compression, ϵ_v is equal to $\Delta e/(1 + e_o)$.

EXAMPLE 8.4

Given:

The compression curve shown in Fig. 8.4a.

Required:

    a. Compute the coefficient of volume change m_v for the stress increment from 20 to 40 kPa.
    b. Determine the constrained modulus D.

Solution:

    a. From Fig. 8.4a, the ϵ_v corresponding to σ_v' of 20 kPa is 23.7% and the ϵ_v corresponding to 40 kPa is 31.4%. Use Eq. 8-6.

$$m_v = \frac{0.314 - 0.237}{40 - 20} = 0.00385 \text{ per kPa}$$

    As with a_v, the units of m_v are the reciprocal of stress.
    b. The constrained modulus is the reciprocal of m_v, or

$$D = E_{oed} = 260 \text{ kPa}$$

EXAMPLE 8.5

Given:

The results of Examples 8.3 and 8.4.

Required:

Show that $m_v = a_v/(1 + e_o)$ for the increment 20 to 40 kPa.

Solution:

From Examples 8.3 and 8.4, $a_v = 0.0145$ per kPa and $m_v = 0.0039$ per kPa. From Fig. 8.4b, $e_o = 2.60$.

$$m_v = \frac{a_v}{1 + e_o} = \frac{0.0145}{1 + 2.6} = 0.0040, \text{ which is close to } 0.0039.$$

When test results are plotted in terms of the void ratio versus the *logarithm* of effective stress (Fig. 8.5b), then the slope of the virgin compression curve is called the *compression index* C_c, or

$$C_c = \frac{-de}{d \log \sigma_v'} = \frac{e_1 - e_2}{\log \sigma_2' - \log \sigma_1'} = \frac{e_1 - e_2}{\log \dfrac{\sigma_2'}{\sigma_1'}} \tag{8-7}$$

EXAMPLE 8.6

Given:

The consolidation test data of Fig. 8.5b.

Required:

Determine the compression index of this soil by (a) Eq. 8-7 and (b) graphically.

Solution:

a. The virgin compression curve of Fig. 8.5b is approximately linear from 10 to 80 kPa. At least we can take the average slope between these two points. Therefore from Eq. 8-7, we have

$$C_c = \frac{2.1 - 1.21}{\log \dfrac{80}{10}} = 0.986$$

Note that C_c is dimensionless.

b. To determine the C_c graphically, we note that

$$\log \frac{\sigma_2'}{\sigma_1'} = \log \frac{100}{10} = \log 10 = 1$$

Therefore if we find the difference in void ratio of the virgin compression curve over *one log cycle*, we automatically have the C_c (because the denominator of Eq. 8-7 is one). If you do this for the log cycle 10 to 100 kPa for example, you find that Δe is slightly less than 1.0 for a line parallel to the average slope between 10 and 80 kPa. Therefore C_c is slightly less than 1.0, which checks the calculation of part (a).

EXAMPLE 8.7

Given:

The consolidation test data of Fig. 8.8a.

Required:

Determine the C_c of tests 9 and 13.

Solution:

We can either use Eq. 8-7 or do this graphically. For test 9, using Eq. 8-7,

$$C_c = \frac{0.88 - 0.64}{\log\dfrac{1500}{400}} = 0.42$$

This is close to what Kaufman and Sherman (1964) obtained (0.44), as shown in Fig. 8.8a. Since the virgin compression curve is not exactly a straight line past σ_p', the value of C_c depends on where you determine the slope.

For test 13, find Δe for the log cycle from 200 to 2000 kPa. $\Delta e = 1.20 - 0.67 = 0.53$; so $C_c = 0.53$.

The slope of the virgin compression curve when the test results are plotted as percent consolidation or vertical strain versus *logarithm* of effective stress (Fig. 8.5a) is called the *modified compression index*, $C_{c\epsilon}$, and it is expressed as

$$C_{c\epsilon} = \frac{\Delta\epsilon_v}{\log\dfrac{\sigma_2'}{\sigma_1'}} \qquad (8\text{-}8)$$

Sometimes $C_{c\epsilon}$ is called the *compression ratio*. The relationship between the modified compression index $C_{c\epsilon}$ and the compression index C_c, is given by

$$C_{c\epsilon} = \frac{C_c}{1 + e_o} \tag{8-9}$$

EXAMPLE 8.8

Given:

The consolidation data of Fig. 8.5a.

Required:

Determine the modified compression index of this soil by (a) Eq. 8-8 and (b) graphically. (c) Check the C_c from Example 8.6 by Eq. 8-9.

Solution:

Do this problem just like Example 8.6.

a. Consider the virgin compression curve to be approximately a straight line over the stress range 10 to 80 kPa. Thus, using Eq. 8-8, we have

$$C_{c\epsilon} = \frac{0.385 - 0.138}{\log \dfrac{80}{10}} = 0.274$$

b. To find $C_{c\epsilon}$ graphically, choose any convenient log cycle; in this case use the cycle 10 to 100 kPa. Then the $\Delta\epsilon_v$ for this cycle is $38 - 10 = 28\%$, or $C_{c\epsilon} = 0.28$, which checks part (a) adequately.

c. Assume $e_o = 2.60$ from Fig. 8.5b. Use Eq. 8-9. Therefore

$$C_c = C_{c\epsilon}(1 + e_o) = 0.274(1 + 2.6) = 0.985$$

which checks the C_c from Example 8.6.

EXAMPLE 8.9

Given:

The void ratio versus log effective pressure data shown in Fig. Ex. 8.9.

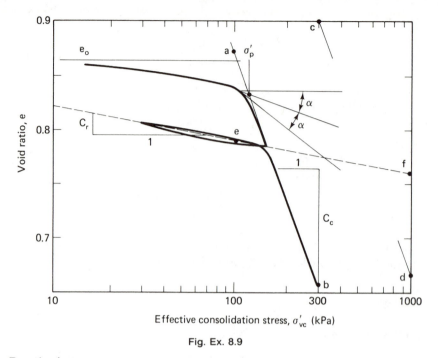

Fig. Ex. 8.9

Required:

Determine (a) the preconsolidation pressure σ_p', (b) the compression index C_c, and (c) the modified compression index $C_{c\epsilon}$.

Solution:

    a. Perform the Casagrande construction according to the procedure outlined in Sec. 8.4, and find $\sigma_p' = 121$ kPa.

    b. By definition (Eq. 8-7),

$$C_c = \frac{\Delta e}{\log \dfrac{\sigma_2'}{\sigma_1'}}$$

Using the points a and b of Fig. Ex. 8.9, $e_a = 0.870$, $e_b = 0.655$, $\sigma_a' = 100$ kPa, and $\sigma_b' = 300$ kPa. Therefore

$$C_c = \frac{e_a - e_b}{\log \dfrac{\sigma_b'}{\sigma_a'}} = \frac{0.870 - 0.655}{\log \dfrac{300}{100}} = \frac{0.215}{0.477} = 0.451$$

A second graphical way is to find Δe over *one* cycle; for example, $\log\frac{1000}{100} = \log 10 = 1$. When this is done, $C_c = \Delta e$. In Fig. Ex. 8.9 the vertical scale is not sufficient for $\Delta\sigma' = 1$ log cycle, but it can be done in two steps, e_a to e_b and e_c to e_d. (To extend the line $\overline{e_a e_b}$ to one full log cycle on the *same* graph, choose e_c at the same pressure as e_b. Then draw the line $\overline{e_c e_d}$ parallel to $\overline{e_a e_b}$. This second line is merely the extension of $\overline{e_a e_b}$ if the graph paper extended lower than shown.) Or,

$$\begin{aligned}\Delta e = C_c &= (e_a - e_b) + (e_c - e_d)\\ &= (0.870 - 0.655) + (0.90 - 0.664)\\ &= 0.215 + 0.236\\ &= 0.451, \text{ or same as above}\end{aligned}$$

c. The modified compression index $C_{c\epsilon}$ is

$$C_{c\epsilon} = \frac{C_c}{1 + e_o} = \frac{0.451}{1 + 0.865} = 0.242$$

To calculate consolidation settlement, Eqs. 8-5, 8-6, or 8-7 and 8-8 may be combined with Eq. 8-4. For example, using Eqs. 8-7 and 8-4 we obtain

$$s_c = C_c \frac{H_o}{1 + e_o} \log\frac{\sigma_2'}{\sigma_1'} \tag{8-10}$$

If the soil is normally consolidated, then σ_1' would be equal to the existing vertical overburden stress σ_{vo}', and σ_2' would include the additional stress $\Delta\sigma_v$ applied by the structure, or

$$s_c = C_c \frac{H_o}{1 + e_o} \log\frac{\sigma_{vo}' + \Delta\sigma_v}{\sigma_{vo}'} \tag{8-11}$$

When computing the settlement by means of the percent consolidation versus log effective stress curve, Eq. 8-8 is combined with Eq. 8-4 to get

$$s_c = C_{c\epsilon} H_o \log\frac{\sigma_2'}{\sigma_1'} \tag{8-12}$$

or, analogous to Eq. 8-11, for normally consolidated clays,

$$s_c = C_{c\epsilon} H_o \log\frac{\sigma_{vo}' + \Delta\sigma_v}{\sigma_{vo}'} \tag{8-13}$$

Other similar settlement equations can be derived using a_v and m_v. In this case the average stress for a given stress increment must be used since the compression curves are nonlinear.

EXAMPLE 8.10

Given:

Test results shown in Fig. 8.4 and 8.5 are representative of the compressibility of a 10 m layer of normally consolidated San Francisco Bay Mud. The initial void ratio is about 2.5.

Required:

Estimate the consolidation settlement of a large fill on the site if the average total stress increase on the clay layer is 10 kPa.

Solution:

First estimate the preconsolidation stress to be about 7 kPa. Since the clay is normally consolidated, $\sigma'_p \simeq \sigma'_{vo}$. Use the results of Examples 8.6 and 8.8. C_c is 0.986 and $C_{c\epsilon}$ is 0.274. Use Eq. 8-11.

$$s_c = 0.986 \left(\frac{10 \text{ m}}{1 + 2.5} \right) \log \frac{7 + 10}{7} = 1.09 \text{ m}$$

Use Eq. 8-13.

$$s_c = 0.274(10 \text{ m}) \log \frac{7 + 10}{7} = 1.06 \text{ m}$$

This slight difference is the result of obtaining data from the small graphs. The settlement would be reported as "about 1 m." With a high water table, the actual settlement would be even slightly less since, with settlements this large, fill that was above the water table would now be submerged. Thus the resulting fill load would be reduced. To take this aspect into account, trial and error computations are required.

There are a couple of reasons for the popularity in engineering practice of using the percent consolidation or vertical strain versus log effective stress curve to compute settlements. First, estimating field settlements is simple. You can read the percent compression directly from the graph, once you have a good estimate of the in situ vertical overburden stress.

Another reason the percent consolidation versus log effective stress plots are popular is that during the consolidation test it is often desirable to know what the shape of the compression curve looks like, so as to be able to obtain an early evaluation of the preconsolidation pressure. The void ratio versus log effective stress curve cannot be plotted during the test as it is necessary to know both the initial and final values of the void ratio. This calculation requires the determination of the dry mass of solids, which can only be determined at the *end* of the test. Therefore the e versus $\log \sigma'_{vc}$ curve cannot be plotted during the test. However the percent consolidation versus log pressure curve can be plotted while the test is being performed. Another advantage is that when the preconsolidation pressure is being approached, the load increments placed on the sample can be reduced so as to define more carefully the transition between the reloading curve and the virgin compression curve. Also, the test can be stopped when two or three points define the straight line portion of the virgin compression curve. Finally, as Ladd (1971a) pointed out, two samples may show very different e versus $\log \sigma'_{vc}$ plots but may have similar vertical strain versus log effective stress curves because of differences in initial void ratio.

EXAMPLE 8.11

Given:

The data of Example 8.10.

Required:

Estimate the settlement *directly* from Fig. 8.5a.

Solution:

If the preconsolidation stress is about 7 kPa, the final stress after loading is 17 kPa. Refer to Fig. 8.5a. At the σ'_p, which is equal to σ'_{vo} since it is normally consolidated, the ϵ_v is about 5.5%. At $\sigma'_v = 17$ kPa, the ϵ_v is about 22%. Therefore the $\Delta\epsilon_v$ is 16.5%, so the estimated settlement will be

$$s_c = 0.165(10 \text{ m}) = 1.65 \text{ m}$$

The settlement is more in this example because the slope of the virgin compression curve is steeper from 7 to 17 kPa than it is from 10 to 80 kPa (see Example 8.6).

All the equations for settlement presented above were for a single compressible layer. When the consolidation properties or the void ratio vary significantly with depth or are different for distinct soil layers, then the total consolidation settlement is merely the sum of the settlements of the individual layers, or

$$s_c = \sum_{i=1}^{n} s_{ci} \qquad (8\text{-}14)$$

where s_{ci} is the settlement of the ith layer of n layers as calculated by Eqs. 8-10 through 8-13.

So far in this section we have only discussed settlement calculations for normally consolidated soils. What happens if the soil is overconsolidated? You will recall that an overconsolidated soil is one in which σ'_{vo} is less than σ'_p. Overconsolidated soils are probably encountered more often in engineering practice than normally consolidated soils. So it is important to know how to make settlement calculations for this important class of soil deposits.

The first thing to do is to check whether the soil is preconsolidated. You do this by comparing the preconsolidation pressure σ'_p to the existing vertical effective overburden pressure, σ'_{vo}. You already know how to calculate σ'_{vo} from Chapter 7. If the soil layer is definitely overconsolidated, then you have to check to see if the stress added by the engineering structure, $\Delta\sigma_v$, plus the σ'_{vo} exceeds the preconsolidation pressure σ'_p. Whether or not it does can make a large difference in the amount of settlement calculated, as is shown in Fig. 8.10.

If you have the case shown in Fig. 8.10a, that is, if $\sigma'_{vo} + \Delta\sigma_v \leq \sigma'_p$, then use either Eq. 8-11 or 8-13, but with the recompression indices C_r or $C_{r\epsilon}$ in place of C_c and $C_{c\epsilon}$, respectively. The *recompression index*, C_r, is defined just like C_c, except that it is for the average slope of the recompression part of the e versus $\log\sigma'_{vc}$ curve (Fig. 8.7). If the data are plotted in terms of ϵ_v versus $\log\sigma'_{vc}$, then the slope of the recompression curve is called the *modified recompression index* $C_{r\epsilon}$ (sometimes called the *recompression ratio*). C_r and $C_{r\epsilon}$ are related just like C_c and $C_{c\epsilon}$ (Eq. 8-9), or

$$C_{r\epsilon} = \frac{C_r}{1 + e_o} \qquad (8\text{-}15)$$

EXAMPLE 8.12

Given:

The void ratio versus log effective pressure data shown in Fig. Ex. 8.9.

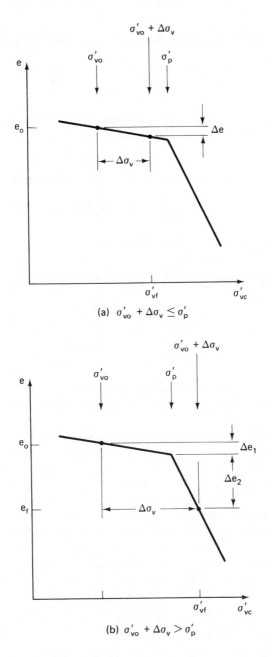

Fig. 8.10 Principle of settlement calculations for overconsolidated soils.

Required:

Calculate (a) the recompression index C_r and (b) the modified recompression index C_{re}.

Solution:

a. The recompression index C_r is found in a similar manner to the C_c (Eq. 8-7). Using the points e and f over 1 log cycle, we find that

$$C_r = e_e - e_f = 0.790 - 0.760 = 0.030$$

b. The modified recompression index C_{re} is found from Eq. 8-15.

$$C_{re} = \frac{C_r}{1 + e_o} = \frac{0.030}{1 + 0.865} = 0.016$$

Note that neither of these terms has units.

———————

To calculate settlements of overconsolidated clays, Eqs. 8-11 and 8-13 become

$$s_c = C_r \frac{H_o}{1 + e_o} \log \frac{\sigma'_{vo} + \Delta\sigma_v}{\sigma'_{vo}} \tag{8-16}$$

$$s_c = C_{re} H_o \log \frac{\sigma'_{vo} + \Delta\sigma_v}{\sigma'_{vo}} \tag{8-17}$$

when $\sigma'_{vo} + \Delta\sigma_v \leq \sigma'_p$. Since C_r is usually much less than C_c, the settlements occurring when $\sigma'_{vo} + \Delta\sigma_v \leq \sigma'_p$ are much less than if the soil were normally consolidated.

If the added stress caused by the structure exceeds the preconsolidation stress, then much larger settlements would be expected. This is because the compressibility of the soil is much greater on the virgin compression curve than on the recompression curve as was shown, for example, in Fig. 8.7. For the case, then, where $\sigma'_{vo} + \Delta\sigma_v > \sigma'_p$ the settlement equation consists of two parts: (1) the change in void ratio or strain on the recompression curve from the original in situ conditions of (e_o, σ'_{vo}) or $(\epsilon_{vo}, \sigma'_{vo})$ to σ'_p; and (2) the change in void ratio or strain on the virgin compression curve from σ'_p to the final conditions of (e_f, σ'_{vf}) or $(\epsilon_{vf}, \sigma'_{vf})$. Note that $\sigma'_{vf} = \sigma'_{vo} + \Delta\sigma_v$. These two parts are shown graphically in Fig. 8.10b. The complete settlement equation then becomes

$$s_c = C_r \frac{H_o}{1 + e_o} \log \frac{\sigma'_{vo} + (\sigma'_p - \sigma'_{vo})}{\sigma'_{vo}}$$
$$+ C_c \frac{H_o}{1 + e_o} \log \frac{\sigma'_p + (\sigma'_{vo} + \Delta\sigma_v - \sigma'_p)}{\sigma'_p} \tag{8-18a}$$

This equation reduces to

$$s_c = C_r \frac{H_o}{1 + e_o} \log \frac{\sigma'_p}{\sigma'_{vo}} + C_c \frac{H_o}{1 + e_o} \log \frac{\sigma'_{vo} + \Delta\sigma_v}{\sigma'_p} \qquad (8\text{-}18b)$$

In terms of the modified indices, we have

$$s_c = C_{r\epsilon} H_o \log \frac{\sigma'_p}{\sigma'_{vo}} + C_{c\epsilon} H_o \log \frac{\sigma'_{vo} + \Delta\sigma_v}{\sigma'_p} \qquad (8\text{-}19)$$

Both Eqs. 8-18 and 8-19 give the same results. One could argue that in the right-hand term of Eq. 8-18 the void ratio corresponding to the precon-solidation pressure on the true virgin compression curve should be used. Although this is technically correct, it doesn't make any significant difference in the answer.

Sometimes the degree of overconsolidation varies throughout the compressible layer. You could apply Eq. 8-16 or 8-17 to the part where $\sigma'_{vo} + \Delta\sigma_v < \sigma'_p$ and Eq. 8-18 or 8-19 to the part where $\sigma'_{vo} + \Delta\sigma_v > \sigma'_p$. In practice, however, it is usually easier to simply divide the entire stratum into several layers, apply the appropriate equation to calculate the average settlement for each layer, and then sum up the settlements by Eq. 8-14.

What is the best way to get C_r and $C_{r\epsilon}$ for use in Eqs. 8-16 through 8-19? Because of sample disturbance, the slope of the initial recompression portion of the laboratory consolidation curve (Fig. 8.7) is too steep and would yield values that are too large for these indices. Leonards (1976) offers the reasons why in situ values are generally smaller than those obtained from laboratory measurements: (1) disturbance during sampling, storage, and preparation of test specimens; (2) recompression of gas bubbles in the voids; and (3) errors in test procedures and methods of interpreting test results. This latter item includes the problem of reproducing the in situ state of stress in the specimen. Leonards recommends that the σ'_{vo} be applied to the specimen and that it be innundated and allowed to come to equilibrium for at least 24 hours before starting the incremental loading. Any tendency to swell should be controlled. Then the consolidation test is continued with relatively large load increments. To reproduce as closely as possible the in situ stress state, Leonards recommends that the sample be consolidated to slightly less than the σ'_p and then be allowed to rebound. This is the first cycle shown in Fig. 8.11. If you don't have a good idea of the σ'_p, then consolidate initially to $\sigma'_{vo} + \Delta\sigma_v$ only, which is presumably less than σ'_p. The determination of C_r or $C_{r\epsilon}$ is over the range of $\sigma'_{vo} + \Delta\sigma_v$, as shown in Fig. 8.11. It is common practice to take the average slope of the two curves. From the typical test results shown in Fig. 8.11, you can see that the actual values of the recompression index depend on the stress at which the rebound-reload cycle starts, especially whether it starts at a stress less than or greater than the σ'_p. See the difference in

Vertical effective stress (log scale)

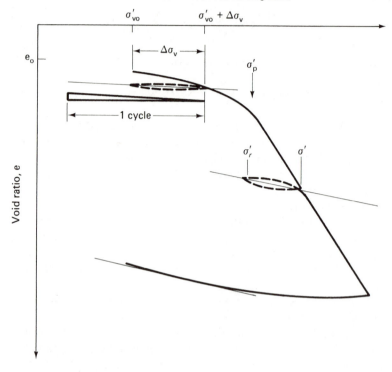

Fig. 8.11   Typical consolidation curve showing the recommended procedure for determining the C_r (after Leonards, 1976).

slopes of the rebound curves shown in the figure. The C_r also depends on the OCR to which rebounding and reloading take place, for example, the ratio of σ'/σ'_r in Fig. 8.11. The final consideration affecting the value of C_r is the presence of gas bubbles in the pores of the soil. Use of back pressure (Chapter 11) can sometimes take care of this problem.

EXAMPLE 8.13

Given:

The data in Example 8.1 and Fig. 8.7 is representative of a layer of silty clay 10 m thick.

Required:

Estimate the consolidation settlement if the structural loads at the surface will increase the average stress in the layer by 35 kPa.

Solution:

From Example 8.1, we know that the σ'_{vo} is 80 kPa and the σ'_p is about 130 kPa; e_o is about 0.84. Since the applied stress is 35 kPa, the $\sigma'_{vo} + \Delta\sigma_v = 115$ kPa < 130 kPa. Therefore use Eq. 8-16. To get C_r, we will take the average slope of the two curves DE and EF near the bottom of Fig. 8.7. C_r is approximately 0.03. Now use Eq. 8-16.

$$s_c = 0.03 \frac{10 \text{ m}}{1 + 0.84} \log \frac{80 + 35}{80} = 0.026 \text{ m or } 26 \text{ mm}$$

From the preceding discussion, the C_r in this example is probably too large since we determined it from a recycle well beyond the σ'_p. It is therefore very likely that the settlements in the field will be less than the prediction of 26 mm.

EXAMPLE 8.14

Given:

The data in Example 8.13, except that the structural engineer made an error in computing the loads; the correct loads now will produce an average stress increase of 90 kPa in the silty clay layer.

Required:

Estimate the consolidation settlement due to the new loads.

Solution:

Now the applied stress is much greater than $\sigma'_{vo} + \Delta\sigma_v$, or $80 + 90 = 170$ > 130 kPa. Therefore we must use Eq. 8-18. In addition to the C_r, we need the compression index C_c. From Fig. 8.7 we find that C_c is about 0.15. Substitution into Eq. 8-18b gives

$$s_c = 0.03 \frac{10 \text{ m}}{1 + 0.84} \log \frac{130}{80} + 0.25 \frac{10 \text{ m}}{1 + 0.84} \log \frac{80 + 90}{130}$$

$$= 0.034 \text{ m} + 0.158 \text{ m}$$

$$= 0.193 \text{ m}$$

As in Example 8.10, this value would be reported as "about 20 cm" due to the uncertainties in sampling, testing, and in estimating the σ_p', the applied stress increase, and C_r and C_c.

8.8 FACTORS AFFECTING THE DETERMINATION OF σ_p'

Brumund, Jonas, and Ladd (1976) discuss three factors which significantly influence the determination of σ_p' from laboratory consolidation tests. We have already mentioned one, the effect of sample disturbance on the shape of the consolidation curve (Fig. 8.7). We showed how the "break" in the curve became less sharp with increasing disturbance. You can see these effects in Fig. 8.12a. With sensitive clays especially (for example, Figs. 8.8d and e), increasing sample disturbance lowers the value of the σ_p'. At the same time the void ratio is decreased (or the strain increased) for any given value of σ_{vc}'. As a consequence, the compressibility at stresses less than the σ_p' are increased, and at stresses greater than the σ_p' the compressibility is decreased.

A load increment ratio (LIR) of unity is used in conventional consolidation testing (for example, ASTM D 2435). The LIR is defined as the change in pressure or the pressure increment divided by the initial pressure before the load is applied. This relationship is as follows:

$$ \text{LIR} = \frac{\Delta\sigma}{\sigma_{\text{initial}}} \tag{8-20} $$

where $\Delta\sigma$ is the incremental stress, and σ_{initial} is the previous stress. An LIR of unity means that the load is doubled each time. This procedure results in evenly spaced data points on the void ratio versus log effective stress curve such as shown in Fig. 8.5b.

Experience with soft, sensitive clays (Fig. 8.8d), has shown that a small stress change or even vibration may drastically alter the soil structure. For such soils a load increment ratio of unity may not accurately define the value of the preconsolidation stress, so an LIR of less than one is often used. The influence of varying the LIR on the compressibility as well as on the σ_p' of a typical clay is shown in Fig. 8.12b. The effect of the duration of the load increment is shown in Fig. 8.12c. The common (ASTM D 2435) procedure is for each increment to be left on the sample

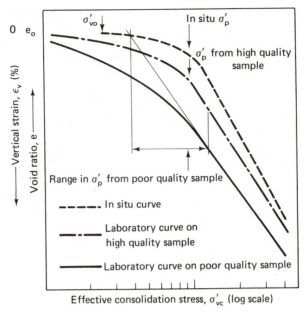

(a)

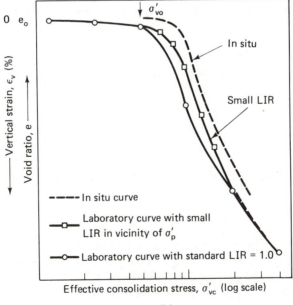

(b)

Fig. 8.12 Factors affecting the laboratory determination of σ_p': (a) effect of sample disturbance; (b) effect of load increment ratio;

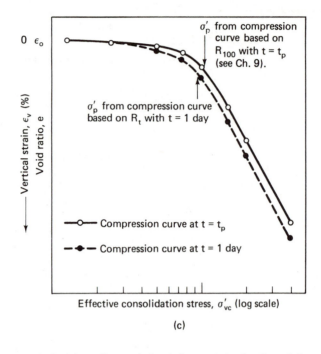

Fig.8.12 (c) effect of load increment duration (after Brumund, Jonas, and Ladd, 1976).

for 24 hours. Note how this procedure affects the σ_p'. Some of the terminology used for these figures will become clearer after you read Chapter 9.

8.9 PREDICTION OF FIELD CONSOLIDATION CURVES

Since the consolidation test really is a reloading of the soil (shown by curve BCD of Fig. 8.7), even with high-quality sampling and testing the actual recompression curve has a slope which is somewhat *less* than the slope of the *field virgin compression curve* (OAD in Fig. 8.7). Schmertmann (1955) developed a graphical procedure to evaluate the slope of the field virgin compression curve. The procedure for this construction technique is illustrated in Fig. 8.13, where typical void ratio versus log effective stress

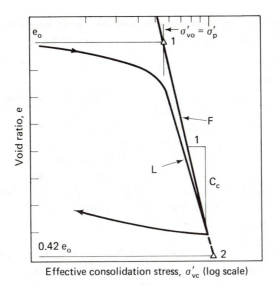

(a)

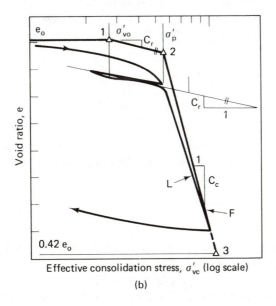

(b)

Fig. 8.13   Illustration of the Schmertmann (1955) procedure to obtain the field virgin compression curve: (a) normally consolidated soil; (b) overconsolidated soil.

curves are plotted. To correct the laboratory virgin compression curve for a normally consolidated soil in the field, proceed as follows:

1. Perform the Casagrande construction and evaluate the preconsolidation pressure σ_p'.
2. Calculate the initial void ratio e_o. Draw a horizontal line from e_o, parallel to the log effective stress axis, to the preconsolidation pressure σ_p'. This defines control point 1, illustrated by triangle 1 in Fig. 8.13a.
3. From a point on the void ratio axis equal to $0.42e_o$, draw a horizontal line, and where the line meets the extension of the laboratory virgin compression curve L, define another control point, as shown by triangle 2. You should note that the coefficient of e_o is not a "magic number," but is a result of many observations on different clays.
4. Connect the two control points by a straight line. The slope of this line, F, defines the compression index C_c that most probably exists in the field. Line F is the *field virgin compression curve*. The Schmertmann correction allows for disturbance of the clay due to sampling, transportation, and storage of the sample plus subsequent trimming and reloading during the consolidation test.

EXAMPLE 8.15

Given:

The e versus log σ data of Fig. Ex. 8.15. This consolidation data is from an undisturbed clay sample taken from the midpoint of a compressible layer 10 m thick. The OCR = 1.0

Required:

Using the Schmertmann procedure determine (a) the slope of the field virgin compression curve. (b) Compute the settlement of this clay layer if the stress is increased from 275 to 800 kPa. Calculate, using both the laboratory and field virgin compression curves. (c) Comment on the difference, if any.

Solution:

a. First, establish the field virgin compression curve according to the Schmertmann procedure outlined above. Perform the Casagrande con-

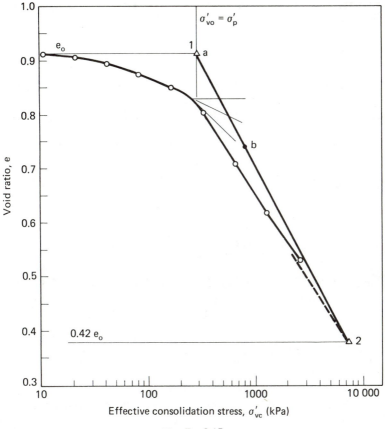

Fig. Ex. 8.15

struction on the curve shown in Fig. Ex. 8.15 to obtain the preconsolidation   pressure. σ'_p is found to be 275 kPa. Draw a horizontal line from
$e_o = 0.912$ to the point where it intersects the preconsolidation pressure to
establish control point 1, shown by triangle 1. Extend the virgin compression curve to 0.42 e_o (0.42 × 0.912) or 0.38, to establish control point 2.
Connecting the two control points 1 to 2 creates the field virgin compression curve.

The value of C_c from the field virgin compression curve is determined
just like you did for the laboratory consolidation curve (see Examples 8.6,
8.7, and 8.9). For the log cycle from 1000 to 10 000 kPa, $e_{1000} = 0.705$ and
$e_{10000} = 0.329$; therefore $C_c = 0.705 - 0.329 = 0.376$. The slope of the
laboratory virgin compression curve is found in the same way and equals
0.31. We'll need this value later.

b. To compute the settlement, we may use either Eq. 8-4 or 8-11. Use Eq. 8-4 first:

$$s_c = \frac{\Delta e}{1 + e_o} H_o$$

The change in void ratio, Δe, is merely the difference in void ratio for $\sigma = 275$ kPa and $\sigma = 800$ kPa. These values are 0.912 at point a and 0.744 at point b in Fig. Ex. 8.15 on the field virgin compression curve. Therefore,

$$s_c = \frac{0.912 - 0.744}{1 + 0.912} \, 10 \text{ m} = 0.88 \text{ m}$$

Using Eq. 8-11:

$$s_c = \frac{C_c}{1 + e_o} H_o \log \frac{\sigma'_{vo} + \Delta\sigma_v}{\sigma'_{vo}}$$

$$= \frac{0.376}{1 + 0.912} (10 \text{ m}) \log \frac{800}{275} = 0.91 \text{ m}$$

The slight difference in the calculated values of the consolidation settlement s_c is due to small errors in reading data points from Fig. Ex. 8.15.

If we calculate the consolidation settlement using the laboratory virgin compression curve to establish C_c, we would obtain (Eq. 8-11)

$$s_c = \frac{0.31}{1 + 0.912} (10 \text{ m}) \log \frac{800}{275} = 0.75 \text{ m, or 16\% lower}$$

c. Comment on the difference. Sixteen percent could be significant in some cases, especially if the proposed structure is particularly sensitive to settlements. Ladd (1971a) has found the Schmertmann correction will increase compression indices about 15% for fairly good samples of soft to medium clay. Since the procedure is simple, it would seem prudent to use it to make the best possible estimates of field compressibility. On the other hand, beware of too much precision in settlement calculations. When foundation engineers present their results in an engineering report, the expected settlement would be stated as "approximately 0.9 m", because including more significant figures would imply more than the actual precision.

The Schmertmann procedure for an overconsolidated soil is illustrated in Fig. 8.13b. If it is suspected that an overconsolidated soil is being tested, then it is good practice to follow the test procedure suggested in Sec. 8.7 and Fig. 8.11. A cycle of partial unloading and reloading is shown in Fig. 8.13b and in Figs. 8.8a, b, and c. The average slope of the rebound-reload curve establishes C_r. The remaining steps in the Schmertmann procedure are as follows:

1. Calculate the initial void ratio e_o. Draw a horizontal line from e_o, parallel to the log effective stress axis, to the existing vertical overburden pressure σ'_{vo}. This establishes control point 1, as shown by triangle 1 in Fig. 8.13b.
2. From control point 1, draw a line parallel to the rebound-reload curve to the preconsolidation pressure σ'_p. This will establish control point 2, as shown by triangle 2 in Fig. 8.13b.
3. In a manner similar to that used for the normally consolidated soil, draw a horizontal line from a void ratio equal to $0.42e_o$. Where this line intersects the laboratory virgin compression curve L, establish a third control point, as shown by triangle 3 in Fig. 8.13b. Connect control points 1 and 2, and 2 and 3 by straight lines. The slope of the line F joining control points 2 and 3 defines the compression index C_c for the field virgin compression curve. The slope of the line joining control points 1 and 2 of course represents the recompression index C_r. An example of a field compression curve is shown in Fig. 8.8c.

EXAMPLE 8.16

Given:

The void ratio versus pressure data shown below. The initial void ratio is 0.725, and the existing vertical effective overburden pressure is 130 kPa.

Void Ratio	Pressure (kPa)
0.708	25
0.691	50
0.670	100
0.632	200
0.635	100
0.650	25
0.642	50
0.623	200
0.574	400
0.510	800
0.445	1600
0.460	400
0.492	100
0.530	25

Required:

  a. Plot the data as e versus log σ'_{vc}.
  b. Evaluate the overconsolidation ratio.
  c. Determine the field compression index using the Schmertmann procedure.
  d. If this consolidation test is representative of a 12 m thick clay layer, compute the settlement of this layer if an additional stress of 220 kPa were added.

Solution:

  a. The data is plotted in Fig. Ex. 8.16.

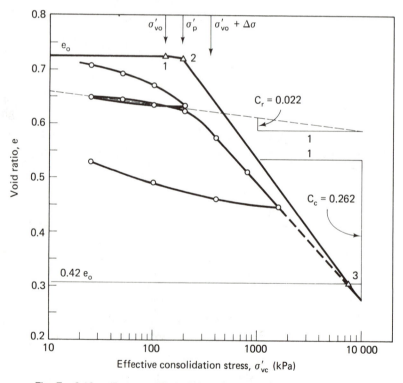

Fig. Ex. 8.16   (Data modified slightly from Soderman and Kim, 1970.)

b. The given value of σ'_{vo} is plotted on the graph, and the Casagrande construction performed to evaluate σ'_p. A value of 190 kPa is found.

$$\text{OCR} = \frac{\sigma'_p}{\sigma'_{vo}} = \frac{190}{130} = 1.46$$

Thus the soil is slightly overconsolidated.

c. Using the Schmertmann procedure for overconsolidated clays as given previously, control points 1, 2, and 3 are established, as shown in Fig. Ex. 8.16. The values of C_r and C_c are evaluated directly from Fig. Ex. 8.16 over one log cycle. $C_r = 0.611 - 0.589 = 0.022$, and $C_c = 0.534 - 0.272 = 0.262$. (Note that $C_r \simeq 10\%$ of C_c.)

d. Using Eq. 8-18b, the settlement is computed:

$$s_c = \frac{C_r}{1 + e_o} H_o \log \frac{\sigma'_p}{\sigma'_{vo}} + \frac{C_c}{1 + e_o} H_o \log \frac{\sigma'_{vo} + \Delta\sigma}{\sigma'_p}$$

$$= \frac{0.022}{1 + 0.725} (12\text{ m}) \log \frac{190}{130} + \frac{0.262}{1 + 0.725} (12\text{ m}) \log \frac{130 + 220}{190}$$

$$= 0.025\text{ m} + 0.484\text{ m}$$

$$= 0.509\text{ m} \simeq 0.5\text{ m}$$

8.10 SOIL PROFILES

In Table 8-1 we list some of the causes of preconsolidation in soil deposits. In this section, we study some typical soil profiles from various parts of the world and indicate their preconsolidation stresses as well as their effective vertical overburden stresses with depth. These overburden stress profiles were calculated just like those of Chapter 7, using the densities and thicknesses of the soil layers as well as the depths to the water table. To perform a detailed settlement analysis, typical profiles such as these are established for the proposed site and are based on subsurface investigations, undisturbed sampling, and laboratory testing. The typical soil profiles are presented in Figs. 8.14 through 8.18.

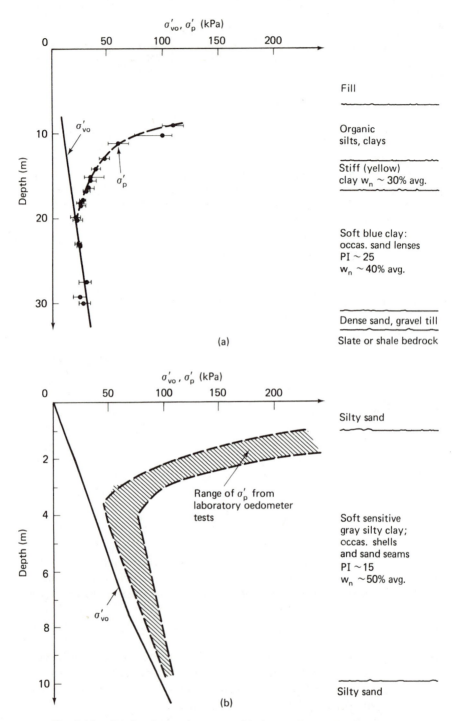

Fig. 8.14   Overburden and preconsolidation stress profiles for marine clays of the Boston area: (a) Mystic power station (after Casagrande and Fadum, 1944); (b) I-95 test section, Portsmouth, NH (after Ladd, 1972).

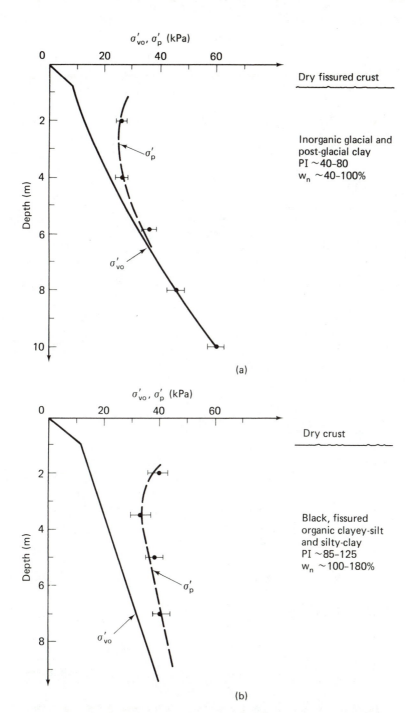

Fig. 8.15 Overburden and preconsolidation profiles for two Swedish clays: (a) Skå-Edeby test field near Stockholm (after Holm and Holtz, 1977); (b) Kalix test site (after Holtz and Holm, 1979).

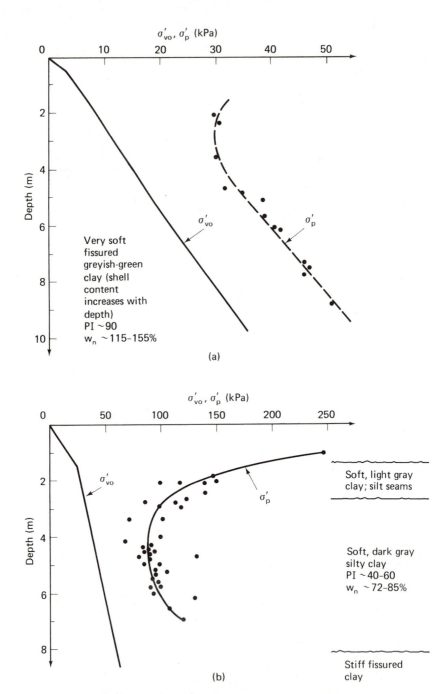

Fig. 8.16 Overburden and preconsolidation stress profiles for marine clays near Bangkok, Thailand: (a) Bangkok-Siracha highway (after Eide and Holmberg, 1972); (b) Asian Institute of Technology (after Moh, Brand, and Nelson, 1972).

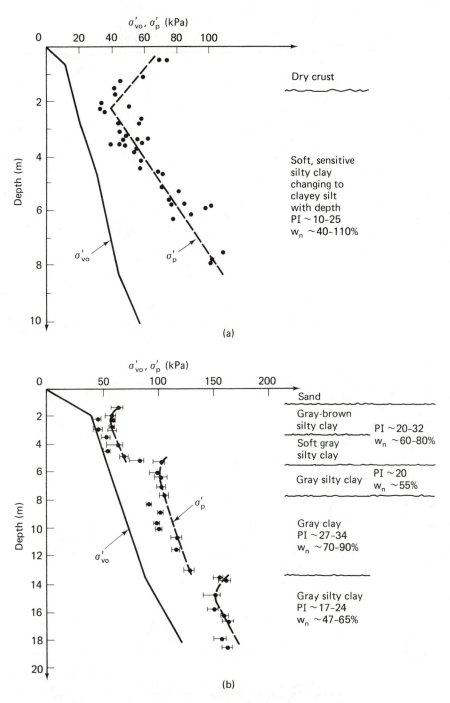

Fig. 8.17 Overburden and preconsolidation stress profiles for Lake Champlain deposits (Laurentian or Leda clays) of Eastern Canada: (a) Saint-Alban, Québec, test fills (after Leroueil, et al., 1978a); (b) C. F. S. Gloucester test (after Bozozuk and Leonards, 1972).

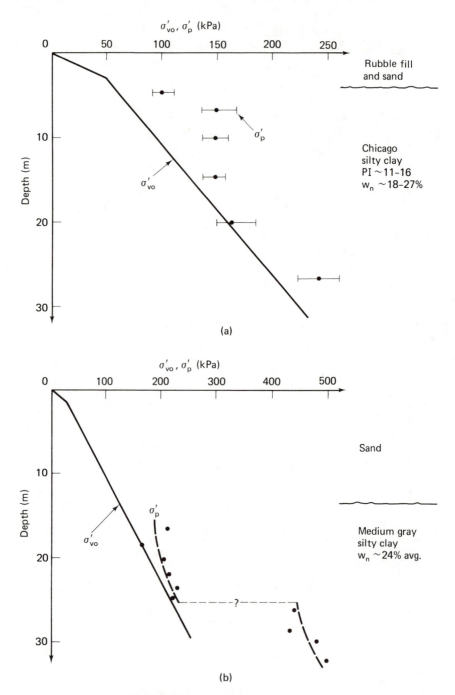

Fig. 8.18 Overburden and preconsolidation stress profiles for glacial lake clays (reworked till?) of the Chicago area: (a) Chicago "Loop" (after data from Prof. J. O. Osterberg's graduate soil mechanics class, Northwestern University, 1966); (b) Hammond, Indiana (after Osterberg, 1963).

8.11 APPROXIMATE METHODS AND TYPICAL VALUES OF COMPRESSION INDICES

Because of the time and expense involved in consolidation testing, it is sometimes desirable to be able to relate the compression indices to the simple classification properties of soils. These relationships are also commonly used for preliminary designs and estimates and for checking the validity of test results.

Table 8-2 is a list of some published equations for the prediction of compression indices (Azzouz, Krizek, and Corotis, 1976).

TABLE 8-2 Some Empirical Equations for C_c and $C_{c\epsilon}$*

Equation	Regions of Applicability
$C_c = 0.007 \, (LL - 7)$	Remolded clays
$C_{c\epsilon} = 0.208 e_o + 0.0083$	Chicago clays
$C_c = 17.66 \times 10^{-5} w_n^2 + 5.93 \times 10^{-3} w_n - 1.35 \times 10^{-1}$	Chicago clays
$C_c = 1.15(e_o - 0.35)$	All clays
$C_c = 0.30(e_o - 0.27)$	Inorganic, cohesive soil; silt, some clay; silty clay; clay
$C_c = 1.15 \times 10^{-2} w_n$	Organic soils—meadow mats, peats, and organic silt and clay
$C_c = 0.75(e_o - 0.50)$	Soils of very low plasticity
$C_{c\epsilon} = 0.156 e_o + 0.0107$	All clays
$C_c = 0.01 w_n$	Chicago clays

*As summarized by Azzouz, Krizek, and Corotis (1976).
Note: w_n = natural water content.

Terzaghi and Peck (1967) proposed the following equation, based on research on undisturbed clays of low to medium sensitivity:

$$C_c = 0.009 \, (LL - 10) \qquad (8\text{-}21)$$

which has a reliability range of about ±30%. This equation is widely used, despite its wide reliability range, to make initial consolidation settlement estimates. The equation should not be used where the sensitivity of the clay is greater than 4, if the LL is greater than 100, or if the clay contains a high percentage of organic matter. Some typical values of the compression index, based on our experience and the geotechnical literature, are listed in Table 8-3.

Often, C_r is assumed to be 5% to 10% of C_c. Typical values of C_r range from 0.015 to 0.035 (Leonards, 1976). The lower values are for clays of lower plasticity and low OCR. Values of C_r outside the range of 0.005 to 0.05 should be considered questionable.

TABLE 8-3   Typical Values of the Compression Index C_c

Soil	C_c
Normally consolidated medium sensitive clays	0.2 to 0.5
Chicago silty clay (CL)	0.15 to 0.3
Boston blue clay (CL)	0.3 to 0.5
Vicksburg buckshot clay (CH)	0 5 to 0.6
Swedish medium sensitive clays (CL-CH)	1 to 3
Canadian Leda clays (CL-CH)	1 to 4
Mexico City clay (MH)	7 to 10
Organic clays (OH)	4 and up
Peats (Pt)	10 to 15
Organic silt and clayey silts (ML-MH)	1.5 to 4.0
San Francisco Bay Mud (CL)	0.4 to 1.2
San Francisco Old Bay clays (CH)	0.7 to 0.9
Bangkok clay (CH)	0.4

8.12 STRESS DISTRIBUTION

In the previous sections of this chapter when we calculated settlements, the increase in stress $\Delta\sigma$ caused by an applied load was given. In this section, we shall show you how to estimate the stress increase in the soil due to boundary or surface loads.

Suppose a very large area such as a subdivision or shopping mall is to be filled with several metres of select compacted material. In this instance, the loading is *one dimensional*, and the stress increase felt at depth would be 100% of the applied stress at the surface. However, near the edge or end of the filled area you might expect a certain amount of attenuation of stress with depth because no stress is applied beyond the edge. Likewise, with a footing of limited size, the applied stress would dissipate rather rapidly with depth.

One of the simplest methods to compute the distribution of stress with depth for a loaded area is to use the *2 to 1 (2:1) method*. This is an empirical approach based on the assumption that the area over which the load acts increases in a systematic way with depth. Since the same vertical force is spread over an increasingly larger area, the unit stress decreases with depth, as shown in Fig. 8.19. In Fig. 8.19a, a strip or continuous footing is seen in elevation view. At a depth z, the enlarged area of the footing increases by $z/2$ on each side. The width at depth z is then $B + Z$, and the stress σ_z at that depth is

$$\sigma_z = \frac{\text{load}}{(B+z) \times 1} = \frac{\sigma_o(B \times 1)}{(B+z) \times 1} \qquad (8\text{-}22)$$

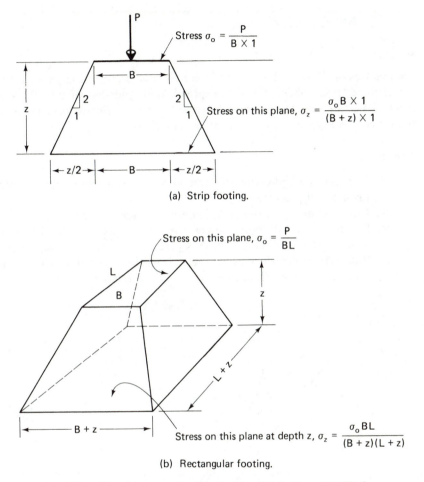

(a) Strip footing.

(b) Rectangular footing.

Fig. 8.19 The 2:1 approximation for the distribution of vertical stress with depth.

where σ_o is the surface or contact stress.

By analogy, a rectangular footing of width B and length L would have an area of $(B + Z)(L + Z)$ at a depth z, as shown in Fig. 8.19b. The corresponding stress at depth z would be

$$\Delta\sigma_z = \frac{\text{load}}{(B + z)(L + z)} = \frac{\sigma_o BL}{(B + z)(L + z)} \qquad (8\text{-}23)$$

Example 8.17 illustrates the use of the 2:1 method.

EXAMPLE 8.17

Given:

Two metres of fill ($\rho = 2.04$ Mg/m^3) are compacted over a large area. On top of the compacted fill, a 3 × 4 m spread footing loaded with 1400 kN is placed. Assume that the average density of the soil prior to placement of the fill is 1.68 Mg/m^3. The water table is very deep.

Required:

a. Compute and plot the effective vertical stress profile with depth prior to fill placement.
b. Compute and plot the added stress, $\Delta\sigma$, due to the fill.
c. Compute the additional stress with depth due to the 3 × 4 m footing when the footing base is placed 1 m below the top of the filled ground surface. Use the 2:1 method. (Assume weight of footing plus backfill equals weight of soil removed.)

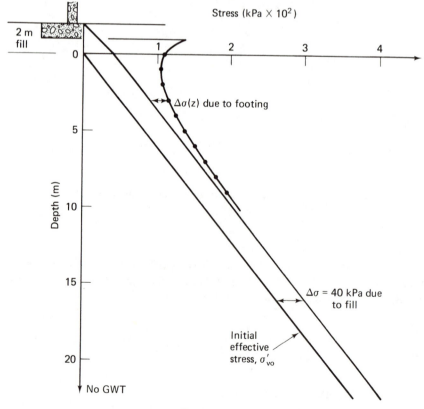

Fig. Ex. 8.17a

Fig. Ex. 8.17b

(1)	(2)	(3)	(4)	(5)
z (m)	(B + z)	(L + z)	Area (m²)	$\Delta\sigma(z)$ (kPa)
0	3	4	12	117
1	4	5	20	70
2	5	6	30	47
3	6	7	42	33
4	7	8	56	25
5	8	9	72	19
6	9	10	90	16
7	10	11	110	13
8	11	12	132	11
9	12	13	156	9
10	13	14	182	8

Note: z taken below bottom of footing.

Fig. Ex. 8.17b

Solution:

a. Just as you did in Chapter 7, the initial effective stress distribution is calculated and plotted in Fig. Ex. 8.17a. The stress is zero at zero depth and 330 kPa at a depth of 20 m ($\rho g z = 1.68 \times 9.81 \times 20 = 330$ kPa).

b. The added stress due to the 2 m fill is $2 \times 2.04 \times 9.81 = 40$ kPa. This is shown in Fig. Ex. 8.17a by the line parallel to the in situ vertical effective stress line. Notice that at any depth, the additional stress due to the fill is a constant 40 kPa because the fill is large in areal extent and thus 100% of its influence is felt throughout.

c. The contact stress σ_o between the footing and the soil equals the column load, 1400 kN, divided by the footing area, 3×4 m, or 12 m², or

$$\sigma_o = \frac{\text{load}}{\text{area}} = \frac{1400 \text{ kN}}{12 \text{ m}^2} = 117 \text{ kN/m}^2 \text{ or kPa}$$

Using the 2:1 method, a tabulation of how the stress changes with depth z is shown in Fig. Ex. 8.17b. The change in stress, $\Delta\sigma(z)$, in column 5 is added to the change in stress due to the fill in Fig. Ex. 8.17a. It can be seen that the stress due to the footing diminishes quite rapidly with depth.

The *theory of elasticity* is also used by foundation engineers to estimate stresses within soil masses. The soil does not necessarily have to be elastic for the theory to work, at least for vertical stresses; only the ratio of stress to strain should be constant. As long as the added stresses are well below failure, the strains are still approximately proportional to stresses.

In 1885, Boussinesq developed equations for the state of stress within a homogeneous, isotropic, linearly elastic half-space for a *point load* acting perpendicular to the surface. The value of the vertical stress is

$$\sigma_z = \frac{Q(3z^3)}{2\pi(r^2 + z^2)^{5/2}} \tag{8-24}$$

where Q = point load,
z = depth from ground surface to the place where σ_z is desired, and
r = horizontal distance from point load to the place where σ_z is desired.

Equation 8-24 may also be written as

$$\sigma_z = \frac{Q}{z^2} N_B \tag{8-25}$$

where N_B is an influence factor which combines the constant terms in Eq. 8-24 and is a function of r/z.

These terms are illustrated in Fig. 8.20a; values of N_B versus r/z are shown in Fig. 8.20b. Boussinesq also derived equations for the radial, tangential, and shear stress; these can be found in most advanced textbooks on soil mechanics. Note that the equation for σ_z is independent of the material; the modulus does not enter into the equation at all.

By integrating the point load equation along a line, the stress due to a *line load* (force per unit length) may be found. In this case, the value of the

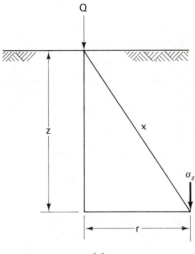

(a)

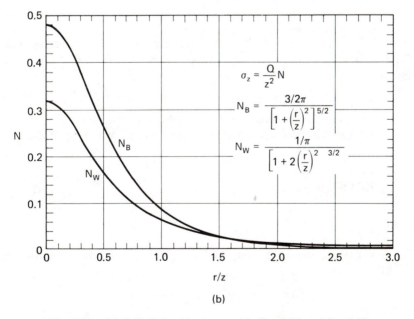

$$\sigma_z = \frac{Q}{z^2} N$$

$$N_B = \frac{3/2\pi}{\left[1 + \left(\frac{r}{z}\right)^2\right]^{5/2}}$$

$$N_W = \frac{1/\pi}{\left[1 + 2\left(\frac{r}{z}\right)^2\right]^{3/2}}$$

(b)

Fig. 8.20   (a) Definition of terms used in Eq. 8-25 and Eq. 8-26; (b) relationship between N_B, N_W, and r/z for a point load (after Taylor, 1948).

vertical stress is

$$\sigma_z = \frac{2P}{\pi} \frac{z^3}{x^4} \tag{8-26}$$

where P = line load, and
  $x = (z^2 + r^2)^{1/2}$ (see Fig. 8.20a).
Equations for the horizontal and shear stress are also available.

The next logical step is to integrate a line load over a finite area. Newmark (1935) performed the integration of Eq. 8.26 and derived the following equation for the vertical stress under the corner of a *uniformly loaded rectangular area*:

$$\sigma_z = q_o \frac{1}{4\pi} \left[\frac{2mn(m^2 + n^2 + 1)^{1/2}}{m^2 + n^2 + 1 + m^2 n^2} \times \frac{(m^2 + n^2 + 2)}{(m^2 + n^2 + 1)} \right.$$

$$\left. + \arctan \frac{2mn(m^2 + n^2 + 1)^{1/2}}{m^2 + n^2 + 1 - m^2 n^2} \right] \tag{8-27}$$

where q_o = surface or contact stress,
  $m = x/z$, $\tag{8-28}$
  $n = y/z$, and $\tag{8-29}$
  x, y = length and width of the uniformly loaded area, respectively.
The parameters m and n are interchangeable. Fortunately Eq. 8-27 may be rewritten as

$$\sigma_z = q_o I \tag{8-30}$$

where I = an influence value which depends on m and n.
  Values of I for various values of m and n are shown in Fig. 8.21.

EXAMPLE 8.18

Given:

The 3 × 4 m rectangular footing of Example 8.17 is loaded uniformly by 117 kPa.

Required:

  a. Find the vertical stress under the corner of the footing at a depth of 2 m.

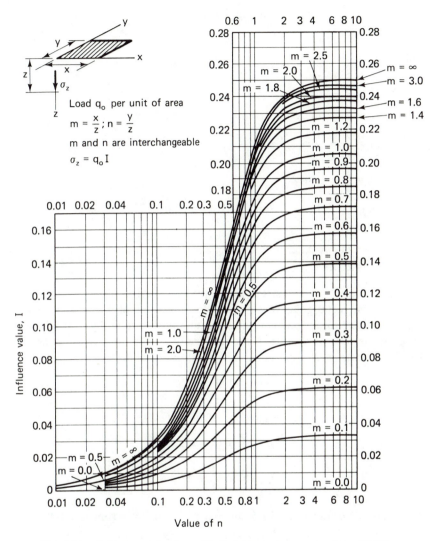

Load q_o per unit of area

$m = \dfrac{x}{z}$; $n = \dfrac{y}{z}$

m and n are interchangeable

$\sigma_z = q_o I$

Value of n

Fig. 8.21   Influence value for vertical stress under corner of a uniformly loaded rectangular area (after U.S. Navy, 1971).

  b. Find the vertical stress under the center of the footing at a depth of 2 m.

  c. Compare results with Fig. Ex. 8.17a.

Solution:

  a. $x = 3$ m

  $y = 4$ m

  $z = 2$ m: therefore from Eqs. 8-28 and 8-29,

  $$m = \frac{x}{z} = \frac{3}{2} = 1.5$$

  $$n = \frac{y}{z} = \frac{4}{2} = 2$$

  From Fig. 8.21, find $I = 0.223$. From Eq. 8-30,

  $\sigma_z = q_o I$

  $= 117 \times 0.223$

  $= 26$ kPa

  b. To compute the stress under the center, it is necessary to divide the 3×4 m rectangular footing into four sections of 1.5×2 m in size. Find the stress under one corner and multiply this value by 4 to take into account the four quadrants of the uniformly loaded area. We can do this because, for an elastic material, superposition is valid.

$$x = 1.5 \text{ m}$$

$$y = 2 \text{ m}$$

$$z = 2 \text{ m; then}$$

$$m = \frac{x}{z} = \frac{1.5}{2} = 0.75$$

$$n = \frac{y}{z} = \frac{2}{2} = 1$$

The corresponding value of I from Fig. 8.21 is 0.159. From Eq. 8-30,

$$\sigma_z = 4q_o I = 4 \times 117 \times 0.159 = 74 \text{ kPa}$$

Thus the vertical stress under the center for this case is about three times that under the corner. This seems reasonable since the center is loaded from all sides but under the corner it is not.

  c. At a depth of 2 m below the 3×4 m footing, the vertical stress according to the 2:1 theory is 47 kPa (see Fig. Ex. 8.17b). This value represents the *average* stress beneath the footing at -2 m. The average of the corner and center stress by elastic theory is $(26 + 74.2)/2 = 50.1$ kPa.

Thus the 2:1 method underestimates the vertical stress at the center but overestimates σ_z at the corners.

Suppose we want to find the vertical stress at some depth z *outside* the loaded area. Under these conditions we merely fabricate other uniformly loaded rectangles, all with corners above the point where the vertical stress is desired, and subtract and add their stress contributions as necessary.

EXAMPLE 8.19

Given:

A 5×10 m area uniformly loaded with 100 kPa.

Required:
  a. Find the stress at a depth of 5 m under point A in Fig. Ex. 8.19.
  b. Find the stress at point A if the right half of the 5×10 m area were loaded with an additional 100 kPa.

Solution:

  a. Refer to Fig. Ex. 8.19 and the numbered points as shown. Add the rectangles in the following manner (+ for loaded areas and − for unloaded areas): $+A123 - A164 - A573 + A584$ result in the loaded rectangle we want 8627. Find four separate influence values from Fig. 8.21 for each rectangle at a depth of 5 m, then add and subtract the computed stresses. Note that it is necessary to add rectangle A584 because it was subtracted twice as part of rectangles A164 and A573.

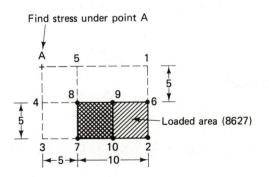

Fig. Ex. 8.19

The computations are shown in the following table.

Item \ Area	+A123	−A164	−A573	+A584
x	15	15	10	5
y	10	5	5	5
z	5	5	5	5
$m = x/z$	3	3	2	1
$n = y/z$	2	1	1	1
I	0.238	0.209	0.206	0.180
σ_z	23.8	− 20.9	− 20.6	+ 18.0

Total $\sigma_z = 23.8 - 20.9 - 20.6 + 18.0 = 0.3$ kPa

b. When rectangle $789\overline{10}$ is loaded with 100 kPa and rectangle $962\overline{10}$ is loaded with 200 kPa, repeat part (a) above to obtain the stress under point A at 5 m depth for the entire rectangle 8627 loaded with 100 kPa. Next, a second set of four rectangles would have to be calculated just as for part (a) above but only rectangle $962\overline{10}$ would be loaded with +100 kPa; the others would be −100 kPa. The total σ_z equals 0.3 kPa from part (a) plus $23.8 - 21.0 - 23.2 + 20.6$ or 0.5 kPa.

Thus it is possible to find the stress at any depth z, in or around a uniformly loaded area or even under a step loaded area, by using the procedures outlined in Examples 8.18 and 8.19. Remember that a new set of calculations is required for each depth where σ_z is desired.

Similar procedures are available for vertical stresses under *uniformly loaded circular areas*. Use Fig. 8.22 to obtain influence values in terms of x/r and z/r, where $z =$ depth,

$r =$ radius of uniformly loaded area,
$x =$ horizontal distance from the center of the circular area, and
$q_0 =$ surface contact pressure, in kPa.

EXAMPLE 8.20

Given:

A circular tank 3.91 m in diameter is uniformly loaded with 117 kPa.

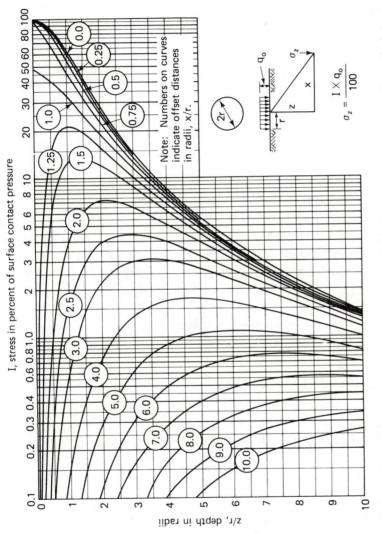

Fig. 8.22   Influence values, expressed in percentage of surface contact pressure, q_o, for vertical stress under uniformly loaded circular area (after Foster and Ahlvin, 1954, as cited by U.S. Navy, 1971).

Required:

  a. Compute the stress under the center of the tank at a depth of 2 m below the tank.
  b. Compute the stress under the edge of the tank, also a depth of 2 m.

Solution:

  a. Refer to Fig. 8.22.

$$z = 2 \text{ m}$$
$$r = 3.91/2 = 1.95 \text{ m}$$
$$x = 0; \text{ then}$$
$$z/r = 2/1.95 = 1.02$$
$$x/r = 0/1.95 = 0.$$

Find $I = 0.63$. Using Eq. 8-30 we obtain,

$$\sigma_z = q_o I = 117 \times 0.63 = 74 \text{ kPa}$$

(This compares exactly with $\sigma_z = 74$ kPa at the center for a 3×4 m rectangular area in Example 8.18. In both cases, the area is 12 m².)

  b. Again, refer to Fig. 8.22. For the edge of the circular loaded area:

$$z = 2 \text{ m}$$
$$r = 1.95 \text{ m}$$
$$x = r = 1.95 \text{ m}$$
$$z/r = 2/1.95 = 1.02$$
$$x/r = 1.0$$

Find $I = 0.33$; then using Eq. 8-30,

$$\sigma_z = q_o I = 117 \times 0.33 = 39 \text{ kPa}$$

(This compares with $\sigma_z = 26$ kPa at a corner for a 3×4 m uniformly loaded rectangular area. In both cases, the area of the loaded area is the same.)

———————

Another useful integration of the Boussinesq equations is the trapezoidal loading shown in Fig. 8.23, which models the loading caused by a *long embankment*. Influence values are in terms of the dimensions a and b, as shown in the figure. If the embankment is not infinitely long, then use Fig. 8.24 together with Fig. 8.21 to represent different load configurations.

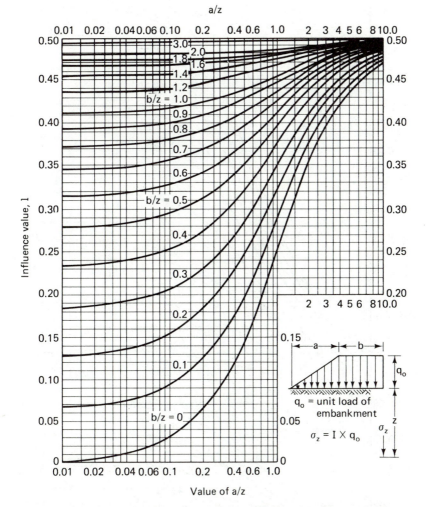

Fig. 8.23 Influence values for vertical stress under a very long embankment; length = ∞ (from U.S. Navy, 1971, after Osterberg, 1957).

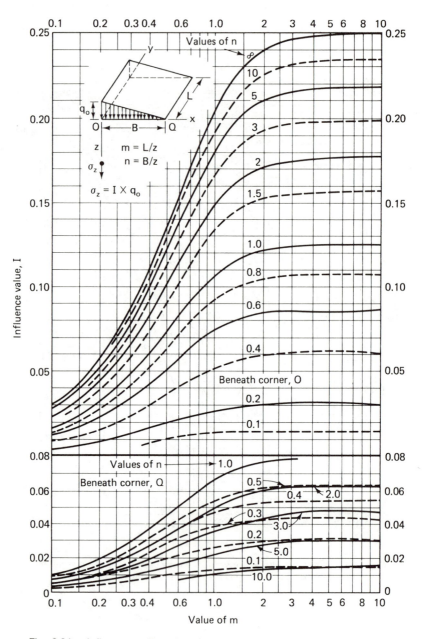

Fig. 8.24 Influence values for vertical stress under the corners of a triangular load of limited length (after U.S. Navy, 1971).

EXAMPLE 8.21

Given:

A highway embankment, as shown in Fig. Ex. 8.21. Assume the average density of the material in the embankment is 2.0 Mg/m³.

Required:

Compute the vertical stress under the centerline at depths of 3 and 6 m.

Solution:

First, calculate the applied surface stress q_o and the dimensions of the embankment in terms of a and b.

$$q_o = \rho g h = 2.0 \text{ Mg/m}^3 \times 9.81 \text{ m/s}^2 \times 3 \text{ m} = 59 \text{ kPa}$$

From Fig. 8.23, and Fig. Ex. 8.21,

$$b = 5 \text{ m}$$
$$a = 2 \times 3 \text{ m} = 6 \text{ m}$$

Next, calculate the vertical stress for $z = 3$ m.

$$a/z = 6/3 = 2$$
$$b/z = 5/3 = 1.67$$

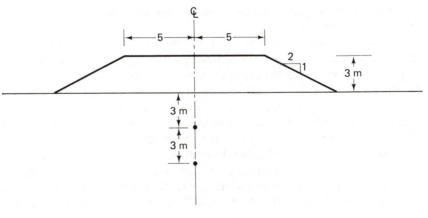

Fig. Ex. 8.21

From Fig. 8.23, $I = 0.49$,

$$\sigma_z = q_o I = 59 \text{ kPa} \times 0.488 = 29 \text{ kPa}$$

for one-half of the embankment, or 58 kPa for the entire embankment. Thus at this shallow depth σ_z is almost the same as the contact stress.

Finally, calculate the vertical stress for $z = 6$ m.

$$a/z = 6/6 = 1$$
$$b/z = 5/6 = 0.83$$

From Fig. 8.23, $I = 0.44$,

$$\sigma_z = q_o I = 59 \text{ kPa} \times 0.44 \times 2 = 52 \text{ kPa}$$

Now and then it becomes necessary to compute the vertical stress due to an irregularly shaped loaded area at various points inside and/or outside an area. To facilitate computations, Newmark (1942) developed *influence charts* from which the vertical stress (and even the horizontal and shear stresses) may be computed. These influence charts are based on Boussinesq's theory, although similar charts have been prepared for the Westergaard theory, to be discussed shortly. Examples of influence charts may be found in foundation engineering textbooks, for example, Leonards (1962) and Peck, Hanson, and Thornburn (1974). Figure 8-25 shows the Newmark influence chart for the computation of vertical stresses due to a loaded area. Think of the chart as a contour map that shows a volcano, the top of which is located at the center (O) of the influence chart. If it were possible to look normal to a three-dimensional surface of the chart, you would see that each of the "areas" or "blocks" has the *same surface area*. We see only the projection on the contour map; the blocks grow smaller as the center is approached.

The charts are scaled with respect to depth so that they may be used for a structure of any size, in the following manner. On the chart is the line OQ. This line represents the distance below the ground surface z for which the vertical stress σ_v is desired, and this distance is used as the scale for a drawing of the loaded area. The vertical stress is computed by merely counting the number of areas or blocks on the chart, *within* the boundary of the loaded area that is drawn to the proper scale and then placed upon the chart. The number of areas is multiplied by an influence value I, specified on the chart, and by the contact pressure to obtain the stress at the desired depth. The point at which the vertical stress is desired is placed *over the center* of the chart. Example 8.22 illustrates the use of the chart.

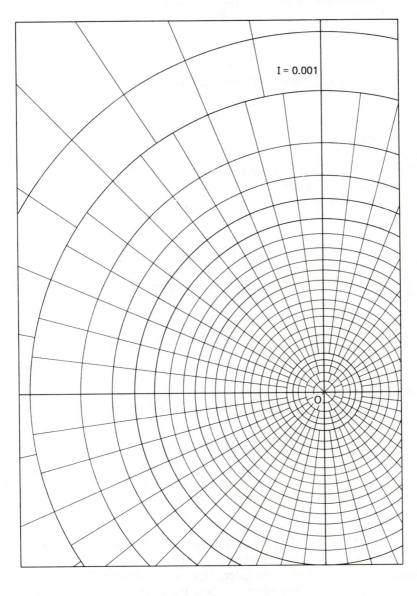

I = 0.001

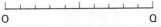

O   Q

Scale of distance OQ =
depth z at which stress is computed

Fig. 8.25   Influence chart for vertical stress on horizontal planes (after
Newmark, 1942).

EXAMPLE 8.22

Given:

A uniform stress of 250 kPa is applied to the loaded area shown in Fig. Ex. 8.22a.

Required:

Compute the stress at a depth of 80 m below the ground surface due to the loaded area under point O'.

Solution:

Draw the loaded area such that the length of the line $\overline{OQ}$ is scaled to 80 m. For example, the distance $\overline{AB}$ in Fig. Ex. 8.22a is 1.5 times the distance $\overline{OQ}$. $\overline{OQ} = 80$ m and $\overline{AB} = 120$ m. Next, place point O', the point where the stress is required, over the center of the influence chart (as shown in Fig. Ex. 8.22b to a slightly smaller scale). The number of blocks (and partial blocks) are counted under the loaded area. In this case, about eight blocks are found. The vertical stress at 80 m is then indicated by

$$\sigma_v = q_o I \times \text{No. of blocks} \qquad (8\text{-}31)$$

where q_o = surface or contact stress, and
I = influence value per block (0.02 in Fig. Ex. 8.22b).
Therefore,

$$\sigma_v = 250 \text{ kPa} \times 0.02 \times 8 \text{ blocks} = 40 \text{ kPa}$$

To compute the stress at other depths, the process is repeated by making other drawings for the different depths, changing the scale *each time* to correspond to the distance $\overline{OQ}$ on the influence chart.

All of the preceding stress distribution solutions were integrations of the original Boussinesq equations for vertical stress in a homogeneous isotropic linearly elastic half-space. Natural soil deposits do not approach these ideal material conditions. In fact, many important sedimentary soil deposits were formed by the aggradation of alternate horizontal layers of silts and clays. These deposits are called *varved clays*, and the solution for stresses at a point developed by Westergaard (1938) may be more applicable. In this theory, an elastic soil is interspersed with infinitely thin but perfectly rigid layers that allow only vertical movement but no lateral movement. Westergaard's solution for the vertical stress for a *point load*

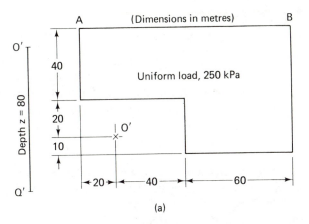

A (Dimensions in metres) B

O′

40

Uniform load, 250 kPa

Depth z = 80

20

O′
×

10

Q′

◄20►◄── 40 ──►◄──── 60 ────►

(a)

Fig. Ex. 8.22a   (After Newmark, 1942.)

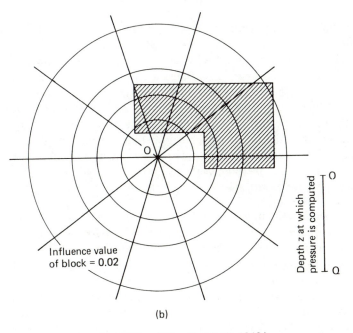

O

Influence value
of block = 0.02

Depth z at which
pressure is computed

O

Q

(b)

Fig. Ex. 8.22b   (After Newmark, 1942.)

(for Poisson's ratio $\nu = 0$) is

$$\sigma_z = \frac{Q}{z^2 \pi} \frac{1}{\left[1 + 2\left(\dfrac{r}{z}\right)^2\right]^{3/2}} \tag{8-32}$$

where the terms were defined in Fig. 8.20 and Eq. 8-24. *Poisson's ratio, ν,* is defined as the the ratio of the horizontal strain, ϵ_h, to the vertical strain, ϵ_v, or

$$\nu = \frac{\epsilon_h}{\epsilon_v} \tag{8-33}$$

Typical values of ν for silts and sands range from 0.2 for loose materials to 0.4 for dense materials. Values for saturated clays vary from about 0.40 to 0.5. The theoretical maximum for a saturated clay undergoing no volume change when stressed (undrained) is 0.5.

Equation 8-32 may be written as

$$\sigma_z = \frac{Q}{z^2} N_W \tag{8-34}$$

where N_W is an influence factor combining terms in Eq. 8-32 and is a function of r/z. Values of N_W are plotted in Fig. 8.20b.

The Boussinesq and Westergaard theories are compared in Fig. 8.20b. For r/z less than 1.5, Boussinesq indicates values larger than Westergaard. When $r/z \geqslant 1.5$, both theories provide about the same results. Which theory should you use? From a philosophical point of view, both theories are based on assumptions which are far from reality. It often boils down to a matter of personal preference, even though the assumptions of the Westergaard theory probably are closer to reality for a layered soil deposit. The 2:1 method, crude as it may be, is probably used about as often in practice as the solutions from the theory of elasticity for estimating vertical stresses.

A graph similar to Fig. 8.21 for influence values for vertical stress under a corner of a *uniformly loaded rectangular area* has been prepared for the Westergaard case (for Poisson's ratio = 0) and is shown as Fig. 8.26. You use it as you would use Fig. 8.21.

Tables 8-4 through 8-6 present the influence values for vertical stress under the center of a *square load*, under the center of an infinitely long *strip load*, and under the corner of a *uniformly loaded rectangular area*, respectively. These tables present influence coefficients for both the Boussinesq and Westergaard assumptions. You may find these charts useful in engineering practice.

It must be pointed out that once you have found the vertical stresses from the equations and charts provided in this section, they must be *added* to the existing in situ overburden effective stress, as was done in Example 8.17. This procedure is necessary because the elastic solutions consider the half-space to be weightless and only the stress due to an external loading is

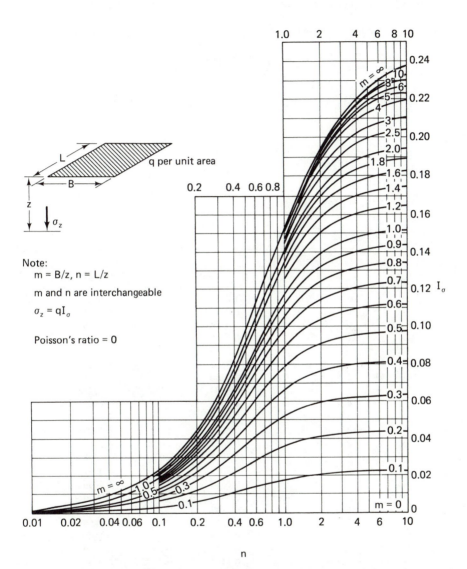

Fig. 8.26   Influence values for vertical stress under corners of a uniformly loaded rectangular area for the Westergaard theory (after Duncan and Buchignani, 1976).

TABLE 8-4 Influence Values for Vertical Stress Under the Center of a Square Uniformly Loaded Area*

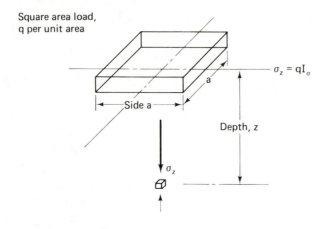

a/z	I_σ	
	Boussinesq	Westergaard
∞	1.0000	1.0000
20	0.9992	0.9365
16	0.9984	0.9199
12	0.9968	0.8944
10	0.9944	0.8734
8	0.9892	0.8435
6	0.9756	0.7926
5	0.9604	0.7525
4	0.9300	0.6971
3.6	0.9096	0.6659
3.2	0.8812	0.6309
2.8	0.8408	0.5863
2.4	0.7832	0.5328
2.0	0.7008	0.4647
1.8	0.6476	0.4246
1.6	0.5844	0.3794
1.4	0.5108	0.3291
1.2	0.4276	0.2858
1.0	0.3360	0.2165
0.8	0.2410	0.1560
0.6	0.1494	0.0999
0.4	0.0716	0.0477
0.2	0.0188	0.0127
0	0.0000	0.0000

*After Duncan and Buchignani (1976).

TABLE 8-5 Influence Values for Vertical Stress Under the Center of an Infinitely Long Strip Load*

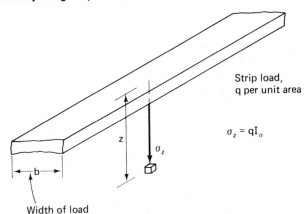

b/z	I_σ	
	Boussinesq	Westergaard
∞	1.000	1.000
100	1.000	0.990
10	0.997	0.910
9	0.996	0.901
8	0.994	0.888
7	0.991	0.874
6.5	0.989	0.864
6.0	0.986	0.853
5.5	0.983	0.835
5.0	0.977	0.824
4.5	0.970	0.807
4.0	0.960	0.784
3.5	0.943	0.756
3.0	0.920	0.719
2.5	0.889	0.672
2.0	0.817	0.608
1.5	0.716	0.519
1.2	0.624	0.448
1.0	0.550	0.392
0.8	0.462	0.328
0.5	0.306	0.216
0.2	0.127	0.089
0.1	0.064	0.045
0	0.000	0.000

*After Duncan and Buchignani (1976).

TABLE 8-6 Influence Values for Vertical Stress Under Corner of a Uniformly Loaded Rectangular Area*

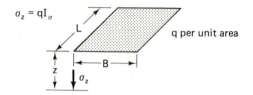

$\sigma_z = qI_\sigma$

L

B

z

σ_z

q per unit area

Boussinesq Case

B/z	L/z							
	0.1	0.2	0.4	0.6	0.8	1.0	2.0	∞
0.1	0.005	0.009	0.017	0.022	0.026	0.028	0.031	0.032
0.2	0.009	0.018	0.033	0.043	0.050	0.055	0.061	0.062
0.4	0.017	0.033	0.060	0.080	0.093	0.101	0.113	0.115
0.6	0.022	0.043	0.080	0.107	0.125	0.136	0.153	0.156
0.8	0.026	0.050	0.093	0.125	0.146	0.160	0.181	0.185
1.0	0.028	0.055	0.101	0.136	0.160	0.175	0.200	0.205
2.0	0.031	0.061	0.113	0.153	0.181	0.200	0.232	0.240
∞	0.032	0.062	0.115	0.156	0.185	0.205	0.240	0.250

Westergaard Case

B/z	L/z							
	0.1	0.2	0.4	0.6	0.8	1.0	2.0	∞
0.1	0.003	0.006	0.011	0.014	0.017	0.018	0.021	0.022
0.2	0.006	0.012	0.021	0.028	0.033	0.036	0.041	0.044
0.4	0.011	0.021	0.039	0.052	0.060	0.066	0.077	0.082
0.6	0.014	0.028	0.052	0.069	0.081	0.089	0.104	0.112
0.8	0.017	0.033	0.060	0.081	0.095	0.105	0.125	0.135
1.0	0.018	0.036	0.066	0.089	0.105	0.116	0.140	0.152
2.0	0.021	0.041	0.077	0.104	0.125	0.140	0.174	0.196
∞	0.022	0.044	0.082	0.112	0.135	0.152	0.196	0.250

*After Duncan and Buchignani (1976).

considered. Further, for sites where a layered subsoil exists, that is, where there are large variations in the modulus of elasticity, other solutions must be used to take into account the relative rigidities of the layers. Solutions to these kinds of stress distributions may be found in Harr (1966) and Poulos and Davis (1974). These references also supply equations and charts for estimating the horizontal and shear stresses in elastic media.

PROBLEMS

8-1. For the e versus log σ curves of Fig. 8.8a, compute the compression indices. Explain why H is possible to get slightly different answers than those shown at the bottom of the figure.

8-2. Verify the values of the preconsolidation stress shown in Fig. 8.8a.

8-3. Determine the overconsolidation ratio (OCR) for the five fine-grained soils of Fig. 8.8a.

8-4. Verify that the values for the preconsolidation stress and the virgin compression index shown in Fig. 8.8b are correct.

8-5. What is the OCR of the clay till in Fig. 8.8c?

8-6. Estimate the preconsolidation stress for (a) the undisturbed Leda clay in Fig. 8.8d, (b) undisturbed Mexico City clay in Fig. 8.8e (c) undisturbed Chicago clay in Fig. 8.8f, and (d) the swelling clays from Texas in Fig. 8.8g.

8-7. Determine the compression indices for the four soils of Problem 8-6.

8-8. The pressure versus void ratio data determined from a consolidation test on an undisturbed clay specimen are as follows:

Pressure (kPa)	Void Ratio
20	0.953
40	0.948
80	0.938
160	0.920
320	0.878
640	0.789
1280	0.691
320	0.719
80	0.754
20	0.791
0	0.890

(a) Plot the pressure versus void ratio curve on both arithmetic and semilogarithmic graphs.

(b) Determine the equations for the virgin compression curve and for the rebound curve for unloading starting at 1280 kPa.

(c) What are the corresponding modified compression and rebound indices for this soil?

(d) Estimate the stress to which this clay has been preconsolidated. (After A. Casagrande.)

8-9. A building is to be constructed on a 6 m thick stratum of the clay for which consolidation data are given in Problem 8-8. The average existing overburden pressure on this clay stratum is 120 kPa. The average pressure on the clay after construction of the building is 270 kPa.

(a) Estimate the decrease in thickness of the clay stratum caused by full consolidation under the building load.

(b) Estimate the decrease in thickness due to the building load if the clay had never been preconsolidated under a load greater than the existing overburden.

(c) Show on the e versus log σ' plot of Problem 8-8 the values of Δe used for making the estimates in parts (a) and (b). (After A. Casagrande.)

8-10. The compression curve for a certain clay is a straight line on the semilogarithmic plot, and it passes through the point $e = 1.21$, $\sigma'_{vc} = 50$ kPa, and $e = 0.68$, $\sigma'_{vc} = 800$ kPa. Determine an equation for this relationship. (After Taylor, 1948.)

8-11. Show that Eqs. 8-9 and 8-15 are valid.

8-12. The following consolidation test data were obtained from undis-

Stress (kPa)	Dial Reading (mm)	Void Ratio
0	12.700	2.855
5	12.352	2.802
10	12.294	2.793
20	12.131	2.769
40	11.224	2.631
80	9.053	2.301
160	6.665	1.939
320	4.272	1.576
640	2.548	1.314
160	2.951	1.375
40	3.533	1.464
5	4.350	1.589

turbed San Francisco Bay Mud. For this clay, LL = 88, PL = 43, ρ_s = 2.70 Mg/m^3, and w_n = 105.7%. Initially, the specimen height was 2.54 cm and its volume was 75.14 cm^3. Plot the data as percent consolidation versus log pressure. Evaluate the preconsolidation pressure and the modified virgin compression index.

8-13. Plot the data of Problem 8-12, on a void ratio versus log pressure graph. Evaluate the preconsolidation pressure and the virgin compression index. Do these values agree with what you found in Problem 8-12? Comments?

8-14. The initial water content of the sample in Problem 8-12 is 105.7%, and the density of the solids, ρ_s, is 2.70 Mg/m^3. Compute the wet and dry density and degree of saturation of the consolidation test sample if the dry weight of the sample is 52.8 g. If the final water content is 59.6%, compute the degree of saturation and dry density at the end of consolidation.

8-15. A 7.5 m thick layer of soft San Francisco Bay Mud is to be loaded with a granular fill 3 m thick, on the average. The total density of the fill is about 1.9 Mg/m^3. Assume the test data in Problem 8-12 is typical of the clay layer, and that the layer is normally consolidated. What consolidation settlement will take place due to the weight of the fill? Make these calculations using (a) the $C_{c\epsilon}$ determined in Problem 8-12, (b) the C_c determined in Problem 8-13, and (c) directly from the percent consolidation versus log pressure diagram you plotted in Problem 8-12.

8-16. Assume the laboratory test results in Problem 8-12 are typical of another San Francisco Bay Mud site, but where the clay is slightly overconsolidated. The present vertical effective overburden stress is calculated to be about 12 kPa, and the thickness of the clay is 4 m. At this location, the granular fill (ρ = 1.9 Mg/m^3) will be only about 1 m thick. Estimate the consolidation settlement due to the weight of the fill.

8-17. What settlement would you expect at the overconsolidated site in Problem 8-16 if the fill to be constructed were 3 m thick? Do this problem (a) directly from the percent consolidation plot and (b) using Eq. 8-18 or 8-19. How do the results compare?

8-18. Plot the following data and determine the preconsolidation pressure
and the modified compression index.

% Consolidation (compression is +)	Pressure (kPa)
0.08	5
0.10	10
0.11	20
0.24	40
0.89	80
1.74	160
3.81	320
7.32	640
7.30	320
7.12	160
6.55	80
6.67	160
6.91	320
7.59	640
11.50	1280
15.83	2560
20.16	5120
19.68	1280
18.75	160
17.51	40
13.95	5

Specimen height is 25.4 mm, $w_n = 29.3\%$, $\rho_d = 1.50$ Mg/m^3. Sample
is from a depth of -10.7 m.

8-19. At the site where the sample of Problem 8-18 was taken, the soil
profile consists of about 6 m of sand and rubble fill and then 8.5 m
of clay. The water table is about 2 m below the ground surface.
Average densities of the sand and rubble fill are 1.5 Mg/m^3 above
the water table and 1.65 Mg/m^3 below the water table. Estimate the
consolidation settlement if the average stress increase in the compressible layer is (a) 50 kPa, (b) 100 kPa, and (c) 250 kPa. Use both
Eq. 8-19 and your percent compression plot from Problem 8-18, and
compare the results. Comments?

8-20. Plot the following void ratio versus pressure data, and evaluate the
compression index and the recompression index. Determine the
preconsolidation stress.

Void Ratio, e	Pressure (kPa)
1.079	0
1.059	10
1.049	20
1.029	40
1.001	80
0.959	160
0.940	200
0.881	300
0.821	400
0.690	800
0.530	2000
0.570	500
0.620	160
0.717	20

8-21. Use the consolidation data from Problem 8-20 to compute the settlement of a structure that adds 150 kPa to the already existing overburden pressure of 120 kPa at the middle of a 6 m thick layer.

8-22. What would be the settlement of the same structure in Problem 8-21 if the overconsolidation ratio of the clay were 1.0 and $\sigma'_{vo} + \Delta\sigma_v = 270$ kPa at the middepth of the clay layer? Show your work and assumptions on the e versus log σ curve of Problem 8-20.

8-23. The consolidation curve of Fig. Ex. 8.9 is typical of a compressible layer 5 m thick. If the existing overburden pressure is 50 kPa, compute the settlement due to an additional stress of 150 kPa added by a structure.

8-24. For the test data of Problem 8-12, construct the field virgin compression curve using the Schmertmann procedure for an OCR of unity.

8-25. Do Problem 8-24 for an OCR = 2.0.

8-26. At the midpoint of a 7 m thick soil layer, the void ratio is 2.4. Find this point on the field virgin compression curve determined in Problem 8-24. What is the corresponding pressure? If this pressure is *doubled* over the entire site, compute the consolidation settlement of the layer.

8-27. Show that the field virgin compression curve shown on Fig. 8.8c ($C_c = 0.19$) is correct.

8-28. Show that the point of intersection where the laboratory and field virgin compression curves meet for the percent consolidation versus log σ' graph is equal to $0.58e_o/(1 + e_o)$. This intersection is equivalent to the $0.42e_o$ point on the e versus log σ' graph.

8-29. Using the appropriate empirical relationship from Sec. 8.11, estimate the compression and recompression indices and both modified indices for as many of the clays of Fig. 8.8 as you can. How well do the empirical relationships agree with the laboratory data?

8-30. Do Problem 8-29 for the clays in Problems 8-8, 8-12, 8-18, and 8-20. Again, how good is the agreement?

8-31. Compare the stress distribution with depth for (a) a point load of 1000 kN and (b) a 1000 kN load applied over an area of 3×3 m. Plot the results.

8-32. If you used the Boussinesq (or Westergaard) theory for Problem 8-31, do the problem again but use the Westergaard (or Boussinesq) theory instead. Comment on the differences between the two theories.

8-33. Compute the data and draw a curve of σ_z/Q versus depth for points directly below a point load Q. On the same plot draw curves of σ_z/Q versus depth for points directly below the center of square footings with breadths of 5 m and 15 m, respectively, each carrying a uniformly distributed load Q. On the basis of this plot, make a statement relative to the range within which loaded areas may be considered to act as point loads. (After Taylor, 1948.)

8-34. The center of a rectangular area at ground surface has Cartesian coordinates $(0,0)$, and the corners have coordinates $(6,15)$. All dimensions are in metres. The area carries a uniform pressure of 150 kPa. Estimate the stresses at a depth of 20 m below ground surface at each of the following locations: $(0,0)$, $(0,15)$, $(6,0)$, $(6,15)$, and $(10,25)$; obtain values by both Boussinesq and Westergaard methods, and also determine the ratio of the stresses as indicated by the two methods. (After Taylor, 1948.)

8-35. Compare the results of Problem 8-34 with the 2:1 method. Comments?

8-36. Calculate the stress distribution with depth at a point 3 m from the corner (along the longest side) of a rectangularly loaded area 10 by 30 m with a uniform load of 60 kPa. Do by (a) the Boussinesq theory, (b) the Westergaard theory, and (c) the 2:1 method.

8-37. How far apart must two 20 m diameter tanks be placed such that their stress overlap is not greater than 10% of the contact stress at depths of 10, 20, and 30 m?

8-38. Compute the stresses for the data of Example 8.19, parts (a) and (b), using the Newmark chart, Fig. 8.25.

8-39. Work Example 8.21, using superposition of the results of Figs. 8.24 and 8.21. How does your answer compare with the solution for Example 8.21?

8-40. Given the data of Example 8.22. Instead of a load on the surface, compute the depth of an excavation to cause a reduction in stress at the bottom of the excavation of 250 kPa if $\rho = 2$ Mg/m³. The excavation plan area is shown in Fig. Ex. 8.22a.

8-41. For the excavation of Problem 8-40, *estimate* the stress change at a depth of 65 m below the bottom of the excavation at point O'.

8-42. Is the 2:1 method usable for excavations? Why?

8-43. A strip footing 3 m wide is loaded on the ground surface with a pressure equal to 150 kPa. Calculate the stress distribution at depths of 3, 6, and 12 m under the center of the footing. If the footing rested on a normally consolidated cohesive layer whose LL was 84 and whose PL was 50, estimate the settlement of the footing.

8-44. How would the estimated settlement under the center of a 3 × 3 m square footing compare with the settlement of the 3 m wide strip footing in the previous problem, assuming soil conditions were the same? Assume the footing is flexible enough to provide uniform contact pressure to the soil.

8-45. How much difference in the computed settlements is there in Problem 8-44 if the Westergaard theory is used instead of Boussinesq theory?

8-46. A large oil storage tank 100 m in diameter is to be constructed on the soil profile shown in Fig. P8-46. Average depth of the oil in the tank is 20 m, and the specific gravity of the oil is 0.92. Consolidation tests from the clay layer are similar to those given in Problem 8-18. Estimate the maximum total and differential consolidation settlement of the tank. Neglect any settlements in the sand. Work this problem assuming (a) conditions at the middepth of the clay are typical of the entire clay layer, and (b) dividing the clay layer into four or five thinner layers, computing the settlement of each thin layer and summing up by Eq. 8-14. Hint: See Example 9.12.

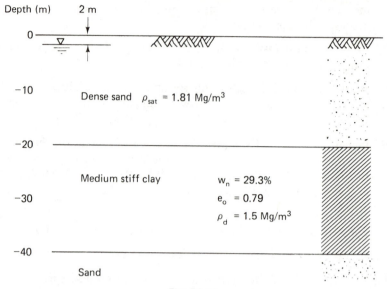

Fig. P8-46

8-47. Estimate the ultimate consolidation settlement under the centerline of a 15 × 15 m mat foundation. The mat is 1 m thick reinforced concrete, and the average stress on the surface of the slab is 76 kPa. The soil profile is shown in Fig. P8-47. Oedometer tests on samples of the clay provide these average values:

$$\sigma_p' = 130 \text{ kPa}, \qquad C_c = 0.40, \qquad C_r = 0.03$$

Neglect any settlements due to the sand layer.

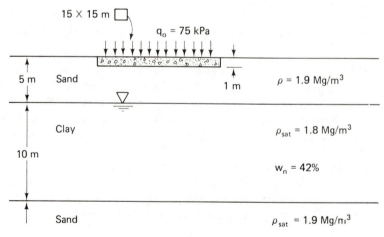

Fig. P8-47

8-48. Three uniformly distributed loads of 100 kPa each are applied to
10 × 10 m square areas on the soil profile shown in Fig. P8-48.
Undisturbed samples of the clay were taken prior to construction,
and consolidation tests indicated that the average preconsolidation
stress is about 110 kPa, the average compression index is 0.50, and
the average recompression index is 0.02. Estimate the total consolidation settlement for the clay layer only under the center of the
middle loaded area.

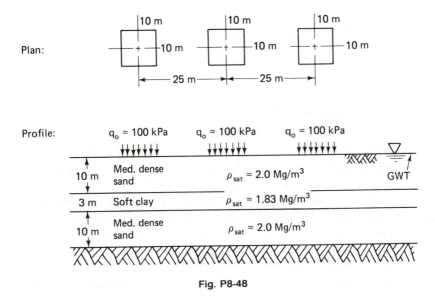

Fig. P8-48

8-49. A series of oil storage tanks are to be constructed near Mystic River
power station in Boston, MA. The typical tank is 20 m in diameter,
and it exerts an average foundation stress of about 100 kPa. The soil
profile at the site is very similar to that shown in Fig. 8.14a. Estimate
both the total and differential consolidation settlement under the
average tank. Hint: See Example 9.12.

8-50. A new highway to Siracha, Thailand, is to be constructed east of
Bangkok, across a region of deep deposits of very soft marine clay. A
typical soil profile is shown in Fig. 8.16a. The average $C_{c\epsilon} = 0.8$
below the drying crust. The proposed embankment is 15 m wide at
the top, has three horizontal to one vertical side slope and is 2 m
high. Estimate the ultimate consolidation settlement of the centerline
of the embankment.

nine

Time Rate of Consolidation

9.1 INTRODUCTION

In Chapter 8 we showed how to calculate the magnitude of consolidation settlement that a clay layer under a structure will undergo before it reaches equilibrium with the external load. We described how the pore water pressure in excess of hydrostatic dissipates with time (consolidation) and how the effective stress ultimately becomes equal to the applied load. It was mentioned that the rate of consolidation would depend, among other things, on the permeability of the soil.

Sometimes consolidation is called *primary consolidation* to distinguish it from the other time-dependent component of total settlement, *secondary compression*. You will recall from Sec. 8.2 that secondary compression occurs after essentially all of the excess pore water pressure has dissipated, that is, it occurs at constant effective stress. In some soils, especially inorganic clays, primary consolidation is the largest component of total settlement, whereas secondary compression constitutes a major part of the total settlement of peats and other highly organic soils. In this chapter, the theories for estimating the time rate of both primary consolidation and secondary compression of fine-grained soils are discussed.

Why is it important to know how fast a structure will settle under the applied load? For example, if the design life of a structure is 50 years, and it is estimated that it will take 500 years for all the settlement to occur, then the foundation engineer would expect only minor settlement problems during the life of the structure. On the other hand, if the settlement is expected to take about the time required to build the structure, then most if not all of the settlement will have occurred by the time the structure is completed. If the structure is sensitive to rapid settlements (for example, reinforced concrete frames or concrete pavement), then structural damage could result. Most structures on clay foundations experience gradual settlements during their lifetimes, which may or may not impair the performance of the structures. This chapter presents procedures for estimating the *rate* of foundation settlement. The engineer then can decide what effect, if any, the settlement may have on the structural integrity as well as the intended use of the structure.

The following notation is introduced in this chapter.

Symbol	Dimension	Unit	Definition
C_α	—	—	Secondary compression index (Eq. 9-15)
$C_{\alpha\epsilon}$	—	—	Modified secondary compression index (Eq. 9-16)
c_v	$L^2 T^{-1}$	m^2/s	Coefficient of consolidation (Eq. 9-3)
H_{dr}	L	m	Length of drainage path (Eq. 9-5)
R_0	L	mm	Initial dial reading (Eq. 9-13)
R_n	L	mm	Dial reading at time $t, n = 1, 2, \ldots$
$s(t)$	L	m	Consolidation settlement at time t
T	—	—	Time factor (Eq. 9-5)
U_{avg} or U	—	(%)	Degree of consolidation
U_z	—	— (or %)	Consolidation ratio (Eq. 9-9)
Z	—	—	Depth factor (Eq. 9-9)

9.2 THE CONSOLIDATION PROCESS

It is useful to return to the spring analogy as presented in Chapter 8 (Fig. 8.2). Figure 9.1a shows a spring with a piston and a valve in a single cylinder. A pressure versus depth diagram is shown in Fig. 9.1b. The soil, represented by the spring, is at equilibrium with an initial effective stress σ'_{vo}. For the time being, we shall assume that all of the applied load on the piston, $\Delta\sigma$, is initially transferred to the excess pore water pressure Δu (excess above hydrostatic or initial u_o). This is the case for one-dimensional loading, but later we shall see that this is not true for three-dimensional loading.

With time, water is squeezed out through the valve, and the excess pore water pressure decreases. Thus there is a gradual transfer of stress from the pore water to the soil skeleton and a concurrent increase in effective stress. Figure 9.1c shows the initial effective stress σ'_{vo}, the change (increase) in effective stress, $\Delta\sigma'$, and the pore pressure still to be dissipated, Δu, at $t = t_1$. The vertical dashed lines, labeled t_1, t_2, etc., represent time from the start of load application. These lines are called

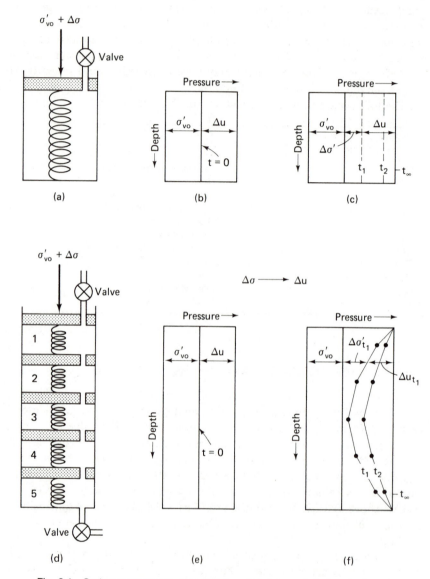

Fig. 9.1 Spring analogy for consolidation: (a)–(c), model of a single soil layer; (d)–(f), model of multiple soil layers.

isochrones because they are lines of equal times. (*Isobars* are either arctic taverns or contour lines of equal pressure found on a weather map; *isopachs* are lines of equal thickness of a geologic deposit; and *isotachs* are lines of equal velocity on wind maps.) Finally, at $t \to \infty$ all of the excess pore water pressure Δu will be dissipated, and the effective stress will equal the initial stress σ'_{vo} plus the applied stress increment $\Delta \sigma$. During this time, the piston will have settled an amount that is directly related to how much water is squeezed out of the cylinder.

A typical soil layer is much more complex than the simple model shown in Figs. 9.1a–c. Let us increase the number of springs, pistons, and valves as shown Fig. 9.1d. As before, we can show the initial effective stress σ'_{vo}, within the soil layer, and the corresponding induced pore water pressure Δu, due to the external load on the pistons, $\Delta \sigma$, in Fig. 9.1e. Let us allow drainage to occur through each piston and valve so that we have both internal as well as top and bottom drainage. In order for the water to be squeezed out of cylinders 2, 3, and 4, it is necessary for some of the water in cylinders 1 and 5 to escape beforehand. Likewise, before the water can be squeezed out of the soil in cylinder 3, it is necessary that some of the water in cylinders 2 and 4 be squeezed out first, etc. Because all valves are open, upon application of the external stress $\Delta \sigma$, water will start to flow out immediately from the top and bottom cylinders. This will result in an immediate reduction of the excess pore water pressure and an increase in effective stress in cylinders 1 and 5, etc. As shown in Fig. 9.1f, with time the pore pressure isochrones move to the right, and they are segmented lines because of the finite number of pistons and valves. With an infinite number of pistons, the isochrones would be smooth curves which rather accurately represent what is physically occurring with time in a consolidating soil deposit. At the center of a *doubly drained* layer, which is modeled by Figs. 9.1d–f, it can be seen that the decrease in the induced pore water pressure, for example at t_1, is small compared to the change at the top and bottom of the layer. This is because the *drainage path* for the center cylinder is considerably longer than for cylinders 1 and 5. As a result, it takes a longer time for the center of a doubly drained layer (or the bottom of a singly drained layer) to dissipate its excess pore pressure.

The flow of water out of the cylinders (soil voids) is physically due to the gradient i, which equals $h/l = (\Delta u / \rho_w g) / \Delta z$. The slope of the segmented isochrones in Fig. 9.1f is $\Delta u / \Delta z$. At the exact center of the clay layer, the flow is zero because the gradient $\Delta u / \Delta z$ is zero. At the ends, the gradient approaches infinity and thus the flow is the largest right at the drainage surfaces.

The process just described is called *consolidation*. The amount of settlement the spring-piston system (or clay layer) experiences is directly related to how much water has squeezed out of the cylinders (or voids in the clay). How much water has squeezed out and thus the change in void ratio of the clay is in turn directly proportional to the amount of excess pore water pressure that has dissipated. Thus the *rate* of settlement is directly related to the *rate* of excess pore pressure dissipation. What we need in order to predict the rate of settlement of a foundation is an equation or theory that predicts the pore pressure and void ratio at any point in time and space in the consolidating clay layer. Then the change in thickness or settlement of the layer after any time of loading can be determined by integration of the equation over the thickness of the clay layer. The theory of consolidation which is most commonly used in soil mechanics is a one-dimensional theory. It was first developed by Terzaghi in the 1920's and its derivation and solution are summarized in the following sections.

9.3 TERZAGHI'S ONE-DIMENSIONAL CONSOLIDATION THEORY

In this section, we will present the Terzaghi (1925) one-dimensional consolidation equation and discuss some of the assumptions necessary to derive the equation. A detailed derivation and the solution to the equation is given in Appendix B-2. In order to use the Terzaghi theory with some confidence, it is important that you understand the assumptions and therefore the limitations of the theory.

The compressible soil layer is assumed to be both homogeneous and completely saturated with water, and the mineral grains in the soil and the water in the pores are completely incompressible. Darcy's law is considered to govern the egress of water from the soil pores, and usually both drainage and compression are assumed to be one dimensional. Usually drainage is provided at both the top and bottom of the compressible layer, but we could just as easily assume drainage at only one surface. The Terzaghi theory is a *small strain* theory in that the applied load increment produces only small strains in the soil; therefore both the coefficient of compressibility, a_v, and the Darcy coefficient of permeability, k, remain essentially constant during the consolidation process. If a_v is a constant over the increment of applied stress, then there is a *unique* relationship between the change in void ratio, Δe, and the change in effective stress, $\Delta \sigma'$. This implies also that there is *no secondary compression*; if secondary compression occurs, then the relationship between Δe and $\Delta \sigma'$ would not be unique, by definition. Recall that secondary compression is the change in void ratio that occurs at constant effective stress.

The derivation of the Terzaghi equation considers the volume of water flowing out of a differential compressible soil element. From Darcy's law, we know the quantity of flow depends on the hydraulic gradient as well as on the permeability of the soil. The hydraulic gradient causing flow can be related to the excess pore water pressure in the element by $u/\rho_w g$. Since the water is assumed incompressible, by continuity the volume change in the element must be the difference in flow in and out of the element in a differential time dt. This part of the equation can be written as

$$\frac{-k}{\rho_w g} \frac{\partial^2 u}{\partial z^2} dz \, dt$$

where z is the space or depth variable in the soil element. Everything else is as previously defined. Partial differentials must be used because u is a function of both the position z and time t.

The other part of the equation is obtained by relating the volume change or change in void ratio of the soil skeleton to the change in effective stress by means of the coefficient of compressibility a_v, which we determined in the oedometer test. (Thus a_v is really the stress-strain relationship or "modulus" of our soil.) From the effective stress principle, we can equate the change in effective stress to the change in pore pressure. In other words as long as the total stress is constant, as the excess pore pressure dissipates with time, there is a concurrent increase in effective stress, or $\Delta\sigma' = -\Delta u$. As before, u is a function of both z and t. This half of the equation is usually written as

$$\frac{-a_v}{1 + e_o} \frac{\partial u}{\partial t} dt \, dz$$

Putting the two parts together, we obtain

$$\frac{-k}{\rho_w g} \frac{\partial^2 u}{\partial z^2} dz \, dt = \frac{-a_v}{1 + e_o} \frac{\partial u}{\partial t} dt \, dz \tag{9-1}$$

Rearranging, we obtain

$$c_v \frac{\partial^2 u}{\partial z^2} = \frac{\partial u}{\partial t} \tag{9-2}$$

where

$$c_v = \frac{k}{\rho_w g} \frac{1 + e_o}{a_v} \tag{9-3}$$

The coefficient c_v is called the *coefficient of consolidation* because it contains the material properties that govern the consolidation process. If you perform a dimensional analysis of Eq. 9-3, you will find that c_v has dimensions of $L^2 T^{-1}$ or m^2/s.

Equation 9-2 is the *Terzaghi one-dimensional consolidation equation*. It could just as easily be written in three dimensions, but most of the time in engineering practice one-dimensional consolidation is assumed. Basically, the equation is a form of the diffusion equation from mathematical physics. Many physical diffusion phenomena are described by this equation, for example, heat flow in a solid body. The "diffusion constant" for the soil is the c_v. Note that we called the c_v a constant. It really isn't, but we must assume it is, that is, that k, a_v, and e_o are constants, in order to make the equation linear and easily solvable.

So how do we solve the Terzaghi consolidation equation? Just like we solve all other second-order partial differential equations with constant coefficients. There are a variety of ways; some are mathematically exact; others are only approximate. For example, Harr (1966) presents an approximate solution by using the method of finite differences. Taylor (1948), following Terzaghi (1925), gives a mathematically rigorous solution in terms of a Fourier series expansion, and this is what we do in detail in Appendix B-2. Here we shall just give an outline of the solution. First, the boundary and initial conditions for the case of one-dimensional consolidation are:

1. There is complete drainage at the top and bottom of the compressible layer.
2. The initial *excess* hydrostatic pressure $\Delta u = u_i$ is equal to the applied increment of stress at the boundary, $\Delta\sigma$.

We can write these boundary and initial conditions as follows:

When $z = 0$ and when
$$z = 2H, u = 0$$
When $t = 0$, $\Delta u = u_i = \Delta\sigma = (\sigma_2' - \sigma_1')$

We usually take the thickness of the consolidating layer to be $2H$, so that the *length of the longest drainage path* is equal to H or H_{dr}. Of course at $t = \infty$, $\Delta u = 0$, or complete dissipation of the pore pressure will have occurred.

Terzaghi (1925) was obviously familiar with the early work on heat transfer, and he adapted those closed-form solutions to the consolidation problem. The solution comes out in terms of a Fourier series expansion of the form

$$u = (\sigma_2' - \sigma_1') \sum_{n=0}^{\infty} f_1(Z)f_2(T) \qquad (9\text{-}4)$$

where Z and T are dimensionless parameters (see also Taylor, 1948). The first term, Z, is a geometry parameter, and it is equal to z/H. The second term, T, is known as the *time factor*, and it is related to the coefficient of consolidation c_v by

$$T = c_v \frac{t}{H_{dr}^2} \qquad (9\text{-}5)$$

where $t =$ time, and
  $H_{dr} =$ length of the longest drainage path.

We have already mentioned that c_v has dimensions of $L^2 T^{-1}$ or units of m^2/s (or equivalent).

From Eq. 9-3, the time factor can also be written as

$$T = \frac{k(1 + e_o)}{a_v \rho_w g} \frac{t}{H_{dr}^2} \qquad (9\text{-}6)$$

Note that t has the same time units as k. That is, if k is in centimetres per second, then t must be in seconds. The drainage path for double drainage would be equal to half the thickness H of the clay layer, or $2H/2 = H_{dr}$. If we had only a singly drained layer, the drainage path would still be H_{dr}, but then it would be equal to the thickness H of the layer.

The progress of consolidation after some time t and at any depth z in the consolidating layer can be related to the void ratio at that time and the final change in void ratio. This relationship is called the *consolidation ratio*, and it is expressed as

$$U_z = \frac{e_1 - e}{e_1 - e_2} \qquad (9\text{-}7)$$

where e is some intermediate void ratio, as shown on Fig. 9.2. What we are looking at graphically in that figure is the ratio of ordinates corresponding to AB and AC. In terms of stresses and pore pressures, Eq. 9-7 becomes

$$U_z = \frac{\sigma' - \sigma_1'}{\sigma_2' - \sigma_1'} = \frac{\sigma' - \sigma_1'}{\Delta\sigma'} = \frac{u_i - u}{u_i} = 1 - \frac{u}{u_i} \qquad (9\text{-}8)$$

where σ' and u are intermediate values corresponding to e in Eq. 9-7, and u_i is the initial *excess* pore pressure induced by the applied stress $\Delta\sigma'$. You should satisfy yourself that these equations are correct from the relationships shown in Fig. 9.2 and from $\Delta\sigma' = -\Delta u$. (See also Appendix B-2.)

From Eqs. 9-7 and 9-8, it is evident that U_z is zero at the start of loading, and it gradually increases to 1 (or 100%) as the void ratio decreases from e_1 to e_2. At the same time, of course, as long as the total stress remains constant, the effective stress increases from σ_1' to σ_2' as the excess hydrostatic stress (pore water pressure) dissipates from u_i to zero. The consolidation ratio U_z is sometimes called the *degree or percent consolidation*, and it represents conditions *at a point* in the consolidating layer. It is now possible to put our solution for u in Eq. 9-4 in terms of the

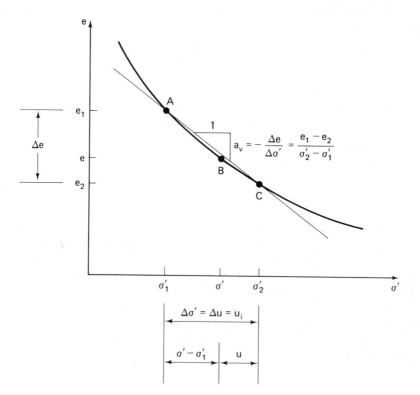

Fig. 9.2 Laboratory compression curve. Note: $\sigma' - \sigma'_1 = (\sigma'_2 - \sigma'_1) - u = u_i - u.$

consolidation ratio, Eq. 9-8, or

$$U_z = 1 - \sum_{n=0}^{\infty} f_1(Z)f_2(T) \tag{9-9}$$

The solution to this equation is shown graphically in Fig. 9.3 in terms of the dimensionless parameters already defined. The tedious calculations involved in solving Eq. 9-9 are no longer necessary. From Fig. 9.3 it is possible to find the amount or degree of consolidation (and therefore u and σ') for any real time after the start of loading and at any point in the consolidating layer. All you need to know is the c_v for the particular soil deposit, the total thickness of the layer, and boundary drainage conditions. With these items, the time factor T can be calculated from Eq. 9-5. It is applicable to any one-dimensional loading situation where the soil properties can be assumed to be the same throughout the compressible layer.

Figure 9.3 also is a picture of the *progress of consolidation*. The *isochrones* (lines of constant T) in Fig. 9.3 represent the degree or percent

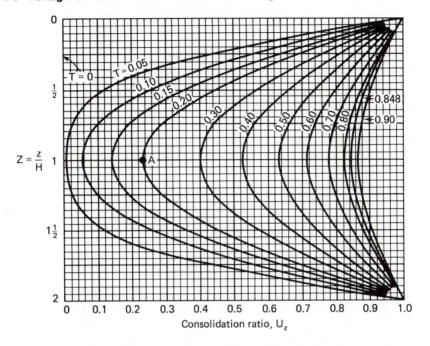

Fig. 9.3 Consolidation for any location and time factor in a doubly drained layer (after Taylor, 1948).

consolidation for a given time factor throughout the compressible layer. For example, the percent consolidation at midheight of a doubly drained layer (total thickness $= 2H$) for a time factor equal to 0.2 is approximately 23% (see point A in Fig. 9.3). However at the same time (and time factor) at other locations within the soil layer, the degree of consolidation is different. At 25% of the depth, for example, $z/H = \frac{1}{2}$ and $U_z = 44\%$. Similarly, near the drainage surfaces at $z/H = 0.1$, for the same time factor, because the gradients are much higher, the clay is already 86% consolidated, which means that at that depth and time, 86% of the original excess pore pressure has dissipated and the effective stress has increased by a corresponding amount.

EXAMPLE 9.1

Given:

A 12 m thick layer of Chicago clay is *doubly drained*. (This means that a very pervious layer compared to the clay exists on top of and under the 12 m clay layer.) The coefficient of consolidation $c_v = 8.0 \times 10^{-8}$ m^2/s.

Required:

Find the degree or percent consolidation for the clay 5 yr after loading at depths of 3, 6, 9, and 12 m.

Solution:

First, compute the time factor. From Eq. 9-5,

$$T = \frac{c_v t}{H_{dr}^2}$$

$$= \frac{8.0 \times 10^{-8} \text{ m}^2/\text{s} \ (3.1536 \times 10^7 \text{ s/yr}) \ (5 \text{ yr})}{(6)^2} = 0.35$$

Note that $2H = 12$ m and $H_{dr} = 6$ m since there is double drainage.
Next, from Fig. 9.3 we obtain (by interpolation) for $T = 0.35$:

At $z = 3$ m, $z/H = 0.50$, $U_z = 61\%$
At $z = 6$ m, $z/H = 1.0$, $U_z = 46\%$
At $z = 9$ m, $z/H = 1.50$, $U_z = 61\%$
At $z = 12$ m, $z/H = 2.0$, $U_z = 100\%$

EXAMPLE 9.2

Given:

The soil conditions of Example 9.1.

Required:

If the structure applied an average vertical stress increase of 100 kPa to the clay layer, estimate the excess pore water pressure remaining in the clay after 5 yr for the depths in the clay layer of 3, 6, 9, and 12 m.

Solution:

Assuming one-dimensional loading, the induced excess pore water pressure at the beginning of consolidation is 100 kPa. From Eq. 9-8,

$$U_z = 1 - \frac{u}{u_i}$$

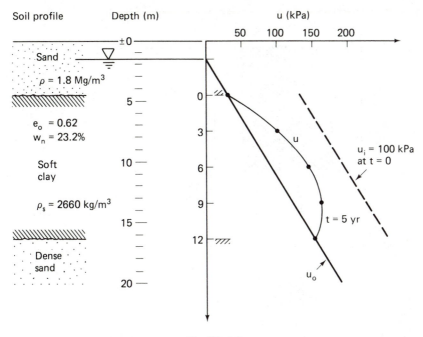

Fig. Ex. 9.2

or

$$u = u_i(1 - U_z)$$

From the solution in Example 9.1 we obtain:

At $z = 3$ m,	$U_z = 61\%$,	$u = 39$ kPa
At $z = 6$ m,	$U_z = 46\%$,	$u = 54$ kPa
At $z = 9$ m,	$U_z = 61\%$,	$u = 39$ kPa
At $z = 12$ m,	$U_z = 100\%$,	$u = 0$ kPa

Figure Ex. 9.2 shows these values versus depth. Note that they are *excess* pore pressures, that is, they are above the hydrostatic water pressure.

In most cases, we are not interested in how much a given point in a layer has consolidated. Of more practical interest is the *average degree or percent consolidation* of the entire layer. This value, denoted by U or U_{avg}, is a measure of how much the entire layer has consolidated and thus it can be directly related to the *total settlement* of the layer at a given time after loading. Note that U can be expressed as either a decimal or a percentage.

To obtain the average degree of consolidation over the entire layer corresponding to a given time factor we have to find the area under the T

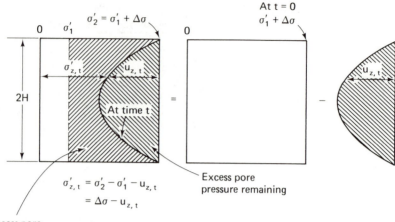

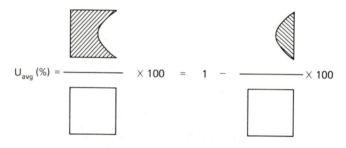

Fig. 9.4   Average degree of consolidation, U_{avg}, defined.

curve of Fig. 9.3. (Actually we obtain the area outside the T curve as shown in Fig. 9.4.) How the integration is done mathematically is shown in Appendix B-2. Table 9-1 presents the results of the integration for the case where a *linear* distribution of excess pore water pressure is assumed.

The results in Table 9-1 are shown graphically in Fig. 9.5. In Fig. 9.5a the relationship is shown arithmetically, whereas in Fig. 9.5b, the relationship between U and T is shown semilogarithmically. Another form of the relationship is found in Fig. 9.5c, where U is plotted versus $\sqrt{T}$. As discussed in the next section, Figs. 9.5b and 9.5c have been found to show certain characteristics of the theoretical U-T relationship to better advantage than Fig. 9.5a. Note that as T becomes very large, U asymptotically approaches 100%. This means that, theoretically, consolidation never stops but continues infinitely. It should also be pointed out that the solution for U versus T is dimensionless and applies to all types of problems where $\Delta\sigma = \Delta u$ varies *linearly* with depth. Solutions for cases

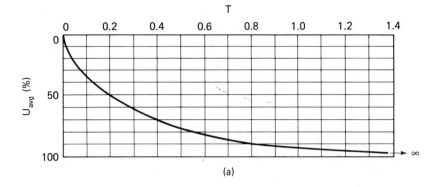

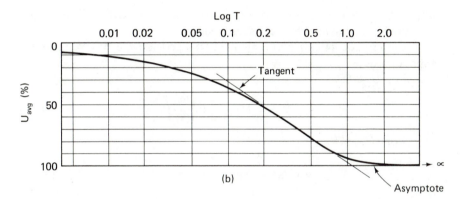

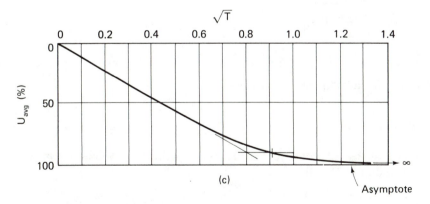

Fig. 9.5 U_{avg} versus T: (a) arithmetic scale; (b) log scale; (c) square root scale.

TABLE 9-1

U_{avg}	T
0.1	0.008
0.2	0.031
0.3	0.071
0.4	0.126
0.5	0.197
0.6	0.287
0.7	0.403
0.8	0.567
0.9	0.848
0.95	1.163
1.0	∞

where the initial pore pressure distribution is sinusoidal, half sine, and triangular are presented by Leonards (1962).

Casagrande (1938) and Taylor (1948) provide the following useful approximations:

For $U < 60\%$,

$$T = \frac{\pi}{4} U^2 = \frac{\pi}{4} \left(\frac{U\%}{100} \right)^2 \tag{9-10}$$

For $U > 60\%$,

$$T = 1.781 - 0.933 \log (100 - U\%) \tag{9-11}$$

EXAMPLE 9.3

Given:

$T = 0.05$ for a compressible clay deposit.

Required:

Average degree of consolidation and the percent consolidation at the center and at $z/H = 0.1$.

Solution:

From Table 9-1 and Fig. 9.5, $U_{avg} = 26\%$. Therefore the clay is 26% consolidated, on the average. From Fig. 9.3 you can see that the center of the layer is less than 0.5% consolidated, while at the "10%" depth ($z/H =$ 0.1) the clay is 73% consolidated. But, *on the average* throughout the layer, the clay is 26% consolidated.

What does the average consolidation mean in terms of settlements? U_{avg} can be expressed as

$$U_{avg} = \frac{s(t)}{s_c} \tag{9-12}$$

where $s(t)$ is the settlement at any time, and s_c is the final or ultimate consolidation (primary) settlement at $t = \infty$.

EXAMPLE 9.4

Given:

The data of Example 9.3.

Required:

Find the settlement when U_{avg} is 26%, if the final consolidation settlement is 1 m.

Solution:

From Eq. 9-12, $s(t) = U_{avg}(s_c)$. Therefore

$$s(t) = 26\% \,(1 \text{ m}) = 0.26 \text{ m}$$

EXAMPLE 9.5

Given:

The soil profile and properties of Examples 9.1 and 9.2.

Required:

Compute the time required for the clay layer to settle 0.25 m.

Solution:

To compute the average degree of consolidation, you first must estimate the consolidation settlement s_c, as you did in Chapter 8. For Chicago clay, a reasonable value of C_c is about 0.25 (Tables 8-2 and 8-3). From Fig. Ex. 9.2, $H_o = 12$ m and $e_o = 0.62$. Determine ρ for the soft clay and calculate σ'_{vo} at the middepth of layer from Eqs. 7-14c and 7-15. Assume the clay is normally consolidated. So

$$\sigma'_{vo} = 1.8 \times 9.81 \times 1.5 + (1.8 - 1) \times 9.81 \times 3$$
$$+ (2.02 - 1) \times 9.81 \times 6$$
$$= 110 \text{ kPa}$$

From Eq. 8-11,

$$s_c = 0.25 \frac{12 \text{ m}}{1 + 0.62} \log \frac{110 + 100}{110} = 0.52 \text{ m}$$

The average degree of consolidation U_{avg} when the clay layer settles 0.25 m is (Eq. 9-12):

$$U_{avg} = \frac{s(t)}{s_c} = \frac{0.25 \text{ m}}{0.52 \text{ m}} = 0.48, \text{ or } 48\%$$

To obtain T we can use either Table 9-1 or Fig. 9.5. Or since $U_{avg} < 60\%$, we can use Eq. 9-10.

$$T = \frac{\pi}{4} (0.48)^2 = 0.182$$

From Eq. 9.5, $t = TH_{dr}^2/c_v$, where $H_{dr} = 6$ m for double drainage; or

$$t = \frac{0.182 \times (6 \text{ m})^2}{8 \times 10^{-8} \text{ m}^2/\text{s} \times 3.1536 \times 10^7 \text{ s/yr}}$$
$$= 2.6 \text{ yr}$$

EXAMPLE 9.6

Given:

The data of Examples 9.1 and 9.5.

Required:

How much time would be required for a settlement of 0.25 m to occur if the clay layer were singly drained?

Solution:

Use Eq. 9-5 directly.

$$c_v = 8 \times 10^{-8} \text{ m}^2/\text{s} \times 3.1536 \times 10^7 \text{ s/yr} = 2.523 \text{ m}^2/\text{yr}$$

$$t = \frac{TH_{dr}^2}{c_v}$$

where $H_{dr} = 12$ m for single drainage.

$$t = \frac{0.182 \times (12 \text{ m})^2}{2.523 \text{ m}^2/\text{yr}} = 10.4 \text{ yr}$$

or *four times as long* as with double drainage.

EXAMPLE 9.7

Given:

A 10 m thick clay layer with *single* drainage settles 9 cm in 3.5 yr. The coefficient of consolidation for this clay was found to be 0.544×10^{-2} cm^2/s.

Required:

Compute the ultimate consolidation settlement, and find out how long it will take to settle to 90% of this amount.

Solution:

From Eq. 9-5 solve for T:

$$T = \frac{tc_v}{H^2}$$

$$= \frac{3.5 \text{ yr} (0.544 \times 10^{-2}) \text{ cm}^2}{(100 \text{ m}^2) \text{ s}} \left(\frac{1 \text{ m}^2}{10\,000 \text{ cm}^2} \right) \left(3.1536 \times 10^7 \frac{\text{s}}{\text{yr}} \right)$$

$$= 0.6$$

From Table 9-1 we see that the average degree of consolidation is between 0.8 and 0.9. Therefore we can use either Eq. 9-11 or Fig. 9.5a, or we can interpolate from Table 9-1. Using Eq. 9-11, we have

$$0.6 = 1.781 - 0.933 \log (100 - U\%)$$

$$1.27 = \log (100 - U\%)$$

or

$$U = 81.56\%, \text{ or } 82\%$$

Thus if 9 cm of settlement represents 82% of the total settlement, then the total consolidation settlement is (Eq. 9-12):

$$s_c = \frac{s(t)}{U_{avg}} = \frac{9 \text{ cm}}{0.82} = 11 \text{ cm}$$

For the time for 90% settlement to occur, find $T = 0.848$ for $U_{avg} = 0.9$, from Table 9-1. Using Eq. 9-5 and solving for t, we find that:

$$t = \frac{TH_{dr}^2}{c_v} = \frac{0.848 \, (10 \text{ m})^2}{0.544 \times 10^{-2} \text{ cm}^2/\text{s}} \quad \frac{10\,000 \text{ cm}^2}{\text{m}^2}$$

$$= 1.559 \times 10^8 \text{ s} \, \frac{\text{yr}}{3.1536 \times 10^7 \text{ s}}$$

$$= 4.94 \text{ yr}$$

EXAMPLE 9.8

Given:

The data of Example 9.7.

Required:

Find the variation in the degree of consolidation throughout the layer when $t = 3.5$ yr.

Solution:

When $t = 3.5$ yr, the corresponding time factor $= 0.6$, from Example 9.7. Find the curve for $T = 0.6$ in Fig. 9.3. (For a layer with single drainage, we use the top half or bottom half, depending on where the layer is drained. Assume for this problem that the layer is drained at the *top*.) The curve for $T = 0.6$ represents the degree of consolidation at any depth z. Since $T = 0.6$ and using Eq. 9-5 we find that this isochrone shows the variation of U_z for $t = 3.5$ yr. It can be seen that at the bottom of the layer, where $z/H = 1$, $U_z = 71\%$. At midheight of the 10 m thick layer, where $z/H = 0.5$, $U_z = 79.5\%$. Thus the degree of consolidation varies through the depth of the clay layer, but the *average* degree of consolidation for the entire layer is 82% (Example 9.7). Another interesting point about Fig. 9.3 is that the area to the left of the curve $T = 0.6$ represents 82% of the area of the entire graph, $2H$ versus U_z, whereas the area to the right of the curve $T = 0.6$ represents 18%, or the amount of consolidation yet to take place. (See also Fig. 9.4.)

9.4 DETERMINATION OF THE COEFFICIENT OF CONSOLIDATION c_v

How do we obtain the coefficient of consolidation c_v? This coefficient is the only part of the solution to the consolidation equation that takes into account the soil properties which govern the rate of consolidation. In Chapter 8 we described the procedure for performing consolidation or oedometer tests to obtain the compressibility of the soil. We mentioned that each load increment usually remains on the test specimen an arbitrary length of time, until (we hope) essentially all of the excess pore pressure has dissipated. Deformation dial readings are obtained during this process, and the coefficient of consolidation c_v is determined from the time-deformation data.

The curves of actual deformation dial readings versus real time for a given load increment often have very similar shapes to the theoretical $U\text{-}T$ curves shown in Fig. 9.5. We shall take advantage of this observation to determine the c_v by so-called "curve-fitting methods" developed by Casagrande and Taylor. These empirical procedures were developed to fit approximately the observed laboratory test data to the Terzaghi theory of consolidation. Many factors such as sample disturbance, load increment ratio (LIR), duration, temperature, and a host of test details have been found to strongly affect the value of c_v obtained by the curve-fitting procedures (Leonards and Ramiah, 1959; Leonards, 1962). But research by Leonards and Girault (1961) has shown that the Terzaghi theory is applicable to the laboratory test if large LIR's (Eq. 8-20), usually around unity, are used.

The curve-fitting procedures outlined in this section will enable you to determine values of the coefficient of consolidation c_v from laboratory test data. In addition, the procedures will allow you to separate the secondary compression from the primary consolidation.

Probably the easiest way to illustrate the curve-fitting methods is to work with time-deformation data from an actual consolidation test. We will use the data for the load increment from 10 to 20 kPa for the test shown in Fig. 8.5. This data is shown in Table 9-2 and plotted in Figs. 9.6a, b, and c. Note how similar the shapes of these curves are to the theoretical curves of Figs. 9.5a, b, and c.

TABLE 9-2 Time-Deformation Data for Load Increment 10 to 20 kPa (Fig. 8.5)

Elapsed Time (min)	$\sqrt{t}$ ($\sqrt{\text{min}}$)	Dial Reading, R (mm)	Displacement (mm)
0	0	6.627	0
0.1	0.316	6.528	0.099
0.25	0.5	6.480	0.147
0.5	0.707	6.421	0.206
1	1.0	6.337	0.290
2	1.41	6.218	0.409
4	2.0	6.040	0.587
8	2.83	5.812	0.815
15	3.87	5.489	1.138
30	5.48	5.108	1.519
60	7.75	4.775	1.852
120	10.95	4.534	2.093
240	15.5	4.356	2.271
480	21.9	4.209	2.418
1382	37.2	4.041	2.586

(a) Casagrande's Logarithm of Time Fitting Method

In this method, the deformation dial readings are plotted versus the *logarithm of time*, as shown in Fig. 9.6b and to larger scale in Fig. 9.7. The idea is to find R_{50} and thus t_{50}, which is the time for 50% consolidation, by approximating R_{100}, the dial reading corresponding to the time for 100% primary consolidation, t_{100} or t_p. Refer to Fig. 9.5b, the theoretical U-T curve, for a moment. Note that the intersection of the tangent and the asymptote to the theoretical curve defines $U_{\text{avg}} = 100\%$. The time for 100% consolidation, of course, occurs at $t = \infty$. Casagrande (1938) suggested that R_{100} could be approximated rather arbitrarily by the intersection of the two corresponding tangents to the laboratory consolidation curve (Fig. 9.7). Later research (for example, Leonards and Girault, 1961) has shown this procedure defines to a good approximation the dial reading at which the excess pore water pressure approaches zero, especially when the LIR is large and the preconsolidation stress is exceeded by the applied load increment. Once R_{100} is defined, then it is fairly easy to determine R_{50} and t_{50}, once we find R_o, the initial dial reading.

How do we determine R_o, the dial reading corresponding to zero percent consolidation, on a semilog plot? Since T is proportional to U_{avg}^2 up to $U = 60\%$ (Eq. 9-10), the first part of the consolidation curve must be a parabola. To find R_o, choose any two times, t_1 and t_2, in the ratio of 4 to 1, and note their corresponding dial readings. Then mark off a distance above R_1 equal to the difference $R_2 - R_1$; this defines the corrected zero

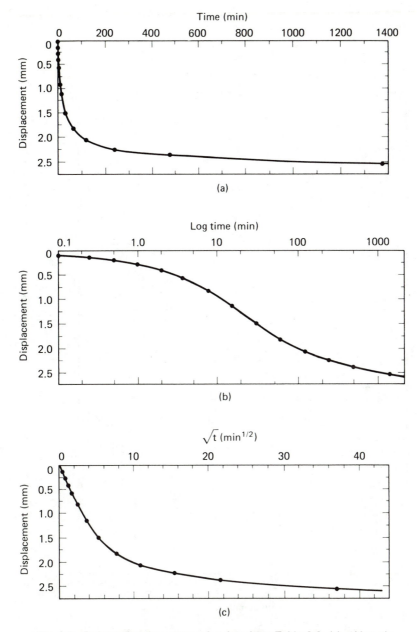

Fig. 9.6 Deformation-time curves for data from Table 9-2: (a) arithmetic scale; (b) log time scale; (c) square root of time scale.

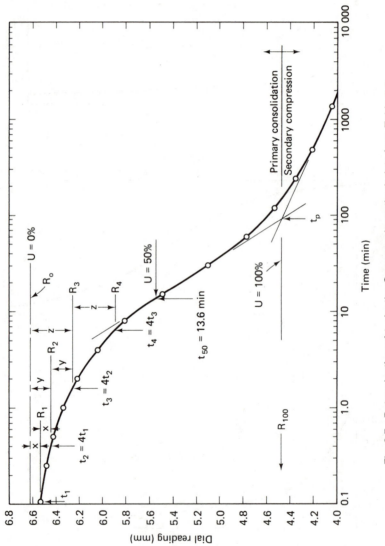

Fig. 9.7   Determination of t_{50} by the Casagrande method; data from Table 9-2.

point R_o. In equation form,

$$R_o = R_1 - (R_2 - R_1) \tag{9-13a}$$

Several trials are usually advisable to obtain a good average value of R_o, or

$$R_o = R_2 - (R_3 - R_2) \tag{9-13b}$$

and

$$R_o = R_3 - (R_4 - R_3) \tag{9-13c}$$

In Fig. 9.7, three different trials are shown for determining R_o from R_1, R_2, R_3, and R_4. The distances x, y, and z are marked off above the ordinates corresponding to times t_2, t_3, and t_4, respectively. You should satisfy yourself that both the graphical procedure and using Eqs. 9-13 (a, b, c) indicate about the same value for R_o (6.62 mm in this case).

Once the initial and 100% primary consolidation points have been determined, find t_{50} by subdividing the vertical distance between R_o and R_{100} [or $R_{50} = \frac{1}{2}(R_o - R_{100})$]. Then t_{50} is simply the time corresponding to the dial reading R_{50}. In Fig. 9.7, $t_{50} = 13.6$ min. To evaluate c_v, we use Eq. 9-5 with $T_{50} = 0.197$ (Table 9-1). We also need the average height of the specimen during the load increment. At the beginning of this increment, H_o was 21.87 mm. From the data of Table 9-2,

$$H_f = H_o - \Delta H = 21.87 - 2.59 = 19.28 \text{ mm}$$

Thus the average height of specimen during the increment is 20.58 mm (2.06 cm). Remember that in the standard oedometer test the specimen is doubly drained, so use $H_{dr} = 2.06/2$ in Eq. 9-5. Thus we have

$$c_v = \frac{TH_{dr}^2}{t} = \frac{T_{50}H_{dr}^2}{t_{50}}$$

$$= \frac{0.197\left(\dfrac{2.06}{2}\right)^2 \text{cm}^2}{13.6 \text{ min}\left(60\ \dfrac{\text{s}}{\text{min}}\right)}$$

$$= 2.56 \times 10^{-4}\ \frac{\text{cm}^2}{\text{s}}\left(3.1536 \times 10^7\ \frac{\text{s}}{\text{yr}}\right)\left(\frac{\text{m}^2}{10^4\ \text{cm}^2}\right)$$

$$= 0.81 \text{ m}^2/\text{yr}$$

Recall that the Casagrande fitting procedure found R_{50} and thus t_{50} by approximating R_{100}. This procedure did not find t_{100} since the time for any other degree of consolidation must be obtained from the classical consolidation theory in which $t_{100} = \infty$. But the procedure does define a t called t_p (for "primary") which is a practical time required to obtain a

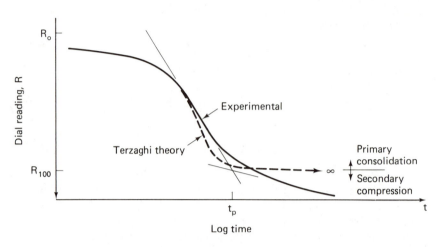

Fig. 9.8   Terzaghi consolidation theory and a typical experimental curve used to define t_p.

good usable value of R_{100}. Often, in practice, t_p is called t_{100}. The deviation of the experimental curve from the theoretical curve is shown in Fig. 9.8. Differences in the curves are the result of secondary compression and other effects such as the rate of effective stress increase (Leonards, 1977) not considered by the Terzaghi theory.

(b) Taylor's Square Root of Time Fitting Method

Taylor (1948) also developed a procedure for evaluating c_v, using the square root of time. As with Casagrande's fitting method, the procedure is based on the similarity between the shapes of the theoretical and experimental curves when plotted versus the square root of T and t. Refer to Fig. 9.5c and compare it with Fig. 9.6c. Note that in Fig. 9.5c the theoretical curve is a straight line to at least $U \simeq 60\%$ or greater. Taylor observed that the abscissa of the curve at 90% consolidation was about 1.15 times the abscissa of the extension of the straight line (Fig. 9.5c). He thus could determine the point of 90% consolidation on the laboratory time curve.

We will use the same data as before (Table 9-2) to illustrate the $\sqrt{t}$ fitting method. These data are plotted in Fig. 9.9. Usually a straight line can be drawn through the data points in the initial part of the compression curve. The line is projected backward to zero time to define R_o. The common point at R_o may be slightly lower than the initial dial reading (at zero time) observed in the laboratory due to immediate compression of the specimen and apparatus. Draw a second line from R_o with all abscissas 1.15 times as large as corresponding values on the first line. The intersection of this second line and the laboratory curve defines R_{90} and is the point of 90% consolidation. Its time is, of course, t_{90}.

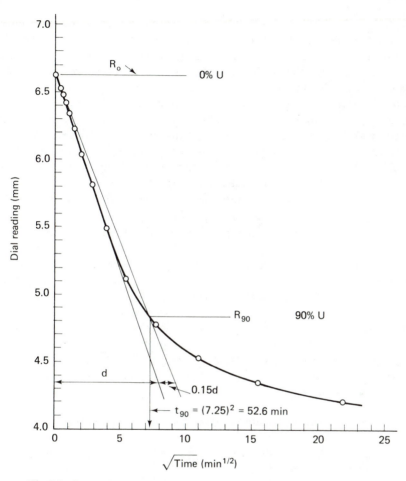

Fig. 9.9 Determination of c_v, using Taylor's square root of time method; data from Table 9-2.

The coefficient of consolidation is, as before, determined by using Eq. 9-5. From Table 9-1, $T_{90} = 0.848$. The average height of specimen is also used, as before. Therefore

$$c_v = \frac{0.848 \, (2.06/2)^2 \, \text{cm}^2}{52.6 \, \text{min} \, (60 \, \text{s/min})}$$

$$= 2.85 \times 10^{-4} \, \text{cm}^2/\text{s or } 0.90 \, \text{m}^2/\text{yr}$$

This value is reasonably close to the value obtained using the Casagrande method. Because both fitting methods are approximations of theory, you should not expect them to agree exactly. Often c_v as determined by the $\sqrt{t}$ method is slightly greater than c_v by the log t fitting method.

You should also note that c_v is not a constant for a test on a given soil, but it depends greatly on the load increment ratio and whether the preconsolidation stress has been exceeded or not (Leonards and Girault, 1961). For load increments less than the preconsolidation stress, consolidation occurs quite rapidly, and c_v values can be rather high. However, determinations of t_p for these increments is often difficult because the time-settlement curves do not have the "classical" shapes of Figs. 9.7 and 9.9. For undisturbed clays c_v is usually a minimum for increments near the preconsolidation pressure (Taylor, 1948). For design, this minimum value is often used. However, for some situations it may be more appropriate to use the c_v for the anticipated load increment in the field.

A strong advantage of the $\sqrt{t}$ fitting method is that t_{90} can be determined without going too far beyond t_p. If dial readings are plotted as you go during the test, then it is possible to add the next increment of load as soon as t_{90} is reached. Not only is the time for testing significantly reduced compared to when the conventional 24 h increments are used, but also the contribution of secondary compression to the e versus log σ' curve can be effectively minimized (see Leonards, 1976).

By now you should have noticed that the data do not exactly coincide with the initial starting point in either of Figs. 9.7 or 9.9; that is, R_o does not equal exactly the initial reading of Table 9-2. The reason for the difference between the initial laboratory dial reading and R_o, the "corrected dial reading" corresponding to 0% consolidation, is due to several factors. These may include:

1. Vertical elastic compression of the soil specimen, porous stones, and apparatus.
2. Lateral expansion of the soil specimen if it is not trimmed exactly to the diameter of the ring.
3. Deformation associated with lateral expansion of the oedometer ring.

You will have the opportunity to use the two curve-fitting methods to determine c_v in the problems at the end of this chapter.

9.5 DETERMINATION OF THE COEFFICIENT OF PERMEABILITY

You may recall from Fig. 7.6 that the coefficient of permeability, k, of the soil may also be obtained indirectly from the consolidation test. If you take Eq. 9-3 and solve for k, you obtain

$$k = \frac{c_v \rho_w g a_v}{1 + e_o} \tag{9-14}$$

The value of e_o is the void ratio at the start of the time rate readings for a given load increment.

EXAMPLE 9.9

Given:

The time-deformation data for the load increment 10 to 20 kPa of the test in Fig. 8.4. From Table 9-2 and Fig. 9.7, a c_v value of 0.81 m^2/yr $(2.56 \times 10^{-4}$ cm^2/s) can be determined.

Required:

Compute the coefficient of permeability, assuming the temperature of the water is 20°C.

Solution:

It is first necessary to compute the coefficient of compressibility from Eq. 8-5 and using Fig. 8.4b:

$$a_v = \frac{e_1 - e_2}{\sigma_2' - \sigma_1'} = \frac{2.12 - 1.76}{(20 - 10) \text{ kPa}}$$

$$= 0.036/\text{kPa} = 3.6 \times 10^{-5} \frac{\text{m}^2}{\text{N}}$$

From Eq. 9-14,

$$k = \frac{c_v \rho_w g a_v}{1 + e_o}$$

$$= \frac{2.56 \times 10^{-4} \frac{\text{cm}^2}{\text{s}} \times 1000 \frac{\text{kg}}{\text{m}^3} \times 9.81 \frac{\text{m}}{\text{s}^2} \times 3.6 \times 10^{-5} \frac{\text{m}^2}{\text{N}} \frac{1 \text{ m}}{100 \text{ cm}}}{1 + 2.12}$$

$$= 2.9 \times 10^{-7} \frac{\text{cm}}{\text{s}} = 2.9 \times 10^{-9} \frac{\text{m}}{\text{s}}$$

Note that the e used in the equation is the void ratio at the start of the load increment rather than the original or in situ void ratio.

TABLE 9-3 Typical Values of the Coefficient of Consolidation c_v

Soil	c_v cm²/s, ×10⁻⁴	c_v m²/yr
Boston blue clay (CL) (Ladd and Luscher, 1965)	40 ± 20	12 ± 6
Organic silt (OH) (Lowe, Zaccheo, and Feldman, 1964)	2–10	0.6–3
Glacial lake clays (CL) (Wallace and Otto, 1964)	6.5–8.7	2.0–2.7
Chicago silty clay (CL) (Terzaghi and Peck, 1967)	8.5	2.7
Swedish medium sensitive clays (CL-CH) (Holtz and Broms, 1972)		
1. laboratory	0.4–0.7	0.1–0.2
2. field	0.7–3.0	0.2–1.0
San Francisco Bay Mud (CL)	2–4	0.6–1.2
Mexico City clay (MH) (Leonards and Girault, 1961)	0.9–1.5	0.3–0.5

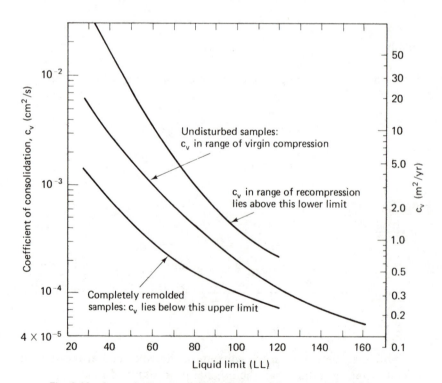

Fig. 9.10 Approximate correlations of the coefficient of consolidation c_v with the liquid limit (after U.S. Navy, 1971).

9.6 TYPICAL VALUES OF c_v

Typical values of the coefficient of consolidation c_v for a variety of soils are listed in Table 9-3. Approximate correlations of c_v with the liquid limit are presented in Fig. 9.10.

9.7 EVALUATION OF SECONDARY SETTLEMENT

Thus far we have discussed how to compute the consolidation or primary settlement s_c, and how it varies with time. The other two components of total settlement as given by Eq. 8-1 were the immediate settlement s_i and secondary settlement s_s. Immediate settlement is calculated from elastic theory, and it involves an evaluation of the modulus of elasticity and Poisson's ratio of the compressible soils. In addition, the contact stress distribution in the soil under the loaded area must be considered. The estimation of immediate settlements is treated in foundation engineering textbooks and will not be discussed herein.

Secondary compression is a continuation of the volume change that started during primary consolidation, only it usually occurs at a much slower rate. Secondary compression is different from primary consolidation in that it takes place at a *constant effective stress*, that is, after essentially all the excess pore pressure has dissipated. This component of settlement seems to result from compression of the bonds between individual clay particles and domains, as well as other effects on the microscale which are not yet clearly understood. Another complicating factor is that in the field it is difficult to separate secondary compression from consolidation settlement, especially if the consolidating clay layer is relatively thick. Parts of the layer near the drainage surfaces may be fully consolidated, and therefore undergoing "secondary" compression, while portions near the center of the layer are still in "primary." Both types of settlements contribute to the total surface settlement, and separating the effects in order to predict the final surface settlement is not a simple matter. However, in this section we shall present a practical working hypothesis, acceptable for engineering practice, for estimating secondary compression, and we shall show you how to make estimates of secondary settlement for some simple cases.

There is, unfortunately, a lot of confusion in the geotechnical literature as to the best way to describe the magnitudes and rates of secondary compression. In this section, we shall follow Raymond and Wahls (1976) and Mesri and Godlewski (1977), who define the *secondary compression index* C_α as

$$C_\alpha = \frac{\Delta e}{\Delta \log t} \qquad (9\text{-}15)$$

where $\Delta e =$ the change in void ratio along a part of the void ratio versus the *logarithm* of time curve between times t_1 and t_2, and
$\Delta t =$ the time between t_2 and t_1.

This definition is analogous, of course, to the primary compression index C_c, defined as $\Delta e / \Delta \log \sigma'$ (Eq. 8-7). In addition, we will define the *modified secondary compression index* $C_{\alpha\epsilon}$, analogous to Eq. 8-9, as

$$C_{\alpha\epsilon} = \frac{C_\alpha}{1 + e_p} \qquad (9\text{-}16)$$

where $C_\alpha =$ the secondary compression index, Eq. 9-15,
$e_p =$ the void ratio at the start of the *linear portion* of the e versus $\log t$ curve. (One could also use e_o, the in situ void ratio, with no appreciable loss of accuracy.)

Sometimes $C_{\alpha\epsilon}$ is called the *secondary compression ratio*, or the *rate of secondary consolidation*. As Ladd et al. (1977) note, $C_{\alpha\epsilon} = \Delta\epsilon / \Delta \log t$.

The secondary compression index, C_α, and the modified secondary compression index, $C_{\alpha\epsilon}$, can be determined from the slope of the straight line portion of the dial reading versus log time curve which occurs after primary consolidation is complete (see, for example, Fig. 9.7). Usually the ΔR is determined over one log cycle of time. The corresponding change in void ratio is calculated from the settlement equation (Eq. 8-3) since you know the height of specimen for that increment and e_o.

To provide a working hypothesis for estimating secondary settlements, we shall make the following *assumptions* about the behavior of fine-grained soils in secondary compression. These assumptions, based on the work of Ladd (1971a) and others and summarized by Raymond and Wahls (1976), are as follows:

1. C_α is independent of time (at least during the time span of interest).
2. C_α is independent of the thickness of the soil layer.
3. C_α is independent of the LIR, as long as some primary consolidation occurs.
4. The ratio C_α / C_c is approximately constant for many normally consolidated clays over the normal range of engineering stresses.

Typical dial reading versus log time behavior curves illustrating these assumptions for a normally consolidated clay are shown in Fig. 9.11. You

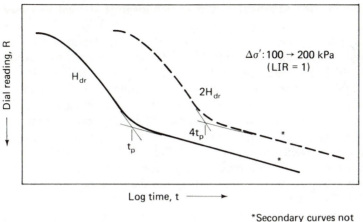

(a) Effect of drainage distance.

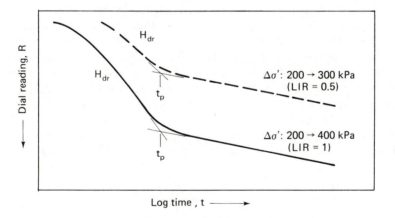

(b) Effect of load increment ratio and consolidation stress

Fig. 9.11 Typical secondary compression behavior from the working hypothesis by Raymond and Wahls (1976).

can see that the rate of secondary compression expressed in terms of settlement (ΔR) per log cycle is assumed to be independent of the thickness of the specimen as well as the load increment. There is some effect however of the consolidation stress, and as Mesri and Godlewski (1977) point out, C_α is strongly dependent on the final effective stress.

The working hypothesis is useful as a first approximation for estimating secondary settlements. However you should expect some aberrations in the actual long-term settlement response of the foundation because the assumptions are admittedly an oversimplification of real behavior. For example, the secondary compression curves of Fig. 9.11 may not be actually parallel or even have a constant slope. There is some evidence that C_α may change with time, both in the laboratory (Mesri and Godlewski, 1977) and in the field (Leonards, 1973). Also, the duration and therefore the magnitude of secondary settlement is a function of the time required for completion of primary consolidation (t_p), and from earlier work in this chapter you know that the thicker the consolidating layer, the longer is the time required for primary consolidation. Even though the strain at the end of primary consolidation for both thin and thick layers is about the same (as shown in Fig. 9.11a), there is limited evidence (Aboshi, 1973) that the slopes may not be parallel and that C_α may decrease as the thickness of the soil layer increases.

Assumptions 3 and 4 are approximately correct. Assumption 3 was verified by Leonards and Girault (1961) and Mesri and Godlewski (1977), except that the load increment must be sufficient to go well beyond the preconsolidation stress. The fourth assumption, that the ratio C_α/C_c is approximately a constant, has also been verified by Mesri and Godlewski (1977) for a wide variety of natural soils. Their work is summarized in Table 9-4. The average value of C_α/C_c is about 0.05, and in no case did they find a value of that ratio to exceed 0.1. The range for inorganic soils is 0.025 to 0.06, while the range for organic soils and peats is somewhat higher. They also showed that this ratio holds at any time, effective stress, and void ratio during secondary compression. The only exception, as shown by Leonards and Girault (1961, Fig. 3) seems to be the load increment that straddles the preconsolidation stress, σ_p'. Obviously, many questions remain to be answered concerning the topic of secondary compression.

TABLE 9-4　Values of C_α/C_c for Natural Soils*

Soil	C_α/C_c
Organic silts	0.035–0.06
Amorphous and fibrous peat	0.035–0.085
Canadian muskeg	0.09–0.10
Leda clay (Canada)	0.03–0.06
Post-glacial Swedish clay	0.05–0.07
Soft blue clay (Victoria, B.C.)	0.026
Organic clays and silts	0.04–0.06
Sensitive clay, Portland, ME	0.025–0.055
San Francisco Bay Mud	0.04–0.06
New Liskeard (Canada) varved clay	0.03–0.06
Mexico City clay	0.03–0.035
Hudson River silt	0.03–0.06
New Haven organic clay silt	0.04–0.075

*Modified after Mesri and Godlewski (1977).

If, for some reason, you do not want to or cannot determine C_α from laboratory test data, you can use the C_α/C_c data of Table 9-4 for similar soils, or simply use an average C_α/C_c value of 0.05, which is acceptable for preliminary calculations. Mesri (1973) has provided another method to obtain the secondary compression index, actually the modified secondary compression index, and it is shown in Fig. 9.12. Here the $C_{\alpha\epsilon}$ is plotted versus natural water content of the soil.

We will illustrate how to estimate secondary settlement in Examples 9.10, 9.11, and 9.12.

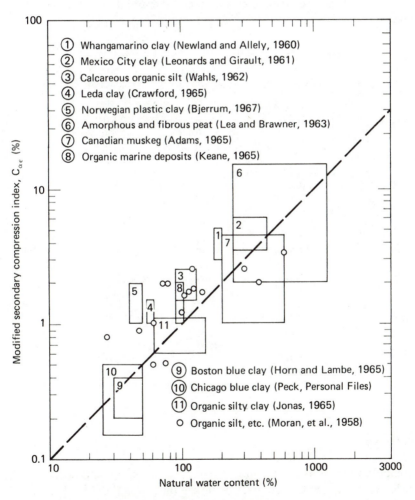

Fig. 9.12 Modified secondary compression index versus natural water content. (After Mesri, 1973. See Mesri, 1973 for details of the references included in this figure.)

EXAMPLE 9.10

Given:

The data for Problem 8-12 plus the following time rate of consolidation data for the load increment of 40 to 80 kPa. (This load increment represents the anticipated load in the field.) Assume the consolidation settlement, s_c, is 30 cm and that it occurs after 25 yr. The thickness of the compressible layer is 10 m. The initial void ratio e_o is 2.855, and the initial height of the test specimen is 25.4 mm, and the initial dial reading is 12.700 mm.

(1) Dial Reading (mm)	(2) Elapsed Time (min)	(3) Void Ratio
11.224	0	2.631
11.151	0.1	2.620
11.123	0.25	2.616
11.082	0.5	2.609
11.019	1.0	2.600
10.942	1.8	2.588
10.859	3.0	2.576
10.711	6	2.553
10.566	10	2.531
10.401	16	2.506
10.180	30	2.473
9.919	60	2.433
9.769	100	2.410
9.614	180	2.387
9.489	300	2.368
9.373	520	2.350
9.223	1350	2.327
9.172	1800	2.320
9.116	2850	2.311
9.053	4290	2.301

Required:

Compute the amount of secondary compression that would occur from 25 to 50 years after construction. Assume the time rate of deformation for the load range in the test approximates that occurring in the field.

Solution:

The solution to this problem requires an evaluation of C_α (Eq. 9-15). So a void ratio versus $\log t$ curve must be plotted from the given data. We can readily calculate the void ratio at any height or thickness of the specimen during the consolidation test by using the following method. By definition, $e = V_v/V_s$ and, for a constant specimen area, $e = H_v/H_s$, which is the ratio of height of voids to the height of solids. Then, from the phase diagram (Fig. Ex. 9.10a), the void ratio at any dial reading R may be obtained from

$$e = \frac{H_v}{H_s} = \frac{H_o - H_s}{H_s} = \frac{H_o - (R_o - R) - H_s}{H_s}$$

$$= \frac{(H_o - H_s) - (R_o - R)}{H_s} \tag{9-17}$$

where H_v = the height of voids at time t,
H_s = the height of solids,
H_o = the original height of specimen,
R_o = the initial dial reading, and
R = dial reading at time t.

From the phase diagram and the initial conditions of this problem,

$$H_s = \frac{H_o}{1 + e_o} = \frac{25.4}{1 + 2.855} = 6.589 \text{ mm}$$

For the load increment 40 to 80 kPa, the initial dial reading is 11.224; the dial reading R_o at the very beginning of the test (corresponding to specimen height H_o) is 12.700. Thus for the beginning of this load increment, e from Eq. 9-17 is

$$e = \frac{(25.4 - 6.589) - (12.700 - 11.224)}{6.589} = 2.631$$

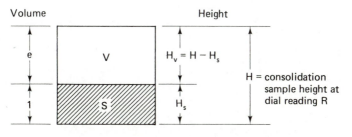

Fig. Ex. 9.10a For initial conditions, $e = e_o$, $H = H_o$, and $R = R_o$.

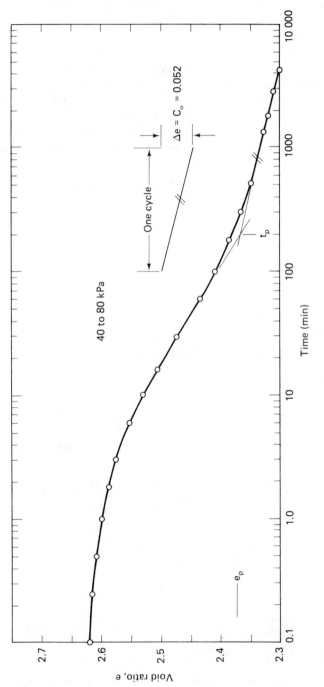

Fig. Ex. 9.10b

This value of e at $R = 11.224$ is shown in column 3 of the given data. The remainder of column 3 can be calculated by substituting the other values of R into Eq. 9-17.

Next, plot the void ratio, column 3, and the elapsed time, column 2, on semilog paper, as shown in Fig. Ex. 9.10b. C_α is then found to be 0.052. Note that $C_\alpha \equiv \Delta e$ when $\Delta \log t$ covers one full log cycle. The corresponding modified secondary compression index $C_{\alpha\epsilon}$ (Eq. 9-16) is $0.052/(1 + e_p)$ $= 0.052/(1 + 2.372) = 0.0154$; e_p is obtained from Fig. Ex. 9.10b at the end of primary consolidation.

To calculate secondary settlement s_s, use our basic settlement equation, Eq. 8-4:

$$s = \frac{\Delta e}{1 + e_o} H_o \qquad (8\text{-}4)$$

Now, however, Δe is a function of *time* and not stress. Substituting Δe from Eq. 9-15 into Eq. 8-4 and using e_p for e_o, we obtain

$$s_s = \frac{C_\alpha}{1 + e_p}(H_o)(\Delta \log t) \qquad (9\text{-}18)$$

$$= \frac{0.052}{1 + 2.372}(10 \text{ m}) \log \frac{50}{25}$$

$$= 0.046 \text{ m} = 4.6 \text{ cm}$$

Thus $s = s_c + s_s = 30 + 4.6 = 34.6$ cm in 50 years. This excludes any immediate settlement s_i that may also have occurred.

The secondary settlement may also be computed by means of Eqs. 8-4 and 9-16, where

$$s_s = C_{\alpha\epsilon} H_o (\Delta \log t) \qquad (9\text{-}19)$$

$$= 0.0154 (10 \text{ m}) \log \frac{50}{25}$$

$$= 0.046 \text{ m, as before}$$

A detailed example illustrating the computations for both s_c and s_s is presented at the end of this chapter.

EXAMPLE 9.11

Given:

Data given in Example 9.10 (Problem 8-12). The initial water content of the specimen is 105.7%.

Required:

From the data listed in Table 9-4 and Fig. 9.12, estimate the (a) C_α and (b) $C_{\alpha\epsilon}$. (c) Compare with the values calculated in Example 9.10.

Solution:

From Problem 8-12, the value of C_c is 1.23 and the value of $C_{c\epsilon}$ is 0.32.

    a. For San Francisco Bay Mud, use an average value of C_α/C_c of 0.05. Therefore

$$C_\alpha = 0.05(C_c) = 0.05(1.23) = 0.062$$

    b. From Eq. 9-16, $C_{\alpha\epsilon} = C_\alpha/1 + e_p$. From Fig. Ex. 9.10b, $e_p = 2.372$. Therefore

$$C_{\alpha\epsilon} = \frac{0.062}{1 + 2.372} = 0.018$$

    A second way to estimate the modified secondary compression index is to use Fig. 9.12, where $C_{\alpha\epsilon}$ is plotted versus natural water content. For our example, the initial water content was 105.7%. From Fig. 9.12, a value of $C_{\alpha\epsilon}$ of about 0.01 (or higher) is obtained if you use the dashed line.

    c. Compare with the calculated values. From Example 9.10, $C_\alpha = 0.052$ and $C_{\alpha\epsilon} = 0.015$. The agreement using the approximate values is certainly acceptable for preliminary design estimates.

9.8 COMPREHENSIVE EXAMPLE OF A TIME RATE OF SETTLEMENT PROBLEM

EXAMPLE 9.12

Given:

A brown silty sand fill 5 m thick was placed over a 15 m thick layer of compressible gray silty clay. Underlying the clay layer is brown sandy gravel. The soil profile is shown in Fig. Ex. 9.12a. Assume for this problem that the settlement of the fill and the sandy gravel is small compared to the

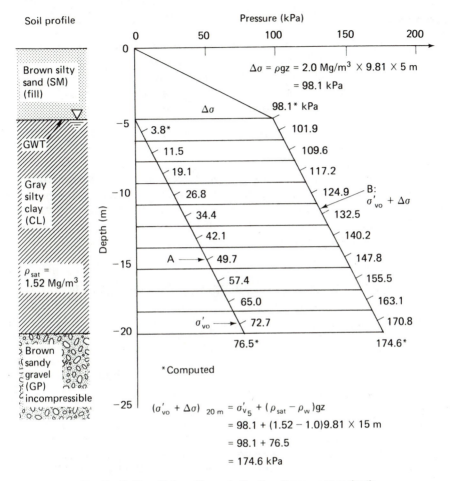

Fig. Ex. 9.12a Soil profile and effective stress versus depth.

settlement of the silty clay layer. Properties of the normally consolidated silty clay layer are:

Initial void ratio, $e_o = 1.1$.
Compression index, $C_c = 0.36$.
Secondary compression index, $C_\alpha = 0.06$.
Saturated density, $\rho_{sat} = 1.52$ Mg/m³.
Coefficient of consolidation, $c_v = 0.858$ m²/yr.
The density of the silty sand fill, ρ, is 2.0 Mg/m³, and the ground water table is at the interface of the fill and clay, at -5 m.

Required:

> Part I. Compute the consolidation settlement of the silty clay layer
> due to the weight of the 5 m of new fill.
> Part II. Compute the time rate of settlement.
> Part III. Compute and plot $\sigma'(z)$ when $U = 50\%$.
> Part IV. Compute the secondary settlement.

Solution:

Part I: The procedure for this part is (1) compute the initial effective overburden pressures of soil layers, (2) compute the increase in vertical stress due to external load, (3) compute the final effective vertical stress, and (4) compute settlement.

1. Initial effective stress: $\sigma'_{vo} = \rho'gz$, prior to fill placement.

$$\sigma'_{vo}(0 \text{ m}) = 0$$

$$\sigma'_{vo}(-15 \text{ m}) = (1.52 - 1.0) \text{ Mg/m}^3 \times 9.81 \text{ m/s}^2 \times 15 \text{ m}$$
$$= 76.5 \text{ kPa}$$

2. Increase in σ due to fill $= \rho g h_{\text{fill}}$.

$$\Delta\sigma_{\text{fill}} = 2.0 \text{ Mg/m}^3 \times 9.81 \text{ m/s}^2 \times 5 \text{ m} = 98.1 \text{ kPa}$$

3. Final effective stress $=$ initial effective stress $+ \Delta\sigma_{\text{fill}}$.

$$\sigma'_v \text{ (top of silty clay)} = 0 + \Delta\sigma = 98.1 \text{ kPa}$$

$$\sigma'_v \text{ (bottom of silty clay)} = 76.5 + 98.1 = 174.6 \text{ kPa}$$

The above stresses are plotted in Fig. Ex. 9.12a. Curve A represents the initial vertical effective overburden stresses σ'_{vo} prior to the placement of the 5 m of silty sand fill. Curve B represents the final vertical effective overburden stress due to the fill after complete consolidation of the silty clay layer has taken place. Curve B equals $\sigma'_{vo} + \Delta\sigma$, where $\Delta\sigma$ is the increase in pressure due to the fill. We will assume that $\Delta u = \Delta\sigma$ (one-dimensional compression) and that the load is placed all at once. (Actually 5 m of fill may take days to weeks to place and compact but, for purposes of our example, let us assume it was placed instantaneously in one load increment.)

4. Recall that the silty clay is normally consolidated. Thus the consolidation settlement of the layer is given by Eq. 8-11.

$$s_c = C_c \frac{H_o}{1 + e_o} \log \frac{\sigma'_{vo} + \Delta\sigma_v}{\sigma'_{vo}} \tag{8-11}$$

For the middepth of the layer, $\sigma'_{vo} = 38.3 \text{ kPa}$ and $\sigma'_{vo} + \Delta\sigma_v = 136.4 \text{ kPa}$.

Thus

$$s_c = 0.36 \frac{15 \text{ m}}{1 + 1.1} \log \frac{136.4}{38.3} = 1.42 \text{ m}$$

Since the silty clay layer is 15 m thick, it is prudent to divide the total thickness into thinner layers to improve the accuracy of results. A 1.5 m thick layer is chosen. The settlement of each of these layers is summed up to obtain the total consolidation settlement of the silty clay layer. To assist in the computations, Fig. Ex. 9.12b is employed, and the corresponding pressures σ'_{vo} and $\sigma'_{vo} + \Delta\sigma$ are indicated at the *middepth* of each of the

(1)	(2)	(3)	(4)	(5)	(6)	(7)	(8)	(9)	(10)	(11)
DEPTH BELOW GROUND SURFACE	AVERAGE DEPTH BELOW GROUND SURFACE	SOIL TYPE	σ'_{vo}	$\Delta\sigma$	$\dfrac{\sigma'_{vo} + \Delta\sigma}{\sigma'_{vo}}$	LOG COL. (6)	C_c	$\dfrac{C_c}{1 + e_o}$	THICKNESS OF DEPTH INCREMENT	SETTLEMENT
m	m		kPa	kPa					m	m
5	5.75	(CL)	3.8	98.1	$\dfrac{101.9}{3.8}$	1.43	0.36	$\dfrac{0.36}{1 + 1.1}$	1.5	0.367
6.5	7.25		11.5		$\dfrac{109.6}{11.5}$	0.979				0.252
8	8.75		19.1		$\dfrac{117.2}{19.1}$	0.788				0.203
9.5	10.25		26.8		$\dfrac{124.9}{26.8}$	0.668				0.172
11	11.75		34.4		$\dfrac{132.5}{34.4}$	0.586				0.151
12.5	13.25		42.1		$\dfrac{140.2}{42.1}$	0.522				0.134 ←6th
14	14.75		49.7		$\dfrac{147.8}{49.7}$	0.473				0.122
15.5	16.25		57.4		$\dfrac{155.5}{57.4}$	0.433				0.111
17	17.75		65.0		$\dfrac{163.1}{65.0}$	0.400				0.103
18.5	19.25		72.7		$\dfrac{170.8}{72.7}$	0.371				0.095
20									$s_c = \Sigma \Delta H = 1.71$ m	
									$\epsilon_v \cong 11\%$	

Fig. Ex. 9.12b Settlement computations.

layers. For example, the average depth of the sixth layer is -13.25 m; $\sigma'_{vo} = 42.1$ kPa while $\sigma'_{vo} + \Delta\sigma = 140.2$ kPa. These values are simply scaled off of Fig. Ex. 9.12a. Inserting the appropriate values into Eq. 8-11, we obtain

$$s_{c_{(12.5-14\,\text{m})}} = 0.36 \frac{1.5\ \text{m}}{1 + 1.1} \log \frac{140.2}{42.1} = 0.134\ \text{m}$$

The computations for each layer are listed in Fig. Ex. 9.12b. The total consolidation settlement of the 15 m thick layer is 1.71 m, or about 11% strain. This value of s_c is about 20% larger than the settlement calculated above for a single 15 m thick layer. All things considered, it probably is a more accurate prediction. Although the answer is given to two decimal places, there is seldom justification for such precision in settlement computations. An estimate of "approximately 1.7 m" would usually be of sufficient accuracy. Research (for example, Holtz and Broms, 1972; Leonards, 1977) has shown that consolidation settlements can be predicted within a range of about 20%.

Note that a settlement of 1.7 m means that 1.7 m of the fill would settle below the ground water table, which is at the original ground surface, and a decrease in the density of the fill due to buoyancy would result. Thus the actual settlement will be somewhat less than 1.7 m. This condition has been ignored in this example.

In part II of this example you are asked to compute the time rate of settlement. We can construct Table Ex. 9.12a incorporating U_{avg}, T, s_c, and t by using Eqs. 9-5, 9-12, and Table 9-1. We usually assume H remains constant at 15 m (see Sec. 9.3 and Appendix B-2). Since the clay layer has double drainage, the value of H_{dr} in Eq. 9-5 is 15 m/2, or 7.5 m. The c_v is given as 0.858 m^2/yr.

TABLE EX. 9.12a   Time Rate of Settlement

(1) U_{avg}	(2) T	(3) s_c (m)	(4) t (yr)
0.1	0.008	0.17	0.52
0.2	0.031	0.34	2.03
0.3	0.071	0.51	4.65
0.4	0.126	0.68	8.26
0.5	0.197	0.86	12.92
0.6	0.287	1.03	18.82
0.7	0.403	1.20	26.42
0.8	0.567	1.37	37.17
0.9	0.848	1.54	55.59
0.95	1.163	1.62	76.25
1.00	∞	1.71	∞

Values of T for given values of U_{avg} from Table 9-1 are substituted into this equation and solved for t, column 4 in Table Ex. 9.12a. The settlement in column 3 is obtained from Eq. 9-12 by multiplying the total consolidation settlement for this example, 1.71 m, by column 1. The data in columns 3 and 4 are plotted (similar to Fig. 9.6a) in Fig. Ex. 9.12c.

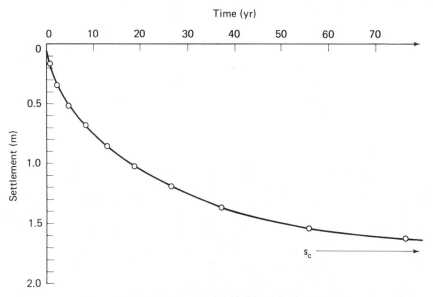

Fig. Ex. 9.12c   Data from Table Ex. 9.12a.

In engineering practice, only *estimates* of the time rate of settlement can be made because of the great dependence that the rate of settlement has on the drainage path. If there were *continuous* layers of permeable soil, for example thin sand seams, interbedded in the 15 m clay layer, then the rate of settlement would be significantly greater (see for instance, Example 9.6). Another factor is our inability to accurately predict c_v. If possible, estimates should be field checked, especially for important jobs.

In part III of this example you are asked to determine the effective stress with depth when $U_{avg} = 50\%$. The computations start with the evaluation of U_z from Fig. 9.3 and the construction of Table Ex. 9.12b.

The depths in column 1 of Table Ex. 9.12b represent evenly spaced elevations within the 15 m thick clay layer. Column 2 is the ratio of the depth to layer thickness. The time factor for U_{avg} of 50% is found from Table 9-1 to be 0.197 (use 0.2 for convenience). Using Fig. 9.3 along the time factor curve for $T = 0.2$ and the various ratios of z/H in column 2 of Table Ex. 9.12b, find the values of the degree of consolidation U_z at these ratios. For example, at $z/H = 0$ (and 2.0), the value of U_z is 1.0, or 100%

TABLE EX. 9.12b   Isochrone Data for $U_{avg} = 50\%$

(1) Depth (m)	(2) z/H	(3) U_z	(4) $\Delta\sigma_z'$ (kPa)
−5	0	1.00	98.1
−6.88	0.25	0.70	68.7
−8.75	0.5	0.455	44.6
−10.63	0.75	0.285	28.0
−12.5	1.0	0.23	22.6
−14.38	1.25	0.285	28.0
−16.25	1.5	0.455	44.6
−18.13	1.75	0.70	68.7
−20	2.0	1.00	98.1

consolidation at the top and bottom of the clay layer. At a ratio of 0.25 z/H (and 1.75 z/H) the degree of consolidation is 70%, etc. These values are placed in column 3. The effective stress is found by multiplying U_z in column 3 by the amount of $\Delta\sigma$, the weight of added fill, or 98.1 kPa. A plot of the isochrone for $U_{avg} = 50\%$ is shown in Fig. Ex. 9.12d. You should compare this figure with Fig. Ex. 9.12a. From Table Ex. 9.12a, you can see that it takes about 13 years to develop this isochrone. The isochrone represents the dividing line between the amount of $\Delta\sigma$ that has gone into effective stress and the amount of pore pressure in the clay layer that remains to be dissipated. If the clay layer were sampled and a consolidation test were performed at a depth of −12.5 m (the middle of the clay layer), the value of the preconsolidation pressure, σ_p', would be 60.9 kPa ($\sigma_{vo}' + \Delta\sigma$, 38.3 + 22.6 kPa). This value is obtained from Fig. Ex. 9.12d.

There are practical implications to be derived from part III. If a foundation engineer wanted to reduce the consolidation settlement of a structure, the site could be *preloaded* with fill and the fill removed later. The time the preload should be applied may be calculated as in this example. After the fill was removed, if the stress distribution of the new structure was about the same or less than the 50% isochrone shown in Fig. Ex. 9.12d, then the consolidation settlement would be calculated by using the recompression index C_r, and the settlements would be substantially less (Sec. 8.7).

Part IV, the final part, illustrates the computation for time rate of secondary compression. First plot the consolidation settlement data in Table Ex. 9.12a, s_c versus log time, shown in Fig. Ex. 9.12e. Note that this is a theoretical settlement-log time relationship. Solving for the secondary

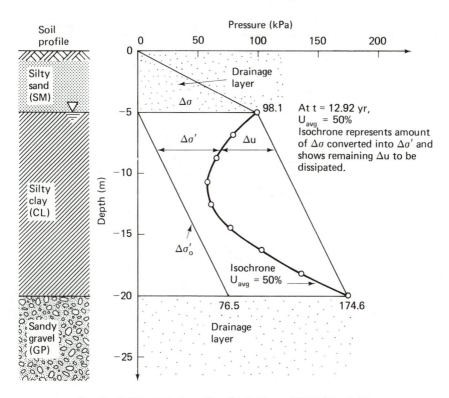

Fig. Ex. 9.12d   Data from Fig. Ex. 9.12a and Table Ex. 9.12b.

settlement (Eq. 9-18) for one log cycle, we obtain

$$s_s = \frac{C_\alpha}{1 + e_p}(H_o)(\Delta \log t)$$

$$= \frac{0.06}{1 + 1.1}(15 \text{ m})(1)$$

$$= 0.429 \text{ m/log cycle of time}$$

This slope is shown in Fig. Ex. 9.12e. This same rate of secondary compression starts at point a on the theoretical settlement-time curve. Point a corresponds to the settlement at 100% primary consolidation ($s_c = 1.71$ m). Note that the primary consolidation curve has been extrapolated slightly to point a. Thus, from Fig. Ex. 9.12e, the total settle-

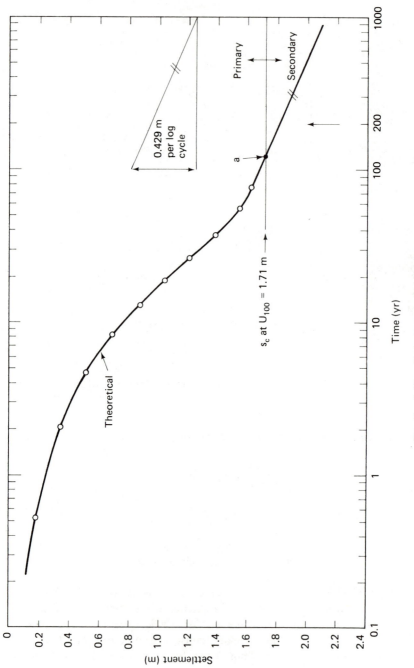

Fig. Ex. 9.12e   Data from Table Ex. 9.12a.

ment at the end of, say, 200 years is expected to be about 1.8 m. To be more precise than this in predicting settlements is beyond our ability to accurately evaluate soil properties and field drainage conditions.

The preceding example illustrates some of the computational details of a time rate of settlement analysis for the simple case of one-dimensional loading and for a normally consolidated clay. If the loaded area were of limited extent, then you would have to take into account the stress distribution with depth at key points under the loaded area. You would use the techniques discussed in Sec. 8.12 to establish curve B in a diagram, similar to Fig. Ex. 9.12a. You might even have to do this for several sections under the foundation, for example in the center, at the edge, and under the corner of the loaded area.

Because the clay in Example 9.12 was normally consolidated, computation of the consolidation settlement (part I) was relatively straightforward. If the clay had been overconsolidated, that is, if $\sigma'_{vo} < \sigma'_p$, then you would use Eqs. 8-16 through 8-19, as the case may be. Sometimes the upper part of the layer is overconsolidated, and the lower part is normally consolidated, and you have to take this into account in your computations. Another complication that often occurs is that the soil and consolidation properties (C_c, e_o) vary throughout the soil profile. In that case, when you break up the profile into smaller layers, as we did in Fig. Ex. 9.12a, the layers will not necessarily be evenly spaced. In this case, tables, such as illustrated in Fig. Ex. 9.12b, are very helpful for making the actual computations.

Procedures for handling complex settlement problems in engineering practice are treated in depth in foundation engineering textbooks.

When soil permeability and therefore c_v varies within the compressible layer, or when boundary layers impede drainage, the problem of time rate of consolidation becomes very complex and numerical techniques such as found in Scott (1963) and Harr (1966) are called for.

PROBLEMS

9-1. The time factor for a clay layer undergoing consolidation is 0.2. What is the degree of consolidation (consolidation ratio) at the center and at the quarter points (that is, $z/H = 0.25$ and 0.75)? What is the average degree of consolidation for the layer?

9-2. If the final consolidation settlement for the clay layer of Problem 9-1 is expected to be 1.0 m, how much settlement has occurred when the time factor is (a) 0.2 and (b) 0.7?

9-3. If the clay layer of Example 9.1 were singly drained, would there be any difference in the calculated U_z values? If so, how much difference?

9-4. Plot a graph of excess pore pressure versus depth, similar to Fig. Ex. 9.2, for the soil and loading conditions given in Example 9.2, but for the case of single drainage. Assume that under the clay there is impervious shale instead of a dense sand.

9-5. For the soil and loading conditions of Examples 9.1 and 9.2, estimate how long it would take for 0.1, 0.25, and 0.4 m of settlement to occur. Consider both single and double drainage.

9-6. By evaluation of the series expression (Eq. B-2-23 in Appendix B-2) for the solution to the consolidation equation, determine the average degree of consolidation U to the nearest 0.001 for time factors 0.2, 0.5, 0.9 and infinity. Verify your computations by referring to Table 9-1 and Fig. 9.5a. Also check by Eqs. 9-10 and 9-11. (After Taylor, 1948.)

9-7. How much difference would there be in the (a) computed ultimate settlement and (b) the time required for 90% consolidation for the soil conditions of Example 9.7 if the clay layer were doubly drained?

9-8. A deposit of Swedish clay is 12 m thick, on the average, and apparently drained on the bottom. The coefficient of consolidation for the clay was estimated to be 1×10^{-4} cm^2/s from laboratory tests. A settlement analysis based on oedometer tests predicted an ultimate consolidation settlement under the applied load in the field to be 1.2 m. (a) How long would it take for settlements of 40 and 70 cm to occur? (b) How much settlement would you expect to occur in 5 yr? 10 yr? 50 yr? (c) How long will it take for the ultimate settlement of 1.2 m to occur?

9-9. A conventional laboratory consolidation test on a 20 mm thick sample gave a time for 90% consolidation equal to 12 min. Calculate c_v in cm^2/s, m^2/s, and ft^2/d.

9-10. List the assumptions of the Terzaghi one-dimensional consolidation theory. List them in the order of their importance in terms of (a) mathematical convenience and (b) practical engineering significance.

9-11. The time rate of settlement data shown below is for the increment from 20 to 40 kPa from the test in Fig. 8.5. The initial sample height is 2.54 cm, and there are porous stones on the top and at the bottom of the sample. Determine c_v by (a) the log time-fitting procedure and (b) the square root of time procedure. (c) Compare the results of (a) and (b).

Elapsed Time (min)	Dial Reading (mm)
0	4.041
0.1	3.927
0.25	3.879
0.5	3.830
1	3.757
2	3.650
4	3.495
8	3.282
15	3.035
30	2.766
60	2.550
120	2.423
240	2.276
505	2.184
1485	2.040

9-12. An oedometer test (Taylor, 1948) was conducted on a sample of soft Chicago silty clay. The specimen had a dry weight of 329.99 g and a density of solids of 2.70 Mg/m³. The area of the ring was 93.31 cm². An old-fashioned dial indicator was used, which measured ten-thousandths of an inch (10^{-4} in. per division), and the incremental stresses applied to the specimen were recorded in kgf/cm². Direct measurements of the thickness of the specimen were as follows:

1.254 in. when under 1/8 kg/cm² (dial reading 2843)
1.238 in. when under 1/2 kg/cm² (dial reading 2694)
1.215 in. when under   1 kg/cm² (dial reading 2458)

Dial readings in 10^{-4} in. recorded during the test are listed in Table P9-12.

(a) Plot the e versus $\log \sigma'$ and/or the ϵ versus $\log \sigma'$ curve for this test. Determine the preconsolidation stress and the appropriate compression index.
(b) Plot dial reading versus $\sqrt{t}$ for each increment and determine c_v. Plot c_v versus $\log \sigma'$.
(c) Same as part (b), only use the Casagrande log time-fitting method.
(d) For two increments, one before the preconsolidation stress and one after the preconsolidation stress, compare the values of c_v as determined by the two fitting procedures.

TABLE P9-12   Dial Readings in 10^{-4} in.*

Elapsed Time (min)	Loading Increment (kg/cm²)						
	$\frac{1}{8}$ to $\frac{1}{4}$	$\frac{1}{4}$ to $\frac{1}{2}$	$\frac{1}{2}$ to 1	1 to 2	2 to 4	4 to 8	8 to 16
0	2843	2796	2694	2458	1500	3100	3102
$\frac{1}{4}$	2834	2780	2664	2421	1451	3047	3040
1	2829	2768	2647	2379	1408	2999	2985
$2\frac{1}{4}$	2824	2761	2629	2337	1354	2946	2931
4	2820	2751	2610	2288	1304	2896	2873
$6\frac{1}{4}$	2817	2742	2592	2239	1248	2841	2822
9	2813	2735	2576	2190	1197	2791	2768
$12\frac{1}{4}$	2811	2729	2562	2142	1143	2743	2728
16	2809	2724	2553	2098	1093	2701	2690
$20\frac{1}{4}$	2808	2720	2546	2044	1043	2660	2658
25	2807	2717	2540	2013	999	2630	2636
$30\frac{1}{4}$	2806	2715	2533	1969	956	2602	—
36	2805	2713	2529	1937	922	2575	2602
$42\frac{1}{4}$	2804	2710	—	1905	892	—	—
60	2803	2709	2517	1837	830	2525	2568
100	2802	2706	2508	1740	765	2496	2537
200	2801	2702	2493	1640	722	2471	2518
400	2799	2699	2478	1585	693	2446	2499
1440	2796	2694	2458	1500	642	2399	2468
(reset to)	—	—	—	—	3100	3102	—

*After Taylor (1948).

9-13. A consolidation test is performed on the specimen with these characteristics:

> Height of specimen = 38.10 mm
> Area of specimen = 90.1 cm²
> Wet weight of specimen = 621.5 g
> Dry weight of specimen = 475.1 g
> Density of solids = 2.80 Mg/m³

The consolidation data (after A. Casagrande) are summarized in Table P9-13.

(a) Plot the effective stress versus void ratio curve for both arithmetic and semilogarithmic scales.

(b) Estimate the preconsolidation pressure.

(c) Compute the compression index for virgin consolidation.

(d) Plot the time curve for the load increment from 256 to 512 kg for both arithmetic and semilogarithmic scales.

(e) Compute the coefficient of compressibility a_v, the coefficient of permeability, and the coefficient of consolidation c_v, for the load increment from 256 kg to 512 kg.

TABLE P9-13 Consolidation Test Data*

Temp. (°C)	Date	Time	Load (kg)	Elapsed Time (min)	Dial Reading (mm)
	5/16/84		0		0
			16		0.787
			32		1.176
			64		1.854
			128		2.896
			256		4.204
23.0	5/22/84	0933	512	Sudden	4.305
				0.10	4.343
				1.00	4.460
				4.00	4.663
				10.00	4.890
				28—	5.235
				72—	5.481
				182—	5.598
22.7		1733		480—	5.669
22.6		2240			5.700
23.4	5/23/84	1055			5.730
22.8	5/24/84	1100			5.753
	5/24/84		1024		7.366
	5/30/84		1024		7.447
			512		7.239
			256		6.949
			128		6.612
			32		5.878
	6/7/84		0.27		4.115
	6/30/84		0.27		3.693

*Modified after A. Casagrande.

9-14. A certain compressible layer has a thickness of 4 m. After 1 yr when the clay is 50% consolidated, 8 cm of settlement has occurred. For a similar clay and loading conditions, how much settlement would occur at the end of 1 yr and 4 yr if the thickness of this new layer were 40 m?

9-15. In a laboratory consolidation test on a representative sample of cohesive soil, the original height of a doubly drained sample was 25.4 mm. Based on the log time versus dial reading data, the time for 50% consolidation was 9 min. The laboratory sample was taken from a soil layer which is 12 m thick in the field, doubly drained, and is subjected to a similar loading. (a) How long will it take until the layer consolidates 50%? (b) If the final consolidation settlement is predicted to be 18 cm, how long will it take for a settlement of 4 cm to take place?

9-16. A layer of normally consolidated clay 3.5 m thick has an average void ratio of 1.3. Its compression index is 0.6 and its coefficient of consolidation is 1 m^2/yr. When the existing vertical pressure on the clay layer is doubled, what change in thickness of the clay layer will result?

9-17. The settlement analysis for a proposed structure indicates that 6 cm of settlement will occur in 4 yr and that the ultimate total settlement will be about 25 cm. The analysis is based on the assumption that the compressible clay layer is drained at both its top and bottom surfaces. However it is suspected that there may not be drainage at the bottom surface. For the case of single drainage, estimate (a) the ultimate total settlement and (b) the time required for 6 cm of settlement. (After Taylor, 1948.)

9-18. The structure of Problem 9-17 was constructed and performed essentially as expected during the first 4 yr (that is, the settlement of the building was about 6 cm). The owner decides to build a duplicate of the first structure nearby. During foundation investigations, it is discovered that the clay layer under the new building would be about 20% thicker than under the first structure. Otherwise, the properties of the clay are the same. Estimate for the new structure (a) the ultimate total settlement, and (b) the settlement in 4 yr. (After Taylor, 1948.)

9-19. A certain doubly drained clay layer has an expected ultimate settlement s_c of 15 cm. The clay layer, which is 17 m thick, has a coefficient of consolidation of 5×10^{-3} cm^2/s. Plot the s_c-time relationship to (a) an arithmetic time scale and (b) a semilog time scale.

9-20. Given the same soil data as for Problem 9-19. After 3 yr, an identical load is placed, causing an additional 15 cm of consolidation settlement. Compute and plot the time rate of settlement under these conditions, assuming that the load causing consolidation settlement is placed instantaneously.

9-21. Given the same data as for Problem 9-19. The load causing the 15 cm ultimate settlement was placed over a period of 2 yr. Although we haven't discussed how to handle this kind of problem, describe the approach you would use to compute the time history of settlement.

9-22. A specimen of clay in a special oedometer (with drainage at the top only) has a height of 2.065 cm when fully consolidated under a pressure of 50 kPa. A pressure transducer is located at the base of the sample to measure the pore water pressure. (a) When another stress increment of 50 kPa is applied, what would you expect the initial reading on the transducer to be? (b) If, after 15 min had elapsed, the transducer records a pressure of 25 kPa, what would you expect it to read 45 min later (total elapsed time of 1 h)? (After G. A. Leonards.)

9-23. The total consolidation settlement for a compressible layer 7 m thick is estimated to be about 30 cm. After about 6 mo (180 d) a point 2 m below the top of the singly drained layer has a degree of consolidation of 60%. (a) Compute the coefficient of consolidation of the material in m^2/d. (b) Compute the settlement for 180 d.

9-24. A 20 m thick normally consolidated clay layer has a load of 100 kPa applied to it over a large areal extent. The clay layer is located below a granular fill ($\rho = 2.0$ Mg/m^3) 3 m thick. A dense sandy gravel is found below the clay. The ground water table is located at the top of the clay layer, and the submerged density of the soil is 0.90 Mg/m^3. Consolidation tests performed on 2.20 cm thick doubly drained samples indicate $t_{50} = 9$ min for a load increment close to that of the loaded clay layer. Compute the effective stress in the clay layer at a depth of 18 m below the ground surface 4 yr after application of the load.

9-25. Given the same data as for Problem 9-24. At $t = 4$ yr, what is the average degree of consolidation for the clay layer?

9-26. Again, given the same data as for Problem 9-24. If the clay layer were *singly drained* from the top only, compute the effective stress at a depth of 18 m below the ground surface and 4 yr after placement of the external load. Comments?

9-27. Determine the average coefficient of permeability, corrected to 20°C, of a clay specimen for the following consolidation increment:

$\sigma_1 = 150$ kPa, $e_1 = 1.30$
$\sigma_2 = 300$ kPa, $e_2 = 1.18$
Height of specimen $= 20$ mm
Drainage at both top and bottom faces
Time required for 50% consolidation $= 20$ min
Test temperature $= 23$°C

(After A. Casagrande.)

9-28. The following data were obtained from a consolidation test on an undisturbed clay sample:

$\sigma_1 = 165$ kPa, $e_1 = 0.895$

$\sigma_2 = 310$ kPa, $e_2 = 0.732$

The average value of the coefficient of permeability of the clay in this pressure increment range is 3.5×10^{-9} cm/s. Compute and plot the decrease in thickness with time for a 10 m layer of this clay which is drained (a) on the upper surface only and (b) on the upper surface, and at a depth of 3 m by a thin horizontal sand layer that provides free drainage. (After A. Casagrande.)

9-29. Given the data of Problem 9-11. Evaluate (a) the secondary compression index and (b) the modified secondary compression index if

$$e_o = 2.60$$
$$H_o = 2.54 \text{ cm}$$
$$\rho_s = 2.75 \text{ Mg/m}^3$$
At $t = 0$, $e = 1.74$, $H = 1.928$ cm
At $t = 1485$ min, $e = 1.455$, $H = 1.728$ cm
Weight of technician = 7 stone; moon phase = full

9-30. Show that $C_{\alpha\epsilon} = \dfrac{C_\alpha}{1 + e_p} = \dfrac{\Delta\epsilon}{\Delta \log t}$ is valid.

9-31. Show that $s_s = \dfrac{C_\alpha}{1 + e_p}(H_o)\Delta \log t$ is true.

9-32. Estimate the secondary compression per log cycle of time for Problem 9-24.

9-33. The liquid limit of a soil is 80. Estimate the value of the modified secondary compression index.

9-34. Given the information of Example 9.12. What would be the settlement of the silty clay layer if *one* layer were chosen instead of the 10 layers actually used? What would be the settlement if 2, 5, and 7 layers represented the 15 m thick layer? Plot s_c versus the number of layers.

9-35. Given the data and information of Example 9.12. After 10 yr of consolidation, an additional areal fill load of 49 kPa is placed on the site. Compute the amount of additional settlement, and prepare a plot of settlement similar to Fig. Ex. 9.12c.

9-36. Given the data and information of Example 9.12. Subsequent subsurface investigation reveals a thin pervious layer at a depth of -9.5 m (that is, 4.5 m into the silty clay layer), which provides drainage throughout the silty clay. Compute the time rate of settlement for these new conditions, and plot the results on a graph similar to Fig. Ex. 9.12c.

ten

The Mohr Circle, Failure Theories, Stress Paths

10.1 INTRODUCTION

Before we discuss the stress-deformation and shear strength properties of soils, we need to introduce some new definitions and concepts about stress and failure. From Chapters 8 and 9, you know something about the load-settlement-time characteristics of cohesive soils, at least those due to one-dimensional loading. In this chapter and the next, we shall describe the reaction of sands and clays to types of loading other than one dimensional.

If the load or stress in a foundation or earth slope is increased until the deformations become unacceptably large, we say that the soil in the foundation or slope has "failed." In this case we are referring to the *strength* of the soil, which is really the maximum or ultimate stress the material can sustain. In geotechnical engineering, we are generally concerned with the *shear strength* of soils because, in most of our problems in foundations and earthwork engineering, failure results from excessive applied shear stresses.

The following notation is introduced in this chapter.

Symbol	Dimension	Unit	Definition
a	$ML^{-1}T^{-2}$	kPa	Intercept of the K_f line on the p-q diagram (Eq. 10-23)
c	$ML^{-1}T^{-2}$	kPa	Intercept of the Mohr failure envelope (Eq. 10-8)
CD	—	—	Consolidated drained (triaxial test)
CU	—	—	Consolidated undrained (triaxial test)

Symbol	Dimension	Unit	Definition
p	$ML^{-1}T^{-2}$	kPa	$(\sigma_v + \sigma_h)/2$ (Eq. 10-19)
q	$ML^{-1}T^{-2}$	kPa	$(\sigma_v - \sigma_h)/2$ (Eq. 10-18)
UU	—	—	Unconsolidated undrained (triaxial test)
α	—	(degree)	An angle
α_f	—	(degree)	Angle of the failure plane
β	—	(degree)	Arctan (q/p) (Eq. 10-21)
γ	—	(%)	Shear strain (angle of rotation in DSS test)
δ	L	m	Horizontal displacement
θ	—	(degree)	An angle
σ	$ML^{-1}T^{-2}$	kPa	Normal stress
σ_1	$ML^{-1}T^{-2}$	kPa	Major principal stress
σ_2	$ML^{-1}T^{-2}$	kPa	Intermediate principal stress
σ_3	$ML^{-1}T^{-2}$	kPa	Minor principal stress
σ_{ff}	$ML^{-1}T^{-2}$	kPa	Normal stress on the failure plane at failure (Eq. 10-7)
τ	$ML^{-1}T^{-2}$	kPa	Shear stress
τ_{ff}	$ML^{-1}T^{-2}$	kPa	Shear stress on the failure plane at failure (Eq. 10-7)
ψ	—	(degree)	Slope of the K_f line on the p-q diagram (Eq. 10-23)
ϕ	—	(degree)	Slope of the Mohr failure envelope (sometimes called the *angle of internal friction*) (Eq. 10-8)

Note: A prime on an angle or stress denotes *effective stress*.

10.2 STRESS AT A POINT

As we mentioned when we discussed effective stresses in Chapter 7, the concept of stress at a point in a soil is really fictitious. The point of application of a force within a soil mass could be on a particle or in a void. Clearly, a void cannot support any force, but if the force were applied to a particle, the stress could be extremely large. Thus when we speak about stress in the context of soil materials we are really speaking about a force per unit area, in which the area under consideration is the gross cross-sectional or engineering area. This area contains both grain-to-grain contacts as well as voids. The concept is similar to the "engineering area" used in seepage and flow problems (Chapter 7).

Consider a soil mass that is acted upon by a set of forces F_1, $F_2, \ldots, F_n$, as shown in Fig. 10.1. For the time being, let's assume that these

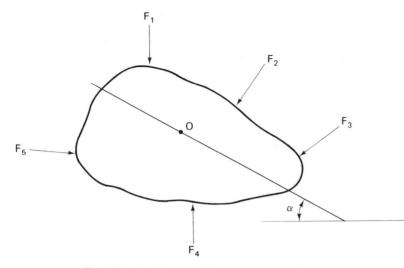

Fig. 10.1 A soil mass acted upon by several forces.

forces act in a two-dimensional plane. We could resolve these forces into components on a small element at any point within the soil mass, such as point O in that figure. The resolution of these forces into normal and shear components acting, for example, on a plane passing through point O at an angle α from the horizontal is shown in Fig. 10.2, which is an expanded view of a small element at point O. Note that for convenience our sign convention has *compressive forces and stresses positive* because most normal stresses in geotechnical engineering are compressive. This convention then requires that a *positive* shear stress produce *counterclockwise* couples on our element. Put another way: *positive* shears produce *clockwise* moments about a point just *outside* the element, as shown by the insert in Fig. 10.2. *Clockwise* angles are also taken to be *positive*. These conventions are the *opposite* of those normally assumed in structural mechanics.

To begin, let's assume that the distance AC along the inclined plane in Fig. 10.2 has unit length, and that the figure has a unit depth perpendicular to the plane of the paper. Thus the vertical plane BC has the dimension of $1 \sin \alpha$, and the horizontal dimension AB has a dimension equal to $1 \cos \alpha$. At equilibrium, the sum of the forces in any direction must be zero. So summing in the horizontal and vertical directions, we obtain

$$\Sigma F_h = H - T\cos\alpha - N\sin\alpha = 0 \qquad (10\text{-}1a)$$

$$\Sigma F_v = V + T\sin\alpha - N\cos\alpha = 0 \qquad (10\text{-}1b)$$

Dividing the forces in Eq. 10-1 by the areas upon which they act, we obtain the normal and shear stresses. (We shall denote the horizontal

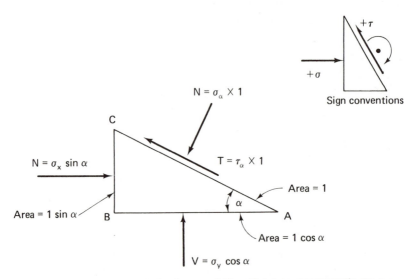

Fig. 10.2 Resolution of the forces of Fig. 10.1 into components on a small element at point O. Sign conventions are shown in the small inset figure.

normal stress by σ_x and the vertical normal stress by σ_y; the stresses on the α-plane are the normal stress σ_α and the shear stress τ_α.)

$$\sigma_x \sin \alpha - \tau_\alpha \cos \alpha - \sigma_\alpha \sin \alpha = 0 \tag{10-2a}$$

$$\sigma_y \cos \alpha + \tau_\alpha \sin \alpha - \sigma_\alpha \cos \alpha = 0 \tag{10-2b}$$

Solving Eqs. 10-2a and 10-2b simultaneously for σ_α and τ_α, we obtain

$$\sigma_\alpha = \sigma_x \sin^2 \alpha + \sigma_y \cos^2 \alpha = \frac{\sigma_x + \sigma_y}{2} + \frac{\sigma_x - \sigma_y}{2} \cos 2\alpha \tag{10-3}$$

$$\tau_\alpha = (\sigma_x - \sigma_y) \sin \alpha \cos \alpha = \frac{\sigma_x - \sigma_y}{2} \sin 2\alpha \tag{10-4}$$

If you square and add these equations, you will obtain the equation for a *circle* with a radius of $(\sigma_x - \sigma_y)/2$ and its center at $[(\sigma_x + \sigma_y)/2, 0]$. When this circle is plotted in τ-σ space, as shown in Fig. 10.3b for the element in Fig. 10.3a, it is known as the *Mohr circle of stress* (Mohr, 1887). It represents the state of stress *at a point at equilibrium*, and it applies to any material, not just soil. Note that the scales for τ and σ have to be the same to obtain a circle from these equations.

Since the vertical and horizontal planes in Fig. 10.2 and Fig. 10.3a have no shearing stresses acting on them, they are by definition *principal planes*. Thus the stresses σ_x and σ_y are really *principal stresses*. You may recall from your study of strength of materials that principal stresses act on planes where $\tau = 0$. The stress with the largest magnitude is called the

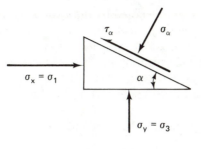

(a)

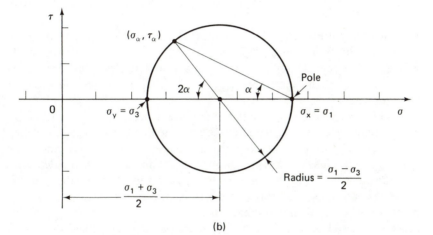

(b)

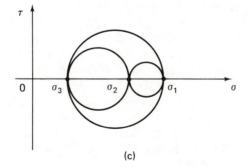

(c)

Fig. 10.3 The Mohr circle of stress: (a) element at equilibrium; (b) the Mohr circle; (c) Mohr circles including σ_2.

major principal stress, and denoted by the symbol σ_1. The smallest principal stress is called the *minor principal stress*, σ_3, and the stress in the third dimension is the *intermediate principal stress*, σ_2. In Fig. 10.3b, σ_2 is neglected since our derivation was for two-dimensional (plane stress) conditions. We could, however, construct two additional Mohr circles for σ_1 and σ_2 and σ_2 and σ_3 to make a complete Mohr diagram, as shown in Fig. 10.3c.

Now we can write Eqs. 10-3 and 10-4 in terms of principal stresses.

$$\sigma_\alpha = \frac{\sigma_1 + \sigma_3}{2} + \frac{\sigma_1 - \sigma_3}{2} \cos 2\alpha \qquad (10\text{-}5)$$

$$\tau_\alpha = \frac{\sigma_1 - \sigma_3}{2} \sin 2\alpha \qquad (10\text{-}6)$$

Here we have arbitrarily assumed that $\sigma_x = \sigma_1$ and $\sigma_y = \sigma_3$. You should verify that the coordinates of $(\sigma_\alpha, \tau_\alpha)$ in Fig. 10.3b can be determined by Eqs. 10-5 and 10-6. From these equations, also verify that the coordinates of the center of the circle are $[(\sigma_1 + \sigma_3)/2, 0]$, and that the radius is $(\sigma_1 - \sigma_3)/2$.

It is now possible to calculate the normal stress σ_α and shear stress τ_α on any plane α, as long as we know the principal stresses. In fact, we could almost as easily derive equations for the general case where σ_x and σ_y are not principal planes. These equations are known as the *double angle equations*, and they are the ones generally presented in strength of materials textbooks. The analytical procedure is sometimes awkward to use in practice because of the double angles; we perfer to use a graphical procedure based on a unique point on the Mohr circle called the *pole* or the *origin of planes*. This point has a very useful property: *any straight line drawn through the pole will intersect the Mohr circle at a point which represents the state of stress on a plane inclined at the same orientation in space as the line*. This concept means that if you know the state of stress, σ and τ, on some plane in space, you can draw a line parallel to that plane through the *coordinates* of σ and τ on the Mohr circle. The pole then is the point where that line intersects the Mohr circle. Once the pole is known, the stresses on *any plane* can readily be found by simply drawing a line from the pole parallel to that plane; the coordinates of the point of intersection with the Mohr circle determine the stresses on that plane. A few examples will illustrate how the pole method works.

EXAMPLE 10.1

Given:

Stresses on an element as shown in Fig. Ex. 10.1a.

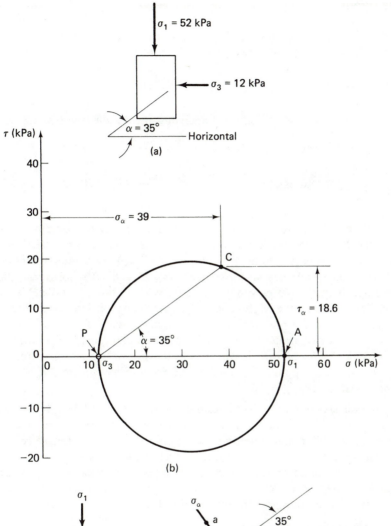

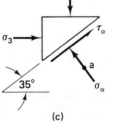

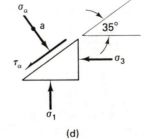

Fig. Ex. 10.1

Required:

The normal stress σ_α and the shear stress τ on the plane inclined at $\alpha = 35°$ from the horizontal reference plane.

Solution:

1. Plot the Mohr circle to some convenient scale (see Fig. Ex. 10.1b).

$$\text{center of circle} = \frac{\sigma_1 + \sigma_3}{2} = \frac{52 + 12}{2} = 32 \text{ kPa}$$

$$\text{radius of circle} = \frac{\sigma_1 - \sigma_3}{2} = \frac{52 - 12}{2} = 20 \text{ kPa}$$

2. Establish the origin of planes or the pole. It is probably easier to use the horizontal plane upon which σ_1 acts. The state of stress on this plane is indicated by point A in Fig. Ex. 10.1b. Draw a line parallel to the plane upon which this state of stress (σ_1, 0) acts (the horizontal plane) through the point representing σ_1 and 0. By definition, the pole P is where this line intersects the Mohr circle. [By coincidence, it intersects at (σ_3, 0)]. A line through the pole inclined at an angle $\alpha = 35°$ from the horizontal plane would be parallel to the plane on the element in Fig. Ex. 10.1a, and this is the plane on which we require the normal and shear stress. The intersection is at point C in Fig. Ex. 10.1b, and we find that $\sigma_\alpha = 39$ kPa and $\tau_\alpha = 18.6$ kPa.

You should verify these results by using Eqs. 10-5 and 10-6. Note that τ_α is positive since point C occurs above the abscissa. Thus the sense of τ_α on the 35° plane is determined as indicated in Figs. Ex. 10.1c and d, which represent the top and bottom parts of the given element. For both parts, the direction or sense of the shear stress τ_α is equal and opposite (as it should be). However, they are both positive shear stresses, which is consistent with our sign convention (Fig. 10.2).

EXAMPLE 10.2

Given:

The same element and stresses as in Fig. Ex. 10.1a, except that the element is rotated 20° from the horizontal, as shown in Fig. Ex. 10.2a.

Required:

As in Example 10.1, find the normal stress σ_α and the shear stress τ_α on the plane inclined at $\alpha = 35°$ from the base of the element.

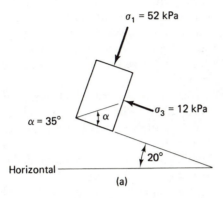

$\sigma_1 = 52$ kPa

$\sigma_3 = 12$ kPa

$\alpha = 35°$

α

$20°$

Horizontal

(a)

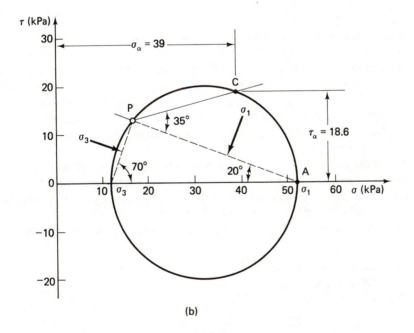

τ (kPa)

30

$\sigma_\alpha = 39$

C

20

P

σ_1

$35°$

10

σ_3

$\tau_\alpha = 18.6$

$70°$

$20°$

0

10 σ_3 20 30 40 50 σ_1 60 σ (kPa)

−10

−20

(b)

Fig. Ex. 10.2

439

Solution:

1. Plot the Mohr circle (Fig. Ex. 10.2b). Since the principal stresses are the same, the Mohr circle will be the same as in Example 10.1.

2. Find the pole of the circle. As in the previous example, draw a line parallel to a plane on which you know the stresses. If we again begin with the major principal plane, this plane is inclined at an angle of 20° to the horizontal. Start at point A, and where this line intersects the Mohr circle, defines the pole P of this circle.

3. Now find the stresses on the α-plane, which as before is inclined at 35° to the base of the element. From line AP, turn an angle in the same direction as in the element, 35°, and the stresses on that plane are defined by the point of intersection of the line with the Mohr circle (in this case at point C). Scale off the coordinates of point C to determine σ_α and τ_α. Note that these stresses are the same as in Example 10.1. Why is this? Because nothing has changed except the orientation in space of the element.

For step 2, we could just as well have used the minor principal plane as our starting point. In this case a line from $(\sigma_3, 0)$ could be drawn at 70° from the horizontal (parallel to the σ_3-plane), and it would intersect the Mohr circle at the same point as before, point P. We now have a check on the step—if we have done everything correctly, we should obtain the same pole. Since line AP is parallel to the major principal plane, we can show the direction of σ_1 right on this line in Fig. Ex. 10.2; similarly, the dashed line from the pole to σ_3 is parallel to the σ_3-plane.

Now you probably can begin to see what is really happening with the pole. It is just a way of relating the Mohr circle of stress to the geometry or orientation of our element in the real world. We could just as well rotate the τ-σ axes to coincide with the directions of the principal stresses in space, but traditionally τ versus σ is plotted with the axes horizontal and vertical.

EXAMPLE 10.3

Given:

The stress shown on the element in Fig. Ex. 10.3a.

Required:

    a. Evaluate σ_α and τ_α when $\alpha = 30°$.
    b. Evaluate σ_1 and σ_3 when $\alpha = 30°$.

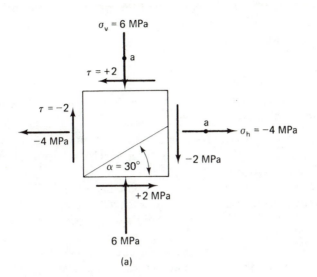

$\sigma_v = 6$ MPa

a

$\tau = +2$

$\tau = -2$

-4 MPa

a

$\sigma_h = -4$ MPa

-2 MPa

$\alpha = 30°$

$+2$ MPa

6 MPa

(a)

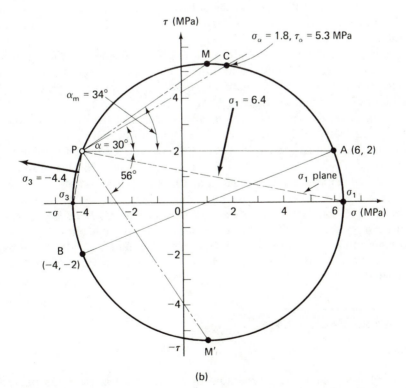

τ (MPa)

$\sigma_\alpha = 1.8$, $\tau_\alpha = 5.3$ MPa

M C

$\alpha_m = 34°$

$\sigma_1 = 6.4$

P

$\alpha = 30°$

A (6, 2)

$\sigma_3 = -4.4$

$56°$

σ_1 plane

σ_3

σ_1

$-\sigma$ -4 -2 0 2 4 6 σ (MPa)

B
$(-4, -2)$

-2

-4

$-\tau$ M'

(b)

Fig. Ex. 10.3

    c. Determine the orientation of the major and minor principal planes.

    d. Find the maximum shear stress and the orientation of the plane on which it acts.

Solution:

Construct the Mohr circle, as shown in Fig. Ex. 10.3, according to the following steps:

    1. Plot the state of stress on the horizontal plane $(6, 2)$ at point A. Note that the shear stress makes a clockwise moment about a and therefore is positive.

    2. In a similar manner, plot point B $(-4, -2)$. The shear stress on the vertical plane is negative since it makes a counterclockwise moment.

    3. Points A and B are two points on a circle (a diameter in this case since their planes are 90° apart); the center of the circle has coordinates of $[(\sigma_x + \sigma_y)/2, 0]$. Construct the Mohr circle with center at $(1, 0)$.

    4. To find the pole, remember that a line drawn parallel to the plane (horizontal in this example) upon which a known state of stress acts, point A, intersects the Mohr circle at the pole P. As a check, you could also draw a line in the vertical direction from point B $(-4, -2)$ and find the same pole.

    5. To find the state of stress on the plane inclined at angle $\alpha = 30°$ from the horizontal, draw the line PC at an angle 30° from the horizontal (see Fig. Ex. 10.3b). The state of stress on this plane is given by the coordinates at point C $(1.8, 5.3)$ MPa.

    6. Lines drawn from P to σ_1 and σ_3 establish the orientation of the major and minor principal planes. The values of σ_1 and σ_3 are determined automatically once the circle is drawn; here they are 6.4 and -4.4 MPa, respectively. Of course σ_1 and σ_3 are perpendicular to their respective planes, which are oriented at 11° and 101° to the horizontal, respectively.

    7. The maximum shear stress can be calculated by Eq. 10-6 when $2\alpha = 90°$. This is $(\sigma_1 - \sigma_3)/2$ or ± 5.4 MPa (see points M or M'). You can also simply scale off the maximum value of τ from the Mohr diagram. The orientation of τ_{max} is the line PM or PM', depending on which mutually perpendicular plane you desire. (Actually $\tau = -5.4$ MPa is the minimum shear stress.)

EXAMPLE 10.4

Given:

Two planes, a and b, are separated by an unknown angle θ. On plane a, $\sigma_a = 10$ kPa and $\tau_a = +2$ kPa. Plane a acts 15° from the horizontal, as shown in Fig. Ex. 10.4a. The stresses on plane b are $\sigma_b = 9$ kPa and $\tau_b = -3$ kPa.

Required:

    a. Find the major and minor principal stresses and their orientation.
    b. Find the stresses on the horizontal plane.
    c. Find the angle between planes a and b.

Solution:

    1. Plot the coordinates of the stresses on planes a and b. If you assume the body or element is in equilibrium, then these coordinates are *on* the circumference of the Mohr circle. To find the center, construct a perpendicular to the line AB, which joins the two points. The intersection of the horizontal σ-axis and the perpendicular bisector to AB is the center of the circle C.
    2. Establish the pole by drawing a line from point A parallel to the plane (15° from the horizontal) upon which the stresses at point A act to where it intersects the Mohr circle. The intersection of this line and the Mohr circle is the pole P.
    3. Lines from the pole P to σ_1 and σ_3 indicate the orientation of the major and minor principal planes. The principal stresses act perpendicular to these planes. The scaled-off value of σ_1 is equal to 10.65 kPa, and σ_3 is found to be 3.61 kPa.
    4. The stresses on the horizontal plane are found by drawing a horizontal line from the pole until it intersects the Mohr circle at point H; the stresses on this plane are (8.6, 3.18) kPa.
    5. To find the angle between the two planes a and b, draw the line PB from the pole B. This line is the actual orientation in space of plane B. The angle θ then represents the true angle between planes A and B, or $\theta = 46°$.

EXAMPLE 10.5

Given:

The stresses on an element shown in Fig. Ex. 10.5a.

Required:

Find the magnitude and direction of the major and minor principal stresses.

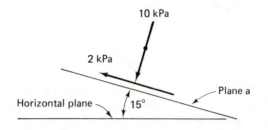

10 kPa

2 kPa

Horizontal plane

15°

Plane a

On plane b: σ_b = 9 kPa, τ_b = −3 kPa, $\theta°$ = ? from plane a

(a)

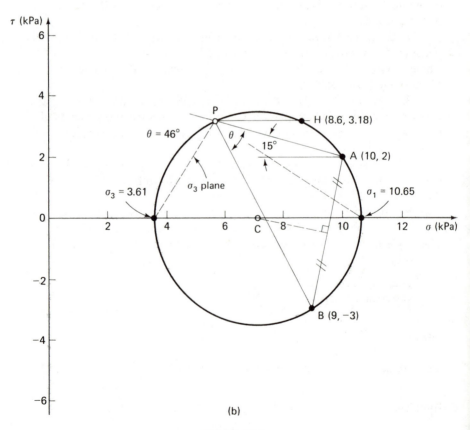

τ (kPa)

6

4

P

H (8.6, 3.18)

θ = 46°

θ

15°

A (10, 2)

2

σ_3 = 3.61

σ_3 plane

σ_1 = 10.65

0

2   4   6   C   8   10   12 σ (kPa)

−2

B (9, −3)

−4

−6

(b)

Fig. Ex. 10.4

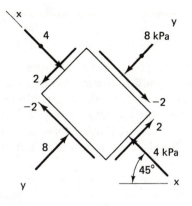

(a)

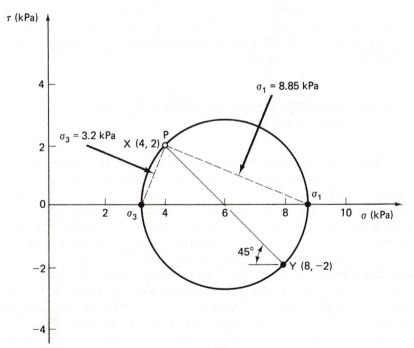

(b)

Fig. Ex. 10.5

Solution:

Refer to Fig. Ex. 10.5b for the following steps.

1. Plot the two points X and Y from the given stress coordinates. These two points lie on the circumference of the circle. Where the line XY intersects the σ-axis establishes the center of the Mohr circle at $(6, 0)$.

2. Locate the pole by drawing a line from point Y parallel to the plane on which the stress at Y acts. This line is at $45°$ from the horizontal, and it intersects the Mohr circle at the pole P, which is the same point as point X.

3. To find the direction of the principal stresses, draw a line from the pole to σ_1 and σ_3; these lines are shown dashed on Fig. Ex. 10.5b. The direction (arrows) of σ_1 and σ_3 are shown in the figure. The values of σ_1 and σ_3 are scaled off the figure and found to be 8.85 kPa and 3.2 kPa, respectively.

By now you can see that the Mohr circle of stress represents the complete two-dimensional state of stress *at equilibrium* in an element or at a point. The pole simply couples the Mohr circle to the orientation of the element in the real world. The Mohr circle and the concept of the pole are very useful in geotechnical engineering; we shall use them throughout the rest of this text.

10.3 STRESS-STRAIN RELATIONSHIPS AND FAILURE CRITERIA

Earlier, in the introduction to Chapter 8, we briefly mentioned some stress-strain relationships. Now we want to elaborate on, as well as illustrate, some of those ideas. The stress-strain curve for mild steel is shown in Fig. 10.4a. The initial portion up to the proportional limit or yield point is *linearly elastic*. This means that the material will return to its original shape when the stress is released, as long as the applied stress is below the yield point. It is possible, however, for a material to have a *nonlinear* stress-strain curve and still be elastic, as shown in Fig. 10.4b. Note that both these stress-strain relationships are independent of time. If time is a variable, then the material is called *visco-elastic*. Some real materials such as most soils and polymers are visco-elastic. Why, then, don't we use a visco-elastic theory to describe the behavior of soils? The problem is that soils have a highly nonlinear stress-strain-time behavior, and unfortunately only a mathematically well-developed linear theory of visco-elasticity is available.

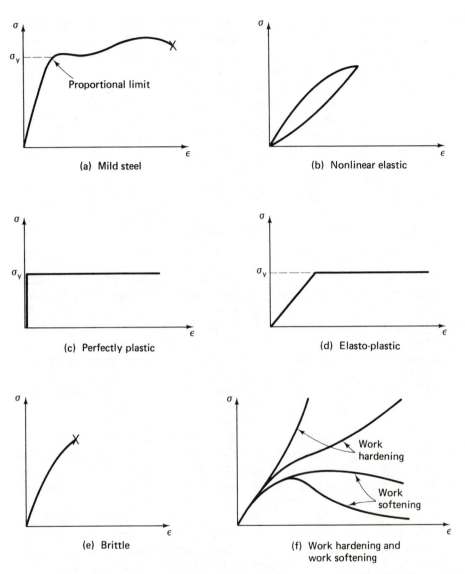

(a) Mild steel

(b) Nonlinear elastic

(c) Perfectly plastic

(d) Elasto-plastic

(e) Brittle

(f) Work hardening and
work softening

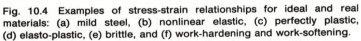

Fig. 10.4 Examples of stress-strain relationships for ideal and real materials: (a) mild steel, (b) nonlinear elastic, (c) perfectly plastic, (d) elasto-plastic, (e) brittle, and (f) work-hardening and work-softening.

Note that so far we've said nothing about failure or yield. Even linearly elastic materials yield, as indicated in Fig. 10.4a, if sufficient stress is applied. At the proportional limit, the material is said to become *plastic* or to *yield plastically*. The behavior of real materials can be idealized by several plastic stress-strain relationships, as shown in Figs. 10.4c, d, and f. *Perfectly plastic* materials (Fig. 10.4c), sometimes called *rigid-plastic*, can be treated relatively easily mathematically, and thus are popular subjects of study by mechanicians and mathematicians. A more realistic stress-strain relationship is *elasto-plastic* (Fig. 10.4d). The material is linearly elastic up to the yield point σ_y; then it becomes perfectly plastic. Note that both perfectly plastic and elasto-plastic materials continue to strain even without any additional stress applied. The stress-strain curve for mild steel can be approximated by an elasto-plastic stress-strain curve, and this theory is very useful in, for example, working, punching, and machining of metals. Sometimes materials such as cast iron, concrete, and a lot of rocks are *brittle*, in that they exhibit very little strain as the stress increases. Then, at some point, the material suddenly collapses or crushes (Fig. 10.4e). More complex but also realistic for many materials are the stress-strain relations shown in Fig. 10.4f. *Work-hardening* materials, as the name implies, become stiffer (higher modulus) as they are strained or "worked." The little hump in the stress-strain curve for mild steel after yield (Fig. 10.4a) is an example of work-hardening. Many soils are also work-hardening, for example, compacted clays and loose sands. *Work-softening* materials (Fig. 10.4f) show a decrease in stress as they are strained beyond a peak stress. Sensitive clay soils and dense sands are examples of work-softening materials.

At what point on the stress-strain curve do we have failure? We could call the yield point "failure" if we wanted to. In some situations, if a material is stressed to its yield point, the strains or deflections are so large that for all practical purposes the material has failed. This means that the material cannot satisfactorily continue to carry the applied loads. The stress at "failure" is often very arbitrary, especially for nonlinear materials. With brittle-type materials, however, there is no question when failure occurs—it's obvious. Even with work-softening materials (Fig. 10.4f), the peak of the curve or the maximum stress is usually defined as failure. On the other hand, with some plastic materials it may not be obvious. Where would you define failure if you had a work-hardening stress-strain curve (Fig. 10.4f)? With materials such as these, we usually define failure at some arbitrary percent strain, for example, 15 or 20%, or at a strain or deformation at which the function of the structure might be impaired.

Now we can also define the *strength* of a material. It is the maximum or yield stress or the stress at some strain which we have defined as "failure."

As suggested by the above discussion, there are many ways of defining failure in real materials; or put another way, there are many *failure criteria*. Most of the criteria don't work for soils, and in fact the one we do use, which is the subject of the next section, doesn't always work so well either. Even so, the most common failure criterion applied to soils is the *Mohr-Coulomb failure criterion*.

10.4 THE MOHR-COULOMB FAILURE CRITERION

Mohr is the same Otto Mohr of Mohr circle fame. Coulomb you know from coulombic friction, electrostatic attraction and repulsion, among other things. Around the turn of this century, Mohr (1900) hypothesized a criterion of failure for real materials in which he stated that materials fail when the *shear stress on the failure plane at failure reaches some unique function of the normal stress on that plane*, or

$$\tau_{ff} = f(\sigma_{ff}) \tag{10-7}$$

where τ is the shear stress and σ is the normal stress. The first subscript f refers to the plane on which the stress acts (in this case the *failure plane*) and the second f means "at failure."

τ_{ff} is called the *shear strength* of the material, and the relationship expressed by Eq. 10-7 is shown in Fig. 10.5a. Figure 10.5b shows an element at failure with the principal stresses that caused failure and the resulting normal and shear stresses on the failure plane.

For the present, we will assume that a failure plane exists, which is not a bad assumption for soils, rocks, and many other materials. Also, we won't worry now about how the principal stresses at failure are applied to the element (test specimen or representative element in the field) or how they are measured.

Anyway, if we know the principal stresses at failure, we can construct (draw, sketch) a Mohr circle to represent this state of stress for this particular element. Similarly, we could conduct several tests to failure or measure failure stresses in several elements at failure, and construct Mohr circles for each element or test at failure. Such a series is plotted in Fig. 10.6. Note that only the top half of the Mohr circles are drawn, which is conventionally done in soil mechanics for convenience only. Since the Mohr circles are determined at failure, it is possible to construct the limiting or failure envelope of the shear stress. This envelope, called the *Mohr failure envelope*, expresses the functional relationship between the shear stress τ_{ff} and the normal stress σ_{ff} at failure (Eq. 10-7).

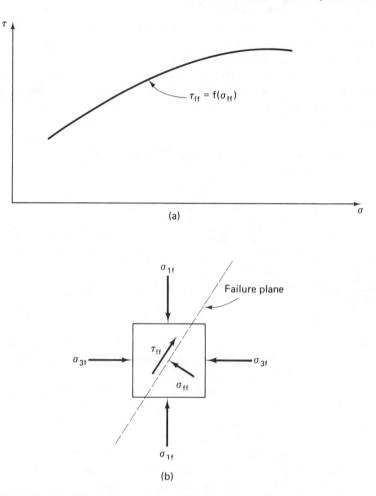

Fig. 10.5 (a) Mohr failure criterion; (b) element at failure, showing the principal stresses and the stresses on the failure plane.

Note that any Mohr circle lying below the Mohr failure envelope (such as circle A in Fig. 10.6) represents a stable condition. Failure occurs only when the combination of shear and normal stress is such that the Mohr circle is *tangent* to the Mohr failure envelope. Note also that circles lying above the Mohr failure envelope (such as circle B in Fig. 10.6) cannot exist. The material would fail before reaching these states of stress. If this envelope is unique for a given material, then the point of tangency of the Mohr failure envelope gives the stress conditions on the failure plane at failure. Using the pole method, we can therefore determine the angle of the failure plane from the point of tangency of the Mohr circle and the Mohr

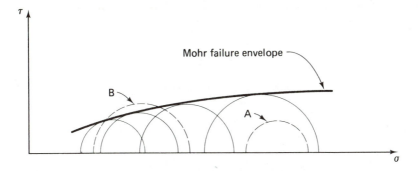

Fig. 10.6   The Mohr circles at failure define the Mohr failure envelope.

failure envelope. The hypothesis, that the point of tangency defines the angle of the failure plane in the element or test specimen, is the *Mohr failure hypothesis*. You should distinguish this hypothesis from the Mohr failure theory. The Mohr failure hypothesis is illustrated in Fig. 10.7a for the element at failure shown in Fig. 10.7b. Stated another way: the Mohr failure hypothesis states that the point of tangency of the Mohr failure envelope with the Mohr circle at failure determines the inclination of the failure plane.

Another thing you should note from Fig. 10.7a is that even though in soil mechanics we commonly draw only the top half of the Mohr circle, there is a bottom half, and also a "bottom-half" Mohr failure envelope. This also means, if the Mohr failure hypothesis is valid, that it is equally likely that a failure plane will form at an angle of $-\alpha_f$, as shown in Fig. 10.7a. In fact, it is the nonuniform stress conditions on the ends of a test specimen and small inhomogeneities within the specimen itself that we think cause a single failure plane to often form in a test specimen. Ever wonder why a cone forms at failure in the top and bottom of a concrete cylinder when it is failed in compression? Shear stresses between the testing machine and specimen caps cause nonuniform stresses to develop within the specimen. If everything is homogeneous and uniform stress conditions are applied to a specimen, then multiple failure planes form at conjugate angles, $\pm\alpha_f$, as shown in Fig. 10.7c.

Now we are going to involve Monsieur Dr. Coulomb in our story. In addition to his famous experiments with cats' fur and ebony rods, M. Coulomb (1776) was also concerned with military defense works such as revetments and fortress walls. At that time, these constructions were built by rule of thumb, and unfortunately for the French military defenses many of these works failed. Coulomb became interested in the problem of the lateral pressures exerted against retaining walls, and he devised a system for analysis of earth pressures against retaining structures that is still used

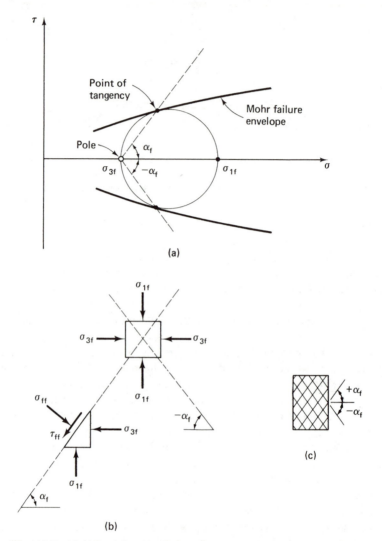

Fig. 10.7 (a) Mohr failure hypothesis for determining the angle of the failure plane in the (b) element; (c) conjugate failure planes.

today. One of the things he needed for design was the shearing strength of the soil. Being also interested in the sliding friction characteristics of different materials, he set up a device for determining the shear resistance of soils. He observed that there was a stress-independent component of shear strength and a stress-dependent component. The stress-dependent component is similar to sliding friction in solids, so he called this component the *angle of internal friction*, denoting it by the symbol ϕ. The other component seemed to be related to the intrinsic *cohesion* of the material and it is commonly denoted by the symbol c. Coulomb's equation is, then,

$$\tau_f = \sigma \tan \phi + c \tag{10-8}$$

where τ_f is the shear strength of the soil, σ is the applied normal stress, and ϕ and c are called the *strength parameters* of the soil as defined above. This relationship gives a straight line and is, therefore, easy to work with. As is explained in the next chapter, neither ϕ nor c are inherent properties of the material; on the contrary they are dependent on the conditions operative in the test. We could, much as M. Coulomb probably did, plot the results of a shear test on soil to obtain the strength parameters ϕ and c (Fig. 10.8). Note that either strength parameter could be zero for any particular stress condition; that is, $\tau = c$ when $\phi = 0$, or $\tau = \sigma \tan \phi$ when $c = 0$. As we shall see in Chapter 11, these relationships are valid for certain specific test conditions for some soils.

Although who first did so is unknown, it would seem reasonable to combine the Coulomb equation, Eq. 10-8, with the Mohr failure criterion, Eq. 10-7. Engineers traditionally prefer to work with straight lines since anything higher than a first-order equation (straight line) is too complicated! So the natural thing to do was to straighten out that curved Mohr failure envelope, or at least approximate the curve by a straight line over some given stress range; then the equation for that line in terms of the Coulomb strength parameters could be written. Thus was born the *Mohr-Coulomb strength criterion*, which is by far the most popular strength criterion applied to soils. The Mohr-Coulomb criterion can be written as

$$\tau_{ff} = \sigma_{ff} \tan \phi + c \tag{10-9}$$

the terms having been defined previously. This is a simple, easy-to-use criterion that has many distinct advantages over other failure criteria. It is the only failure criterion which predicts the stresses on the failure plane at failure, and since soil masses have been observed to fail on rather distinct surfaces, we would like to be able to estimate the state of stress on

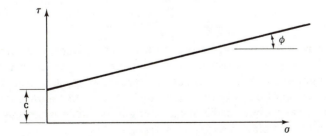

Fig. 10.8   The Coulomb strength equation presented graphically.

potential sliding surfaces. So the Mohr-Coulomb criterion is very useful for analyses of the stability of earth slopes and foundation.

Before we discuss the kinds of tests used to determine the Mohr-Coulomb strength parameters, we should look a little more carefully at some Mohr circles, both before failure and at failure. They have several interesting characteristics that will be useful later on.

First, if we know the angle of inclination of the Mohr failure envelope or have determined it from laboratory tests, then it is possible to write the angle of the failure plane α_f in terms of the slope ϕ of the Mohr failure envelope. To do this, we have to invoke the Mohr failure hypothesis. Then the failure angle measured relative to the plane of the major principal stress is

$$\alpha_f = 45° + \frac{\phi}{2} \tag{10-10}$$

A proof of this equation is requested in one of the problems at the end of the chapter.

Second, let's look at a soil element subjected to principal stresses which are *less* than the stresses required to cause failure. Such a state of stress might be represented by the Mohr circle shown in Fig. 10.9a. In this case τ_f is the *mobilized* shear resistance on the *potential* failure plane, and τ_{ff} is the shear strength available (shear stress on the failure plane at failure). Since we haven't reached failure yet, there is some reserve strength remaining, and this really is a definition of the *factor of safety* in the material. Or

$$\text{factor of safety (F.S.)} = \frac{\tau_{ff}\,(\text{available})}{\tau_f\,(\text{applied})} \tag{10-11}$$

Now, if the stresses increase so that failure occurs, then the Mohr circle becomes tangent to the Mohr failure envelope. According to the Mohr failure hypothesis, failure occurs on the plane inclined at α_f and with shear stress on that plane of τ_{ff}. Note that this is not the largest or maximum shear stress in the element! The maximum shear stress acts on the plane inclined at 45° and is equal to

$$\tau_{\max} = \frac{\sigma_{1f} - \sigma_{3f}}{2} > \tau_{ff} \tag{10-12}$$

Then why doesn't failure occur on the 45° plane? Well, it cannot, because on that plane the shear strength available is greater than $\tau_{\max}$, so failure cannot occur. This condition is represented by the distance from the maximum point on the Mohr circle up to the Mohr failure envelope in Fig. 10.9b. That would be the shear strength available when the normal stress σ_n on the 45° plane was $(\sigma_{1f} + \sigma_{3f})/2$.

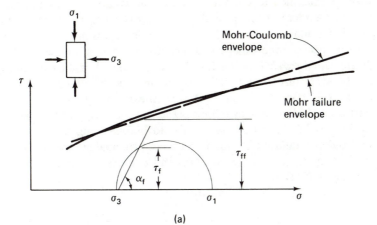

(a)

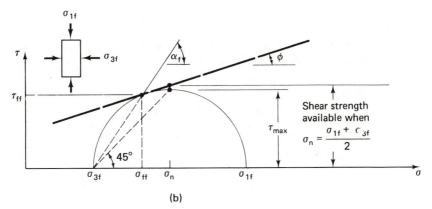

(b)

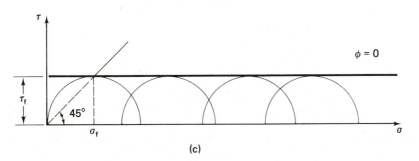

(c)

Fig. 10.9 (a) Stress conditions before failure; (b) stress conditions at failure; (c) Mohr failure envelope for a purely cohesive material (after Hirschfeld, 1963).

The only exception to the above discussion would be when the shear strength is independent of the normal stress; that is, when the Mohr failure envelope is horizontal and $\phi = 0$. This situation is shown in Fig. 10.9c, and it is valid for special conditions which are discussed in Chapter 11. Such materials are called *purely cohesive* for obvious reasons. For the case shown in Figure 10.9c, failure *theoretically* occurs on the 45° plane (it doesn't really, as is explained in Chapter 11). The shear strength is τ_f, and the normal stress on the theoretical failure plane at failure is $(\sigma_{1f} + \sigma_{3f})/2$.

Another useful thing that we should do before going on is to write the Mohr-Coulomb failure criterion in terms of principal stresses at failure, rather than as in Eq. 10-9 in terms of τ_{ff} and σ_{ff}. Look at Fig. 10.10 and note that $\sin \phi = R/D$, or

$$\sin \phi = \frac{\dfrac{\sigma_{1f} - \sigma_{3f}}{2}}{\dfrac{\sigma_{1f} + \sigma_{3f}}{2} + c \cot \phi}$$

or $(\sigma_{1f} - \sigma_{3f}) = (\sigma_{1f} + \sigma_{3f}) \sin \phi + 2c \cos \phi$. If $c = 0$, then $(\sigma_{1f} - \sigma_{3f}) = (\sigma_{1f} + \sigma_{3f}) \sin \phi$, which can be written as

$$\sin \phi = \frac{(\sigma_{1f} - \sigma_{3f})}{(\sigma_{1f} + \sigma_{3f})} \tag{10-13}$$

Rearranging, we have

$$\frac{\sigma_1}{\sigma_3} = \frac{1 + \sin \phi}{1 - \sin \phi} \tag{10-14}$$

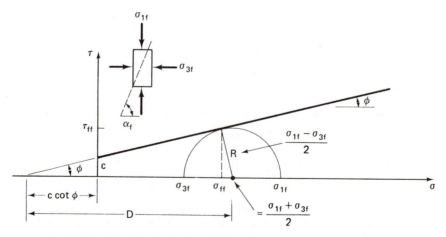

Fig. 10.10 Mohr-Coulomb strength envelope with one Mohr circle at failure.

or the reciprocal is

$$\frac{\sigma_3}{\sigma_1} = \frac{1 - \sin\phi}{1 + \sin\phi} \tag{10-15}$$

Using some trigonometric identities we can express Eqs. 10-14 and 10-15 as

$$\frac{\sigma_1}{\sigma_3} = \tan^2\left(45° + \frac{\phi}{2}\right) \tag{10-16}$$

$$\frac{\sigma_3}{\sigma_1} = \tan^2\left(45° - \frac{\phi}{2}\right) \tag{10-17}$$

Equations 10-14 through 10-17 are called the *obliquity relationships* because the maximum inclination, or obliquity, of the Mohr failure envelope occurs where c is equal to zero. These four equations are, of course, only valid where $c = 0$. Inspection of these equations and Fig. 10.10 shows that the coordinates of the point of tangency of the Mohr failure envelope and the Mohr circle (σ_{ff}, τ_{ff}) are the stresses on the plane of maximum obliquity in the soil element. In other words, the ratio τ_{ff}/σ_{ff} is a maximum on this plane. As we pointed out before, this plane is not the plane of maximum shear stress. On that plane ($\alpha = 45°$), the obliquity will be less than the maximum value since the ratio of τ_{max} to $(\sigma_1 + \sigma_3)/2$ is less than τ_{ff}/σ_{ff}. The obliquity relationships are very useful for evaluating triaxial test data and in theories of lateral earth pressure.

The last factor we should consider is the effect of the intermediate principal stress σ_2 on conditions at failure. Since by definition σ_2 lies somewhere between the major and minor principal stresses, the Mohr circles for the three principal stresses look like those shown in Fig. 10.3c and again in Fig. 10.11. It is obvious that σ_2 can have no influence on the

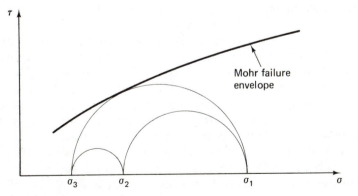

Fig. 10.11 Mohr circles for a three-dimensional state of stress.

conditions at failure for the Mohr failure criterion, no matter what magnitude it has. The intermediate principal stress σ_2 probably does have an influence in real soils, but the Mohr-Coulomb failure theory does not consider it.

10.5 TESTS FOR THE SHEAR STRENGTH OF SOILS

In this section we shall briefly describe some of the more common tests for determining the shearing strength of soils. Some of the tests are rather complicated, and for further details you should consult manuals and books on laboratory testing, especially those by the ASTM (1980), U.S. Army Corps of Engineers (1970), U.S. Bureau of Reclamation (1974), and Bishop and Henkel (1962).

Direct Shear Test

This test is probably the oldest strength test because Coulomb used a type of shear box test more than 200 years ago to determine the necessary parameters for his strength equation. The test in principle is quite simple. Basically, there is a specimen container, or "shear box," which is separated horizontally into halves. One-half is fixed; with respect to that half the other half is either pushed or pulled horizontally. A normal load is applied to the soil specimen in the shear box through a rigid loading cap. The shear load, horizontal deformation, and vertical deformation are measured during the test. Dividing the shear force and the normal force by the *nominal* area of the specimen, we obtain the shear stress as well as the normal stress on the failure plane. Remember that the failure plane is *forced* to be horizontal with this apparatus.

A cross-sectional diagram of the essential features of the apparatus is shown in Fig. 10.12a, while Fig. 10.12b shows some typical test results. The Mohr-Coulomb diagram for conditions at failure appears in Fig. 10.12c. As an example, if we were to test three samples of a sand at the same relative density just before shearing, then as the normal stress σ_n was increased, we would expect from our knowledge of sliding friction a concurrent increase in the shear stress on the failure plane at failure (the shear strength). This condition is shown in the typical shear stress versus deformation curves for a dense sand in Fig. 10.12b for $\sigma_{n1} < \sigma_{n2} < \sigma_{n3}$. When these results are plotted on a Mohr diagram, Fig. 10.12c, the angle of internal friction ϕ can be obtained.

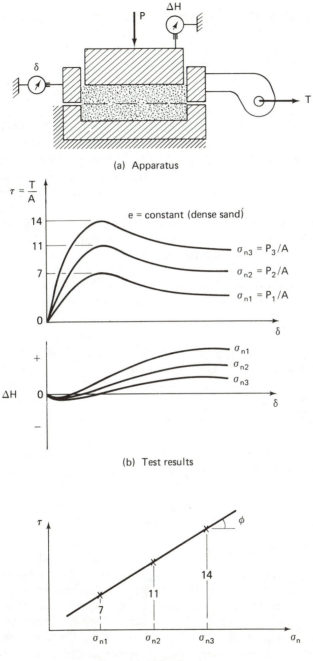

(a) Apparatus

(b) Test results

(c) Mohr diagram

Fig. 10.12 (a) Cross-sectional schematic diagram of direct shear apparatus; (b) typical test results (dense sand); and (c) Mohr diagram for specimens at the same relative density.

Typical results of vertical deformation ΔH for a dense sand are shown in the lower portion of Fig. 10.12b. At first there is a slight reduction in height or volume of the soil specimen, followed by a dilation or increase in height or volume. As the normal stress σ_n increases, the harder it is for the soil to dilate during shear, which seems reasonable.

We do not obtain the principal stresses directly in the direct shear test. Instead, if they are needed, they may be inferred if the Mohr-Coulomb failure envelope is known. Then, as is shown in Example 10.6, the angle of rotation of the principal stresses may be determined. Why is there rotation of the principal planes? Initially, the horizontal plane (potential failure plane) is a principal plane (no shear stress), but after the shearing stress is applied and at failure, by definition, it cannot be a principal plane. Therefore, rotation of the principal planes must occur in the direct shear test. How much do the planes rotate? It depends on the slope of the Mohr failure envelope, but it is fairly easy to determine, as is shown in Example 10.6, if you make some simple assumptions.

EXAMPLE 10.6

Given:

The initial and failure conditions in a direct shear test, as shown in Fig. Ex. 10.6.

Required:

Plot the Mohr circles for both initial conditions and at failure, assuming ϕ is known. Find the principal stresses at failure and their angles of rotation at failure.

Solution:

The Mohr circles for both initial conditions and at failure are shown on the right side of Fig. Ex. 10.6. At failure, you know the normal stress on the failure plane, σ_{ff}, is the same as the initial normal stress, σ_n. Since ϕ is known (assume c is small or zero), from the Mohr failure hypothesis (Fig. 10.7) the shear stress on the failure plane at failure is determined by the point of tangency of the Mohr circle at failure. The center of the failure circle can be found by drawing a perpendicular to the Mohr failure envelope from the point of tangency. The radial distance is, of course, equal to $[(\sigma_1 - \sigma_3)/2]_f$. Another way to find the Mohr circle at failure is

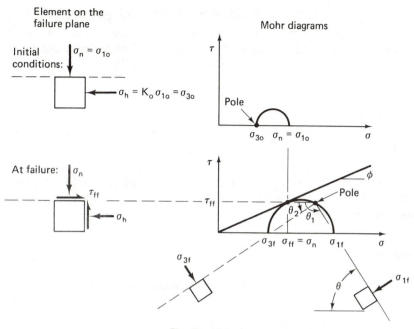

Fig. Ex. 10.6

graphically by trial and error. Find the only circle which is tangent at (σ_{ff}, τ_{ff}) and whose diameter lies on the σ-axis. Once the failure circle is drawn, the values of σ_{1f} and σ_{3f} can be scaled off. From the pole method, the angles of rotation of these stresses are readily found, as shown in Fig. Ex. 10.6.

There are, of course, several advantages and disadvantages of the direct shear test. Primarily, the test is inexpensive, fast, and simple, especially for granular materials. We do observe shear planes and thin failure zones in nature, so it seems alright to actually shear a specimen of soil along some plane to see what the stresses are on that plane. Disadvantages include the problem of controlling drainage—it is very difficult if not impossible, especially for fine-grained soils. Consequently, the test is not so suitable for other than completely drained conditions. When we force the failure plane to occur, how can we be sure that it is the weakest direction or even at the same critical direction as occurs in the field? We don't know. Another flaw in the direct shear test is that there are rather serious stress concentrations at the sample boundaries, which lead to highly nonuniform stress conditions within the test specimen itself. And finally, as shown by Example 10.6, an uncontrolled rotation of principal planes and stresses occurs between the start of the test and failure. To accurately model the in situ loading conditions, the amount of this rotation should be known and accounted for, but it isn't. The Mohr circles for the direct shear test are further illustrated by Example 10.7.

EXAMPLE 10.7

Given:

A direct shear test is run on a medium dense sandy silt, with the normal stress $\sigma_n = 65$ kPa. $K_o = 0.5$. At failure, the normal stress is still 65 kPa and the shear stress is 41 kPa.

Required:

Draw the Mohr circles for the initial conditions and at failure and determine:

    a. The principal stresses at failure.
    b. The orientation of the failure plane.
    c. The orientation of the major principal plane at failure.
    d. The orientation of the plane of maximum shear stress at failure.

Solution:

    a. The initial conditions are shown in Fig. Ex. 10.7 by circle i. Since $K_o = 0.5$, the initial horizontal stress is 32.5 kPa. The normal stress on the specimen is held constant at 65 kPa during the test, so σ_{1i} is also σ_{ff}. Since the shear stress at failure is 41 kPa, the failure point (as in Fig. 10.12c) is plotted as point F. The ϕ is determined to be 32°. What happens between the initial Mohr circle i and at failure f is unknown. The construction of circle f was described in Example 10.6. The center of circle f is found to be at (91 kPa, 0). So $\sigma_{1f} = 139$ kPa and $\sigma_{3f} = 43$ kPa.
    b. The state of stress at failure point F is (65, 41) kPa, and the failure plane is assumed to be horizontal, a good assumption for the direct shear test.
    c. A line drawn horizontally from the known state of stress at point F intersects the Mohr circle at P, the pole. Line $\overline{P\sigma_{1f}}$ indicates the orientation of the major principal plane. It makes an angle of about 60.5° with the horizontal.

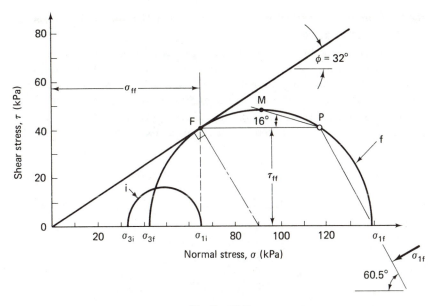

Fig. Ex. 10.7

d. Line $\overline{PM}$ is the orientation of the plane of maximum shear stress; it is about 16° from the horizontal.

Note that in this example, if we didn't assume that the Mohr failure envelope passed through the origin of the Mohr diagram, more than one test at different σ_{1i}'s would be required to establish the Mohr envelope.

Triaxial Test

During the early history of soil mechanics, the direct shear test was the most popular shear test. Then, about 1930, A. Casagrande while at M.I.T. began research on the development of cylindrical compression tests in an attempt to overcome some of the serious disadvantages of the direct shear test. Now this test, commonly called the *triaxial test*, is by far the more popular of the two. The triaxial test is much more complicated than the direct shear but also much more versatile. We can control drainage quite well, and there is no rotation of σ_1 and σ_3. Stress concentrations still exist, but they are significantly less than in the direct shear test. Also, the failure plane can occur anywhere. An added advantage: we can control the stress paths to failure reasonably well, which means that complex stress paths in the field can more effectively be modeled in the laboratory with the triaxial test. Stress paths are explained in the next section.

The principle of the triaxial test is shown in Fig. 10.13a. The soil specimen is usually encased in a rubber membrane to prevent the pressurized cell fluid (usually water) from penetrating the pores of the soil. Axial load is applied through a piston, and often the volume change of the specimen during a drained test or the induced pore water pressure during an undrained test is measured. As mentioned above, we can control the

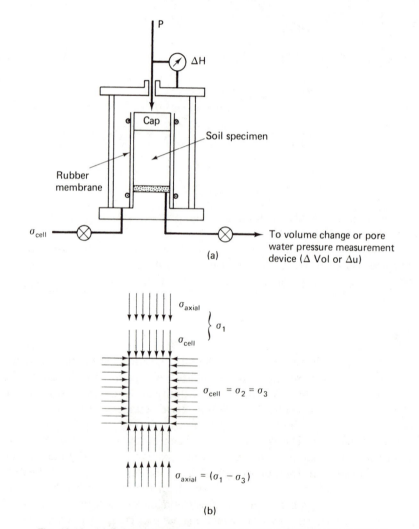

Fig. 10.13 (a) Schematic diagram of the triaxial apparatus; (b) assumed stress conditions on the triaxial specimen.

drainage to and from the specimen, and it is possible, with some assumptions, to control the stress paths applied to the specimen. Basically, we assume the stresses on the boundary of the specimen are principal stresses (Fig. 10.13b). This is not really true because of some small shear stresses acting on the ends of the specimen. Also, as mentioned before, the failure plane is not forced—the specimen is free to fail on any weak plane or, as sometimes occurs, to simply bulge.

You will note that the σ_{axial} in Fig. 10.13b is the difference between the major and minor principal stresses; it is called the *principal stress difference* (or sometimes, wrongly, the deviator stress). Note also that for the conditions shown in the figure, $\sigma_2 = \sigma_3 = \sigma_{cell}$. Sometimes we will assume that $\sigma_{cell} = \sigma_1 = \sigma_2$ for special types of stress path tests. Common triaxial stress paths are discussed in the next section.

The triaxial test is far more complex than the direct shear test; entire books have been written on test details and interpretation of the results (see, for example, Bishop and Henkel, 1962). Most of the data and test results described in Chapter 11 were derived from triaxial tests.

Drainage conditions or paths followed in the triaxial test are models of specific critical design situations required for the analysis of stability in engineering practice. These are commonly designated by a two-letter symbol. The first letter refers to what happens *before shear*—that is, whether the specimen is consolidated. The second letter refers to the drainage conditions *during shear*. The three permissible drainage paths in the triaxial test are as follows:

Drainage Path Before Shear-During Shear	Symbol
Unconsolidated-Undrained	UU
Consolidated-Undrained	CU
Consolidated-Drained	CD

For reasons explained in Chapter 11, the unconsolidated-drained test defies interpretation and is therefore meaningless. Triaxial test results for the three drainage paths are described in detail in Chapter 11.

EXAMPLE 10.8

Given:

A conventional consolidated-drained (CD) triaxial test is conducted on a sand. The cell pressure is 100 kPa, and the applied axial stress at failure is 200 kPa.

Required:

   a. Plot the Mohr circles for both the initial and failure stress conditions.

   b. Determine ϕ (assume $c = 0$).

   c. Determine the shear stress on the failure plane at failure τ_{ff}, and find the theoretical angle of the failure plane in the specimen. Also determine the orientation of the plane of maximum obliquity.

   d. Determine the maximum shear stress at failure τ_{max} and the angle of the plane on which it acts; calculate the available shear strength on this plane and the factor of safety on this plane.

Solution:

   a. Refer to Fig. 10.13b and Fig. Ex. 10.8. The initial conditions are shown at the top of Fig. Ex. 10.8 for the conventional triaxial test. The initial stress is equal to the cell pressure σ_{cell}, and it is equal in all directions (hydrostatic). Therefore the Mohr circle for the initial stress conditions is a *point* at 100 kPa, as shown in the Mohr diagram of Fig. Ex. 10.8. At failure, the $\sigma_{axial} = (\sigma_1 - \sigma_3)_f = 200$ kPa. So

$$\sigma_{1f} = (\sigma_1 - \sigma_3)_f + \sigma_{3f} = 200 + 100 = 300 \text{ kPa}$$

Now we can plot the Mohr circle at failure; $\sigma_{1f} = 300$ and $\sigma_{3f} = 100$. The center is at $(\sigma_1 + \sigma_3)/2 = 200$, and the radius is $(\sigma_1 - \sigma_3)/2 = 100$. The circle at failure is shown in Fig. Ex. 10.8.

   b. We find ϕ graphically to be 30°. We can also use Eq. 10-13 if we prefer an analytical solution. Thus

$$\phi = \arcsin \frac{\sigma_{1f} - \sigma_{3f}}{\sigma_{1f} + \sigma_{3f}} = \arcsin \frac{200}{400} = 30°$$

   c. From the Mohr failure hypothesis, the coordinates of the point of tangency of the Mohr failure envelope and the Mohr circle at failure are (σ_{ff}, τ_{ff}). From Eq. 10-9, we know that $\tau_{ff} = \sigma_{ff} \tan \phi$, but unlike the direct shear test we don't know σ_{ff} in the triaxial test. Look carefully at Fig. 10.10. The small angle near the top of the Mohr circle is ϕ (by a theorem from high school geometry). Therefore, since $c = 0$, $D - \sigma_{ff} = R \sin \phi$. Solving for σ_{ff}, we obtain

$$\sigma_{ff} = \frac{\sigma_{1f} + \sigma_{3f}}{2} - \frac{\sigma_{1f} - \sigma_{3f}}{2} \sin \phi$$

$$= 200 - 100 \sin 30° = 150 \text{ kPa}$$

$$\tau_{ff} = \sigma_{ff} \tan \phi = 150 \tan 30° = 86.6 \text{ kPa}$$

The theoretical angle of inclination of the failure plane can be found

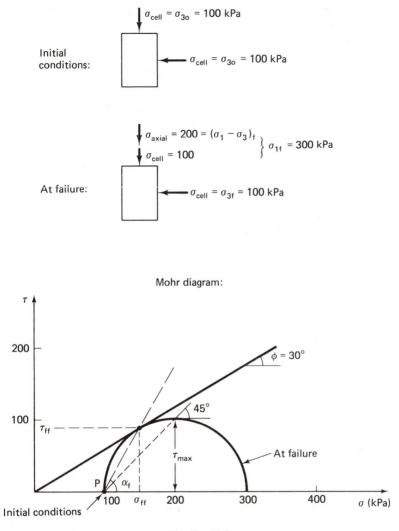

Fig. Ex. 10.8

graphically by the pole method or analytically. From the stress conditions at failure shown in Fig. Ex. 10.8, the pole is at (100, 0), and α_f can be measured to be 60°. For the analytical solution, use Eq. 10-10.

$$\alpha_f = 45° + \frac{\phi}{2} = 60°$$

The plane of maximum obliquity is oriented at this angle also since the maximum inclination of the Mohr failure envelope is 30° and the point of tangency determines the condition of maximum obliquity. In other words, the ratio τ_{ff}/σ_{ff} is a maximum at this point on the Mohr circle and on the plane in the specimen inclined at $60°$ from the horizontal.

    d. $\tau_{max} = R = \dfrac{\sigma_{1f} - \sigma_{3f}}{2} = 100$ kPa. From the pole, the plane of τ_{max} is inclined at $45°$ from the horizontal. The obliquity at this orientation is $\tau_{max}/\sigma_{45°} = 100/200 = 0.5$. The maximum obliquity (part c) is $86.6/150 = 0.58 > 0.5$. The available τ (see Fig. 10.9b) can be determined from

$$\tau_{available} = \sigma_n \tan \phi = \frac{\sigma_{1f} + \sigma_{3f}}{2} \tan \phi$$

$$= 200 \tan 30° = 115.5 \text{ kPa}$$

which is greater than $\tau_{max} = 100$ kPa. Therefore the factor of safety on the $45°$ plane (Eq. 10-11) is

$$\text{F.S.} = \frac{\tau_{available}}{\tau_{max}} = \frac{115.5}{100} = 1.16$$

Note that the factor of safety on the $\alpha_f = 60°$ plane is

$$\text{F.S.} = \frac{\tau_{available}}{\tau_{ff}} = \frac{86.6}{86.6} = 1$$

Special Laboratory Tests

Other types of laboratory strength tests that you may hear about include *hollow cylinder tests*, *plane strain tests*, and so called *true triaxial* or *cuboidal shear tests*. These tests are schematically illustrated in Fig. 10.14. In the common triaxial test, the intermediate principal stress can only be equal to either the major or minor principal stress—nothing in between. With these other tests it is possible to vary σ_2, which probably models the stress conditions in real problems more accurately. Today, however, these tests are primarily used for research rather than for practical engineering applications.

    A couple of other tests of the direct shear type must also be mentioned. *Torsional* or *ring shear* tests (Fig. 10.15a) have been developed so that the test specimen may be sheared to very large deformations. This approach is sometimes necessary to obtain the *residual* or *ultimate shear strength* of certain materials, which is easier to obtain with a ring shear device than by repeatedly reversing the direct shear box. A more common

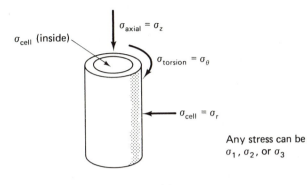

$\sigma_{axial} = \sigma_z$

σ_{cell} (inside)

$\sigma_{torsion} = \sigma_\theta$

$\sigma_{cell} = \sigma_r$

Any stress can be
σ_1, σ_2, or σ_3

(a)

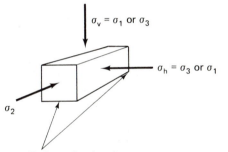

$\sigma_v = \sigma_1$ or σ_3

$\sigma_h = \sigma_3$ or σ_1

σ_2

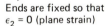

Ends are fixed so that
$\epsilon_2 = 0$ (plane strain)

(b)

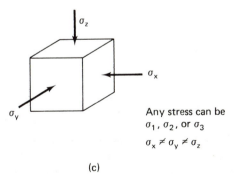

σ_z

σ_x

σ_y

Any stress can be
σ_1, σ_2, or σ_3

$\sigma_x \neq \sigma_y \neq \sigma_z$

(c)

Fig. 10.14 Schematic diagrams for the: (a) hollow cylinder test; (b) plane strain test; and (c) true triaxial or cuboidal shear test.

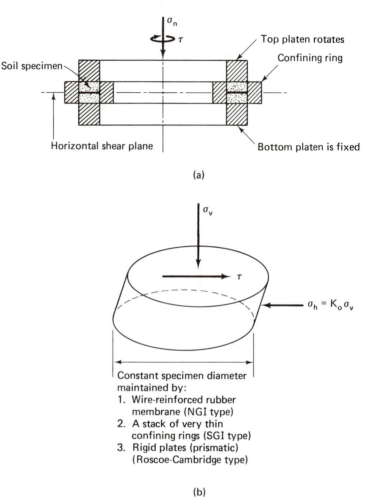

(a)

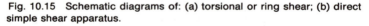

(b)

Fig. 10.15 Schematic diagrams of: (a) torsional or ring shear; (b) direct simple shear apparatus.

test used in both Scandinavia and in the United States for static and dynamic testing is the *direct simple shear* (DSS) test (Fig. 10.15b). In this test, a fairly homogeneous state of shear stress is applied, thereby avoiding the stress concentrations which exist in the ordinary direct shear apparatus. Since stress conditions in the DSS test are not the same as those shown in Examples 10.6 and 10.7 for the direct shear box, they are described in Example 10.9.

EXAMPLE 10.9

Given:

The DSS test.

Required:

Illustrate the stress conditions in the test, and draw the Mohr circles for both initial and failure conditions.

Solution:

The initial conditions for the DSS test shown in Fig. Ex. 10.9a are the same as those for the direct shear box test shown in Fig. Ex. 10.6 and Fig. Ex. 10.7. The sides of the soil sample are forced to rotate through an angle γ by the application of a horizontal shear stress, τ_{hv}. These stress conditions are shown in Fig. Ex. 10.9b. Note the absence of complementary shear stresses on the *outside* of the soil sample; this is necessary for simple shear. *Inside* the sample, however, the applied stress system is assumed to be *pure* shear, and complementary stresses are necessary for equilibrium. With the application of τ_{hv}, and σ_v and σ_h constant, the Mohr circle enlarges about the same center as the initial Mohr circle i. At failure, the Mohr circle is just tangent to the Mohr failure envelope, and the Mohr circle looks like circle f of Fig. Ex. 10.9c.

For this condition, the pole P is found by extending a line from $(\sigma_v, -\tau_{hv})$ horizontally (the plane on which these stresses act) to where it intersects the Mohr circle. Lines drawn from the pole represent the orientations of different states of stress within the soil sample. The line PM represents the plane of maximum (absolute value) shear stress; the line PF represents the orientation of the failure plane—it is *not* horizontal as in the direct shear test. The line $\overline{P\sigma_{1f}}$ represents the orientation of the σ_1 planes when τ_{hv} is negative on the horizontal surface. When (and if) the sign of τ_{hv} becomes positive on the horizontal plane, as in a *cyclic simple shear* test, then the pole is located at P' for that part of the circle. The line $\overline{P'\sigma_{1f}}$ becomes the new orientation of the principal plane with a negative θ, the angle of principal stress rotation.

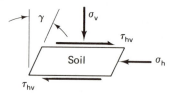

(a) Initial conditions

(b) With application of shear stresses on top and bottom only

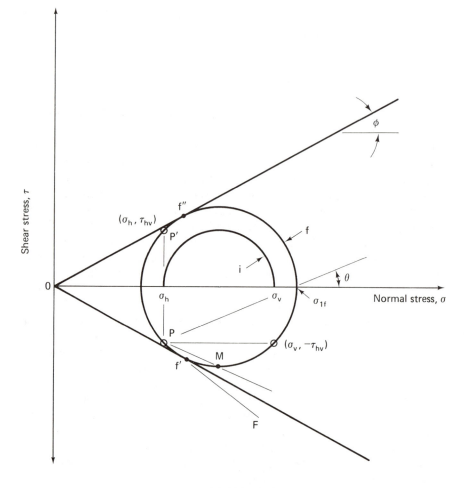

(c) Mohr circles

Fig. Ex. 10.9

Field Tests

Because of all the problems associated with sampling and laboratory testing, it sometimes is better to measure the strength directly in the field. The most common field tests for soft clays are the *vane shear test* and the *Dutch cone penetrometer*. The latter test is also very effective when used for sandy soils. The *standard penetration test* (SPT) is used for granular materials and sometimes for cohesive soils, but it is less accurate in especially soft clays. The *Iowa borehole shear test* has been developed for use in loess soils. The *pressuremeter* and *screw plate* tests are also becoming increasingly popular for determining, among other things, the strength and deformation properties of soils. Field test equipment and test methods are described briefly in Chapter 11 and in detail in most textbooks on foundation engineering. Ladd, et al. (1977) have a good discussion of the applicability of both laboratory and field tests in geotechnical engineering practice.

10.6 STRESS PATHS

As you know from the first part of this chapter, states of stress at a point in equilibrium can be represented by a Mohr circle in a τ-σ coordinate system. Sometimes it is convenient to represent that state of stress by a *stress point*, which has the coordinates $(\sigma_1 - \sigma_3)/2$ and $(\sigma_1 + \sigma_3)/2$, as shown in Fig. 10.16. For many situations in geotechnical engineering, we

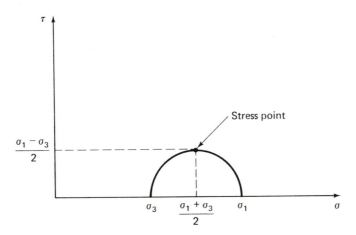

Fig. 10.16   A Mohr circle of stress and corresponding stress point.

assume σ_1 and σ_3 act on vertical and horizontal planes, so the coordinates of the stress point become $(\sigma_v - \sigma_h)/2$ and $(\sigma_v + \sigma_h)/2$, or simply q and p, respectively; or

$$q = \frac{\sigma_v - \sigma_h}{2} \tag{10-18}$$

$$p = \frac{\sigma_v + \sigma_h}{2} \tag{10-19}$$

Both q and p could, of course, be defined in terms of the principal stresses. By convention, q is considered positive when $\sigma_v > \sigma_h$; otherwise it is negative.

We often want to show successive states of stress which a test specimen or a typical element in the field undergoes during loading or unloading. A diagram showing the successive states with a series of Mohr circles could be used (Fig. 10.17a), but it might be confusing, especially if the stress path were complicated. Therefore it is simpler to show only the *locus* of the stress points. This locus is called the *stress path*, and it is plotted on what we call a *p-q diagram* (Fig. 10.17b). Note that both p and q could be defined either in terms of total stresses or effective stresses. As before, a prime mark is used to indicate effective stresses. So from Eqs. 10-18 and 10-19 and the effective stress equation (Eq. 7-13), we know that $q' = q$ while $p' = p - u$, where u is the excess hydrostatic or pore water pressure.

Although the concept of a stress path has been around for a long time, Prof. T. W. Lambe of M.I.T. demonstrated its usefulness as a teaching device (Lambe and Whitman, 1969) and developed the method into a practical engineering tool for the solution of stability and deformation problems (Lambe, 1964 and 1967; Lambe and Marr, 1979). Very often in geotechnical engineering practice, if you understand the complete stress

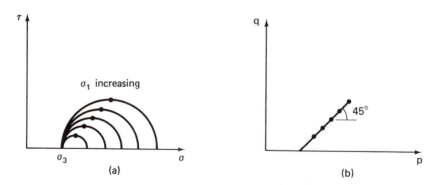

Fig. 10.17 (a) Successive Mohr circles; (b) stress path for constant σ_3 and increasing σ_1 (after Lambe and Whitman, 1969).

path of your problem, you are well along the way towards the solution of that problem.

A simple case to illustrate stress paths is the common triaxial test in which σ_3 remains fixed as we increase σ_1. Some Mohr circles for this test are shown in Fig. 10.17a along with their stress points. The corresponding stress path shown in Fig. 10.17b is a straight line at an angle of 45° from the horizontal because the stress point represents the state of stress on the

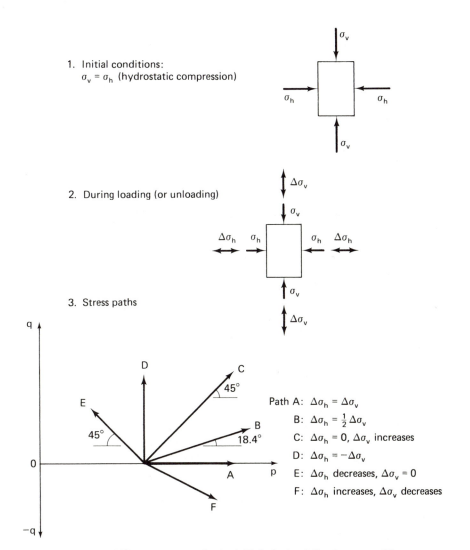

1. Initial conditions:
   $\sigma_v = \sigma_h$ (hydrostatic compression)

2. During loading (or unloading)

3. Stress paths

Path A: $\Delta\sigma_h = \Delta\sigma_v$
   B: $\Delta\sigma_h = \frac{1}{2}\Delta\sigma_v$
   C: $\Delta\sigma_h = 0$, $\Delta\sigma_v$ increases
   D: $\Delta\sigma_h = -\Delta\sigma_v$
   E: $\Delta\sigma_h$ decreases, $\Delta\sigma_v = 0$
   F: $\Delta\sigma_h$ increases, $\Delta\sigma_v$ decreases

Fig. 10.18   Different stress paths for initially hydrostatic stress conditions (after Lambe and Whitman, 1969).

plane oriented 45° from the principal planes. (Note that this is the plane of maximum shear stress.)

Some examples of stress paths are shown in Figs. 10.18 and 10.19. In Fig. 10.18 the initial conditions are $\sigma_v = \sigma_h$, an equal-all-around or hydrostatic state of stress. Those in Fig. 10.19, where the initial vertical stress is not the same as the initial horizontal stress, represent a non-hydrostatic

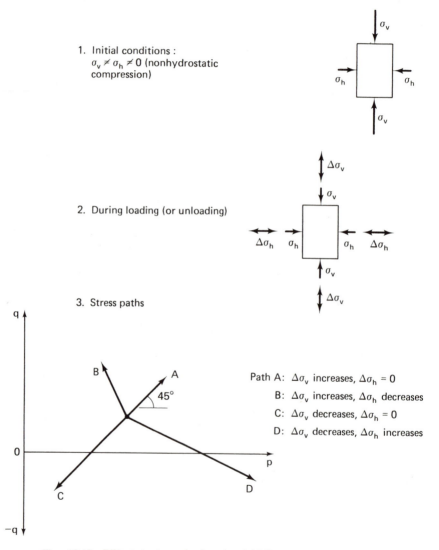

Fig. 10.19 Different stress paths for initially non-hydrostatic stress conditions (after Lambe and Whitman, 1969).

state of stress. You should verify that each stress path in Figs. 10.18 and 10.19 has in fact the direction as indicated in the figures. We shall show you how to do this in Example 10.10.

EXAMPLE 10.10

Given:

Figures 10.18 and 10.19.

Required:

Verify that stress paths A, B, and C of Fig. 10.18 and A and D of Fig. 10.19 are correct as shown.

Solution:

The initial conditions for all stress paths in Fig. 10.18 are $p_o = (\sigma_v + \sigma_h)/2 = \sigma_v = \sigma_h$ and $q_o = 0$. Final conditions are (Eqs. 10-18 and 10-19)

$$q_f = \frac{(\sigma_v + \Delta\sigma_v) - (\sigma_h + \Delta\sigma_h)}{2}$$

$$p_f = \frac{(\sigma_v + \Delta\sigma_v) + (\sigma_h + \Delta\sigma_h)}{2}$$

For stress path A, $\Delta\sigma_v = \Delta\sigma_h$; so

$$q_f = \frac{\sigma_v + \Delta\sigma_v - \sigma_v - \Delta\sigma_v}{2} = 0$$

$$p_f = \frac{\sigma_v + \Delta\sigma_v + \sigma_v + \Delta\sigma_v}{2} = \sigma_v + \Delta\sigma_v$$

Thus the stress path A moves out on the p-axis by an amount $\Delta\sigma_v = \Delta\sigma_h$.
For stress path B, $\Delta\sigma_h = \frac{1}{2}\Delta\sigma_v$; so

$$q_f = \frac{\sigma_v + \Delta\sigma_v - \sigma_v - \frac{1}{2}\Delta\sigma_v}{2} = \frac{1}{4}\Delta\sigma_v$$

$$p_f = \frac{\sigma_v + \Delta\sigma_v + \sigma_v + \frac{1}{2}\Delta\sigma_v}{2} = \sigma_v + \frac{3}{4}\Delta\sigma_v$$

These values are the (p, q) coordinates of the end of stress path B. Thus the q and p both increase by an amount $\Delta q = \frac{1}{4}\Delta\sigma_v$ and $\Delta p = \frac{3}{4}\Delta\sigma_v$, which means that the stress path has a slope of $\frac{1}{3}$ or is inclined at $18.4°$ as shown in Fig. 10.18.

For stress path C, $\Delta\sigma_h = 0$ and $\Delta\sigma_v$ increases by some amount.

$$q_f = \frac{\sigma_v + \Delta\sigma_v - \sigma_v}{2} = \frac{1}{2}\Delta\sigma_v$$

$$p_f = \frac{\sigma_v + \Delta\sigma_v + \sigma_v}{2} = \sigma_v + \frac{1}{2}\Delta\sigma_v$$

So $\Delta q = \frac{1}{2}\Delta\sigma_v$ and $\Delta p = \frac{1}{2}\Delta\sigma_v$. Therefore the slope of the stress path must be 1 or inclined at $45°$. This solution also holds for stress path A in Fig. 10.19. Here initial conditions are non-hydrostatic, so

$$q_o = \frac{\sigma_v - \sigma_h}{2}$$

$$p_o = \frac{\sigma_v + \sigma_h}{2}$$

The final coordinates for path A are

$$q_f = \frac{\sigma_v + \Delta\sigma_v - \sigma_h}{2}$$

$$p_f = \frac{\sigma_v + \Delta\sigma_v + \sigma_h}{2}$$

So $\Delta q = \frac{1}{2}\Delta\sigma_v$ and $\Delta p = \frac{1}{2}\Delta\sigma_v$, which is the same as for stress path C in Fig. 10.18.

For stress path D in Fig. 10.19, $\Delta\sigma_v$ decreases while $\Delta\sigma_h$ increases. Initial (p_o, q_o) are the same as path A in this figure, while the final values of (p_f, q_f) are

$$q_f = \frac{(\sigma_v - \Delta\sigma_v) - (\sigma_h + \Delta\sigma_h)}{2}$$

$$p_f = \frac{(\sigma_v - \Delta\sigma_v) + (\sigma_h + \Delta\sigma_h)}{2}$$

So

$$\Delta q = -\frac{1}{2}\Delta\sigma_v - \frac{1}{2}\Delta\sigma_h \quad \text{and} \quad \Delta p = -\frac{1}{2}\Delta\sigma_v + \frac{1}{2}\Delta\sigma_h$$

The actual slope of the stress path depends on the relative magnitudes of $\Delta\sigma_v$ and $\Delta\sigma_h$, but in general it trends down and out as shown in Fig. 10.19.

It is often convenient to consider *stress ratios*. In Chapter 7 we defined a lateral stress ratio K, which is the ratio of horizontal to vertical stress,

$$K = \frac{\sigma_h}{\sigma_v} \qquad (7\text{-}18)$$

In terms of *effective* stresses, this ratio is

$$K_o = \frac{\sigma_h'}{\sigma_v'} \qquad (7\text{-}19)$$

where K_o is called the *coefficient of lateral earth pressure at rest* for conditions of no lateral strain. Finally, we can define a ratio K_f for the stress conditions at failure.

$$K_f = \frac{\sigma_{hf}'}{\sigma_{vf}'} \qquad (10\text{-}20)$$

where σ_{hf}' = the horizontal effective stress at failure, and

σ_{vf}' = the vertical effective stress at failure.

Usually K_f is defined in terms of effective stresses, but it may just as well be in terms of total stresses. Constant stress ratios appear as straight lines on a *p-q* diagram (Fig. 10.20). These lines could also be stress paths for initial conditions of $\sigma_v = \sigma_h = 0$ with loadings of K equal to a constant (that is, constant σ_h/σ_v). Other initial conditions are, of course, possible, such as those shown in Figs. 10.18 and 10.19.

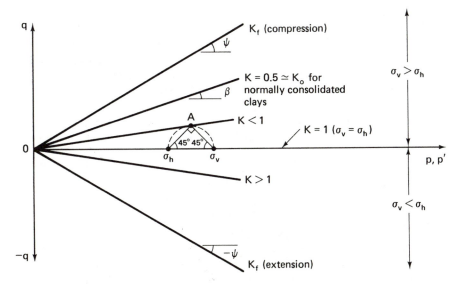

Fig. 10.20 Different constant stress ratios and examples of stress paths, starting from $\sigma_v = \sigma_h = 0$ (after Lambe and Whitman, 1969).

Note that

$$\frac{q}{p} = \tan \beta = \frac{1 - K}{1 + K} \qquad (10\text{-}21)$$

or in terms of K

$$K = \frac{1 - \tan \beta}{1 + \tan \beta} \qquad (10\text{-}22)$$

where β is the slope of the line of constant K when $K < K_f$. At failure, the slope of the K_f line is indicated by the symbol ψ. Note also that for any point where you know p and q, (for example point A in Fig. 10.20), σ_h and σ_v can readily be found graphically; that is, lines at 45° from the stress point intersect the σ-axis at σ_h and σ_v. Finally, there is no reason why σ_v must always be greater than σ_h. It usually is, but there are many important situations in geotechnical engineering where $\sigma_h > \sigma_v$. In these cases, by convention q is negative and $K > 1$, as shown in Fig. 10.20.

Now we shall describe some stress paths which are important in geotechnical engineering. When soils are deposited in a sedimentary environment like a lake or the sea, there is a gradual build up of overburden stress as additional material is deposited from above. As this stress increases, the sediments consolidate and decrease in volume (Chapters 8 and 9). If the area of deposition is relatively large compared with the thickness of the deposit, then it seems reasonable that the compression is essentially one dimensional. In this case the stress ratio would be constant and equal to K_o, and the stress path during sedimentation and consolidation would be similar to path AB in Fig. 10.21. Typical values of K_o for granular materials range from about 0.4 to 0.6, whereas for normally consolidated clays K_o can be a little less than 0.5 up to 0.8 or 0.9. A good average value is about 0.5. (See Sec. 11.7 and 11.11.) When a sample of the soil is taken, stress decrease occurs because the overburden stress σ'_{vo} has to be removed to get at the sample. The stress path follows approximately line BC in Fig. 10.21, and the soil specimen ends up someplace on the hydrostatic ($\sigma_h = \sigma_v$) or $K = 1$ axis. This stress path and its relation to the strength of clays is discussed in Chapter 11.

If, instead of sampling, the overburden stress was decreased by erosion or some other geologic process, an unloading stress path similar to BC would be followed. If the vertical stress continued to be removed, the path could extend to a point well below the p-axis. The soil would then be overconsolidated and K_o would be greater than 1.0.

Sometimes in engineering practice a test specimen is reconsolidated in the laboratory under K_o conditions so as to reinstate the estimated in situ stresses. Such conditions are shown in Fig. 10.19 and at point A in Fig. 10.22. After consolidation, the loading (or unloading) path followed to

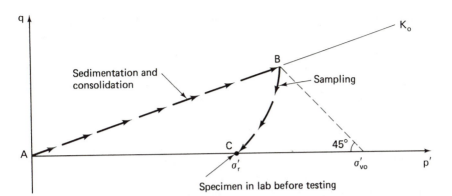

Fig. 10.21 Stress paths during sedimentation and sampling of normally consolidated clay, where $K_o < 1$.

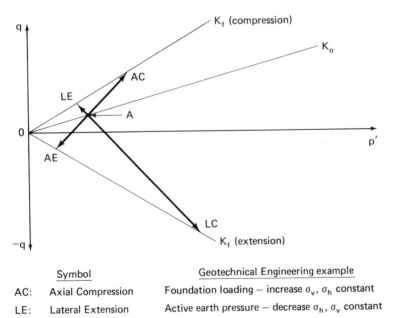

Symbol		Geotechnical Engineering example
AC:	Axial Compression	Foundation loading — increase σ_v, σ_h constant
LE:	Lateral Extension	Active earth pressure — decrease σ_h, σ_v constant
AE:	Axial Extension	Unloading (excavation) — decrease σ_v, σ_h constant
LC:	Lateral Compression	Passive earth pressure — increase σ_h, σ_v constant

Fig. 10.22 Stress paths during drained loadings on normally consolidated clays and sand (after Lambe, 1967).

failure depends on the field loading conditions one wishes to model. Four common field conditions and the laboratory stress paths which model them are shown in Fig. 10.22. Note that these stress paths are for *drained* loading (discussed in the next chapter) in which there is *no* excess pore water pressure; therefore total stresses equal effective stresses and the total stress path (TSP) for a given loading is identical to the effective stress path (ESP).

As suggested by Eq. 10-20, we are often interested in conditions at failure, and it is useful to know the relationship between the K_f line and the Mohr-Coulomb failure envelope. Consider the two Mohr circles shown in Fig. 10.23. The circle on the left, drawn for illustrative purposes only, represents failure in terms of the *p-q* diagram. The identical circle on the right is the same failure circle on the Mohr τ-σ diagram. To establish the slopes of the two lines and their intercepts, several Mohr circles and stress paths, determined over a range of stresses, were used. The equation of the K_f line is

$$q_f = a + p_f \tan\psi \tag{10-23}$$

where a = the intercept on the q-axis, in stress units, and
ψ = the angle of the K_f line with respect to the horizontal, in degrees.

The equation of the Mohr-Coulomb failure envelope is

$$\tau_{ff} = c + \sigma_{ff}\tan\phi \tag{10-9}$$

From the geometries of the two circles, it can be shown that

$$\sin\phi = \tan\psi \tag{10-24}$$

and

$$c = \frac{a}{\cos\phi} \tag{10-25}$$

So, from a *p-q* diagram the shear strength parameters ϕ and c may readily be computed.

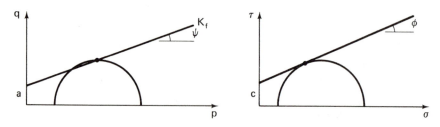

Fig. 10.23 Relationship between the K_f line and the Mohr-Coulomb failure envelope.

Another useful aspect of the *p-q* diagram is that it may be used to show both total and effective stress paths on the same diagram. We said before that for drained loading, the total stress path (TSP) and the effective stress path (ESP) were identical. This is because the pore water pressure induced by loading was approximately equal to zero at all times during shear. However, in general, during *undrained* loading the TSP is not equal to the ESP because excess pore water pressure develops. For axial compression (AC) loading of a normally consolidated clay ($K_o < 1$), a *positive* excess pore water pressure Δu develops. Therefore the ESP lies to the *left* of the TSP because $\sigma' = \sigma - \Delta u$. At any point during the loading, the pore water pressure Δu may be scaled off any horizontal line between the TSP and ESP, as shown in Fig. 10.24.

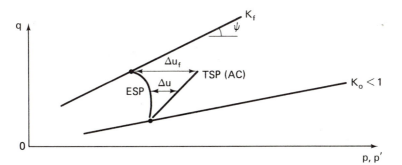

Fig. 10.24 Stress paths during undrained axial compression loading of a normally consolidated clay.

If a clay is overconsolidated ($K_o > 1$), then *negative* pore water pressure ($-\Delta u$) develops because the clay *tends* to expand during shear, but it can't. (Remember: we are talking about undrained loading in which no volume change is allowed.) For AC loading on an overconsolidated clay, stress paths like those shown in Fig. 10.25 will develop. Similarly, we can plot total and effective stress paths for other types of loadings and unloadings, for both normally and overconsolidated soils, and we shall show some of these in Chapter 11.

In most practical situations in geotechnical engineering, there exists a static ground water table; thus an initial pore water pressure u_o, is acting on the element in question. So there are really three stress paths we should consider, the ESP, the TSP, and the $(T - u_o)$SP. These three paths are shown in Fig. 10.26 for a normally consolidated clay with an initial pore water pressure u_o undergoing AC loading. Note that as long as the ground water table remains at the same elevation, u_o does not affect either the ESP or the conditions at failure.

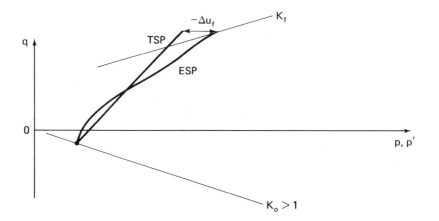

Fig. 10.25   Stress paths during axial compression of a heavily overconsolidated clay.

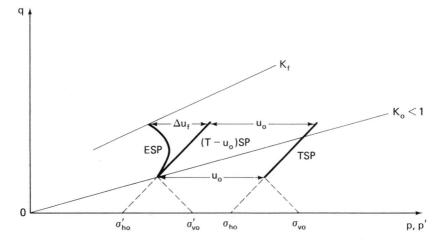

Fig. 10.26   ESP, TSP, and $(T - u_o)$SP for a normally consolidated clay (after Lambe, 1967).

PROBLEMS

10-1. Given an element with stresses as indicated in Fig. P10-1.
   Find: (a) The major and minor principal stresses and the planes on which they act.
   (b) The stresses on a plane inclined at 30° from the horizontal.
   (c) The maximum shear stress and the inclination of the plane on which it acts.

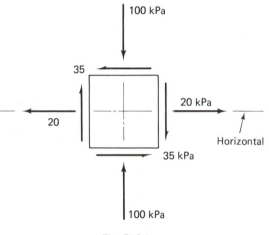

Fig. P10-1

10-2. Work Problem 10-1 with the element rotated 30° clockwise from the horizontal.

10-3. With the element of Problem 10-2 rotated 30°, find the magnitude and direction of the stresses on the *horizontal* plane.

10-4. Work Example 10-3 with the element rotated 20° clockwise from the horizontal. In addition, find the stresses (magnitude and direction) on the vertical plane.

10-5. Equations 10-5 and 10-6 were derived from Fig. 10.2, with σ_x and σ_y as principal stresses. Derive the more general equations for the Mohr circle when σ_x and σ_y are *not* acting on principal planes.

10-6. The state of plane stress in a body is described by the following stresses: $\sigma_1 = 9000 \text{ kN/m}^2$ compression, $\sigma_3 = 2000 \text{ kN/m}^2$ tension. Determine by means of the Mohr circle the normal stress and shear stress on a plane inclined at 10° to the plane on which the minor principal stress acts. Check the results analytically. (After A. Casagrande.)

10-7. At a certain critical point in a steel beam, on a vertical plane the compressive stress is 126 MPa and the shearing stress is 34.5 MPa. There is no normal stress on the longitudinal (horizontal) plane. Find the stresses acting on the principal planes and the orientation of principal planes with the horizontal. (After Taylor, 1948.)

10-8. A soil sample is under a biaxial state of stress. On plane 1, the stresses are $(26, 8)$, while on plane 2, the stresses are $(11.6, -4)$. Find the major and minor principal stresses.

10-9. For the element shown in Fig. P10-9: (a) Find the magnitude of the unknown stresses σ_h and τ_h on the horizontal plane. (b) Find the orientation of the principal stresses; clearly indicate their orientation in a small sketch. (c) Show the orientation of the planes of maximum as well as minimum shear.

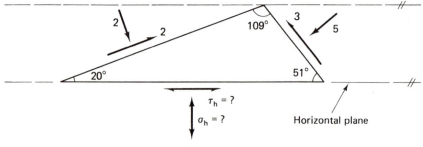

Fig. P10-9

10-10. Given the element with stresses as shown in Fig. P10-10: (a) Find the magnitude and direction of σ_H and τ_H. (b) Find the magnitude and direction of σ_1 and σ_3. Be sure to clearly indicate these stresses and their directions on a separate sketch.

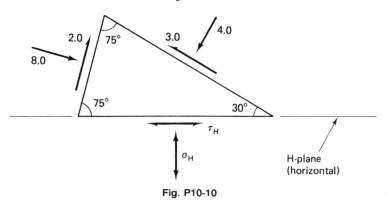

Fig. P10-10

10-11. Given the data of Example 10.5. (a) Find the magnitude and direction of the stresses on the horizontal plane. (b) Find the maximum shear stress and determine the angle between the plane on which it acts and the major principal plane.

10-12. The state of stress on a small element is $\sigma_v = 28$ kPa, $\sigma_h = 14$ kPa, and the shear stress on the horizontal plane is $+4$ kPa. (a) Find the magnitude and directions of the major and minor principal stresses. (b) If the material is a loose sand, can you say whether the element is in a state of failure? If it isn't, how close is it? Why? State your assumptions clearly. (Assume $\phi = 30°$ for the loose sand.)

10-13. Given the vertical and horizontal normal stresses of Problem 10-12. Find the maximum values of shear stress on the horizontal and vertical planes to cause failure in a medium dense sand. Assume the angle of internal friction for the sand is $36°$.

10-14. The state plane stress in a mass of dense cohesionless sand is described by the following stresses:

> Normal stress on horizontal plane = 370 kPa
> Normal stress on vertical plane = 200 kPa
> Shear stress on horizontal and vertical planes = 80 kPa

Determine by means of the Mohr circle the magnitude and direction of the principal stresses. Is this state of stress safe against failure? (After A. Casagrande.)

10-15. At a given point within a sand deposit the major, intermediate, and minor principal stresses are 5, 3, and 2 MN/m^2, respectively. Construct the Mohr diagram, and from it scale the normal and shearing stresses and the obliquity angles on planes at $30°$, $45°$, $60°$, and $75°$ from the major principal plane. (After Taylor, 1948.)

10-16. A 1 m cube within a mass of stressed soil has a stress of 100 kPa on its top and bottom faces, 50 kPa on one pair of vertical faces, and 30 kPa on the other pair of vertical faces. There is no shear stress on any face. Fill in the numerical values for each stress and angle in the following table. (After Taylor, 1948.)

	σ(kPa)	τ(kPa)	α
Major principal plane:			
Intermediate principal plane:			
Minor principal plane:			
Plane of maximum shearing stress:			
Plane of maximum obliquity:			

Note: α is with respect to the horizontal plane.

10-17. In Problem 10-16 what is ϕ, assuming $c = 0$?

10-18. Show that on a plane inclined at $45°$ with respect to the principal planes, the ratio of τ_{max} to $(\sigma_1 + \sigma_3)/2$ is in fact less than τ_{ff}/σ_{ff}. Hint: Assume a ϕ, σ_1, and σ_3.

10-19. Prove that Eq. 10-10 is true, assuming that the Mohr-Coulomb failure criterion is valid. Does it matter if $c = 0$? Hint: Derive the equation first with $c = 0$, then with $c \neq 0$.

10-20. Show that Eq. 10-13 is identical to Eq. 10-14.

10-21. Equation 10-14 is true if $c = 0$. Derive the expression for the principal stress ratio, including the strength parameter c.

10-22. Show that Eq. 10-17 is true from Eq. 10-16.

10-23. Show that Eqs. 10-16 and 10-17 are identical to Eqs. 10-14 and 10-15. (This is a good review of trigonometric identities!)

10-24. In a direct shear test on a specimen of cohesionless sand, the vertical normal stress on the specimen is 300 kN/m^2 and the horizontal shear stress at failure is 200 kN/m^2. (a) Assuming uniform stress distribution within the failure zone and a straight line failure envelope which goes through the origin, determine by means of the Mohr circle the magnitude and direction of the principal stresses at failure. (b) Explain why it is not possible to determine the principal stresses in a direct shear specimen for an applied horizontal shear stress which is not large enough to cause failure. (After A. Casagrande.)

10-25. A specimen of sand is tested in direct simple shear. The stress conditions in the test are as shown in Fig. Ex. 10.9.
Initial conditions:
$$\sigma_v = 4.16 \text{ kg/cm}^2, K_o = 0.5$$
At failure:
$$\sigma_v = 4.16 \text{ kg/cm}^2, \qquad \tau_{hv} = 2.40 \text{ kg/cm}^2$$

(a) Draw the Mohr circles for both initial and final stress conditions.
(b) Show clearly the locations of the poles of these circles.
(c) Determine the magnitude and orientation of the principal stresses at failure.
(d) What is the orientation of the failure plane?
(e) If the shear strain at failure is $10°$ as shown in the figure, what are the stresses σ_s and τ_s on the sides of the specimen at failure? Note: $\tau_s \neq \tau_{hv}$.

10-26. Two conventional CD triaxial compression tests were conducted on a dense angular dry sand at the same void ratio. Test A had a confining pressure of 100 kPa, while in test B the confining pressure was 400 kPa; these stresses were held constant throughout the test. At failure, tests A and B had maximum principal stress differences of 400 and 1700 kPa, respectively.

(a) Plot the Mohr circles for both tests at initial conditions and at failure.

(b) Assuming $c = 0$, determine ϕ.

(c) What is the shear stress on the failure plane at failure for both tests?

(d) Determine the theoretical orientation of the failure plane in each specimen.

(e) What is the orientation of the plane of maximum obliquity?

10-27. Show that stress paths D, E, and F in Fig. 10.18 are correct.

10-28. Show that stress paths B and C in Fig. 10.19 are correct.

10-29. Prove that Eqs. 10-21 and 10-22 are valid.

10-30. A soil sample is subjected to an initial equal-all-around hydrostatic state of stress of 50 kPa. Sketch the stress paths for the loading conditions when (a) σ_h remains constant and σ_v increases to 100 kPa; (b) σ_v is held constant while σ_h increases to 100 kPa; (c) both σ_h and σ_v are increased to 100 kPa; (d) σ_v remains constant while σ_h decreases to 10 kPa; and (e) σ_v is increased by 25 kPa at the same time that σ_h is decreased by 25 kPa.

10-31. Given the same initial conditions as for Problem 10-30, draw the stress paths for loading when (a) $\Delta\sigma_h = \Delta\sigma_v/3$ and (b) $\Delta\sigma_h = \Delta\sigma_v/4$.

10-32. If the initial stress conditions in a soil sample are $\sigma_v = 10$ MPa and $\sigma_h = 5$ MPa, draw the stress paths for σ_v being held constant while (a) σ_h increases to 10 MPa and (b) σ_h decreases to 0 MPa.

10-33. Evaluate the K_o and β for the conditions shown in Fig. 10.21. Are these values reasonable? Why?

10-34. A triaxial sample of loose sand is tested in lateral extension (LE) (see Fig. 10.22). The sample is first consolidated non-hydrostatically, with $\sigma_1 = 15$ kPa and $\sigma_3 = 10$ kPa. The sample is then failed in LE, and the angle of internal friction is 30° ($c = 0$). (a) Draw the Mohr circles for both initial and "at failure" conditions. (b) What will be the major and minor principal stresses at failure? (c) On a p-q diagram draw the stress paths for both the consolidation (see Fig. 10.21) and shearing phases of this test. Show the K_f line.

10-35. Another sample of the same sand tested in Problem 10-34 is tested in lateral compression (LC). Complete parts (a) through (c) requested in Problem 10-34 for this test.

eleven

Shear Strength of Sands and Clays

11.1 INTRODUCTION

The shear strength of soils is a most important aspect of geotechnical engineering. The bearing capacity of shallow or deep foundations, slope stability, retaining wall design and, indirectly, pavement design are all affected by the shear strength of the soil in a slope, behind a retaining wall, or supporting a foundation or pavement. Structures and slopes must be stable and secure against total collapse when subjected to maximum anticipated applied loads. Thus *limiting equilibrium* methods of analysis are conventionally used for their design, and these methods require determination of the ultimate or limiting shear resistance (shear strength) of the soil.

In Chapter 10, we defined the shear strength of a soil as the ultimate or maximum shear stress the soil can withstand. We mentioned that sometimes the limiting value of shear stress was based on a maximum allowable strain or deformation. Very often, this allowable deformation actually controls the design of a structure because with the large safety factors we use, the actual shear stresses in the soil produced by the applied loads are much less than the stresses causing collapse or failure.

The shear strength can be determined in several different ways; we described some of the more common laboratory and field tests in Sec. 10.5. In situ methods such as the vane shear test or penetrometers avoid some of the problems of disturbance associated with the extraction of soil samples from the ground. However, these methods only determine the shear strength indirectly through correlations with laboratory results or backcalculated from actual failures. Laboratory tests, on the other hand, yield the shear strength directly. In addition, valuable information about the stress-strain behavior and the development of pore pressures during shear can often be obtained. In this chapter, we shall illustrate the fundamental stress-deformation and shear strength response of soils with the results of laboratory tests for typical soils. In this way, we hope you can gain some understanding of how soils actually behave when sheared.

A word about the scope of this chapter. It is purposely kept as simple as possible. Only test results of typical "well-behaved" sands and clay soils are illustrated; special soils such as cemented sands, stiff fissured clays, highly sensitive ("quick") clays, and organic soils are not considered in detail herein. Special topics such as strength anisotropy, the Hvorslev parameters, complex stress systems, and creep are not included in this chapter. The approach is admittedly classical and we hope not oversimplified. The interested student may wish to consult advanced textbooks and the geotechnical literature on soil behavior for additional information about the real behavior of sands and clays.

In this chapter we draw heavily on the work of our teachers and colleagues. We gratefully acknowledge the important contributions made by A. Casagrande, R. C. Hirschfeld, C. C. Ladd, K. L. Lee, G. A. Leonards, J. O. Osterberg, S. J. Poulos, and H. B. Seed. Our discussion of shear strength of soils starts with sands and is followed by the strength properties of cohesive soils.

The following notation is introduced in this chapter.

Symbol	Dimension	Unit	Definition
$A, \bar{A}, B$	—	—	Skempton's pore pressure parameters
a	—	—	Henkel's pore pressure parameter
A_o, A_s	L^2	m^2	Initial specimen area and area at some strain, respectively (Eq. 11-8)
C_{sk}, C_v, C_w	$M^{-1}LT^2$	$1/kPa$	Compressibility of the soil skeleton, pore fluid, and water, respectively (Eq. 11-11)
c, c_T	$ML^{-1}T^{-2}$	kPa	"Cohesion" or intercept on τ axis when $\sigma = 0$
E_s	$ML^{-1}T^{-2}$	kPa	Secant modulus
E_t	$ML^{-1}T^{-2}$	kPa	Initial tangent modulus
E_u	$ML^{-1}T^{-2}$	kPa	Undrained modulus of linear deformation (Young's modulus)
ESP	—	—	Effective stress path
e_c	—	(decimal)	Void ratio after consolidation
e_{crit}	—	(decimal)	Critical void ratio
e_{cd}	—	(decimal)	e_{crit}-dense
e_{cl}	—	(decimal)	e_{crit}-loose
e_d	—	(decimal)	e-dense
e_l	—	(decimal)	e-loose
h	—	—	An empirical exponent (Eq. 11-7)
S_t	—	—	Sensitivity (Eq. 11-9)
TSP	—	—	Total stress path
u_o	$ML^{-1}T^{-2}$	kPa	Initial pore water pressure; back pressure

Symbol	Dimension	Unit	Definition
u_r	$ML^{-1}T^{-2}$	kPa	Residual (capillary) pore water pressure after sampling
V_o	L^3	m³	Initial volume (Eq. 11-4)
ΔV	L^3	m³	Change in volume (Eq. 11-4)
ϵ_v	—	(%)	Vertical or axial strain
$\epsilon_1, \epsilon_2, \epsilon_3$	—	(%)	Major, intermediate, and minor principal strains
μ	—	—	Correction factor to vane shear strength
σ_{hc}	$ML^{-1}T^{-2}$	kPa	Total horizontal consolidation stress
σ_{vc}	$ML^{-1}T^{-2}$	kPa	Total vertical consolidation stress
σ_{oct}	$ML^{-1}T^{-2}$	kPa	Octahedral normal stress (Eq. 11-23)
σ'_{3c}	$ML^{-1}T^{-2}$	kPa	Effective consolidation pressure
$\sigma'_{3\,crit}$	$ML^{-1}T^{-2}$	kPa	Critical effective confining pressure
σ'_{3f}	$ML^{-1}T^{-2}$	kPa	Effective confining pressure at failure
σ'_1/σ'_3	—	—	Principal effective stress ratio (Eq. 11-1)
$(\sigma_1 - \sigma_3)$	$ML^{-1}T^{-2}$	kPa	Principal stress difference (Eq. 11-2)
τ_f	$ML^{-1}T^{-2}$	kPa	Undrained shear strength
τ_{oct}	$ML^{-1}T^{-2}$	kPa	Octahedral shear stress (Eq. 11-24)
ϕ'_d	—	(degree)	Angle of internal friction from CD tests
ϕ', ϕ_T	—	(degree)	Angle of internal friction in terms of effective stress and total stress, respectively
ϕ_{ps}, ϕ_{tx}	—	(degree)	Angle of internal friction from plane strain tests and triaxial tests, respectively (Eq. 11-5)

Note: A prime mark on an angle or stress indicates *effective stresses*. Subscripts o, c, and f indicate initial, consolidation, and failure conditions, respectively.

11.2 ANGLE OF REPOSE OF SANDS

If we were to deposit a granular soil by pouring it from a single point above the ground, it would form a conical pile. As more and more granular material was deposited on the pile, the slope for a short period of time might appear to be steeper, but then the soil particles would slip and slide down the slope to the *angle of repose*. This angle of the slope with respect to the horizontal plane would remain constant at some *minimum* value. Refer for a moment to Fig. 6.7 for an illustration of the angle of repose. Since this angle is the steepest *stable* slope for very loosely packed sand, the angle of repose represents the angle of internal friction of the granular material at its *loosest* state.

Sand dunes are an example from nature of the angle of repose. Figure 11.1 shows how both a stationary dune (SD) as well as a migrating dune (MD) are formed. On the leeward side (LS), the slope of the dune

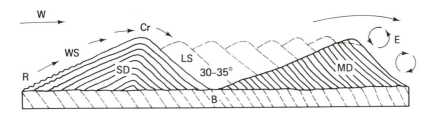

Deposition of sand by wind. Ideal structure of stationary
or fixed dunes (SD) and migrating live dunes (MD). The
arrows indicate the direction of air currents (W). E shows
eddies. WS is the windward slope of the dune, LS the
leeward, or down-wind slope. R marks ripples, and Cr
is the crest of the dune. Dashed lines show the former
positions of live dune MD. B is the base rock. (After
A. Holmes.)

Fig. 11.1 Formation of sand dunes and illustration of the angle of
repose. After *Aerogeology* by Horst F. von Bandat. Copyright © 1962 by
Gulf Publishing Company, Houston, TX. Used with permission. All rights
reserved.

will have an angle (of repose) which varies from 30° to 35°, depending on
factors that are discussed later in this chapter. If the slope on the leeward
side becomes steeper than 30° to 35°, then the slope is unstable and sand
grains will roll down the slope until the angle of repose is reached. An
unstable condition is shown on the slope at the far right-hand side of Fig.
11.1; eventually a smooth slope at the angle of repose will form. The angle
of repose depends on the type of materials and other factors, and it
represents the angle of internal friction or shearing resistance ϕ at its
loosest state. Recall that the terms *loose* or *dense* are only relative terms
(see Sec. 4.9), especially with respect to their behavior in shear. As we shall
soon see, the stress-strain and volume change response depends on the
confining pressure as well as on the relative density.

11.3 BEHAVIOR OF SATURATED SANDS DURING DRAINED SHEAR

To illustrate the behavior of sands during shear, let's start by taking
two samples of sand, one at a very high void ratio, the "loose" sand, and
the other at a very low void ratio, the "dense" sand. We could perform
direct shear tests (Fig. 10.12), but to better measure the volume changes we
shall use the triaxial apparatus, as shown in Fig. 10.13a and Fig. 11.2. We
shall run the two tests under consolidated drained (CD) conditions, which

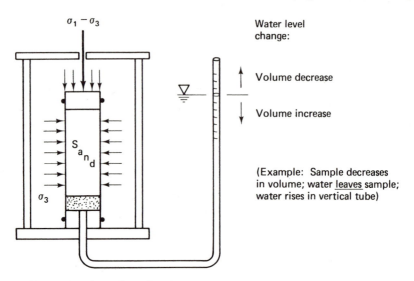

Fig. 11.2   Consolidated-drained triaxial test with volume change measurements.

means we will allow water to freely enter or leave the sample during shear without interference. If we have a saturated sample, we can easily monitor the amount of water that enters or leaves the sample and equate this to the volume change and thus the void ratio change in the sample. Water leaving the sample during shear indicates a volume decrease, and vice versa. In both out tests the confining pressure, σ_c equals σ_3, is held constant and the axial stress is increased until failure occurs. Failure may be defined as:

1. Maximum principal stress difference, $(\sigma_1 - \sigma_3)_{max}$.
2. Maximum principal effective stress ratio, $(\sigma'_1/\sigma'_3)_{max}$.
3. $\tau = [(\sigma_1 - \sigma_3)/2]$ at a prescribed strain.

Most of the time, we will define failure as the *maximum principal stress difference*, which is the same as the *compressive strength* of the specimen. Typical stress-strain curves for loose and dense sand are shown in Fig. 11.3a, while the corresponding stress versus void ratio curves are presented in Fig. 11.3b.

When the loose sand is sheared, the principal stress difference gradually increases to a maximum or ultimate value $(\sigma_1 - \sigma_3)_{ult}$. Concurrently, as the stress is increased the void ratio *decreases* from e_l (e-loose) down to e_{cl} (e_c-loose), which is very close to the *critical void ratio* e_{crit}. Casagrande (1936a) called the ultimate void ratio at which continuous deformation occurs with no change in principal stress difference the *critical void ratio*.

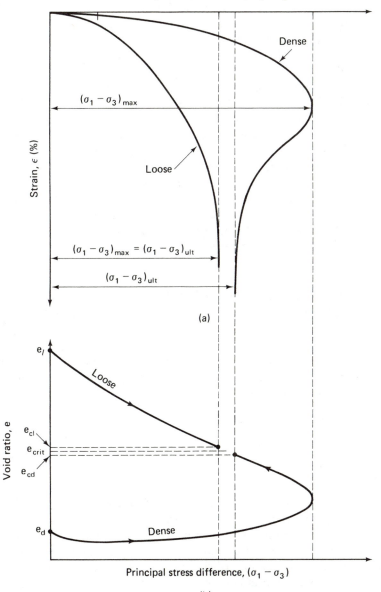

Fig. 11.3   Triaxial tests on "loose" and "dense" specimens of a typical
sand: (a) stress-strain curves; (b) void ratio changes during shear (after
Hirschfeld, 1963).

When the dense specimen is sheared, the principal stress difference reaches a peak or maximum, after which it decreases to a value very close to $(\sigma_1 - \sigma_3)_{ult}$ for the loose sand. The void ratio-stress curve shows that the dense sand decreases in volume slightly at first, then *expands* or *dilates* up to e_{cd} (e_c-dense). Notice that the void ratio at failure e_{cd} is very close to e_{cl}. Theoretically, they both should be equal to the *critical void ratio* e_{crit}. Similarly, the values of $(\sigma_1 - \sigma_3)_{ult}$ for both tests should be the same. The differences are usually attributed to difficulties in precise measurement of ultimate void ratios as well as nonuniform stress distributions in the test specimens (Hirschfeld, 1963). Evidence of this latter phenomenon is illustrated by the different ways in which the samples usually fail. The loose sample just *bulges*, while the dense sample often fails along a distinct plane oriented approximately $45° + \phi'/2$ from the horizontal (ϕ' is, of course, the *effective* angle of shearing resistance of the dense sand). Note that it is at least theoretically possible to set up a sample at an initial void ratio such that the volume change at failure would be zero. This void ratio would, of course, be the critical void ratio e_{crit}.

11.4 EFFECT OF VOID RATIO AND CONFINING PRESSURE ON VOLUME CHANGE

Thus far, in describing the behavior of the two drained triaxial tests on loose and dense sands shown in Fig. 11.3, we have mentioned the following physical quantities:

principal stress difference
strain
volume change
critical void ratio e_{crit} and, indirectly,
relative density (Eqs. 4-2 and 4-3)

We have purposely avoided defining the terms *loose* and *dense* because the volume change behavior during shear depends not only on the initial void ratio and relative density but also on the confining pressure. In this section we shall consider the effect of confining pressure on the stress-strain and volume change characteristics of sands in drained shear.

We can assess the effects of σ_3 (and, remember, in a drained test $\sigma_3 = \sigma_3'$, as the excess pore water pressure is always zero) by preparing several samples at the same void ratio and testing them at different confining pressures. We would find that the shear strength increases with σ_3. A convenient way to plot the principal stress difference versus strain data is to *normalize* it by plotting the *principal stress ratio* σ_1/σ_3 versus strain. For a drained test, of course, $\sigma_1/\sigma_3 = \sigma_1'/\sigma_3'$. At failure, the ratio is $(\sigma_1'/\sigma_3')_{max}$. From Eqs. 10-14 and 10-16,

$$\left(\frac{\sigma_1'}{\sigma_3'}\right)_{max} = \frac{1 + \sin\phi'}{1 - \sin\phi'} = \tan^2\left(45 + \frac{\phi'}{2}\right) \qquad (11\text{-}1)$$

where ϕ' is the *effective* angle of internal friction. The principal stress difference is related to the principal stress ratio by

$$\sigma_1 - \sigma_3 = \sigma_3'\left(\frac{\sigma_1'}{\sigma_3'} - 1\right) \qquad (11\text{-}2)$$

At failure, the relationship is

$$(\sigma_1 - \sigma_3)_f = \sigma_{3f}'\left[\left(\frac{\sigma_1'}{\sigma_3'}\right)_{max} - 1\right] \qquad (11\text{-}3)$$

Let's look first at the behavior of loose sand. Typical drained triaxial test results are shown for loose Sacramento River sand in Fig. 11.4a. The principal stress ratio is plotted versus axial strain for different effective consolidation pressures σ_{3c}'. Note that none of the curves has a distinct peak, and they have a shape similar to the loose curve shown in Fig. 11.3a. The volume change data is also normalized by dividing the volume change ΔV by the original volume V_o to obtain the volumetric strain, or

$$\text{volumetric strain, \%} = \frac{\Delta V}{V_o} \times 100 \qquad (11\text{-}4)$$

To better appreciate what is going on in Fig. 11.4a, let us compute the principal stress difference $(\sigma_1 - \sigma_3)$ at a strain of 5% for $\sigma_{3c}' = 3.9$ MPa and $\sigma_{3c}' = 0.1$ MPa. The principal stress ratios for these conditions are 2.0 and 3.5, respectively, as indicated by the arrows in Fig. 11.4a. Utilizing Eq. 11-2, we obtain the following results:

σ_{3c}' (MPa)	σ_1'/σ_3' —	$(\sigma_1 - \sigma_3)$ (MPa)	σ_1' (MPa)
0.1	3.5	0.25	0.35
3.9	2.0	3.9	7.8

It is interesting to look at the shapes of the volumetric strain versus axial strain curves in Fig. 11.4b. As the strain increases, the volumetric strain decreases for the most part. This is consistent with the behavior of a loose sand, as shown in Fig. 11.3b. However at low confining pressures (for

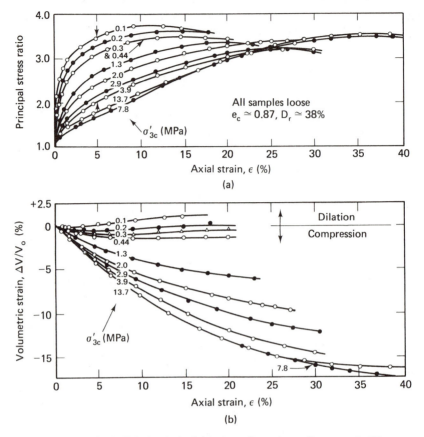

Fig. 11.4   Typical drained triaxial test results on loose Sacramento River sand: (a) principal stress ratio versus axial strain; (b) volumetric strain versus axial strain (after Lee, 1965).

example, 0.1 and 0.2 MPa), the volumetric strain is positive or *dilation* is taking place! Thus even an initially loose sand behaves like a dense sand; that is, it dilates if σ'_{3c} is low enough!

Now, let's look at the behavior of dense sand. The result of several drained triaxial tests on dense Sacramento River sand are presented in Fig. 11.5. Although the results are similar in appearance to Fig. 11.4, there are some significant differences. First, definite peaks are seen in the (σ'_1/σ'_3)-strain curves, which are typical of dense sands (compare with Fig. 11.3a). Second, large *increases* of volumetric strain (dilation) are observed. However, at higher confining pressures, dense sand exhibits the behavior of loose sand by showing a *decrease* in volume or compression with strain.

By testing samples of the same sand at the same void ratios or densities but with different effective consolidation pressures, we can de-

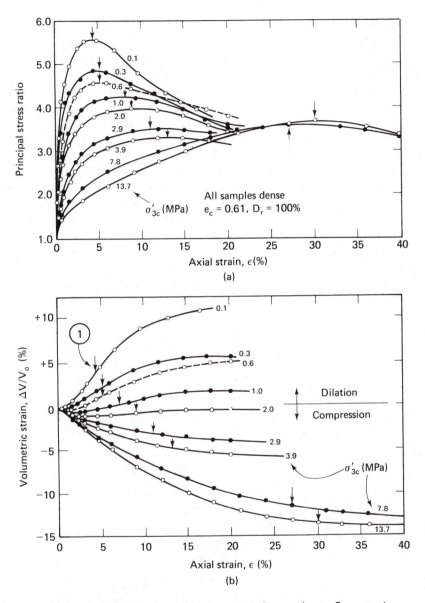

Fig. 11.5 Typical drained triaxial test results on dense Sacramento River sand: (a) principal stress ratio versus axial strain; (b) volumetric strain versus axial strain (after Lee, 1965).

termine the relationship between volumetric strain at failure and void ratio or relative density. We could define failure as either the maximum $(\sigma_1 - \sigma_3)$ or maximum σ_1'/σ_3'. For drained tests, failure occurs at the same strain according to both criteria. Points at failure are shown as small arrows in Fig. 11.5. Volumetric strain at failure versus void ratio at the end of consolidation, from the data in Figs. 11.4b and 11.5b for various confining pressures (other data have been added as well), are shown in Fig. 11.6. For example, point 1 in Fig. 11.5b is plotted as point 1 in Fig. 11.6. It can be seen that for a given confining pressure the volumetric strain decreases (becomes more negative) as the density decreases (void ratio increases). By definition, the critical void ratio is the void ratio *at failure* when the volumetric strain is zero. Thus for the various values of σ_{3c}' in Fig. 11.6, e_{crit} is the void ratio when $\Delta V/V_o = 0$. For example, e_{crit} for $\sigma_{3c}' = 2.0$ MPa is 0.555.

We can see how e_{crit} varies with confining pressure by taking the critical void ratios of Fig. 11.6 and plotting them versus σ_{3c}', as is done in Fig. 11.7. Here we have called σ_{3c}' the *critical* confining pressure $\sigma_{3\ crit}'$

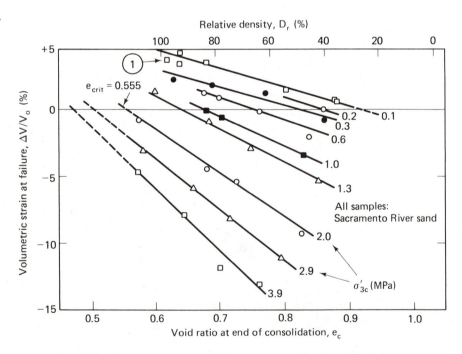

Fig. 11.6 Volumetric strain at failure versus void ratio at end of consolidation for drained triaxial tests at various confining pressures (after Lee, 1965).

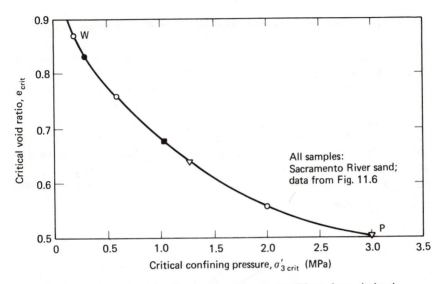

Fig. 11.7 Critical void ratio versus pressure conditions from drained triaxial tests. Data from Fig. 11.6 (after Lee, 1965).

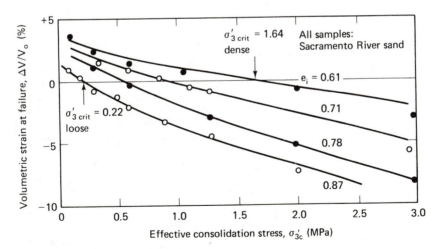

Fig. 11.8 Volumetric strain at failure versus effective consolidation stress for different initial void ratios (after Lee, 1965).

because this is the effective confining pressure at which zero volumetric strain occurs at failure for a given void ratio.

A second and just as interesting approach is to use the data shown in Figs. 11.4b and 11.5b (plus other data at intermediate void ratios) and plot the relationship between volumetric strain at failure and confining pressure for various values of void ratio after consolidation. Such a graph is shown in Fig. 11.8, although the void ratios indicated are initial void ratios and not the void ratios after consolidation. Note that the value of σ'_{3c} at $\Delta V/V_o = 0$ is the critical confining pressure, $\sigma'_{3\,crit}$. Since they are drained tests, $\sigma'_{3c} = \sigma'_{3f}$. This relationship could also be obtained from Fig. 11.6 by noting the values of volumetric strain at constant void ratios and plotting $\Delta V/V_o$ versus σ'_{3c}.

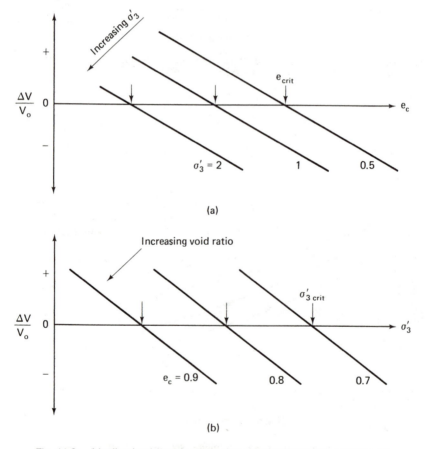

(a)

(b)

Fig. 11.9 Idealized volumetric strain data from drained triaxial tests: (a) $\Delta V/V_o$ versus e_c; (b) $\Delta V/V_o$ versus σ'_3.

We can show the relationships of Figs. 11.6 and 11.8 in Fig. 11.9 (idealized). Since both Figs. 11.6 and 11.8 have a common axis, it is possible to combine them in a single three-dimensional graph known as the *Peacock diagram* (after William Hubert Peacock who first constructed such a diagram in 1967), as shown in Fig. 11.10.

With the Peacock diagram, we are able to predict the behavior of sand at any void ratio after consolidation e_c and at any confining pressure σ'_3. For example, if the effective confining pressure is given at point C in Fig. 11.10, which is higher than $\sigma'_{3\,crit}$ for this given void ratio e_c, then we would expect a decrease in the volume or a minus $\Delta V/V_o$, which is equal to the ordinate BS. On the other hand, if σ'_3 is less than $\sigma'_{3\,crit}$, such as point A for the given value of e_c, then a dilation or positive volume change will take place equal to the ordinate RD. As the void ratio after consolidation varies to-and-fro along the void ratio axis, $\sigma'_{3\,crit}$ varies, and so will the volume changes at failure. For a real sand, the Peacock diagram has curved surfaces. For example, the line KP in Fig. 11.10 should look like one of the curves in Fig. 11.8. The line PW in Fig. 11.10 is also curved. See line PW in Fig. 11.7; here you are looking at a plane on the Peacock diagram where $\Delta V/V_o = 0$.

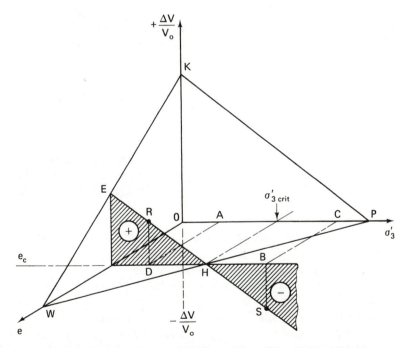

Fig. 11.10 Peacock diagram, which combines Figs. 11.9a and b in an idealized graph to show the behavior of drained triaxial tests on sand.

11.5 BEHAVIOR OF SATURATED SANDS DURING UNDRAINED SHEAR

The main difference between drained and undrained triaxial shear is that in an undrained test no volume change is allowed during axial loading. However, unless the confining pressure just happens to be at $\sigma'_{3\,crit}$, the soil will *tend to change volume* during loading. For example, referring to the Peacock diagram, Fig. 11.10, again, if a soil at e_c is tested *undrained* at a σ'_3 at point C, then the sand sample would *tend* to decrease in volume, but it can't. As a result, a *positive* pore pressure is induced, which causes a *reduction* in the effective stress. The limiting or minimum effective pressure at failure is $\sigma'_{3\,crit}$ because at this pressure $\Delta V/V_o$ is zero. If no tendency towards volume change takes place, then no excess pore pressure is induced. So the maximum possible pore pressure in this example is equal to $\sigma'_{3c} - \sigma'_{3\,crit}$, or the distance $\overline{BH}$ in Fig. 11.10. The Mohr circles at failure for this case are shown in Fig. 11.11a. The dashed circles E represent the effective stress conditions, whereas the solid circle T is in terms of total stresses. Since Eq. 7-13 always holds, the two circles are separated by the value of Δu induced at any time during the test. Since the volume change *tendency* is to reduce, a positive change (increase) in pore pressure is caused, which in turn results in a *reduction* in the effective stress. Thus, for this example, $\Delta u = B - H = \sigma'_{3c} - \sigma'_{3f} = \sigma'_{3c} - \sigma'_{3\,crit}$. The $(\sigma_1 - \sigma_3)_f$ is given by Eq. 11-3 when the confining pressure at failure is $\sigma'_{3\,crit}$.

$$(\sigma_1 - \sigma_3)_f = \sigma'_{3\,crit}\left[\left(\frac{\sigma'_1}{\sigma'_3}\right)_f - 1\right]$$

Also, if we were to run a *drained* test with the confining pressure equal to σ'_{3c} at point C, the drained strength would be much larger than the undrained strength since its Mohr circle must be tangent to the effective Mohr failure envelope. Just look at the relative sizes of the two effective Mohr circles in Fig. 11.11a.

A different response occurs when we run a test with the effective confining pressure less than $\sigma'_{3\,crit}$ such as point A in Fig. 11.10. From the Peacock diagram, we would expect the sample to *tend* to dilate (ordinate RD). Since the specimen is prevented from actually expanding, a *negative* pore pressure is developed which *increases* the effective stress from D (A) towards H ($\sigma'_{3\,crit}$). Thus, as in the previous example, the limiting effective stress is the critical confining pressure $\sigma'_{3\,crit}$. (The situation may arise where the negative pore water pressure approaches -100 kPa or -1 atmosphere, and cavitation takes place, but we will ignore this possibility in this chapter.) The whole point of this exercise is that we may predict the

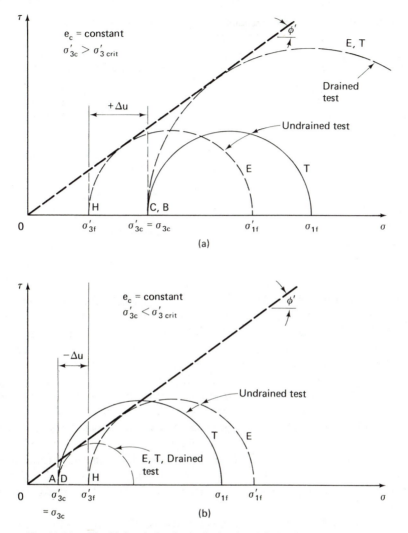

Fig. 11.11 The Mohr circles for undrained and drained triaxial compression tests: (a) case where $\sigma_3'c > \sigma_{3\ \mathrm{crit}}'$; (b) case where $\sigma_3'c < \sigma_{3\ \mathrm{crit}}$.

undrained behavior of sands from the *drained* behavior when we know the volume change *tendencies* as idealized in the Peacock diagram.

The Mohr circle representation for the case where $\sigma_{3c}' < \sigma_{3\ \mathrm{crit}}'$ is presented in Fig. 11.11b. The undrained test starts out at σ_{3c}', point A, and since the induced pore water pressure is negative, the effective confining pressure increases until failure is reached at point H. Note that the effective stress Mohr circles E at failure in Figs. 11.11a and b are the same

TABLE 11-1 A Summary of Concepts Shown in Fig. 11.11

Effective Consolidation Pressure	Mohr Circles		
	Drained, Effective = Total	Undrained, Effective	Undrained, Total
$\sigma'_{3c} > \sigma'_{3\,\mathrm{crit}}$	Larger than undrained	Smaller than drained: Left of total stress circle $\sigma'_{3f} < \sigma'_{3c}$	Smaller than drained: Right of effective stress circle
$\sigma'_{3c} < \sigma'_{3\,\mathrm{crit}}$	Smaller than undrained	Larger than drained: Right of total stress circle $\sigma'_{3f} > \sigma'_{3c}$	Larger than drained: Left of effective stress circle
$\sigma'_{3c} \equiv \sigma'_{3\,\mathrm{crit}}$	All circles would be the same; because no volume change tendencies exist, $\Delta u = 0$ during the test.		

size because, for this void ratio e_c, the effective stress at failure is the same, $\sigma'_{3\,\mathrm{crit}}$. If the effective stress and void ratio are the same, then the samples would have the same compressive strength, $\sigma'_{1f} - \sigma'_{3f}$; thus the circles have the same diameter. Note that the total stress circle T, at failure, is also the same size as the effective stress circle because $(\sigma_1 - \sigma_3)_f$ is the same for both T and E; also T lies to the *left* of E. This case is the opposite of Fig. 11.11a. (The total stress Mohr failure envelopes have been omitted from the figure to simplify things.) Note also that the *drained* Mohr circle for this second case is substantially *smaller* than the effective stress circle for the undrained case. As before, the circle starts at σ'_{3c}, and it must be tangent to the effective Mohr failure envelope. Since the void ratio after consolidation e_c is a constant for all the tests shown in Fig. 11.11, all the effective Mohr circles must be tangent to the effective stress failure envelope.

A summary of the main points just discussed and shown in Fig. 11.11 is presented in Table 11-1. For a more comprehensive treatment of the undrained strength characteristics of sands see Seed and Lee (1967).

EXAMPLE 11.1

Given:

A battery filler (rubber squeeze bulb plus glass tube) contains *dense* sand. The battery filler bulb and sand are completely saturated with water.

Required:

If the bulb is squeezed, describe what happens to the water level in the glass tube. Will it go up, down, or remain the same?

Solution:

Because the sand is dense, it will tend to dilate or expand when sheared. This action will create a slightly negative pressure in the water, which will draw water into the voids and cause the level in the glass tube to move downward.

EXAMPLE 11.2

Given:

The same apparatus as for Example 11.1, only now the bulb is filled with *loose* sand.

Required:

Predict the behavior of the water level in the glass tube when the bulb is squeezed.

Solution:

When loose sand is sheared, the soil will tend to decrease in volume. This action will create a positive pressure in the water, which will squeeze water out of the voids. Thus the water level in the tube will move upward. It follows that if the sand in the battery filler bulb is at its critical void ratio, then upon squeezing (shearing) the bulb, the water level may at first decrease slightly, but with continued squeezing it will return to its original level; that is, no net volume change will occur when the sand is at e_{crit}.

EXAMPLE 11.3

Given:

A CD triaxial test is conducted on a granular soil. At failure, $\sigma_1'/\sigma_3' = 4.0$. The effective minor principal stress at failure was 100 kPa.

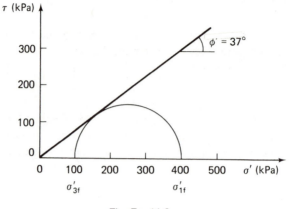

Fig. Ex. 11.3

Required:

    a. Compute ϕ'.
    b. What is the principal stress difference at failure?
    c. Plot the Mohr circle and the Mohr failure envelope.

Solution:

    a. From Eqs. 10-14, 10-16, or 11-1, we know that

$$\frac{\sigma'_{1f}}{\sigma'_{3f}} = \frac{1 + \sin\phi'}{1 - \sin\phi'} = \tan^2\left(45° + \frac{\phi'}{2}\right) = 4.0$$

    Solving for ϕ', we obtain $\phi' = 37°$.
    b. From Eq. 11-3,

$$(\sigma_1 - \sigma_3)_f = \sigma'_3\left(\frac{\sigma'_{1f}}{\sigma'_{3f}} - 1\right) = 100 \text{ kPa } (4 - 1) = 300 \text{ kPa}$$

    c. See Fig. Ex. 11.3.

EXAMPLE 11.4

Given:

Figure 11.6.

Required:

What is the critical void ratio for Sacramento River sand when the confining pressure is 1.5 MPa?

Solution:

From Fig. 11.6, interpolating between the curves for $\sigma_3' = 1.3$ and 2.0 MPa, we find that e_c (for $\sigma_3' = 1.5$) is about 0.61 for Sacramento River sand.

EXAMPLE 11.5

Given:

Figure 11.8.

Required:

What is the critical confining pressure for Sacramento River sand if the void ratio equals 0.75?

Solution:

From Fig. 11.8, we can interpolate between the curves for $e_i = 0.71$ and 0.78 for the value of σ_3' when $\Delta V/V_o$ is zero. We obtain a σ_3' of about 0.7 MPa.

EXAMPLE 11.6

Given:

Figure 11.10, but scaled to the idealized behavior of Sacramento River sand (a combination of Figs. 11.6 and 11.8); σ_3' $_{crit} = 0.4$ MPa and $e_c = e_{crit} = 0.8$.

Required:

Describe both the drained and undrained behavior of this sand if the test void ratios after consolidation at $\sigma_{3c}' = 0.4$ MPa are (a) 0.85 and (b) 0.75.

Solution:

Since σ_{3c}' and e_c are at critical, there is by definition no volume change during shear. Thus our test plots at point H in Fig. 11.10, with the values of σ_3' $_{crit}$ and e_c as given. (You can verify these values in Figs. 11.6 and 11.8.)

a. When $e_c > e_{crit}$ (0.85 > 0.8), then at $\sigma'_{3c} = 0.4$ MPa the coordinates of our test would have to plot *below* the *W0P* plane, which means $\Delta V/V_o$ is negative. During *drained* shear, σ'_3 is constant (no excess pore pressure develops), and the specimen would consolidate and decrease in volume during shear. Its coordinates would be on the extension of plane *WKP*.

In *undrained* shear the specimen would *tend* to decrease in volume, but since it is undrained it cannot. Therefore the specimen would develop positive pore water pressure along with a concurrent decrease in σ'_3. In Fig. 11.10, the test coordinates must remain on the $e = 0.85$ line *and* in the plane *W0P*. The only way this can happen is for σ'_3 to decrease, which makes sense in view of the increase in pore water pressure.

b. When $e_c < e_{crit}$ (0.75 < 0.80) the opposite of (a) will happen: in drained shear, σ'_3 is again constant and equal to 0.4 MPa, so for the coordinates of our test to remain on plane *WKP*, the $\Delta V/V_o$ must increase. In undrained shear, the *tendency* towards volume increase would cause the pore water pressure to decrease and the σ'_3 to increase. This is what happens when our test coordinates remain on plane *W0P*; that is, σ'_3 increases.

EXAMPLE 11.7

Given:

Figure 11.10 is scaled to the behavior of Sacramento River sand (Figs. 11.6 and 11.8), with $e_{crit} = 0.6$ and $\sigma'_{3\ crit} = 1.6$ MPa.

Required:

Describe the behavior, both in drained and undrained shear, if we maintain this void ratio of 0.6 but test the specimen with σ'_{3c} of (a) 1.5 MPa and (b) 1.7 MPa.

Solution:

a. When $\sigma'_{3c} < \sigma'_{3\ crit}$, then the specimen will dilate and a positive $\Delta V/V_o$ will occur. This behavior is similar to what happens to point A in Fig. 11.10. The dilation is measured by the ordinate RD so that the coordinates of our test remain on plane *WKP*.

In undrained shear, the *tendency* will be for dilation which is prevented; we must remain at $e_c = 0.6$ and on plane *W0P*. Therefore σ'_3 must increase, which makes sense physically since the induced pore water pressure tends to decrease.

b. When $\sigma'_{3c} > \sigma'_{3 \text{ crit}}$, the behavior would be similar to path BS in Fig. 11.10 in drained shear. In undrained shear, the tendency towards compression would result in positive excess pore pressure and a decrease in σ'_3.

EXAMPLE 11.8

Given:

A drained triaxial test on sand with $\sigma'_3 = 150$ kPa and $(\sigma'_1/\sigma'_3)_{\text{max}} = 3.7$.

Required:

a. σ'_{1f},
b. $(\sigma_1 - \sigma_3)_f$, and
c. ϕ'.

Solution:

a. $(\sigma'_1/\sigma'_3)_f = 3.7$. Solve for σ'_{1f}. $\sigma'_{1f} = 3.7(150) = 555$ kPa.
b. $(\sigma_1 - \sigma_3)_f = (\sigma'_1 - \sigma'_3)_f = 555 - 150 = 405$ kPa.
c. Assume for sand that $c' = 0$. So, from Eq. 10-13,

$$\phi' = \arcsin\left(\frac{\sigma'_{1f} - \sigma'_{3f}}{\sigma'_{1f} + \sigma'_{3f}}\right) = \arcsin\frac{405}{705} = 35°$$

Note: We could also determine ϕ' graphically from the Mohr circle plotted for failure conditions, as shown in Fig. Ex. 11.8.

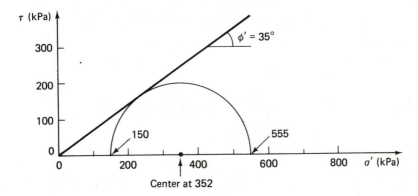

Fig. Ex. 11.8

EXAMPLE 11.9

Given:

Assume the test specimen of Example 11.8 was sheared *undrained* at the same total cell pressure (150 kPa). The induced excess pore water pressure at failure Δu_f is equal to 70 kPa.

Required:

    a. σ'_{1f},
    b. $(\sigma_1 - \sigma_3)_f$,
    c. ϕ in terms of *total* stress, and
    d. the angle of the failure plane α_f.

Solution:

    a., b. Since the void ratio after consolidation would be the same for this test as for Example 11.8, assume ϕ' is the same. You can do this problem either (1) analytically or (2) graphically.
    1. *Analytically*: We know that

$$(\sigma_1 - \sigma_3)_f = \sigma'_{3f}\left[\left(\frac{\sigma'_1}{\sigma'_3}\right)_{max} - 1\right]$$

from Eq. 11-3.

$$\sigma'_{3f} = \sigma_{3f} - \Delta u_f = 150 - 70 = 80 \text{ kPa}$$

So

$$(\sigma_1 - \sigma_3)_f = 80(3.7 - 1) = 216 \text{ kPa}$$

$$\sigma'_{1f} = (\sigma_1 - \sigma_3)_f + \sigma'_{3f} = 216 + 80 = 296 \text{ kPa}$$

These are the answers to parts (a) and (b).
    c. We can write Eqs. 10-13 and 11-1 in terms of total stresses. Using Eq. 10-13,

$$\sin \phi_{total} = \frac{\sigma_1 - \sigma_3}{\sigma_1 + \sigma_3} = \frac{216}{(296 + 70) + 150} = 0.42$$

$$\phi_{total} = 24.8°$$

Using Eq. 11-1,

$$\frac{\sigma_{1f}}{\sigma_{3f}}(\text{no primes}) = \frac{(296 + 70)}{150} = 2.44 = \tan^2\left(45° + \frac{\phi}{2}\right)$$

Solving for ϕ_{total}, we obtain $\phi_{total} = 24.8°$.

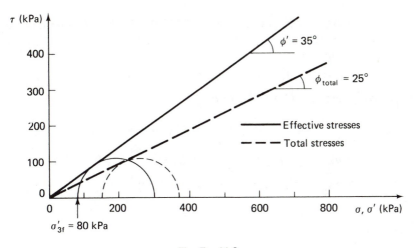

Fig. Ex. 11.9

2. *Graphically*: Plot the Mohr failure envelope with $\phi' = 35°$ on a Mohr diagram (Fig. Ex. 11.9). There is only *one* circle that is tangent to the envelope and with $\sigma'_{3f} = 80$ kPa $(150 - 70)$. Once the circle is drawn (trial and error), σ'_{1f} is automatically determined ($\sigma'_{1f} = 296$ kPa) as is $(\sigma_1 - \sigma_3)_f$, the diameter of the failure circle ($= 216$ kPa).

The Mohr circle at failure in terms of total stresses has the same diameter since $(\sigma_1 - \sigma_3) = (\sigma'_1 - \sigma'_3)$. You can plot the total stress circle starting at $\sigma_{3f} = 150$, the total cell pressure, and determine ϕ_{total}. Compare Figs. Ex. 11.8 and 11.9 with Fig. 11.11a.

d. From Eq. 10-10, $\alpha_f = 45° + \phi'/2 = 62.5°$.

EXAMPLE 11.10

Given:

The same sand as for Example 11.9 except, that the cell pressure is 300 kPa.

Required:

Δu_f.

Solution:

There are several approaches to this problem. Graphically, we could construct a total stress circle tangent to the total failure envelope shown in

Fig. Ex. 11.9 but starting at $\sigma'_{3c} = \sigma_{3f} = 300$ kPa. Then shift your compass or circle maker to the left until the circle is just tangent to the effective Mohr failure envelope.

$$\Delta u_f = \sigma_{3f} - \sigma'_{3f} = 300 \text{ kPa} - 160 = 140 \text{ kPa}$$

Analytically, use Eq. 11-1 and $(\sigma_1/\sigma_3)_{\text{total}}$ from Ex. 11.9.

$$\sigma_{1f} = \sigma_{3f}\left(\frac{\sigma_1}{\sigma_3}\right)_{\text{total}} = 300(2.44) = 732 \text{ kPa}$$

$$\sigma_{1f} - \sigma_{3f} = 732 - 300 = 432 \text{ kPa}$$

From Eq. 11-3 and $(\sigma'_1/\sigma'_3)_f = 3.7$ (Example 11.8),

$$\sigma'_{3f} = \frac{(\sigma_1 - \sigma_3)_f}{(\sigma'_1/\sigma'_3)_f - 1} = \frac{423}{3.7 - 1} = 160 \text{ kPa}$$

$$\Delta u_f = \sigma_{3f} - \sigma'_{3f} = 300 - 160 = 140 \text{ kPa}$$

Check: $\Delta u_f = \sigma_{1f} - \sigma'_{1f} = 732 - 3.7(160) = 140$ kPa

11.6 FACTORS THAT AFFECT THE SHEAR STRENGTH OF SANDS

Since sand is a "frictional" material we would expect those factors that increase the frictional resistance of sand to lead to increases in the angle of internal friction. First, let us summarize the factors that influence ϕ.

1. Void ratio or relative density
2. Particle shape
3. Grain size distribution
4. Particle surface roughness
5. Water
6. Intermediate principal stress
7. Particle size
8. Overconsolidation or prestress

Void ratio, related to the density of the sand, is perhaps the most important single parameter that affects the strength of sands. Generally speaking, for drained tests either in the direct shear or triaxial test apparatus, the lower the void ratio (higher density or higher relative density), the higher the shear strength. The Mohr circles for the triaxial test data presented earlier are shown in Fig. 11.12 for various confining

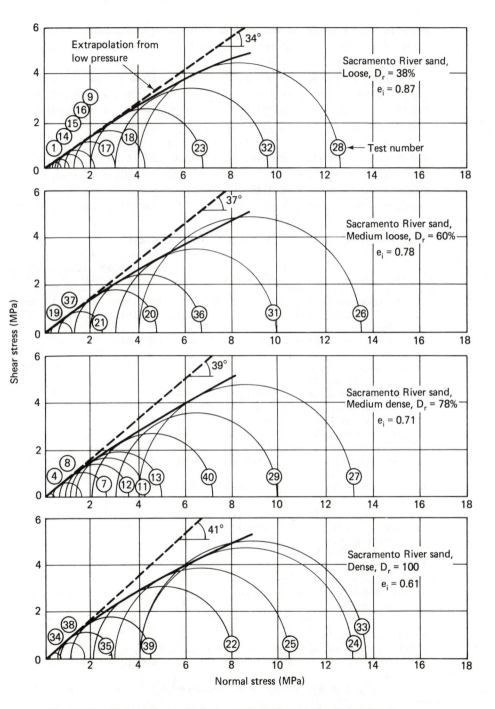

Fig. 11.12   Mohr circles and failure envelopes from drained triaxial tests, illustrating the effects of void ratio or relative density on shear strength (after Lee, 1965; also after Lee and Seed, 1967).

pressures and four initial void ratios. You can see that as the void ratio decreases, or the density increases, the angle of internal friction or angle of shearing resistance ϕ increases.

Another thing you should notice is that the Mohr failure envelopes in Fig. 11.12 are curved; that is, ϕ' is not a constant if the range in confining pressures is large. We usually speak of ϕ' as if it were a constant, but we understand that the Mohr failure envelope really is curved.

The effects of relative density or void ratio, grain shape, grain size distribution, and particle size on ϕ are summarized by Casagrande in Table 11-2. Values were determined by triaxial tests on saturated samples at moderate confining pressures. Generally speaking, with all else constant, ϕ increases with increasing angularity (Fig. 2.5). If two sands have the same

TABLE 11-2   Angle of Internal Friction of Cohesionless Soils*

No.	General Description	Grain Shape	D_{10} (mm)	C_u	Loose e	Loose ϕ(deg)	Dense e	Dense ϕ(deg)
1	Ottawa standard sand	Well rounded	0.56	1.2	0.70	28	0.53	35
2	Sand from St. Peter sandstone	Rounded	0.16	1.7	0.69	31	0.47	37†
3	Beach sand from Plymouth, MA	Rounded	0.18	1.5	0.89	29	—	—
4	Silty sand from Franklin Falls Dam site, NH	Subrounded	0.03	2.1	0.85	33	0.65	37
5	Silty sand from vicinity of John Martin Dam, CO	Subangular to subrounded	0.04	4.1	0.65	36	0.45	40
6	Slightly silty sand from the shoulders of Ft. Peck Dam, MT	Subangular to subrounded	0.13	1.8	0.84	34	0.54	42
7	Screened glacial sand, Manchester, NH	Subangular	0.22	1.4	0.85	33	0.60	43
8‡	Sand from beach of hydraulic fill dam, Quabbin Project, MA	Subangular	0.07	2.7	0.81	35	0.54	46
9	Artificial, well-graded mixture of gravel with sands No. 7 and No. 3	Subrounded to subangular	0.16	68	0.41	42	0.12	57
10	Sand for Great Salt Lake fill (dust gritty)	Angular	0.07	4.5	0.82	38	0.53	47
11	Well-graded, compacted crushed rock	Angular	—	—	—	—	0.18	60

*By A. Casagrande.

†The angle of internal friction of the undisturbed St. Peter sandstone is larger than 60° and its cohesion so small that slight finger pressure or rubbing, or even stiff blowing at a specimen by mouth, will destroy it.

‡Angle of internal friction measured by direct shear test for No. 8, by triaxial tests for all others.

relative density, the soil that is better graded (for example, an SW soil as opposed to an SP soil) has a larger ϕ. (As a reminder, two sands at the same void ratio may not necessarily have the same relative density.) Particle size, at constant void ratio, does *not* seem to influence ϕ significantly. Thus a fine sand and a coarse sand at the same void ratio will probably have about the same ϕ.

Another parameter, not included in Table 11-2, is surface roughness, which is very difficult to measure. It will, however, have an effect on ϕ. Generally, the greater the surface roughness, the greater will be ϕ. It has also been found that wet soils show a 1° to 2° *lower* ϕ than if the sands were dry.

So far we have only discussed results from direct shear or triaxial tests in which $\sigma_2 = \sigma_3$ or σ_1. To investigate the influence of the intermediate principal stress, other types of tests like plane strain or cuboidal shear tests must be used (Fig. 10.14). Research summarized by Ladd, et al. (1977) indicates that ϕ in plane strain is larger than ϕ in triaxial shear by 4° to 9° in dense sands and 2° to 4° for loose sands. A conservative estimate of the plane strain angle of internal friction ϕ_{ps} may be found from triaxial test results ϕ_{tx}, using the following equations (after Lade and Lee, 1976):

$$\phi_{ps} = 1.5\phi_{tx} - 17° \quad (\phi_{tx} > 34°) \tag{11-5a}$$

$$\phi_{ps} = \phi_{tx} \quad\quad\quad (\phi_{tx} \leq 34°) \tag{11-5b}$$

The final factor on our list, overconsolidation or prestress of sands, has been found to not significantly affect ϕ, but it strongly affects the compression modulus of granular materials (Lambrects and Leonards, 1978). Ladd, et al. (1977) discuss the various effects of prestress on the behavior of granular materials.

All the factors mentioned above are summarized in Table 11-3. Some correlations between ϕ' and dry density, relative density, and soil classification are shown in Fig. 11.13. This figure and Table 11-2 are very useful for estimating the frictional characteristics of granular materials. If you

TABLE 11-3 Summary of Factors Affecting ϕ

Factor	Effect
Void ratio e	$e\uparrow, \phi\downarrow$
Angularity A	$A\uparrow, \phi\uparrow$
Grain size distribution	$C_u\uparrow, \phi\uparrow$
Surface roughness R	$R\uparrow, \phi\uparrow$
Water W	$W\uparrow, \phi\downarrow$ slightly
Particle size S	No effect (with constant e)
Intermediate principal stress	$\phi_{ps} \geq \phi_{tx}$ (see Eqs. 11-5a, b)
Overconsolidation or prestress	Little effect

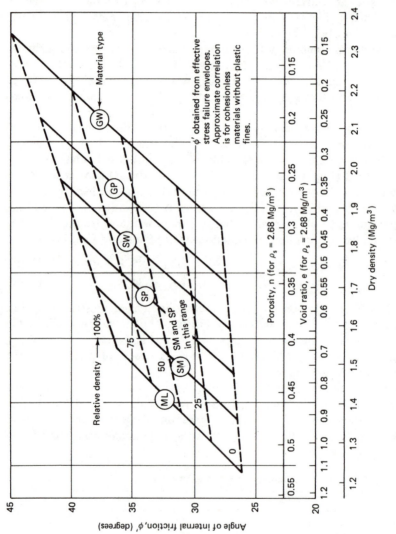

Fig. 11.13 Correlations between the effective friction angle in triaxial compression and the dry density, relative density, and soil classification (after U.S. Navy, 1971).

have a complete visual classification of the materials at your site, together with some idea of the in situ relative density, you already have a pretty good idea about the shear strength behavior of the soils in advance of a laboratory testing program. For small projects, such estimates may be all you need for design.

11.7 THE COEFFICIENT OF EARTH PRESSURE AT REST FOR SANDS

In Sec. 7.6 we defined the coefficient of earth pressure at rest as

$$K_o = \frac{\sigma'_{ho}}{\sigma'_{vo}} \tag{7-19}$$

where σ'_{ho} = the horizontal effective stress in situ, and
σ'_{vo} = the vertical effective stress in situ.

We mentioned that a knowledge of K_o is very important for the design of earth-retaining structures and many foundations; it also influences liquefaction potential, as we shall soon see. Thus, if your assessment of the initial in situ stresses in the soil is inaccurate, you can be way off in your prediction of the performance of such structures.

You already know from Chapter 7 (Sec. 7.5) how to estimate σ'_{vo} from the densities of the overlying materials, the thicknesses of the various layers, and the location of the ground water table. Accurate measurements of σ'_{ho} are not easy, especially in sands. It is virtually impossible to install an earth pressure cell in situ, for example, without causing some disturbance and densification of the sands around the cell, and this changes the stress field at the very point of measurement. Consequently, the approach usually taken is to estimate K_o from theory or laboratory tests, and then calculate σ_{ho} and σ'_{ho} from Eq. 7-19.

The best known equation for estimating K_o was derived by Jáky (1944, 1948), which is a theoretical relationship between K_o and the angle of internal friction ϕ', or

$$K_o = 1 - \sin \phi' \tag{11-6}$$

This relationship, as shown in Fig. 11.14, seems to be an adequate predictor of K_o for normally consolidated sands. Since most of the points lie between 0.35 and 0.5 for these sands, K_o of 0.4 to 0.45 would be a reasonable average value to use for preliminary design purposes.

If the sand has been preloaded, then K_o is somewhat greater. Schmidt (1966, 1967) and Alpan (1967) suggested that the increase in K_o could be

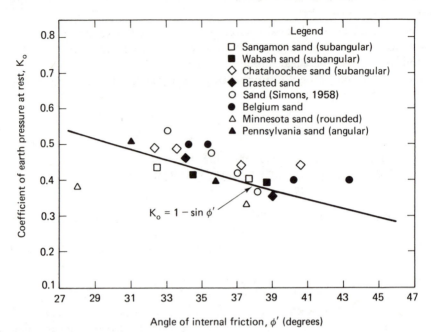

Fig. 11.14   Relationship between K_o and ϕ' for normally consolidated sands (after Al-Hussaini and Townsend, 1975).

related to the overconsolidation ratio (OCR) by

$$K_{o\text{-}oc} = K_{o\text{-}nc}(OCR)^h \qquad (11\text{-}7)$$

where $K_{o\text{-}oc} = K_o$ for the overconsolidated soil,
  $K_{o\text{-}nc} = K_o$ for the normally consolidated soil, and
  h = an empirical exponent.
Values of h range between 0.4 and 0.5 (Alpan, 1967; Schmertmann, 1975) and even as high as 0.6 for very dense sands (Al-Hussaini and Townsend, 1975). Ladd, et al. (1977) pointed out that this exponent itself varies with OCR, and it seems to depend on the direction of the applied stresses. For example, Al-Hussaini and Townsend (1975) found a significantly lower K_o during reloading than during unloading in laboratory tests on a uniform medium sand. Thus K_o appears to be very sensitive to the precise stress history of the deposit.

We shall have more to say about this subject when we discuss K_o for clays.

11.8 LIQUEFACTION AND CYCLIC MOBILITY BEHAVIOR OF SATURATED SANDS

You may recall that we mentioned the phenomenon of liquefaction during our description of the quicksand tank (Sec. 7.8). We described the behavior of the very loose sands in the tank during the upward flow of water (Fig. 7.12a) or when a shock load was applied to the side of the tank (Fig. 7.12c). We also gave a physical explanation for this phenomenon. We said that when loose saturated sands are subjected to strains or shocks, there is a tendency for the sand to decrease in volume. This tendency causes a positive increase in pore pressure which results in a decrease in effective stress within the soil mass. Once the pore pressure becomes equal to the effective stress, the sand loses all its strength, and it is said to be in a state of *liquefaction*.

Examples of liquefaction briefly mentioned in Sec. 7.8 included the failure of Ft. Peck Dam, Montana, and the flow slides that have occurred along the banks of the lower Mississippi River. Here, liquefaction takes place under conditions of large statically induced strains (Casagrande, 1936a, 1950). We will call this statically (monotonic loading) induced condition *liquefaction*. River banks composed of loose uniform fine sands can liquefy when subjected to large strains, such as might be caused by steepening of the banks due to erosion, and the strains produce increased pore pressures. Such a situation is shown in Fig. 11.15. Initially a soil element at A, some distance from the slope, is under a much safer state of initial stress (K_o conditions—discussed in the previous section) than the element at B. As erosion starts at the base of the slope, the soil stresses are increased, the pore water pressure rises, and a limited zone (shown in Fig. 11.15a) can liquefy. As this material *flows* out into the river, additional stresses are applied to the adjacent soils, and they also can liquefy (Fig. 11.15b). In this way, liquefaction *progresses* inland until the material comes to equilibrium on a very flat slope angle (Fig. 11.15c). Some important characteristics of different types of flow slides are listed in Table 11-4. Other types of soils which may be afflicted by flow slides should be added to column 2: hydraulically placed fills of sands and silty sands such as mine waste or tailings dams. As mentioned in Sec. 7.8, these structures are built with very little engineering design or construction supervision, and liquefaction-type failures are relatively common. Notice also in Table 11-4 the character of the strains necessary to start flow (column 3). These strains can be caused by a static increase in stress, like the case of the riverbank erosion leading to progressive liquefaction, or they can be caused by dynamic or vibratory loads. Examples of this second type of

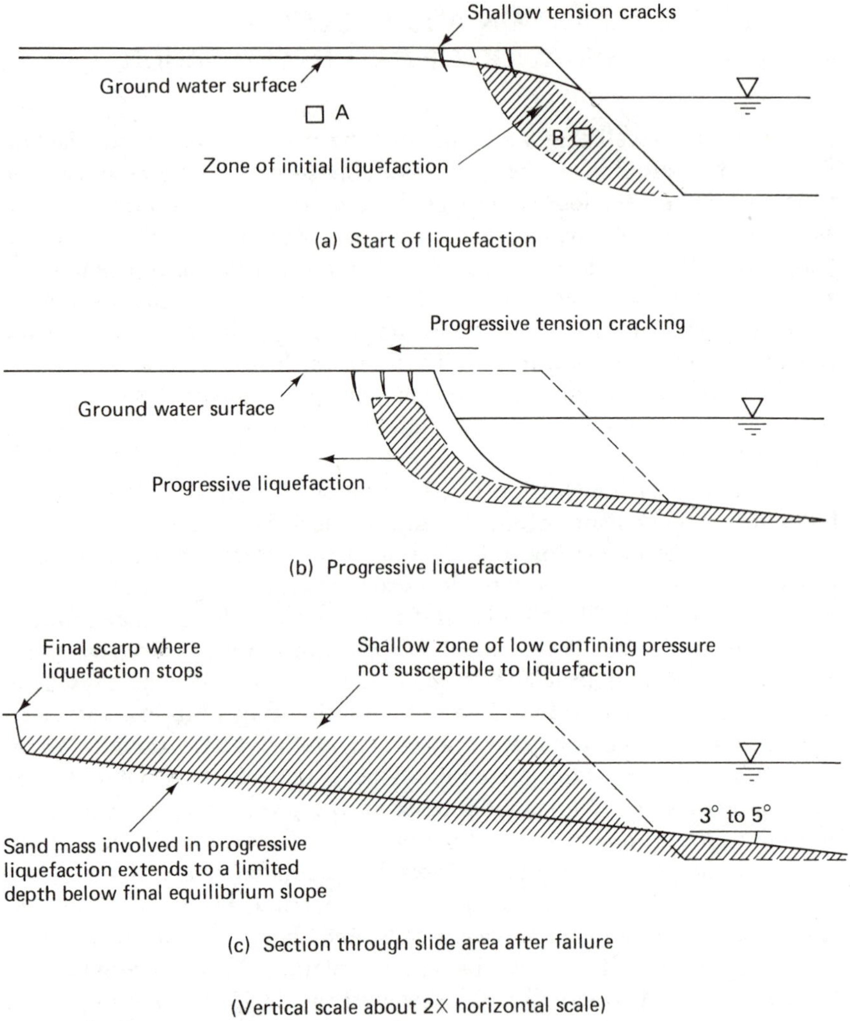

Shallow tension cracks

Ground water surface

□ A

B □

Zone of initial liquefaction

(a) Start of liquefaction

Progressive tension cracking

Ground water surface

Progressive liquefaction

(b) Progressive liquefaction

Final scarp where
liquefaction stops

Shallow zone of low confining pressure
not susceptible to liquefaction

3° to 5°

Sand mass involved in progressive
liquefaction extends to a limited
depth below final equilibrium slope

(c) Section through slide area after failure

(Vertical scale about 2X horizontal scale)

Fig. 11.15   Liquefaction in loose sand adjacent to a waterfront (after
Casagrande, 1975).

loading are pile driving, blasting, traffic, rotating machinery, storm waves,
and earthquakes. Because major earthquakes affect large areas, they can
and have caused large deposits of loose saturated sand to liquefy (for
example, Seed and Idriss, 1967; Seed and Wilson, 1967).

A different kind of liquefaction than that which occurs due to static
stresses is called *cyclic mobility*. Here cyclic loads, like those from an
earthquake, cause a buildup of pore pressures in medium to high density

TABLE 11-4 Flow Slides in Soils*

Sensitivity of Soil to Liquefaction (1)	Soils Which May Be Affected by Indicated Types of Flow Slides (2)	Character of Strain Necessary to Start Flow (3)	Character and Speed of Flow Failure (4)	Examples of Flow Failures (5)
High sensitivity, "type A"	Sand in bulked condition; rock flour	Small strains such as earthquake shocks, explosions, or vibrations simultaneously affecting a large mass	Rapid flow (a few minutes)	Flow failure of railroad embankment in Holland (1918); silt flows in Laurentian Mts.
Low sensitivity, "type B"	River sands; rock flour	Large strains† created simultaneously in a large volume; e.g., shear failure in clay or shale transmitted into overlying sand	Rapid flow (a few minutes)	Ft. Peck Dam (river sands in foundation and hydraulic fill sand in dam)
Low sensitivity, "type C"	River sands; rock flour; varved silts and clays; clays having very great sensitivity to remolding	Large strains created progressively	Progressive liquefaction; up to several hours' duration, depending on mass involved	Mississippi River bank slides; flow slides in Holland; flow slides in "bull's liver" and in varved clays in excavations

*After A. Casagrande (1950).

†Large strains may be invited by intrusive pore pressures created, for example, in a varved clay. However, these intrusive pressures are only indirectly responsible for subsequent liquefaction in silt layers, or layers of super-sensitive clays.

saturated sands and induce measurable strains in samples that ordinarily exhibit a dilative response under static loads; this phenomenon is opposite to the behavior predicted by the Peacock diagram of Fig. 11.10. Thus *cyclic* stresses, if they are large enough and for a sufficient duration (that is, number of cycles), can cause medium dense to very dense saturated sands to liquefy under the right conditions of density and confining pressure. Loose sands, of course, fail at the least number of cycles.

Let's begin our discussion of the behavior of sands under cyclic loading by first studying some test results which showed liquefaction under static stresses. These tests, from Castro (1969), presented in Fig. 11.16, show the results of three CU tests and one CD test, all hydrostatically consolidated to 400 kPa. The relative densities D_r of each specimen after consolidation are also indicated on the figure next to the stress-strain curve for each specimen. The specimens were loaded axially (monotonically) by small dead-load increments of weight applied about every minute to the soil sample.

In test A, the one with the lowest D_r, the peak stress difference of 200 kPa was reached in 15 min, which corresponded to an axial strain of about 1%. Then, when the next small increment of load was applied, the specimen suddenly collapsed—liquefied—and in about 0.2 s the stress decreased from 200 to 30 kPa at 5% strain, where it remained as the specimen continued to flow. Notice how the pore pressure for specimen A remained the same during flow. At this maximum value of pore pressure, the effective minor principal stress was only about 15 kPa, and if you calculate the ϕ' from these stresses (use Eq. 10-13 or 11-1), you get $\phi' = 30°$.

The total and effective Mohr circles at the peak and during flow after liquefaction are shown in Fig. 11.17. Also shown for comparison are the results of the CD test on the same sand at the same D_r. Both tests indicate that $\phi' = 30°$ for this loose sand, although as pointed out by Casagrande (1975), the agreement may be only a coincidence. In any event, the effective stress circle at the peak on maximum stress difference lies below the effective failure envelope.

Figure 11.17 is another good illustration of the very large differences in the strength of sands, depending on the drainage conditions we discussed in the previous sections of this chapter. Here you see the results of CD versus CU tests on the same sand at the same relative density and at the same effective consolidation stress. The differences are even greater when you consider the strength of the sand after liquefaction. In a flow slide, this sand would simply flow out like a very dense liquid, and its equilibrium slope angle might be only a very few degrees.

Now let's look at the results of tests on specimens B and C in Fig. 11.15. Specimen B (Fig. 11.16) at $D_r = 44\%$ also liquefied after a peak stress difference of 250 kPa was reached at about 2% strain—then the

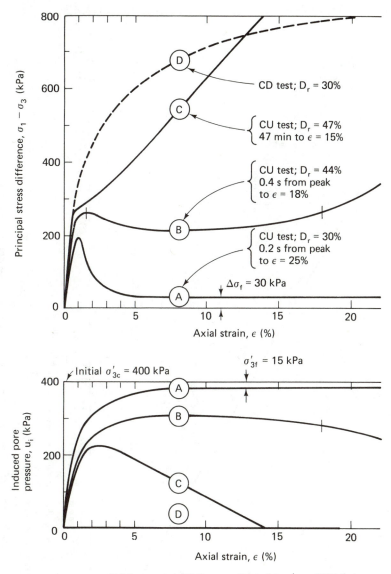

Fig. 11.16   Comparison of three hydrostatically consolidated CU tests and one CD test on banding sand loaded incrementally to failure (after Casagrande, 1975, from Castro, 1969).

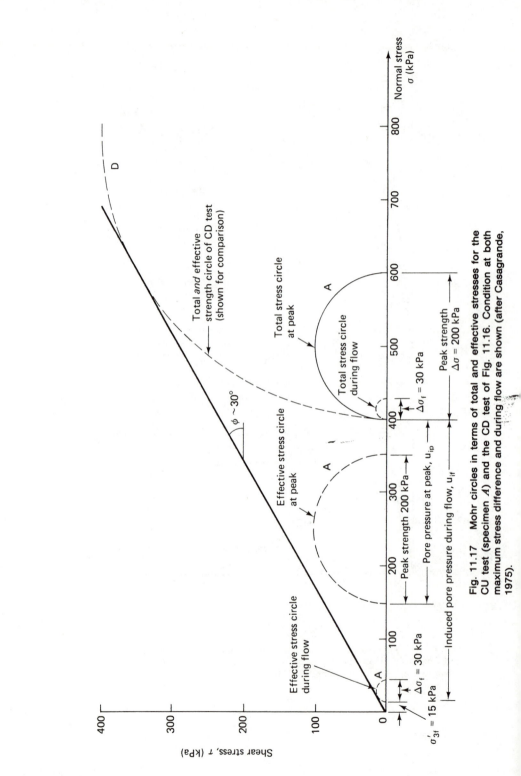

Fig. 11.17 Mohr circles in terms of total and effective stresses for the CU test (specimen *A*) and the CD test of Fig. 11.16. Condition at both maximum stress difference and during flow are shown (after Casagrande, 1975).

specimen flowed rapidly to a strain of 18% and stopped flowing. To get it moving again, additional small weights had to be added to the piston. Notice that the pore pressure induced in this sample during liquefaction was more than 300 kPa, and if you calculate ϕ', you get a value of about 32°. Since it was denser than specimen A, a higher friction angle is reasonable. Specimen C was slightly denser than B, and it never experienced liquefaction. The specimen obviously tried to dilate as the axial stress was increased since the pore pressure *decreased* after the maximum. These tests show how liquefaction occurs under static or monotonic loading. Their liquefaction behavior can be explained by the critical void ratio concept, and the Peacock diagram can be used to predict their behavior.

The occurrences of the 1964 Niigata, Japan, and the Anchorage, Alaska, earthquakes stimulated much research into the problems of ground subsidence and failure due to earthquake (dynamic) loading of saturated sands. Prof. H. B. Seed and his students at the University of California at Berkeley began to study this problem, using both hydrostatically and non-hydrostatically consolidated undrained cyclic triaxial tests to simulate earthquake loadings. Insight may be gained from a series of tests where the variables thought to govern the cyclic behavior of sands are tested in a systematic manner. From the Peacock diagram, it is evident that initial relative density and effective confining pressure are two key parameters. In addition, the magnitude of the cyclic stress and number of cycles to cause failure were studied. Several definitions of failure were used, such as various percent cyclic strains and when the pore pressure ratio ($\Delta u / \sigma'_{3c}$) equaled one.

Typical results from a hydrostatically consolidated cyclic undrained triaxial test on loose sand are shown in Fig. 11.18 (Seed and Lee, 1966). Very little strain developed during the first nine cycles of cyclic stress application, even though the pore pressures gradually increased. Then between the ninth and tenth cycles the pore pressure suddenly increased to a value equal to the confining pressure, and the specimen developed very large strains in the next couple of cycles. It was observed that the specimen was in a fluid condition over a wide range of strains. The suddenness of the collapse or *liquefaction* was also of interest. In several tests, the specimens showed very little strain, after even a relatively large number of cycles—then they would suddenly liquefy after only one or two more cycles were applied. All in all, it was found that loose sands behaved about as expected.

When dense sands were tested, however, the resulting behavior was quite surprising. Typical results of cyclic triaxial tests on the same sand as used in Fig. 11.18 and at the same effective consolidation pressure are shown in Fig. 11.19. Only the relative density is 78% now instead of the

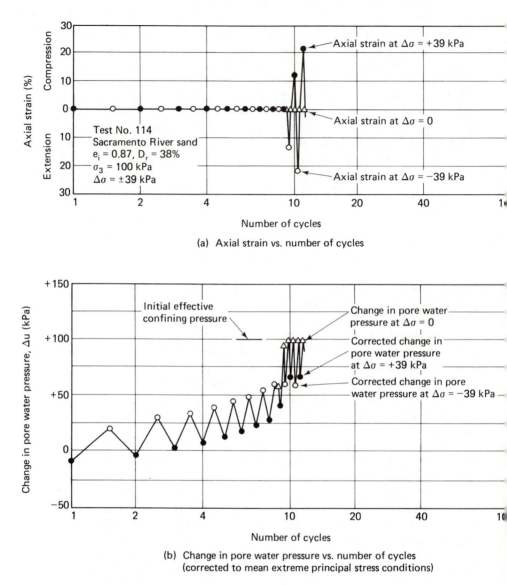

(a) Axial strain vs. number of cycles

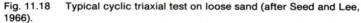

(b) Change in pore water pressure vs. number of cycles
(corrected to mean extreme principal stress conditions)

Fig. 11.18 Typical cyclic triaxial test on loose sand (after Seed and Lee, 1966).

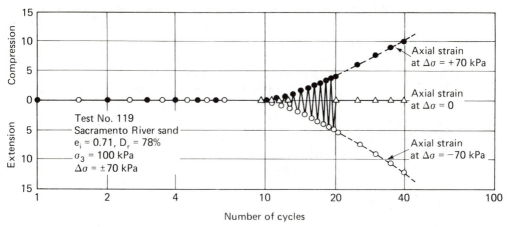

(a) Axial strain vs. number of cycles

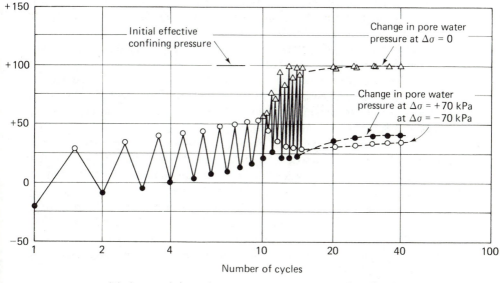

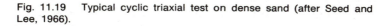

(b) Corrected change in pore water pressure vs. number of cycles

Fig. 11.19   Typical cyclic triaxial test on dense sand (after Seed and Lee, 1966).

previous 38%. During the first 10 or so cycles, very little axial strain occurred, but the induced pore water pressures gradually increased. At cycle 13, the pore pressure momentarily became equal to the total cell pressure when the principal stress difference was zero during the cycle; that is, the effective confining pressure was momentarily zero. Even though the effective stress was zero during part of the cycle, the specimen was still able to withstand additional cyclic stress. As can be seen in Fig. 11.19, the strain amplitude was less than 10% even after 20 cycles, and the sample did not collapse as was the case for the loose sand. Based on other tests, Seed and Lee (1966) also found that the lower the confining pressure, the more easily liquefaction, or *cyclic mobility* as it is now called, would develop. In other words, increasing the effective confining pressure would decrease the potential for cyclic mobility.

The variables that affect the cyclic mobility of saturated sands are shown in Fig. 11.20a, where peak cyclic stress versus log number of cycles is shown. It can be seen that as the peak stress is lowered, more cycles are required to fail the sample. If the relative density and/or the effective confining pressure is increased, it takes a higher cyclic stress to fail the sample for a given number of cycles to failure. Said in another way, it will take a larger number of cycles to cause failure for the same cyclic stress. The definition of cyclic stress is illustrated in Fig. 11.20b for both cyclic triaxial tests and cyclic simple shear tests. Cyclic simple shear tests seem to more closely represent actual field stress conditions. The differences in stress conditions between these two kinds of tests have been the subject of much research (for example, Seed and Peacock, 1971; Finn, et al., 1971; Park and Silver, 1975; and Castro, 1975).

Other factors have been found to influence the results of cyclic testing of saturated sands. The most significant factor is perhaps the method of sample preparation and the resulting soil structure (Mulilis, et al., 1975; Ladd, 1977). Other factors include previous cyclic strain history (from prior earthquakes, for example), the coefficient of earth pressure at rest, K_o, and the overconsolidation ratio of the soil deposit (Seed, 1979). As the value of K_o or OCR increases, a larger number of cycles to cause failure is required for a given cyclic stress. Prior cycling or prestraining causes the same result.

It is difficult to perceive that an initially dense sand could liquefy during cyclic loading. From our previous discussion in this chapter, dense sands *tend* to increase in volume (dilate), which means the pore pressure should decrease and the effective stresses should increase. Is it really possible that the opposite reaction can occur? Further, as was shown in the Peacock diagram (Fig. 11.10), increasing the effective confining stress on an initially dense sand tends to cause "loose"-type behavior; that is, it would increase (rather than decrease) the potential for liquefaction.

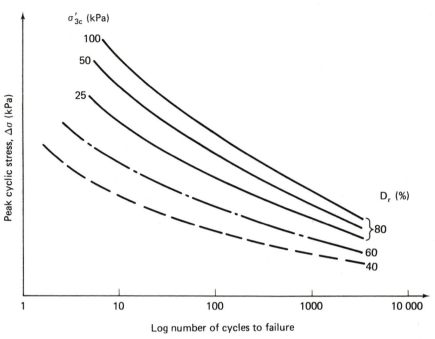

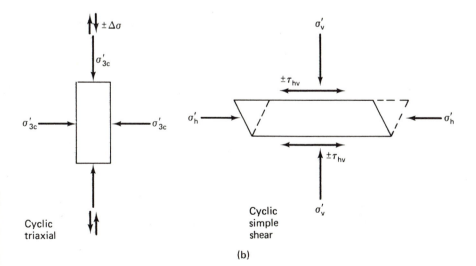

Fig. 11.20 (a) Generalized relationship between peak cyclic stress and number of cycles to cause cyclic mobility failure; effects of initial density and confining pressure are indicated; (b) definition of cyclic stress $\Delta\sigma$ in the triaxial test or τ_{hv} in the cyclic simple shear test.

Additional work by Castro (1969 and 1975) seemed to answer these anomalies. By very careful measurements of the *failed* cyclic triaxial specimens, he found a radical water content and void ratio redistribution in the samples at failure. They were alternately necking down and bulging out at the top, and he found the relative densities varied significantly throughout the specimen. The reasons for this behavior are complex and are discussed at some length by Casagrande (1975) and Castro (1969 and 1975). In any event, Castro's work seemed to explain that we were seeing two basically different phenomena: (1) classical liquefaction of loose sands, which we described earlier and which we all understand, and (2) the phenomenon called *cyclic mobility* which occurs in the laboratory during *cyclic* triaxial or simple shear tests.

These two phenomena are illustrated in Fig. 11.21, which is similar to Fig. 11.7. This figure is like looking upward from beneath the Peacock diagram on the *W0P* plane. The "steady-state line" represents the critical void ratio and effective stress relationship after liquefaction. Soils with void ratios and effective stresses lying above and to the right of the steady-state line are contractive or loose and thus are subject to liquefaction. For example, a sample starting at point *C* when stressed or vibrated

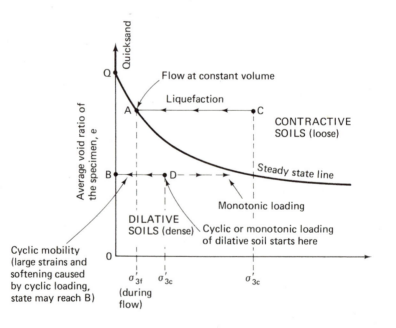

Fig. 11.21   State diagram showing liquefaction potential based on un-drained tests of saturated sands (after Castro and Poulos, 1977).

develops a large amount of positive excess pore pressure and ends up at point A on the steady-state line, where the sample has no further tendency to change volume. On the other hand, a dense dilative specimen originally at point D below the steady-state line, if subjected to cyclic shear, will move towards point B, a condition of zero effective stress. This is the condition of cyclic mobility defined above. If the same sample were loaded monotonically or statically in an ordinary triaxial test, then it would go in the opposite direction towards the steady-state line. This sample would behave just as we described in Sec. 11.5 for dense saturated sands under undrained loading. Note that there is nothing contradictory about this behavior. A state diagram showing the steady-state lines for several typical sands is shown in Fig. 11.22. The differences between liquefaction and cyclic mobility have been summarized by Castro and Poulos (1977) in Table 11-5.

However, the case is not closed on this subject. We can still use the results of the cyclic triaxial test and all our experience with it. Mulilis, et al. (1975) conducted cyclic triaxial tests where relative densities varied from 50% to 90% and the cyclic stress ratios (τ/σ'_{vo}) from 0.2 to 0.5. (These values of density and stress ratios adequately cover most conditions to be expected in the field.) They found that "... there was no apparent effect on nonuniform strains or water content redistribution in the samples prior to the development" of failure (Seed, 1979). After failure, the same nonuniform conditions (necking, bulging) as Castro observed were noticed. These results suggest that the cyclic triaxial test (carefully performed) can in fact be used to evaluate the behavior of field conditions by making appropriate corrections for K_o or by using the cyclic simple shear test with corrections (Seed, 1979).

What can you do to avoid a liquefaction failure? For the static case of natural slopes, monitoring of the field pore water pressures with piezometers may give some indication of impending instability. Observations of erosion and small slides along rivers may also help. If the problem involves earthquakes, it is impossible to control the number of cycles or the applied cyclic stresses. However it may be possible to increase the in situ density by removal and replacement of the loose soil or by compaction of the loose soils by techniques described in Chapter 5. Likewise, the addition of a surcharge fill or berm over a saturated sand layer will increase the effective stresses which should reduce the liquefaction potential (or at least the cyclic mobility!). Finally, it may be possible to permanently lower the ground water table by means of drains and/or pumping, which would reduce the possibility of liquefaction.

Clearly, the problem of a liquefaction failure in a deposit of loose saturated sands is real and should not be overlooked, especially for important structures such as dams and power plants. We don't have all the

TABLE 11-5 Differences Between Liquefaction and Cyclic Mobility*

	Liquefaction	Cyclic Mobility
General	Most likely in uniform, fine, clean, loose sand. Static load can cause liquefaction. Cyclic loads causing shear stresses larger than the steady-state strength also can cause liquefaction.	Any soil in any state can develop cyclic mobility in the laboratory if the cyclic stresses are large enough.
Effect of σ'_{3c} at constant void ratio for $\sigma'_{1c}/\sigma'_{3c}$	Increased σ'_{3c} means larger deformations if liquefaction is induced. The magnitude and/or number of cyclic loads needed to cause liquefaction increases with σ'_{3c}. Cyclic loads smaller than the steady-state strength cannot cause liquefaction but may cause cyclic mobility.	Increased σ'_{3c} means increased cyclic load to cause cyclic mobility. But the cyclic mobility ratio[†] usually decreases with increasing σ'_{3c}.
Effect of $\sigma'_{1c}/\sigma'_{3c}$ at constant void ratio and σ'_{3c}	Smaller additional loads are needed to cause liquefaction as $\sigma'_{1c}/\sigma'_{3c}$ increases. When $\sigma'_{1c}/\sigma'_{3c}$ is large, a soil is more unstable and may, in the extreme, be susceptible to "spontaneous liquefaction."	In soils that have low permeability, increased $\sigma'_{1c}/\sigma'_{3c}$ seems to result in somewhat smaller cyclic mobility stresses, which is a reasonable trend. In clean sands, cyclic mobility stress increases with $\sigma'_{1c}/\sigma'_{3c}$. This unusual result for clean sands is postulated to be due to the substantial test error due to redistribution of void ratio in the laboratory specimens.

*After Castro and Poulos (1977).

[†] $\dfrac{(\sigma_1 - \sigma_3)/2}{\sigma'_{3c}}$, where $(\sigma_1 - \sigma_3)$ is the dynamic principal stress difference, or the cyclic mobility stress.

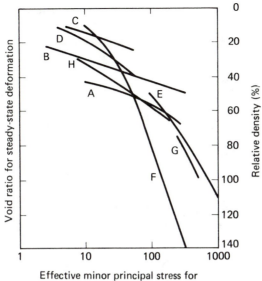

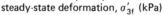

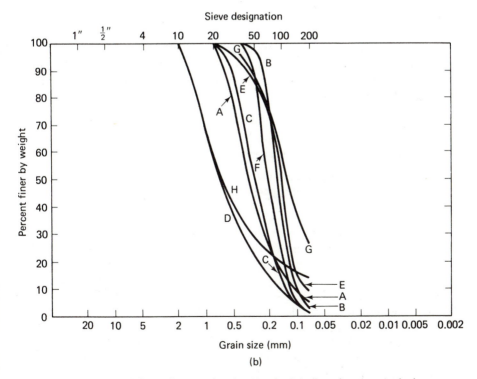

Fig. 11.22   (a) State diagram showing steady-state lines for some typical sands; (b) grain size distributions. All specimens except *G*, which was undisturbed, were prepared by tamping the moist soil in layers (after Castro and Poulos, 1977).

Sand designation	Grain size distribution,			Grain shape	Maximum density (Mg/m^3)	Minimum density (Mg/m^3)
	D_{60} (mm)	C_u	% Finer than 0.074 mm (%)			
A	0.40	3.1	5	Subangular to angular	1.76	1.35
B	0.17	1.8	0	Subrounded	1.76	1.44
C	0.33	2.3	1	Angular	1.73	1.44
D	0.90	5.6	0	Subangular	1.83	1.54
E	0.17	2.1	8	Angular	1.62	1.27
F	0.23	2.0	0	Angular	1.17	0.90
G	0.15	—	26	Angular	—	—
H	0.85	~17	13	Subangular	1.99	1.57

(c)

Fig. 11.22 Continued (c) index properties for sands in (a) and (b).

answers, but fortunately research continues and practical design procedures are maturing (Seed, 1979). In most cases, however, the present designs await the ultimate test of their adequacy under field loading (earthquake) conditions.

11.9 STRESS-DEFORMATION AND STRENGTH CHARACTERISTICS OF SATURATED COHESIVE SOILS

What happens when shear stresses are applied to saturated cohesive soils? Most of the remainder of this chapter concerns this question. But first, let's briefly review what happens when saturated sands are sheared. From our previous discussion, for example, you know that volume changes can take place in a drained test, and that the direction of the volume changes, whether dilation or compression, depends on the relative density as well as the confining pressure. If shear takes place undrained, then the volume change tendencies produce pore pressures in the sand.

Basically, the same things happen when clay soils are sheared. In drained shear, whether the volume changes are dilation or compression depends not only on the density and the confining pressure but also on the stress history of the soil. Similarly, in undrained shear the pore pressures developed depend greatly on whether the soil is normally consolidated or overconsolidated.

Typically, engineering loads are applied much faster than the water can escape from the pores of a clay soil, and consequently excess hydrostatic or pore pressures are produced. If the loading is such that failure does not occur, then the pore pressures dissipate and volume changes develop by the process we call *consolidation* (Chapters 8 and 9). The primary difference in behavior between sands and clays, as mentioned when we discussed the compressibility of soils (Chapter 8), is in the *time* it takes for these volume changes to occur. The time aspect strictly depends on, or is a function of, the difference in permeability between sands and clays. Since cohesive soils have a much lower permeability than sands and gravels, it takes much longer for the water to flow in or out of a cohesive soil mass.

Now, what happens when the loading is such that a shear failure is imminent? Since (by definition) the pore water cannot carry any shear stress, all the applied shear stress must be resisted by the soil structure. Put another way, the shear strength of the soil depends *only on the effective stresses* and not on the pore water pressures. This does not mean that the pore pressures induced in the soil are unimportant. On the contrary, as the total stresses are changed because of some engineering loading, the pore water pressures also change, and until equilibrium of effective stresses occurs instability is possible. These observations lead to two fundamentally different approaches to the solution of stability problems in geotechnical engineering: (1) the *total stress approach* and (2) the *effective stress approach*. In the total stress approach, we allow no drainage to take place during the shear test, and we make the assumption, admittedly a big one, that the pore water pressure and therefore the effective stresses in the test specimen are identical to those in the field. The method of stability analysis is called the *total stress analysis*, and it utilizes the *total* or the *undrained shear strength* τ_f, of the soil. The undrained strength can be determined by either laboratory or field tests. If field tests such as the vane shear, Dutch cone penetrometer, or pressuremeter test are used, then they must be conducted rapidly enough so that undrained conditions prevail in situ.

The second approach to calculate the stability of foundations, embankments, slopes, etc., uses the shear strength in terms of effective stresses. In this approach, we have to measure or estimate the excess hydrostatic pressure, both in the laboratory and in the field. Then, if we know or can estimate the initial and applied total stresses, we may calculate the effective stresses acting in the soil. Since we believe that shear strength and stress-deformation behavior of soils is really controlled or determined by the effective stresses, this second approach is philosophically more satisfying. But, it does have its practical problems. For example, estimating or measuring the pore pressures, especially in the field, is not easy to do. The method of stability analysis is called the *effective stress analysis*, and it utilizes the *drained shear strength* or the shear strength in terms of effective stresses. The drained shear strength is ordinarily only determined by laboratory tests.

You probably recall, when we described triaxial tests in Sec. 10.5, that there are limiting conditions of drainage in the test which model real field situations. We mentioned that you could have consolidated-drained (CD) conditions, consolidated-undrained (CU) conditions, or unconsolidated-undrained (UU) conditions. It is also convenient to describe the behavior of cohesive soils at these limiting drainage conditions. It is not difficult to translate these test conditions into specific field situations with similar drainage conditions.

We mentioned in Sec. 10.5 that the unconsolidated-drained test (UD) is not a meaningful test. First, it models no real engineering design situation. Second, the test cannot be interpreted because drainage would occur during shear, and you could not separate the effects of the confining pressure and the shear stress.

As we did with sands, we shall discuss the shear behavior of cohesive soils with reference to their behavior during triaxial shear tests. You can think of the sample in the triaxial cell as representing a typical soil element in the field under different drainage conditions and undergoing different stress paths. In this manner, we hope you will gain some insight into how cohesive soils behave in shear, both in the laboratory and in the field. Keep in mind that the following discussion is somewhat simplified, and that real soil behavior is much more complicated. Towards the end of the chapter we shall indicate some of these complexities. Our primary references are Leonards (1962), Hirschfeld (1963), and Ladd (1964 and 1971b), as well as the lectures of Professor H. B. Seed and S. J. Poulos.

11.9.1 Consolidated-Drained (CD) Test Behavior

We have already described this test when we discussed the strength of sands earlier in this chapter. Briefly, the procedure is to consolidate the test specimen under some state of stress appropriate to the field or design situation. The consolidation stresses can either be *hydrostatic* (equal in all directions, sometimes called *isotropic*) or *non-hydrostatic* (different in different directions, sometimes called *anisotropic*). Another way of looking at this second case is that a stress difference or (from the Mohr circles) a shear stress is applied to the soil. When consolidation is over, the "C" part of the CD test is complete.

During the "D" part, the drainage valves remain *open* and the stress difference is applied very slowly so that essentially *no* excess pore water pressure develops during the test. Professor A. Casagrande termed this test the S-test (for "slow" test).

In Fig. 11.23, the total, neutral, and effective stress conditions in an axial compression CD test at the end of consolidation, during application of axial load, and at failure are shown. The subscripts v and h refer to vertical and horizontal, respectively; c means consolidation. For conventional axial compression tests, the initial consolidation stresses are hydrostatic. Thus $\sigma_v = \sigma_h = \sigma'_{3c}$ cell pressure, which is usually held constant during the application of the axial stress $\Delta\sigma$. In the axial compression test, $\Delta\sigma = \sigma_1 - \sigma_3$, and at failure $\Delta\sigma_f = (\sigma_1 - \sigma_3)_f$. The axial stress can be applied either by increasing the load on the piston incrementally (*stress-controlled* loading) or through a motor-jack system which deforms the sample at a constant rate (called a *constant rate of strain* test).

Note that at all the times during the CD test, the pore water pressure is essentially zero. This means that the total stresses in the drained test are

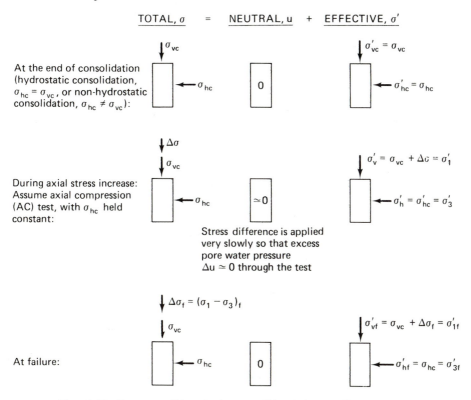

Fig. 11.23 Stress conditions in the consolidated-drained (CD) axial compression triaxial test.

always equal to the effective stresses. Thus $\sigma_{3c} = \sigma'_{3c} = \sigma_{3f} = \sigma'_{3f}$, and $\sigma_{1f} = \sigma'_{1f} = \sigma'_{3c} + \Delta\sigma_f$. If non-hydrostatic consolidation stresses were applied to the specimen, then $\sigma_{1f} = \sigma'_{1f}$ would be equal to $\sigma'_{1c} + \Delta\sigma_f$.

Typical stress-strain curves and volume change versus strain curves for a remolded or compacted clay are shown in Fig. 11.24. Even though the two samples were tested at the same confining pressure, the overconsolidated specimen has a greater strength than the normally consolidated clay. Note also that it has a higher modulus and that failure [the maximum $\Delta\sigma$, which for the triaxial test is equal to $(\sigma_1 - \sigma_3)_f$] occurs at a much lower strain than for the normally consolidated specimen. Note too the analogy to drained behavior of sands. The overconsolidated clay *expands* during shear while the normally consolidated clay *compresses* or consolidates during shear. This is analogous to the behavior described earlier for

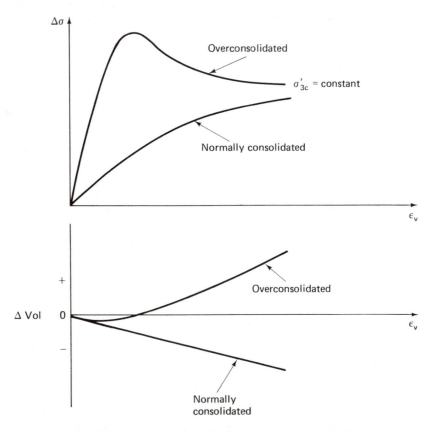

Fig. 11.24 Typical stress-strain and volume change versus strain curves for CD axial compression tests at the same effective confining stress.

sands: normally consolidated clays behave similarly to loose sands, whereas overconsolidated clays behave like dense sands.

In the CD triaxial test, the stress paths are straight lines since we usually keep one of the stresses constant and simply vary the other stress. Typical drained stress paths are shown in Fig. 10.22 for four common engineering situations which can be modeled in the triaxial test. The stress path for the axial compression test illustrated in Fig. 11.23 is the straight line AC.

The Mohr failure envelopes for CD tests of typical clay soils are shown in Figs. 11.25 and 11.26b. The envelope for a remolded clay as well as a normally consolidated undisturbed clay is shown in Fig. 11.25. Even though only one Mohr circle (representing the stress conditions at failure in Fig. 11.23) is shown, the results of three or more CD tests on identical specimens at different consolidation pressures would ordinarily be required to plot the complete Mohr failure envelope. If the consolidation stress range is large or the specimens do not have exactly the same initial water content, density, and stress history, then the three failure circles will not exactly define a straight line, and an average best-fit line by eye is drawn. The slope of the line determines the Mohr-Coulomb strength parameter ϕ', of course, in terms of effective stresses. When the failure envelope is extrapolated to the shear axis, it will show a surprisingly small intercept. Thus it is usually assumed that the c' parameter for normally consolidated non-cemented clays is essentially zero for all practical purposes.

For overconsolidated clays the c' parameter is greater than zero, as indicated by Fig. 11.26b. The overconsolidated portion of the strength envelope (DEC) lies *above* the normally consolidated envelope (ABCF). This portion (DEC) of the Mohr failure envelope is called the *preconsolidation hump*. The explanation for this behavior is shown in the e versus σ'

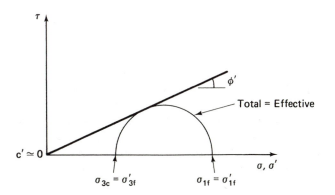

Fig. 11.25 Mohr failure envelope for a normally consolidated clay in drained shear.

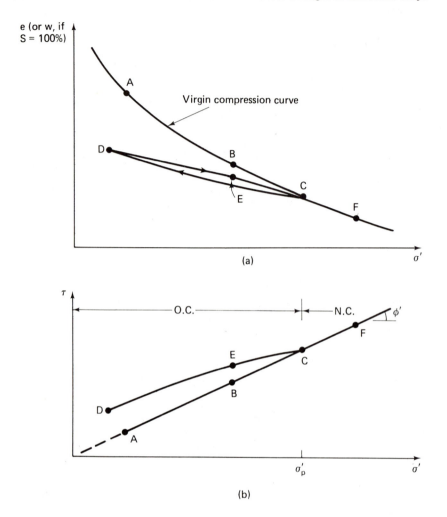

Fig. 11.26   (a) Compression curve; (b) Mohr failure envelope (DEC) for
an overconsolidated clay.

curve of Fig. 11.26a. (Recall from Fig. 8.4 that the virgin compression
curve when plotted arithmetically is concave upward.) Let us assume that
we begin consolidation of a sedimentary clay at a very high water content
and high void ratio. As we continue to increase the vertical stress we reach
point A on the virgin compression curve and conduct a CD triaxial test.
(We could, of course, do the same thing with a CD direct shear test.) The
strength of the sample consolidated to point A on the virgin curve would
correspond to point A on the normally consolidated Mohr failure envelope
in Fig. 11.26b. If we consolidate and test another otherwise identical

specimen which is loaded to point B, then we would obtain the strength, again normally consolidated, at point B on the failure envelope in Fig. 11.26b. If we repeat the process to point C (σ'_p, the preconsolidation stress), then rebound the specimen to point D, then reload it to point E and shear, we would obtain the strength shown at point E in the lower figure. Note that the shear strength of specimen E is greater than specimen B, even though they are tested at exactly the same effective consolidation stresses. The reason for the greater strength of E than B is suggested by the fact that E is at a lower water content, has a lower void ratio, and thus is denser than B, as shown in Fig. 11.26a. If another specimen were loaded to C, rebounded to D, reloaded back past E and C and on to F, it would have the strength as shown in the figure at point F. Note that it is now back on the virgin compression curve and the normally consolidated failure envelope. The effects of the rebounding and reconsolidation have been in effect *erased* by the increased loading to point F. Once the soil has been loaded well past the preconsolidation pressure σ'_p, it no longer "remembers" its stress history.

11.9.2 Typical Values of Drained Strength Parameters

For the Mohr failure envelopes of Figs. 11.25 and 11.26 we did not indicate any numerical values for the effective stress strength parameters ϕ'. Average values of ϕ' for undisturbed clays range from around 20° for normally consolidated highly plastic clays up to 30° or more for silty and sandy clays. The value of ϕ' for compacted clays is typically 25° or 30° and occasionally as high as 35°. As mentioned earlier, the value of c' for normally consolidated non-cemented clays is very small and can be neglected for practical work. If the soil is overconsolidated, then ϕ' would be less, and the c' intercept greater than for the normally consolidated part of the failure envelope (see Fig. 11.26b again). According to Ladd (1971b), for natural overconsolidated non-cemented clays with a preconsolidation stress of less than 500 to 1000 kPa, c' will probably be less than 5 to 10 kPa at low stresses. For compacted clays at low stresses, c' will be much greater due to the prestress caused by compaction. For stability analyses, the Mohr-Coulomb effective stress parameters ϕ' and c' are determined over the range of effective normal stresses likely to be encountered in the field.

It has been observed (for example, Kenney, 1959) that there is not much difference between ϕ' determined on undisturbed or remolded samples at the same water content. Apparently, the development of the maximum value of ϕ' requires so much strain that the soil structure is broken down and almost remolded in the region of the failure plane.

Empirical correlations between ϕ' and the plasticity index for normally consolidated clays are shown in Fig. 11.27. This correlation is based

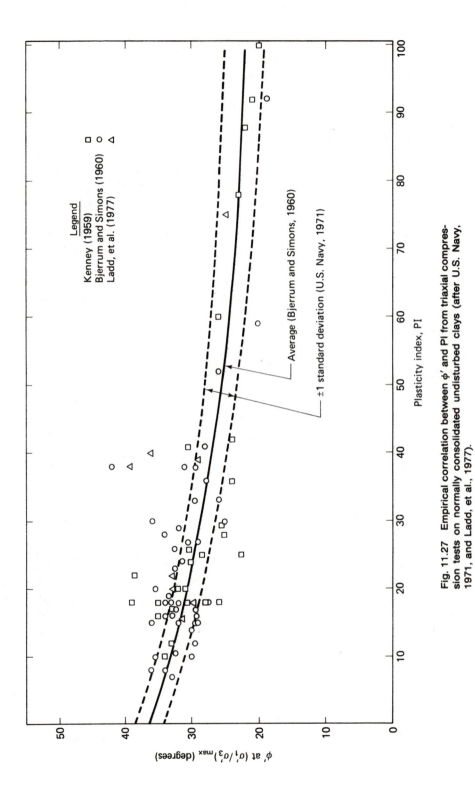

Fig. 11.27 Empirical correlation between φ' and PI from triaxial compression tests on normally consolidated undisturbed clays (after U.S. Navy, 1971, and Ladd, et al., 1977).

Legend
Kenney (1959)   □
Bjerrum and Simons (1960)   ○
Ladd, et al. (1977)   △

Average (Bjerrum and Simons, 1960)

±1 standard deviation (U.S. Navy, 1971)

on work by Kenney (1959), Bjerrum and Simons (1960), U.S. Navy (1971), and Ladd, et al. (1977). Since there is considerable scatter around the "average line," you should use this correlation with considerable caution. However Fig. 11.27 is useful for preliminary estimates and for checking laboratory results.

11.9.3 Use of CD Strength in Engineering Practice

Where do we use the strengths determined from the CD test? As mentioned previously, the limiting drainage conditions modeled in the triaxial test refer to real field situations. CD conditions are the most critical for the long-term steady seepage case for embankment dams and the long-term stability of excavations or slopes in both soft and stiff clays. Examples of CD analysis are shown in Fig. 11.28. How you actually go about making these analyses for stability can be found in textbooks on foundation and embankment dam engineering.

You should be aware that, practically speaking, it is not easy to actually conduct a CD test on a clay in the laboratory. To ensure that no pore pressure is really induced in the specimen during shear for materials with very low permeabilities, the rate of loading must be very slow. The time required to fail the specimen ranges from a day to several weeks (Bishop and Henkel, 1962). Such a long time leads to practical problems in the laboratory such as leakage of valves, seals, and the membrane that surrounds the sample. Consequently, since it is possible to measure the induced pore pressures in a consolidated-undrained (CU) test and thereby calculate the effective stresses in the specimen, CU tests are more practical for obtaining the effective stress strength parameters. Therefore CD triaxial tests are not very popular in most soils laboratories.

11.9.4 Consolidated-Undrained (CU) Test Behavior

As the name implies, the test specimen is first consolidated (drainage valves open, obviously) under the desired consolidation stresses. As before, these can either be hydrostatic or non-hydrostatic consolidation stresses. After consolidation is complete, the drainage valves are closed, and the specimen is loaded to failure in undrained shear. Often, the pore water pressures developed during shear are measured, and both the total and effective stresses may be calculated during shear and at failure. Thus this test can either be a total *or* an effective stress test. This test is sometimes called the *R-test*.

Total, neutral, and effective stress conditions in the specimen during the several phases of the CU test are shown in Fig. 11.29. The symbols are

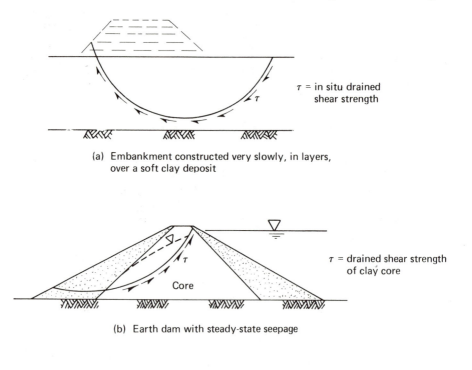

(a) Embankment constructed very slowly, in layers, over a soft clay deposit

τ = in situ drained shear strength

(b) Earth dam with steady-state seepage

τ = drained shear strength of clay core

(c) Excavation or natural slope in clay

τ = in situ drained shear strength

Fig. 11.28 Some examples of CD analyses for clays (after Ladd, 1971b).

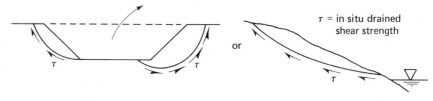

the same as we used before in Fig. 11.23. The general case of unequal consolidation is shown, but typically for routine triaxial testing the specimen is consolidated hydrostatically under a cell pressure which remains constant during shear. Thus,

$$\sigma_{\text{cell}} = \sigma_{vc} = \sigma_{hc} = \sigma'_{1c} = \sigma'_{3c} = \sigma_{3f} \neq \sigma'_{3f}$$

$$\Delta\sigma_f = (\sigma_1 - \sigma_3)_f$$

Like the CD test, the axial stress can be increased incrementally or at a constant rate of strain. At failure, then, the test in Fig. 11.29 is rather conventional in that the axial stress is increased to failure (axial compres-

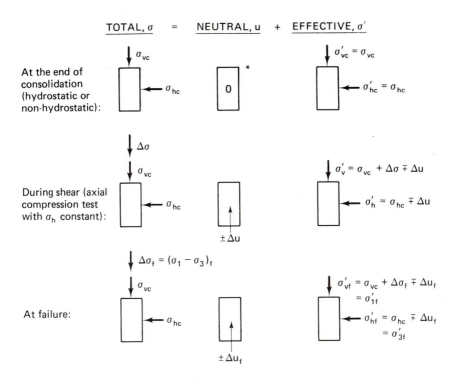

TOTAL, σ = NEUTRAL, u + EFFECTIVE, σ'

At the end of consolidation (hydrostatic or non-hydrostatic):

σ_{vc} 0* $\sigma'_{vc} = \sigma_{vc}$

σ_{hc} $\sigma'_{hc} = \sigma_{hc}$

During shear (axial compression test with σ_h constant):

$\Delta\sigma$

σ_{vc} $\sigma'_v = \sigma_{vc} + \Delta\sigma \mp \Delta u$

σ_{hc} $\sigma'_h = \sigma_{hc} \mp \Delta u$

$\pm \Delta u$

At failure:

$\Delta\sigma_f = (\sigma_1 - \sigma_3)_f$

σ_{vc} $\sigma'_{vf} = \sigma_{vc} + \Delta\sigma_f \mp \Delta u_f$
   $= \sigma'_{1f}$

σ_{hc} $\sigma'_{hf} = \sigma_{hc} \mp \Delta u_f$
   $= \sigma'_{3f}$

$\pm \Delta u_f$

*In practice, to ensure 100% saturation, which is necessary for good measurements of the pore water pressure, a *back pressure* is applied to the pore water. To keep the effective consolidation stresses constant, the total stresses during consolidation are accordingly increased by an amount exactly equal to the applied back pressure, which is the same as raising atmospheric pressure by a constant amount—the effective stresses on the clay do not change.

Example: Initial conditions with back pressure:

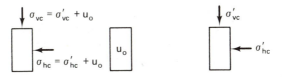

$\sigma_{vc} = \sigma'_{vc} + u_o$

$\sigma_{hc} = \sigma'_{hc} + u_o$ u_o σ'_{vc}

σ'_{hc}

Fig. 11.29 Conditions in specimen during a consolidated-undrained axial compression (CU) test.

sion test). Note that the excess pore water pressure Δu developed during shear can either be positive (that is, increase) or negative (that is, decrease). This happens because the sample tries to either contract or expand during shear. Remember, we are not allowing any volume change (an undrained test) and therefore no water can flow in or out of the specimen during shear. Because volume changes are prevented, the *tendency* towards volume change induces a pressure in the pore water. If the specimen *tends* to contract or consolidate during shear, then the induced pore water pressure is *positive*. It wants to contract and squeeze water out of the pores, but cannot; thus the induced pore water pressure is positive. Positive pore pressures occur in normally consolidated clays. If the specimen *tends* to expand or swell during shear, the induced pore water pressure is *negative*. It wants to expand and draw water into the pores, but cannot; thus the pore water pressure decreases and may even go negative (that is, below zero gage pressure). Negative pore pressures occur in overconsolidated clays. Thus, as noted in Fig. 11.29, the *direction* of the induced pore water pressure Δu is important since it directly affects the magnitudes of the effective stresses.

Also you might note that in actual testing the initial pore water pressure typically is greater than zero. In order to ensure full saturation, a *back pressure* u_o is usually applied to the test specimen (Fig. 11.29). When a back pressure is applied to a sample, the cell pressure must also be increased by an amount equal to the back pressure so that the effective consolidation stresses will remain the same. Since the effective stress in the specimen does not change, the strength of the specimen is not supposed to be changed by the use of back pressure. In practice this may not be exactly true, but the advantage of having 100% saturation for accurate measurement of induced pore water pressures far outweighs any disadvantages of the use of back pressure.

Typical stress-strain, Δu, and σ_1'/σ_3' curves for CU tests are shown in Fig. 11.30, for both normally and overconsolidated clays. Also shown for comparison is a stress-strain curve for an overconsolidated clay at low effective consolidation stress. Note the peak, then the drop-off of stress as strain increases (work-softening material, Fig. 10.4). The pore pressure versus strain curves illustrate what happens to the pore pressures during shear. The normally consolidated specimen develops positive pore pressure. In the overconsolidated specimen, after a slight initial increase, the pore pressure goes "negative"—in this case, negative with respect to the back pressure u_o. Another quantity that is useful for analyzing test results is the principal (effective) stress ratio σ_1'/σ_3'. Note how this ratio peaks early, just like the stress difference curve, for the overconsolidated clay. Similar test specimens having similar behavior on an effective stress basis will have similarly shaped σ_1'/σ_3' curves. They are simply a way of normal-

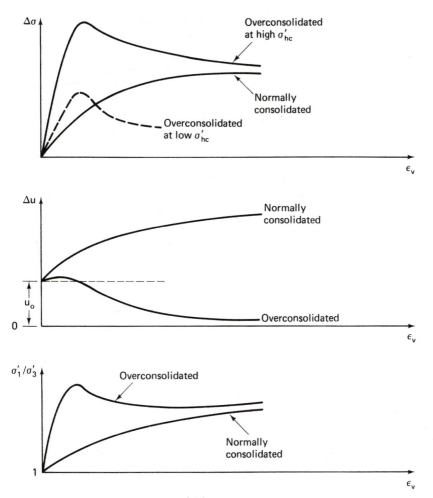

Note: For hydrostatic consolidation, $\sigma_1'/\sigma_3' = 1$ at the start of the test;
for non-hydrostatic consolidation, $\sigma_1'/\sigma_3' > 1$.

Fig. 11.30 Typical σ-ϵ, Δu, and σ_1'/σ_3' curves for normally and overconsolidated clays in undrained shear (CU test).

izing the stress behavior with respect to the effective minor principal stress
during the test. Sometimes, too, the maximum of this ratio is used as a
criterion of failure. However, in this text we will continue to assume failure
occurs at the maximum principal stress difference (compressive strength).

What do the Mohr failure envelopes look like for CU tests? Since we
can get both the total and effective stress circles at failure for a CU test
when we measure the induced pore water pressures, it is possible to define

the Mohr failure envelopes in terms of both total and effective stresses from a series of triaxial tests conducted over a range of stresses, as illustrated in Fig. 11.31 for a normally consolidated clay. For clarity, only one set of Mohr circles is shown. These circles are simply plotted from the stress conditions at failure in Fig. 11.29. Note that the effective stress circle is displaced to the left, towards the origin, for the normally consolidated case, because the specimens develop positive pore pressure during shear and $\sigma' = \sigma - \Delta u$. Note that both circles have the *same diameter* because of our definition of failure at maximum $(\sigma_1 - \sigma_3) = (\sigma_1' - \sigma_3')$. You should verify that this equation is true. Once the two failure envelopes are drawn, the Mohr-Coulomb strength parameters are readily definable in terms of both total (c, ϕ or sometimes c_T, ϕ_T) and effective stresses (c', ϕ'). Again, as with the CD test, the envelope for normally consolidated clay passes essentially through the origin, and thus for practical purposes c' can be taken to be zero, which is also true for the total stress c parameter. Note that ϕ_T is less than ϕ', and often it is about one-half of ϕ'.

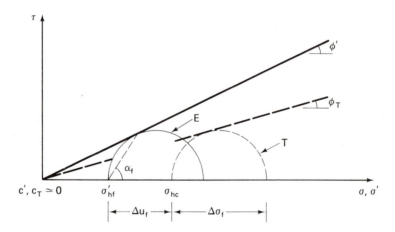

Fig. 11.31 Mohr circles at failure and Mohr failure envelopes for total (T) and effective (E) stresses for a normally consolidated clay.

Things are different if the clay is overconsolidated. Since an overconsolidated specimen tends to expand during shear, the pore water pressure decreases or even goes negative, as shown in Fig. 11.30. Because $\sigma_{3f}' = \sigma_{3f} - (-\Delta u_f)$ or $\sigma_{1f}' = \sigma_{1f} - (-\Delta u_f)$, the effective stresses are *greater* than the total stresses, and the effective stress circle at failure is shifted to the *right* of the total stress circle, as shown in Fig. 11.32. The shift of the effective stress circle at failure to the right sometimes means that the ϕ' is less than ϕ_T. Typically, the complete Mohr failure envelopes are determined by tests on several specimens consolidated over the working

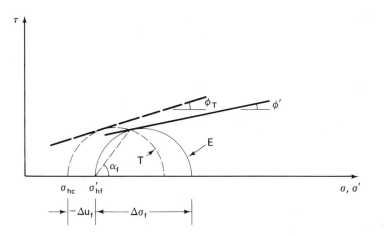

Fig. 11.32 Mohr circles at failure and Mohr failure envelopes for both total (T) and effective (E) stresses for an overconsolidated clay.

stress range of the field problem. Figure 11.33 shows the Mohr failure envelopes over a wide range of stresses spanning the preconsolidation stress. Thus some of the specimens are overconsolidated and others are normally consolidated. You should note that the "break" in the *total* stress envelope (point z) occurs roughly about twice the σ'_p for typical clays (Hirschfeld, 1963). The two sets of Mohr circles at failure shown in Fig.

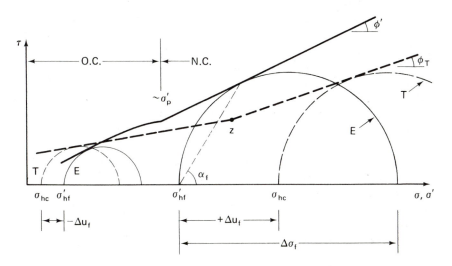

Fig. 11.33 Mohr failure envelopes over a range of stresses spanning the preconsolidation stress σ'_p.

11.33 correspond to the two tests shown in Fig. 11.30 for the "normally consolidated" specimen and the specimen "overconsolidated at low σ'_{hc}."

You may have noticed that an angle α_f was indicated on the effective stress Mohr circles of Figs. 11.31, 11.32, and 11.33. Do you recall the Mohr failure hypothesis wherein the point of tangency of the Mohr failure envelope with the Mohr circle at failure defined the angle of the failure plane in the specimen? If not, reread Sec. 10.4. Since we believe that the shear strength is controlled by the effective stresses in the specimen at failure, the Mohr failure hypothesis is valid in terms of *effective stresses only*.

Stress paths for the two tests of Fig. 11.33 are shown in Fig. 11.34. The tests are quite conventional, hydrostatically consolidated, axial compression tests. Let's look first at Fig. 11.34a, the stress paths for the test on normally consolidated clay. Three stress paths are shown, the effective stress path (ESP), the total stress path (TSP), and the total-u_0 stress paths, $(T - u_0)$ SP. The paths begin on the hydrostatic axis at values of p equal to the total and effective consolidation pressures, respectively. Note that $p = p' + u_0$. The total stress path for axial compression and constant cell pressure is the straight line inclined at 45° as shown. Since positive pore pressures develop in the normally consolidated clay, the ESP lies to the *left* of the TSP because $\sigma' = \sigma - \Delta u$. The situation is directly analogous to that shown in Fig. 10.24. Note that q_f is the same for all three stress paths because we define the failure at the maximum $(\sigma_1 - \sigma_3)$. Figure 11.34a is similar to Fig. 10.24, except the initial consolidation in that case was non-hydrostatic ($K_0 < 1$).

Since the overconsolidated clay was tested in axial compression with a constant hydrostatic cell pressure, the two total stress paths of Fig. 11.34b are exactly like those of Fig. 11.34a—straight lines inclined at 45° to the hydrostatic axis. But the shape of the ESP is significantly different. Look back at the development of pore pressure with axial strain for this test in Fig. 11.30. See how it starts out slightly positive, then goes way negative (actually, less than u_0, as was explained previously). The same thing happens to the ESP in Fig. 11.34b. It goes slightly to the left $(+\Delta u)$ of the $(T - u_0)$ SP at first, then as the pore pressure becomes increasingly negative, the ESP crosses the $(T - u_0)$ SP until maximum q or q_f is reached. Again, because of the way we define failure, q_f is the same for all three stress paths. You may recall that the ESP in Fig. 11.34b for the overconsolidated clay has a shape similar to that shown in Fig. 10.25, except that the latter sample was consolidated with $K_0 > 1$.

If you are still unclear about stress paths, it would be a good idea to reread Sec. 10.6.

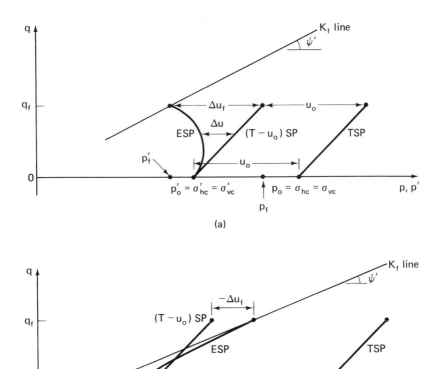

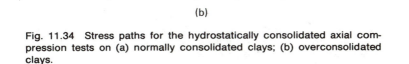

Fig. 11.34 Stress paths for the hydrostatically consolidated axial compression tests on (a) normally consolidated clays; (b) overconsolidated clays.

11.9.5 Typical Values of the Undrained Strength Parameters

Earlier in this section, we gave some typical values for c' and ϕ' determined by CD triaxial tests. The range of values indicated is typical for effective stress strengths determined in CU tests with pore pressure measurements, with the following reservation. In our discussion so far, we have tacitly assumed that the Mohr-Coulomb strength parameters in terms of effective stresses determined by CU tests with pore pressure measurements would be the same as those determined by CD tests. We used the same symbols, c' and ϕ', for the parameters determined both ways. This assumption is not strictly correct. The problem is complicated by alternative definitions of failure. We have used the maximum principal stress difference $(\sigma_1 - \sigma_3)_{max}$ to define failure throughout this chapter, but often in the literature and sometimes in practice you will find failure defined in terms of the maximum principal effective stress ratio $(\sigma_1'/\sigma_3')_{max}$, which is the same as the maximum obliquity (Eqs. 10-14 through 10-17). Depending on how the stress difference and the pore water pressures actually develop with strain, these two definitions may indicate different c's and ϕ's. This is especially true for sensitive clays, as shown in Fig. 11.35.

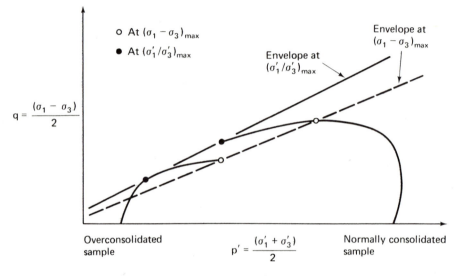

Fig. 11.35 Typical failure envelopes for CU tests on a sensitive clay, illustrating the effect of different failure criteria on the slope and intercept of the Mohr-Coulomb failure envelope (after Ladd, 1971b).

Bjerrum and Simons (1960) studied this problem in some detail, and their results are summarized in Fig. 11.36. Here, ϕ' as defined at $(\sigma_1'/\sigma_3')_{max}$ and $(\sigma_1 - \sigma_3)_{max}$ are plotted versus ϕ_d', the effective stress parameter determined in drained tests. Note that ϕ' from the maximum principal effective stress ratio (the dots) is from 0° to 3° greater than ϕ_d'. Also note that ϕ' at maximum principal stress difference (the squares) is less than both ϕ_d' and ϕ' at the maximum principal effective stress ratio. In one case the difference is about 7°.

The point is that you should be careful when studying published data or engineering test reports to determine exactly how the strength tests were conducted, how failure was defined, and how any reported Mohr-Coulomb parameters were determined.

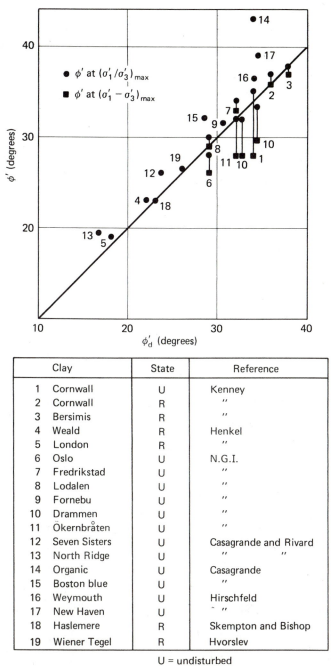

	Clay	State	Reference
1	Cornwall	U	Kenney
2	Cornwall	R	"
3	Bersimis	R	"
4	Weald	R	Henkel
5	London	R	"
6	Oslo	U	N.G.I.
7	Fredrikstad	U	"
8	Lodalen	U	"
9	Fornebu	U	"
10	Drammen	U	"
11	Ökernbråten	U	"
12	Seven Sisters	U	Casagrande and Rivard
13	North Ridge	U	" "
14	Organic	U	Casagrande
15	Boston blue	U	"
16	Weymouth	U	Hirschfeld
17	New Haven	U	" "
18	Haslemere	R	Skempton and Bishop
19	Wiener Tegel	R	Hvorslev

U = undisturbed
R = remolded

Fig. 11.36 Relationship between ϕ'_d determined from CD tests and ϕ' determined from CU tests with pore pressures measured. Two failure criteria are indicated for the undrained tests (after Bjerrum and Simons, 1960).

For the Mohr-Coulomb strength parameters in terms of total stresses, the problem of definition of failure doesn't come up. Failure is defined at the maximum compressive strength $(\sigma_1 - \sigma_3)_{max}$. For normally consolidated clays, ϕ seems to be about half of ϕ'; thus values of 10° to 15° or more are typical. The total stress c is very close to zero. For overconsolidated and compacted clays, ϕ may decrease and c will often be significant. When the failure envelope straddles the preconsolidation stress, proper interpretation of the strength parameters in terms of total stresses is difficult. This is especially true for undisturbed samples which may have some variation in water content and void ratio, even within the same geologic stratum.

In the section on typical values of drained strength parameters, we provided an empirical correlation for ϕ' and PI (Fig. 11.27). These were for normally consolidated undisturbed clays tested in triaxial compression, and in fact most of the tests used to develop this figure were CU tests with pore pressures measured. Figure 11.27 still can be used for preliminary estimates and for checking laboratory test results because the differences in ϕ', depending on how failure is defined, etc., are less than the scatter in the figure.

11.9.6 Use of CU Strength in Engineering Practice

Where do we use the CU strength in engineering practice? As mentioned before, this test, with pore pressures measured, is commonly used to determine the shear strength parameters in terms of both total and effective stresses. CU strengths are used for stability problems where the soils have first become fully consolidated and are at equilibrium with the existing stress system. Then, for some reason, *additional* stresses are applied quickly, with no drainage occurring. Practical examples include rapid drawdown of embankment dams and the slopes of reservoirs and canals. Also, in terms of effective stresses, CU test results are applied to the field situations mentioned in the earlier discussion of CD tests. Some of these practical examples are illustrated in Fig. 11.37.

Just as with CD tests, there are some problems with CU tests on clay. For proper measurement of the pore pressures induced during shear, special care must be taken to see that the sample is fully saturated, that no leaks occur during testing, and that the rate of loading (or rate of strain) is sufficiently slow so that the pore pressures measured at the ends of the specimen are the same as those occurring in the vicinity of the failure plane. As we mentioned, the use of back pressure is common to assure 100% saturation. The effects of the other two factors can be minimized by proper testing techniques, which are described at length by Bishop and Henkel (1962).

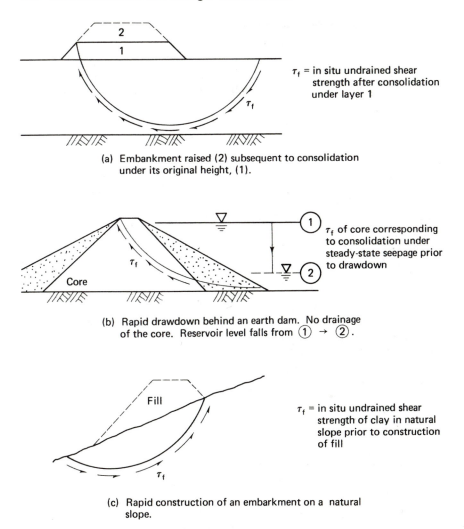

τ_f = in situ undrained shear
strength after consolidation
under layer 1

(a) Embankment raised (2) subsequent to consolidation
under its original height, (1).

τ_f of core corresponding
to consolidation under
steady-state seepage prior
to drawdown

(b) Rapid drawdown behind an earth dam. No drainage
of the core. Reservoir level falls from ① → ②.

τ_f = in situ undrained shear
strength of clay in natural
slope prior to construction
of fill

(c) Rapid construction of an embarkment on a natural
slope.

Fig. 11.37 Some examples of CU analyses for clays (after Ladd, 1971b).

Another problem, not often mentioned, results from trying to determine the long-term or effective stress strength parameters and the short-term or CU-total stress strength parameters from the same test series. The rates of loading or strain required for correct effective stress strength determination may not be appropriate for the short-term or undrained loading situation. The stress-deformation and strength response of clay soils is rate-dependent; that is, usually the faster you load a clay, the stronger it becomes. In the short-term case, the rate of loading in the field may be quite rapid, and therefore for correct modeling of the field situation, the rates of loading in the laboratory sample should be comparable. Thus the two objectives of the CU-effective stress test are really incompatible. The best thing to do, though rarely done in practice, would be to have two sets of tests, one set tested CD modeling the long-term situation and the other CU set modeling the short-term undrained loading.

EXAMPLE 11.11

Given:

A normally consolidated clay was consolidated under a stress of 150 kPa, then sheared undrained in axial compression. The principal stress difference at failure was 100 kPa, and the induced pore pressure at failure was 88 kPa.

Required:

Determine the Mohr-Coulomb strength parameters in terms of both total and effective stresses (a) analytically and (b) graphically. Plot the total and effective Mohr circles and failure envelopes. (c) Compute $(\sigma_1'/\sigma_3')_f$ and $(\sigma_1/\sigma_3)_f$. (d) Determine the theoretical angle of the failure plane in the specimen.

Solution:

To solve this problem we need to assume that both c' and c_T are negligible. Then we can use the obliquity relationships (Eqs. 10-14 through 10-17) to solve for ϕ' and ϕ_T.

a. To use these equations, we need σ_{1f}, σ_{1f}', σ_{3f} and σ_{3f}'. We know $\sigma_{3f} = 150$ kPa and $(\sigma_1 - \sigma_3)_f = 100$ kPa. Therefore

$$\sigma_{1f} = (\sigma_1 - \sigma_3)_f + \sigma_{3f} = 100 + 150 = 250 \text{ kPa}$$

$$\sigma_{1f}' = \sigma_{1f} - u_f = 250 - 88 = 162 \text{ kPa}$$

$$\sigma_{3f}' = \sigma_{3f} - u_f = 150 - 88 = 62 \text{ kPa}$$

From Eq. 10-13,

$$\phi' = \arcsin \frac{100}{224} = 26.5°$$

$$\phi_T = \arcsin \frac{100}{400} = 14.5°$$

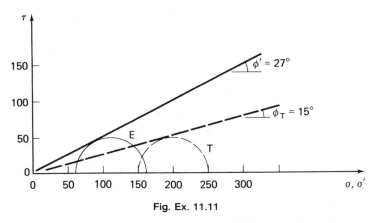

Fig. Ex. 11.11

b. The graphical solution including the failure envelopes is shown in Fig. Ex. 11.11. To plot the total and effective Mohr circles, we would still need to calculate σ_{1f}, σ'_{1f} and σ'_{3f}. The centers of the circles are at $(200, 0)$ for total stresses and at $(112, 0)$ for effective stresses.

c. The stress ratios at failure are

$$\frac{\sigma'_1}{\sigma'_3} = \frac{162}{62} = 2.61$$

$$\frac{\sigma_1}{\sigma_3} = \frac{250}{150} = 1.67$$

Another way to get these values would be to use Eq. 10-14.

$$\frac{\sigma'_1}{\sigma'_3} = \frac{1 + \sin 26.5°}{1 - \sin 26.5°} = \frac{1.45}{0.55} = 2.61$$

$$\frac{\sigma_1}{\sigma_3} = \frac{1 + \sin 14.5°}{1 - \sin 14.5°} = \frac{1.25}{0.75} = 1.67$$

d. Use Eq. 10-10, in terms of *effective* stresses:

$$\alpha_f = 45° + \frac{\phi'}{2} = 58° \text{ from the horizontal}$$

11.9.7 Unconsolidated-Undrained (UU) Test Behavior

In this test, the specimen is placed in the triaxial cell with the drainage valves closed from the beginning. Thus, even when a confining pressure is applied, no consolidation can occur if the sample is 100% saturated. Then, as with the CU test, the specimen is sheared undrained. The sample is loaded to failure in about 10 to 20 min; usually pore water pressures are not measured in this test. This test is a *total stress test* and it yields the strength in terms of total stresses. A. Casagrande first called this test the *Q-test* (for "quick") since the sample was loaded to failure much more quickly than in the *S*-test.

Total, neutral, and effective stress conditions in the specimen during the several phases of the UU test are shown in Fig. 11.38. The symbols are as used before in Fig. 11.23 and 11.29. The test illustrated in Fig. 11.38 is

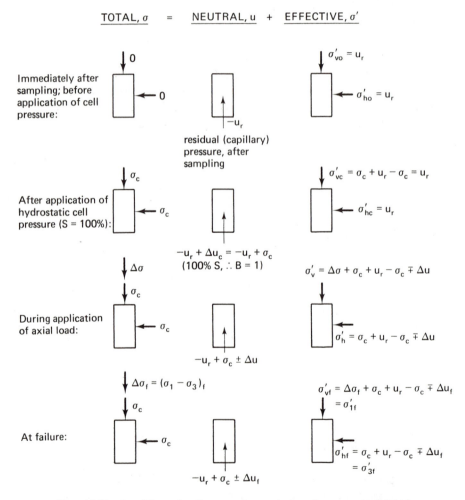

Fig. 11.38 Conditions in the specimen during the unconsolidated-undrained (UU) axial compression test.

quite conventional in that hydrostatic cell pressure is usually applied, and the specimen is failed by increasing the axial load, usually at a constant rate of strain. As with the other tests, the principal stress difference at failure is $(\sigma_1 - \sigma_3)_{max}$.

Note that initially for undisturbed samples, the pore pressure is negative, and it is called the *residual pore pressure* $-u_r$, which results from stress release during sampling. Since the effective stresses initially must be greater than zero (otherwise the specimen would simply disintegrate) and the total stresses are zero (atmospheric pressure = zero gage pressure), the pore pressure must be negative. (See Fig. 10.21 for insight into the sampling process.) When the cell pressure is applied with the drainage valves closed, a positive pore pressure Δu_c is induced in the specimen, which is exactly equal to the applied cell pressure σ_c. All the increase in hydrostatic stress is carried by the pore water because (1) the soil is 100% saturated, (2) the compressibility of the water and individual soil grains is small compared to the compressibility of the soil structure, and (3) there is a unique relationship between the effective hydrostatic stress and the void ratio (Hirschfeld, 1963). Number 1 is obvious. Number 2 means that no volume change can occur unless water is allowed to flow out of (or into) the sample, and we are preventing that from occurring. Number 3 means basically that no secondary compression (volume change at constant effective stress) takes place. You may recall from the discussion of the assumptions of the Terzaghi theory of consolidation (Chapter 9) that the same assumption was required; that is, that the void ratio and effective stress were uniquely related. Thus there can be no change in void ratio without a change in effective stress. Since we prevent any change in water content, the void ratio and effective stress remain the same.

Stress conditions during axial loading and at failure are similar to those for the CU test (Fig. 11.29). They may appear to be complex, but if you study Fig. 11.38 you will see that the UU case is as readily understandable as the CU case.

Typically, stress-strain curves for UU tests are not particularly different from CU or CD stress-strain curves for the same soils. For undisturbed samples, especially the initial portions of the curve (initial tangent modulus), are strongly dependent on the *quality* of the undisturbed samples. Also, the sensitivity (Sec. 2.7) affects the shape of these curves; highly sensitive clays have sharply peaked stress-strain curves. The maximum stress difference often occurs at very low strains, usually less than 0.5%. Some typical UU stress-strain curves are shown in Fig. 11.39.

The Mohr failure envelopes for UU tests are shown in Fig. 11.40 for 100% saturated clays. All test specimens for fully saturated clays are presumably at the same water content (and void ratio), and consequently they will have the same shear strength since there is no consolidation

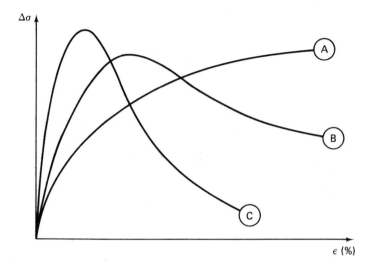

Fig. 11.39 Typical UU stress-strain curves for (A) remolded and some compacted clays, (B) medium sensitive undisturbed clay, and (C) highly sensitive undisturbed clay.

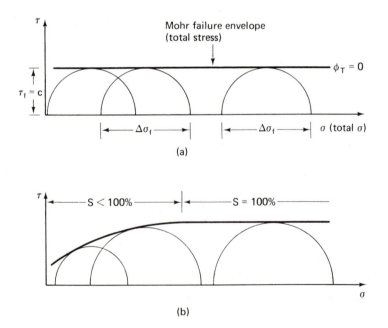

Fig. 11.40 Mohr failure envelopes for UU tests: (a) 100% saturated clay; (b) partially saturated clay.

allowed. Therefore all Mohr circles at failure will have the *same* diameter and the Mohr failure envelope will be a horizontal straight line (see Fig. 10.9c). This is a very important point. If you don't understand it, refer again to Fig. 11.38 to see that in the UU test the effective consolidation stress is the same throughout the test. If all the samples are at the same water content and density (void ratio), then they will have the same strength. The UU test, as previously mentioned, gives the shear strength in terms of total stresses, and the slope ϕ_T of the UU Mohr failure envelope is *equal to zero*. The intercept of this envelope on the τ-axis defines the total stress strength parameter c, or $\tau_f = c$, where τ_f is undrained shear strength.

For partially saturated soils, a series of UU tests will define an initially curved failure envelope (Fig. 11.40b) until the clay becomes essentially 100% saturated due simply to the cell pressure alone. Even though the drainage valves are closed, the confining pressure will compress the air in the voids and decrease the void ratio. As the cell pressure is increased, more and more compression occurs and eventually, when sufficient pressure is applied, essentially 100% saturation is achieved. Then, as with the case for initially 100% saturated clays, the Mohr failure envelope becomes horizontal, as shown on the right side of Fig. 11.40b.

Another way of looking at the compression of partially saturated clays is shown in Fig. 11.41. As the cell pressure is increased incrementally,

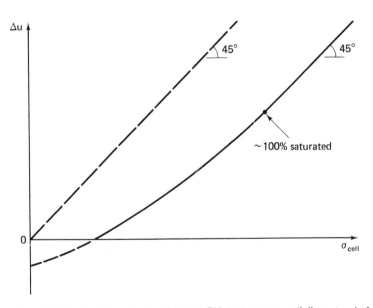

Fig. 11.41 Results obtained from a PH test on a partially saturated compacted clay (after Skempton, 1954, and Hirschfeld, 1963).

the measured increment of pore pressure increases gradually until at some point for every increment of cell pressure added, an equal increment of pore water pressure is observed. At this point, the soil is 100% saturated and the solid (experimental) curve becomes parallel to the 45° line shown in the figure.

In principle, it is possible to measure the induced pore water pressures in a series of UU tests although it is not commonly done. Since the effective stresses at failure are *independent* of the total cell pressures applied to the several specimens of a test series, there is only *one* UU effective stress Mohr circle at failure. This point is illustrated in Fig. 11.42. Note that no matter what the confining pressure (for example, σ_{c1}, σ_{c2}, etc.), there is only one effective stress Mohr circle at failure. The minor effective principal stress at failure (σ'_{hf}) is the same for *all* total stress circles shown in the figure. Since we have only one effective circle at failure, strictly speaking, we need to know both ϕ' and c' in advance in order to draw the Mohr failure envelope in terms of effective stresses for the UU test. We could perhaps measure the angle of the failure plane in the failed UU specimens and invoke the Mohr failure hypothesis, but as was discussed in Sec. 10.4, there are practical problems with this approach. It should also be noted that the angle of inclination of the failure plane α_f shown in Fig. 11.42 is defined by the effective stress envelope. Otherwise, as indicated in Fig. 10.9c and Eq. 10-10, theory would predict α_f to be 45°.

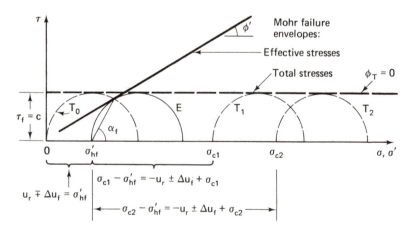

Note: σ'_{hf} is the same for all three total stress circles!

Fig. 11.42 UU test results, illustrating the unique effective stress Mohr circle at failure.

Since the strength ultimately is controlled or governed by the effective stresses, we believe that the physical conditions controlling the formation of a failure plane in the test specimen must in some fashion be controlled by the effective stresses acting in the specimen at failure. Thus Eq. 10-10 should be in terms of ϕ' instead of ϕ_T.

Stress paths for the UU tests of Fig. 11.42 are shown in Fig. 11.43. Behavior is for a normally consolidated clay, and the values of p and q for all three tests are listed in the table below the figure. Refer to Fig. 11.38 if necessary to verify these values. If the clay were overconsolidated, then from your knowledge of CU behavior you would expect the ESP to have a shape similar to those of Fig. 11.34b.

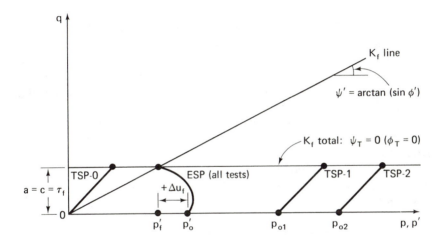

		Initial Conditions		At Failure	
	Test	p_o	q_o	p_f	q_f
Total stresses	0	0	0	$\dfrac{\Delta\sigma_f}{2}$	$\dfrac{\Delta\sigma_f}{2}$
	1	σ_{c1}	0	$\dfrac{\Delta\sigma_f + 2\sigma_{c1}}{2}$	$\dfrac{\Delta\sigma_f}{2}$
	2	σ_{c2}	0	$\dfrac{\Delta\sigma_f + 2\sigma_{c2}}{2}$	$\dfrac{\Delta\sigma_f}{2}$
Effective stresses		p'_o	q_o	p'_f	q_f
	All	u_r	0	$\dfrac{\Delta\sigma_f + 2u_r - 2\Delta u_f}{2}$	$\dfrac{\Delta\sigma_f}{2}$

Fig. 11.43 Stress paths for UU tests on a normally consolidated clay. Same tests as in Fig. 11.42.

11.9.8 Typical Values of UU Strengths

The undrained strength of clays varies widely. Of course, ϕ_T is zero, but the magnitude of τ_f can vary from almost zero for extremely soft sediments to several MPa for very stiff soils and soft rocks. Often, the undrained shear strength at a site is normalized with respect to the vertical effective overburden stress σ'_{vo} at each sampling point. Then the τ_f / σ'_{vo} ratios are analyzed and compared with other data. This point is covered in more detail later in this chapter.

11.9.9 Unconfined Compression Test

We can, theoretically at least, conduct an *unconfined compression test* and obtain the UU-total stress strength. This test is a special case of the UU test with the confining or cell pressure equal to zero (atmospheric pressure). The stress conditions in the unconfined compression test specimen are similar to those of Fig. 11.38 for the UU test, except that σ_c is equal to zero, as shown in Fig. 11.44. If you compare these two figures, you will see that the effective stress conditions at failure are *identical* for both tests. And if the effective stress conditions are the same in both tests, then the strengths will be the same!

Practically speaking, for the unconfined compression test to yield the same strength as the UU test, several assumptions must be satisfied. These are as follows:

1. The specimen must be 100% saturated; otherwise compression of the air in the voids will occur and cause a decrease in void ratio and an *increase* in strength.
2. The specimen must not contain any fissures, silt seams, varves, or other defects; this means that the specimen must be *intact*, homogeneous clay. Rarely are overconsolidated clays intact, and often even normally consolidated clays have some fissures.
3. The soil must be very fine grained; the initial effective confining stress as indicated in Fig. 11.44 is the residual capillary stress which is a function of the residual pore pressure $-u_r$; this usually means that *only clay soils* are suitable for testing in unconfined compression.
4. The specimen must be sheared rapidly to failure; it is a total stress test and the conditions must be undrained throughout the test. If the time to failure is too long, evaporation and surface drying will increase the confining pressure and too high a strength will result. Typical time to failure is 5 to 15 min.

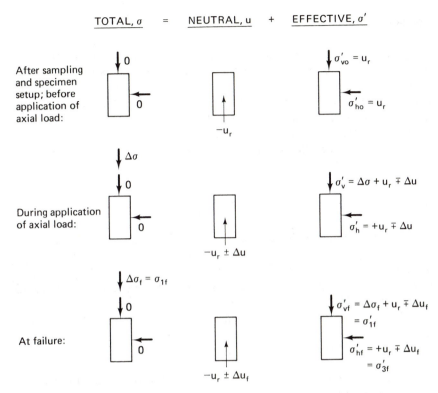

Fig. 11.44 Stress conditions for the unconfined compression test.

Be sure to distinguish between unconfined *compressive* strength $(\sigma_1 - \sigma_3)_f$ and the undrained *shear* strength, which is $\tau_f = \frac{1}{2}(\sigma_1 - \sigma_3)_f$.

EXAMPLE 11.12

Given:

An unconfined compression test was conducted on a soft clay. The specimen was trimmed from the undisturbed tube sample and was 35 mm in diameter and 80 mm high. The load on the load ring at failure was 14.3 N, and the axial deformation was 11 mm.

Required:

Calculate the unconfined compressive strength and the shear strength of the soft clay sample.

Solution:

To calculate the stress at failure, we have to know the area of the specimen A_s. At failure it is *not* equal to the original area A_o, but is somewhat greater. (In compression, the specimen decreases in height and increases in diameter as long as Poisson's ratio (Eq. 8-33) is greater than zero.) So, first we need to determine the actual area of the specimen at failure. Since the specimen is tested in undrained shear, we can assume the volume is unchanged and that the specimen deforms as a right circular cylinder. Thus A_s at any strain ϵ is

$$A_s = \frac{A_o}{1 - \epsilon} \qquad (11\text{-}8)$$

Now we can calculate the area of the specimen. The strain at failure is $\Delta L/L_o = 11 \text{ mm}/80 \text{ mm} = 0.1375$, or 13.8%. Thus $A_s = 1115 \text{ mm}^2$. Now the compressive stress at failure is $14.3 \text{ N}/1115 \text{ mm}^2 = 12.8 \text{ kN/m}^2$ (kPa). If we had simply divided by the original area of the specimen, we would have obtained 14.9 kN/m^2, a significant error.

The shear strength for the unconfined compression test is one-half the compressive strength, or 6.4 kPa.

It should be noted that the actual shear stress on the failure plane at failure τ_{ff} is somewhat less than the undrained shear strength $\tau_f = c$ because τ_{ff} occurs on a failure plane whose inclination is determined by the effective stresses, as explained previously for the UU test. The conditions and the approximate magnitude of associated error is indicated in Fig. 11.45a for the specimen at failure in Fig. 11.45b. The magnitude of the error depends on ϕ', as indicated by the calculations in Example 11.13.

EXAMPLE 11.13

Given:

The stress conditions for the unconfined compression test shown in Figs. 11.45a and 11.45b.

Required:

Find the error in assuming the undrained shear strength $\tau_f = c = \frac{1}{2}\Delta\sigma_f$, rather than τ_{ff} for a normally consolidated clay where $\phi' = 30°$.

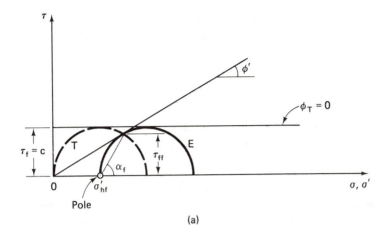

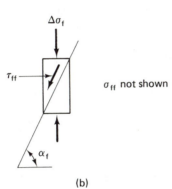

Fig. 11.45 (a) Difference between τ_{ff} and $\tau_f = c$ in an (b) unconfined compression test specimen (after Hirschfeld, 1963).

Solution:

From Eq. 10-6,

$$\tau_{ff} = \frac{\sigma_1 - \sigma_3}{2} \sin 2\alpha_f$$

$$= \frac{\Delta \sigma_f}{2} \sin 2\alpha_f$$

From Eq. 10-10, $\alpha_f = 45° + \phi'/2$. So $\alpha_f = 60°$. Therefore

$$\tau_{ff} = \frac{\Delta \alpha_f}{2} \sin 120° = 0.433 \Delta \sigma_f$$

But $\tau_f = c = 0.5 \Delta \sigma_f$.

Conclusion: $\tau_f = c$ strength is about 15% greater than τ_{ff} for $\phi' = 30°$. Note that the error is less for smaller ϕ' angles. Also note that

$$\tau_f = c = \frac{\Delta \sigma_f}{2} = \frac{(\sigma_1 - \sigma_3)_f}{2} = \tau_{\max}$$

The point illustrated by Example 11.13 is that the actual shear strength on the failure is *overestimated* by the one-half unconfined compressive strength. The magnitude of the error is probably about 15% at most.

Why then does the unconfined compression test apparently work satisfactorily? It is far and away the most common laboratory strength test used today in the United States for the design of shallow and pile foundations in clay. Part of the answer lies in *compensating errors*. Sample disturbance especially tends to reduce the undrained shear strength. Anisotropy also is a factor, as is the assumption of plane strain conditions for most design analyses whereas the real stress conditions are more three dimensional. These factors tend to reduce the undrained shear strength so that the difference between $\tau_f = c$ and τ_{ff} becomes negligible in engineering practice. Several of these points are discussed by Ladd, et al. (1977).

11.9.10 Other Ways to Determine the Undrained Shear Strength

There are other ways besides the unconfined compression test or the UU triaxial test to obtain the undrained shear strength of cohesive soils. Some of the field methods were mentioned briefly at the end of Sec. 10.5; other methods are used exclusively in the laboratory on undisturbed samples. In all the methods, loading to failure is assumed to take place so rapidly that undrained conditions exist. The result obtained from the test is then correlated with the undrained shear strength τ_f.

Table 11-6 and Figs. 11.46 through 11.55 illustrate the methods commonly used for determining τ_f. References are listed in the table if you need additional details about these tests or their interpretation.

Except for the SPT, all the field techniques listed in Table 11-6 were developed in Europe, but there has been increasing interest in them in recent years in North America (Ladd, et al., 1977). The conditions for a reliable unconfined compression test are not often met. More sophisticated laboratory testing techniques are attractive but increasingly expensive. Sampling of poor quality, unfortunately the rule rather than the exception in the United States, can significantly affect the measured shear strength. Some soils such as stiff fissured clays are difficult if not impossible to even

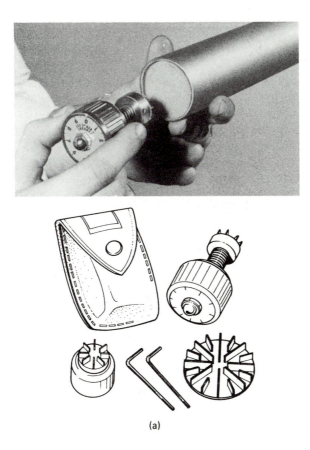

(a)

Diameter (mm)	Height of vanes (mm)	Maximum τ_f (kPa)
19	3	250
25	5	100 (standard)
48	5	20

(b)

Fig. 11.46 Torvane (TV): (a) Standard model shown on its side. The other two vanes, which can be attached to the standard Torvane, are for very soft or very stiff clays. (b) Specifications for the three vanes. (Photograph courtesy of Soiltest, Inc., Evanston, Illinois.)

TABLE 11-6 Laboratory and Field Methods for Determining τ_f

No.	Test	Use	Fig. No.	Remarks
1	Torvane (TV)	Lab, field	11.46	Hand held; calibrated spring; quick; used on tube samples or the sides of exploratory trenches, etc. Sample tested is seen.
2	Pocket penetrometer (PP)	Lab, field	11.47	Same as above, except spring is calibrated in unconfined compressive strength ($= 2\tau_f$).
3	Swedish fall-cone test (SFC)	Lab	11.48	Quick; sample tested is seen; used on tube samples. τ_f depends on cone angle and mass.
4	Vane shear test (VST)	Lab, field	11.49	Various sizes and configurations available for both field and lab use. Height/diameter ratio (H/D) = 2 for field vanes; H/D = 1 for lab vanes. Only lab vane sample is seen.
5	Standard penetration test (SPT)	Field	11.51	A standard "split-spoon" sampler is driven by a 63.5 kg hammer falling 0.76 m. The number of blows required to drive the sampler 0.3 m is called the *standard penetration resistance*, or *blow count*, *N*. Disturbed sample obtained.
6	Dutch cone penetrometer (CPT)	Field	11.52	A 60° cone with a projected area of 10 cm^2 is pushed at 1 to 2 m/min. Point resistance q_c and friction on the friction sleeve f_s are measured either electrically or mechanically.
7	Pressuremeter (PMT)	Field	11.53	A cylindrical probe is inserted in a drill hole (may be self-boring). Lateral pressure is applied incrementally to side of hole.
8	Screw plate compresso-meter (SPC)	Field	11.54	The plate is screwed down to the desired testing depth; hydraulic pressure is applied incrementally and the settlement is observed; continue loading until the bearing capacity of the soil is reached.
9	Iowa borehole shear test (BST)	Field	11.55	Device is lowered into a borehole and expanded against the side walls (σ_n). Then entire mechanism is pulled from ground surface and maximum load measured (τ_f). Stage test results are used to plot Mohr diagram for CD tests. Range of σ_n is from about 30 to 100 kPa.

TABLE 11-6 (cont.)

	Best For	Limitations	References
1	Very soft to stiff clays	Cohesive soils without pebbles, fissures, etc. Test only a small amount of soil near the surface. Only rough calibration with τ_f.	
2	Very soft to stiff clays	Same as above.	
3	Very soft to soft clays	Same as above, except good correlation with τ_f on soft, sensitive clays.	Hansbo (1957)
4	Soft to stiff clays	May overestimate τ_f; see Fig. 11.50 for correction factor for very soft clays. Unreliable readings if vane encounters sand layers, varves, stones, etc., or if vane rotated too rapidly.	Cadling and Odenstad (1950) Bjerrum (1972) Schmertmann (1975) ASTM (1980) D 2573 Ladd, et al. (1977)
5	Granular soils	Very rough correlation with τ_f for cohesive soils. Boulders can cause problems. Results are sensitive to test details.	ASTM (1980) D 1586 de Mello (1971) Schmertmann (1975) Kovacs, et al. (1977)
6	All soil types except very coarse granular soils	Boulders cause problems. Requires local correlation for soft clays.	Sanglerat (1972) ESOPT (1974) Schmertmann (1975) Ladd, et al. (1977) ASTM (1980) D 3441
7	All soil types	Requires a correlation between p_l and τ_f.	Ménard (1956, 1975) Schmertmann (1975) Ladd, et al. (1977) Baguelin, et al. (1978)
8	All soil types except very coarse granular soils	Mostly used to study the compressibility of granular soils. Schwab (1976) found good agreement with the screw plate and the vane shear test in plastic Swedish clays.	Janbu and Senneset (1973) Mitchell and Gardener (1975) Schwab (1976) Schmertmann (1970)
9	Loessial (silty) soils	Cannot be used with soils with 10% or more gravel or caving sands. Uncertain drainage conditions during shear makes the test difficult to interpret. (Is it CD or CU or somewhere in between?)	Wineland (1975) Schmertmann (1975)

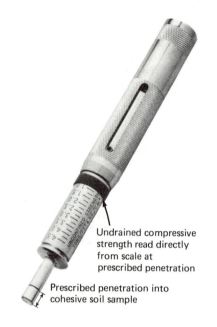

Undrained compressive
strength read directly
from scale at
prescribed penetration

Prescribed penetration into
cohesive soil sample

Fig. 11.47 Pocket penetrometer (PP), a hand-held device which indicates unconfined compressive strength (photograph courtesy of Soiltest,
Inc., Evanston, Illinois).

sample. There are statistical advantages, too, for having lots of indirect
subsurface information obtained rapidly and at relatively low cost compared with a few, expensive laboratory tests on what may not even be the
weakest or the most critical strata at the site. Finally, some soil properties
(K_o, permeability, deformation modulus) can only be determined reliably
in the field.

The major disadvantage of the in situ methods is that τ_f is obtained
only indirectly through correlations with laboratory tests or by backcalcu-
lation from theory or actual failures. Figure 11.50 is an example of the
correction factor that must be applied to the field vane test to obtain
the best estimate of the in situ τ_f. Other correlations are provided in the
references for each test listed in Table 11-6. Especially useful are the
correlations presented by U.S. Navy (1971), de Mello (1971), and Schmertmann (1975) for the SPT and Dutch cone penetrometers. For the pressuremeter test, see Ménard (1975) and Baguelin, et al. (1978). In a real
sense, then, these tests only give an *index* of the actual undrained shear
strength of the soil.

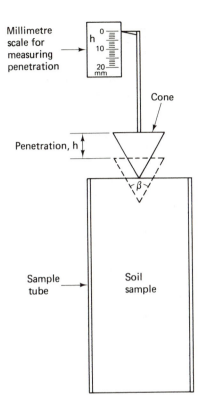

Mass (g)	Cone angle β (degrees)	Range of τ_f (kPa)
400	30	10–250
100	30	25–63
60	60	0.5–11
10	60	0.08–2

Fig. 11.48 Principle of the Swedish fall-cone test (SFC). The undrained shear strength is proportional to the mass of the cone and inversely proportional to the penetration squared. The test must be calibrated (after Hansbo, 1957). Shear strength ranges are tabulated for the four standard cones.

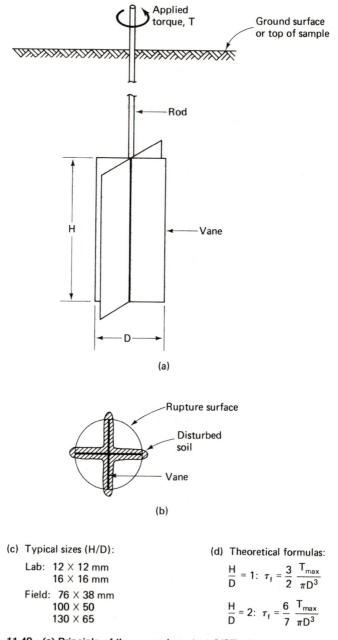

(a)

(b)

(c) Typical sizes (H/D):

    Lab: 12 × 12 mm
           16 × 16 mm
  Field: 76 × 38 mm
          100 × 50
          130 × 65

(d) Theoretical formulas:

$$\frac{H}{D} = 1: \quad \tau_f = \frac{3}{2}\frac{T_{max}}{\pi D^3}$$

$$\frac{H}{D} = 2: \quad \tau_f = \frac{6}{7}\frac{T_{max}}{\pi D^3}$$

Fig. 11.49 (a) Principle of the vane shear test (VST); (b) end view of the vane, showing the probable zone of disturbance and the rupture surface (after Cadling and Odenstad, 1950); (c) typical vane sizes; (d) theoretical formulas for τ_f.

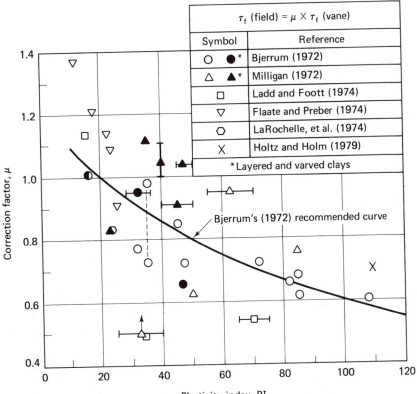

Fig. 11.50 Correction factor for the field vane test as a function of PI, based on embankment failures (after Ladd, 1975, and Ladd, et al., 1977).

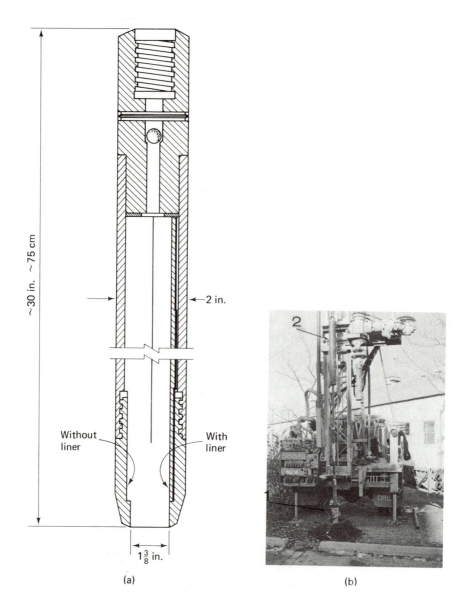

~30 in. ~75 cm

2 in.

Without liner

With liner

$1\frac{3}{8}$ in.

2

(a)

(b)

Fig. 11.51 Standard penetration test (SPT): (a) "split-spoon" sampler; (b) drill rig with sampler being inserted inside hollow stem auger at 1. The sleeve encloses the 63.5 kg hammer. Hammer in the raised position is shown at 2. (Drawing courtesy of Mobile Drilling Co., Indianapolis, Indiana. Photograph by W.D. Kovacs.)

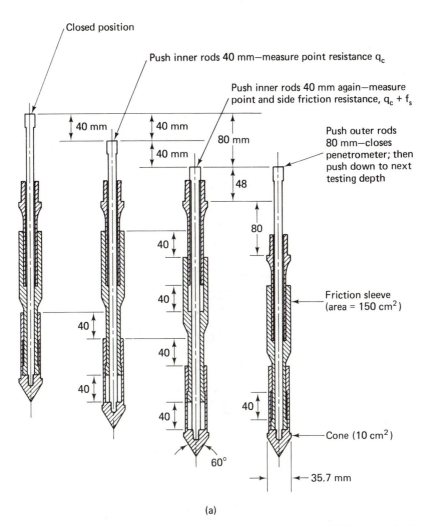

Closed position

Push inner rods 40 mm—measure point resistance q_c

Push inner rods 40 mm again—measure point and side friction resistance, $q_c + f_s$

40 mm 40 mm 80 mm 40 mm

Push outer rods 80 mm—closes penetrometer; then push down to next testing depth

48

80

40

40

Friction sleeve (area = 150 cm^2)

40

40

40

40

40

Cone (10 cm^2)

60°

35.7 mm

(a)

Fig. 11.52 Dutch cone penetrometer (CPT): (a) Begemann (1953) mechanical cone with friction sleeve.

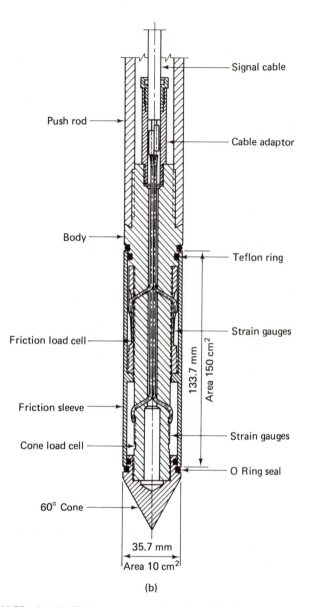

Signal cable

Push rod

Cable adaptor

Body

Teflon ring

Strain gauges

Friction load cell

133.7 mm   Area 150 cm^2

Friction sleeve

Cone load cell

Strain gauges

60° Cone

O Ring seal

35.7 mm

Area 10 cm^2

(b)

Fig. 11.52 (cont.) Dutch cone penetrometer: (b) Cross section of a modern electrical penetrometer with strain-gage load cells to measure both the point resistance and the sleeve friction (after Holden, 1974).

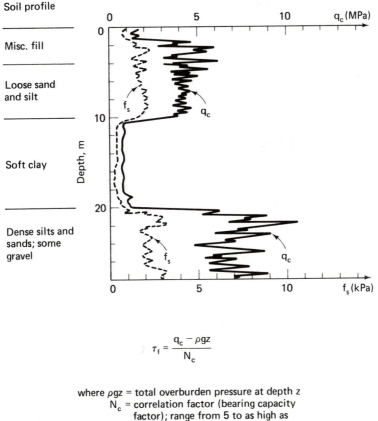

$$\tau_f = \frac{q_c - \rho gz}{N_c}$$

where ρgz = total overburden pressure at depth z
N_c = correlation factor (bearing capacity
factor); range from 5 to as high as
70, depending on the soil deposit

(c)

Fig. 11.52 (cont.) Dutch cone penetrometer: (c) Typical cone penetrometer test results correlated with the soil profile, and formula for calculating τ_f from cone penetrometer results.

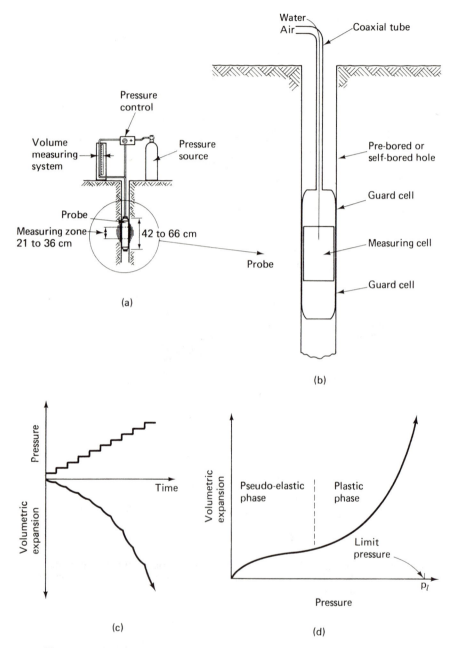

Fig. 11.53 Pressuremeter test (PMT): (a) Schematic diagram of probe and measuring system (after Mitchell and Gardner, 1975). (b) Detail of probe. Typical test results: (c) Pressure and volumetric expansion versus time, and (d) Volumetric expansion versus pressure (after Ménard, 1975).

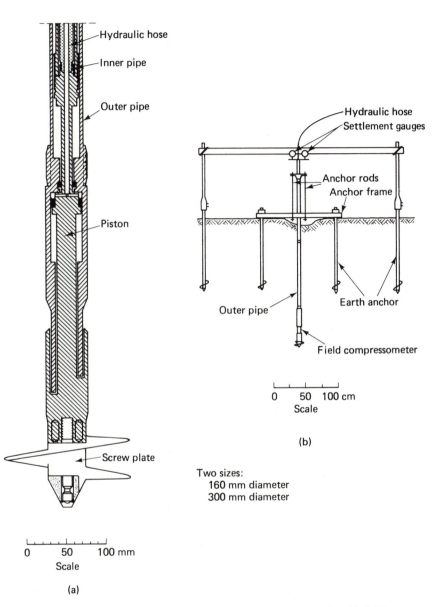

Fig. 11.54 Screw plate compressometer (SPC): (a) principle; (b) field setup (after Janbu and Senneset, 1973).

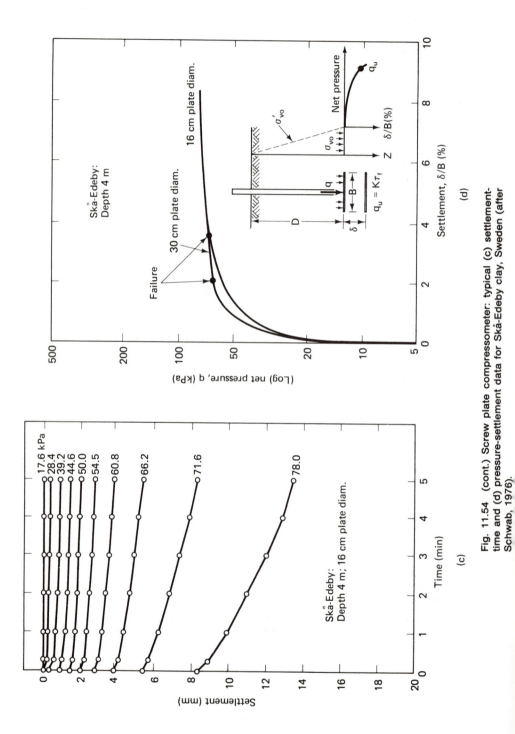

Fig. 11.54 (cont.) Screw plate compressometer: typical (c) settlement-time and (d) pressure-settlement data for Skå-Edeby clay, Sweden (after Schwab, 1976).

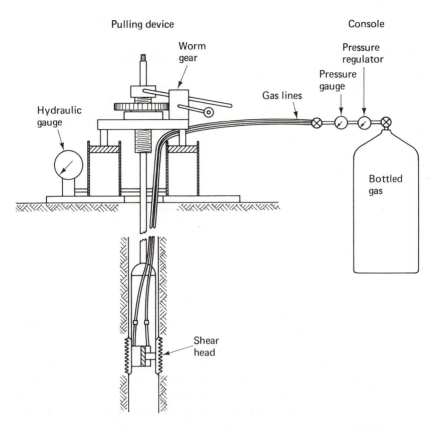

Fig. 11.55 The Iowa borehole shear (BST) device, showing the pressure source and instrumentation console, the pulling device, and the expanded shear head on the sides of a borehole (after Wineland, 1975).

11.9.11 Sensitivity

Earlier, in Sec. 2.7, we very generally defined *sensitivity* as the ratio of the undisturbed or natural strength of a clay to its remolded strength. *Strength* was left purposely vague. Now we can define sensitivity more precisely, at least within the precision limits of the strength measurements themselves. Usually, sensitivity is based on the unconfined compressive strength or the unconfined shear strength $\tau_f = c$, but the laboratory or field vane shear tests or the Swedish fall-cone test could also be used. *Sensitivity S_t* is therefore

$$S_t = \frac{\text{unconfined compressive strength (undisturbed)}}{\text{unconfined compressive strength (remolded)}}$$

$$= \frac{\tau_f(\text{undisturbed})}{\tau_f(\text{remolded})} \qquad (11\text{-}9)$$

It should be noted that the remolded strength determination must be at the *same* water content—the natural water content w_n—as the water content of the undisturbed specimen. Table 11-7 indicates the range of sensitivity values commonly used in the United States, where highly sensitive clays are rare. Sensitive clays exist in other parts of the world, especially eastern Canada and Scandinavia. Other sensitivity scales are available besides those listed in Table 11-7 (for example, Skempton and Northey, 1952; Bjerrum, 1954).

TABLE 11-7 Typical Values of Sensitivity

Condition	Range of S_t	
	U.S.	Sweden
Low sensitive	2–4	< 10
Medium sensitive	4–8	10–30
Highly sensitive	8–16	> 30
Quick	16	> 50
Extra quick	—	> 100
Greased lightning	—	

Figure 2.9 shows what happened to a sample of Leda clay from eastern Canada before and after remolding. Leda clays are often very stiff in their natural state. Their unconfined compressive strengths may be greater than 100 kPa, but their liquidity indices (Eq. 2-23), are often 2 or more. No wonder that their strengths are so low when they are thoroughly remolded! The sample shown in Fig. 2.9 had a sensitivity of about 1500 (Penner, 1963) which definitely qualifies it as extra quick (or even greased lightning!) according to Table 11-7. Note that with such clays, you have to use either a laboratory vane or fall-cone test to obtain the remolded τ_f (Eden and Kubota, 1962).

Correlations between sensitivity and liquidity index have been made by several researchers, as shown in Fig. 11.56.

11.9.12 Use of the Undrained (UU) Shear Strength in Engineering Practice

Like the CD and CU tests, the undrained or UU strength is applicable to certain critical design situations in engineering practice. These situations are where the engineering loading is assumed to take place so rapidly that there is no time for the induced excess pore water pressure to dissipate or for consolidation to occur during the loading period. We also assume that the change in total stress during construction does not affect the in situ undrained shear strength (Ladd, 1971b). Examples shown in Fig. 11.57 include the end of construction of embankment dams and foundations for

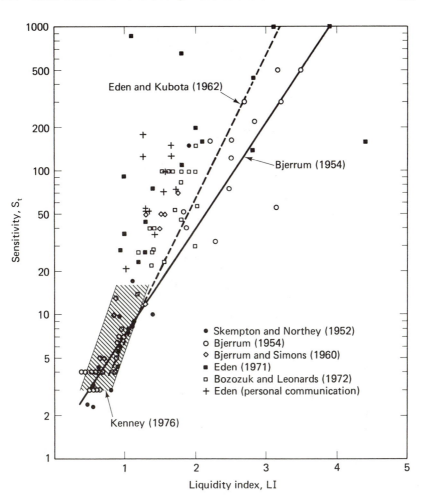

Fig. 11.56 The relationship between sensitivity and liquidity index for Scandinavian, British, Canadian, and some U.S. clays.

embankments, piles, and footings on normally consolidated clays. For these cases, often the most critical design condition is *immediately after* the application of the load (at the *end of construction*) when the induced pore pressure is the greatest but *before* consolidation has had time to take place. Once consolidation begins, the void ratio and the water content naturally decrease and the strength increases. So the embankment or foundation becomes increasingly *safer* with time.

One of the more useful ways to express the undrained shear strength is in terms of the τ_f/σ'_{vo} ratio for normally consolidated clays. Sometimes this is called the *c/p ratio*. In natural deposits of sedimentary clays the

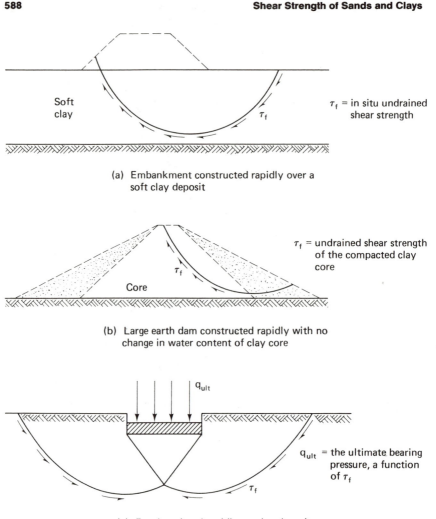

(a) Embankment constructed rapidly over a
soft clay deposit

(b) Large earth dam constructed rapidly with no
change in water content of clay core

(c) Footing placed rapidly on clay deposit

Fig. 11.57 Some examples of UU analyses for clay (after Ladd, 1971b).

undrained shear strength has been found to increase with depth, and thus
it is proportional to the increase in effective overburden stress with depth.
It was first observed by Skempton and Henkel (1953) and confirmed by
Bjerrum (1954) that the τ_f/σ'_{vo} ratio seemed to increase with increasing
plasticity index. Bjerrum's (1954) results are shown in Fig. 11.58 along with
those of several other researchers; in addition, several best-fit correlations
are also shown in the figure. There is a lot of scatter so Fig. 11.58 should
only be used with caution. However, as with Fig. 11.27, such correlations
are useful for preliminary estimates and for checking laboratory data.

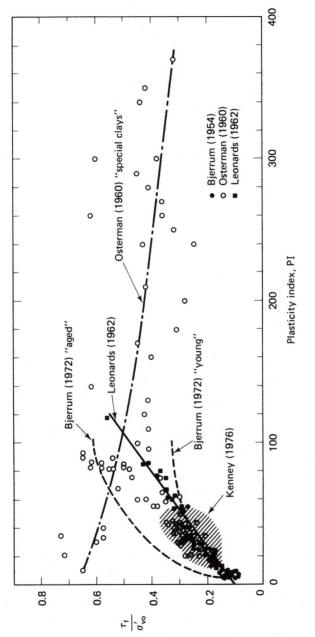

Fig. 11.58 Relationship between the ratio τ_f/σ'_{vo} and plasticity index for normally consolidated clays.

Kenney (1959) and Bjerrum and Simons (1960) presented some theoretical τ_f/σ'_{vo} ratios versus PI based on the correlations of Fig. 11.27, K_o, and the Skempton pore pressure parameter A (to be discussed in Sec. 11.11). These theoretical relationships tended to decrease rather than increase with PI, but the agreement was satisfactory for PI > 30. Kenney (1959) concluded that τ_f/σ'_{vo} was essentially independent of PI after all; rather, it probably depended on the geologic history of the clay.

Bjerrum and Simons (1960) also presented the relationship between τ_f/σ'_{vo} and liquidity index (LI) for some Norwegian marine clays, as shown in Fig. 11.59. As you know from Fig. 11.56, the quick clays are those with very high LI's. Therefore it appears that Norwegian quick clays have a τ_f/σ'_{vo} ratio of about 0.1 to 0.15.

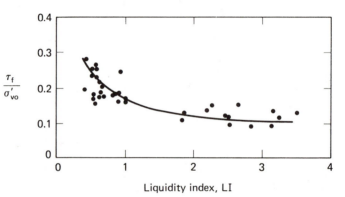

Fig. 11.59 Relationship between τ_f/σ'_{vo} and liquidity index for Norwegian clays (after Bjerrum and Simons, 1960).

You should be aware of the fact that the τ_f/σ'_{vo} ratio depends strongly on the total stress path. This point is discussed by Bjerrum (1972) and Ladd, et al. (1977), among others. In other words, you probably will obtain different values of τ_f/σ'_{vo}, depending on whether you run field vane tests, axial compression or axial extension triaxial tests, or direct simple shear tests.

Sometimes it is better to normalize the undrained shear strength with respect to the effective *consolidation* pressure σ'_{vc} or the preconsolidation stress σ'_p if the clay is slightly overconsolidated. For these soils the undrained strength is really controlled by the effective consolidation pressure rather than the existing effective overburden stress. Bjerrum (1972) hypothesized that the ratio between σ'_p and σ'_{vo} would vary with PI, as shown in Fig. 11.60a. So-called "young" clays are normally consolidated recent sediments, thus they haven't had time to be overconsolidated by any of the factors listed in Table 8-1. On the other hand, "aged" clays are slightly overconsolidated, and Bjerrum found that the amount of overconsolidation increased somewhat with the PI (Fig. 11.60b). The resulting effect on the strength was indicated by the dashed curves labeled "Bjerrum (1972)" in Fig. 11.58.

Recall from the discussion of the vane shear test that Bjerrum (1972) proposed a correction factor for the vane shear test based on a study of actual embankment failures (Fig. 11.50). For convenient reference, this figure is reproduced without all the data points as Fig. 11.60c.

Mesri (1975) discovered a very interesting relationship between all these observations. Combining Figs. 11.60a and 11.60b Mesri obtained Fig. 11.60d, τ_f/σ'_p versus PI, which shows essentially the same behavior for "aged" and "young" clays. Now apply Bjerrum's correction factor μ for the vane shear test to obtain the in situ strengths; the result is Fig. 11.60e. In other words, $(\tau_f/\sigma'_{vc})_{\text{field}}$ is almost a constant equal to 0.22 and independent of PI! There is great uncertainty in such a conclusion because of the scatter in the empirical relationships upon which it is based, and the relationships shown in Figs. 11.60d and 11.60e may be only a coincidence.

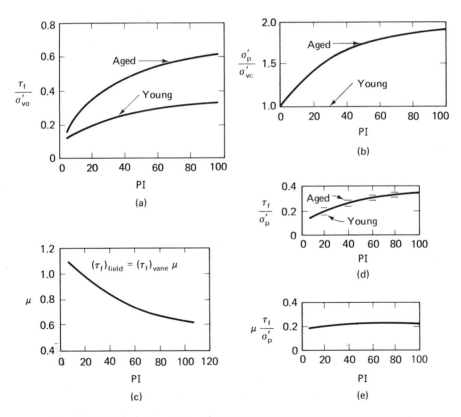

Fig. 11.60 (a) τ_f/σ'_{vo} and (b) σ'_p/σ'_{vo} for normally consolidated late glacial clays (after Bjerrum, 1972); (c) Bjerrum's (1972) correction factor for the vane shear test; (d) τ_f/σ'_p from (a) and (b); (e) $\mu(\tau_f/\sigma'_p)$ (after Mesri, 1975).

Still the possibility that in situ τ_f/σ_p' may well exist within a rather narrow range for soft sedimentary clays has tremendous practical implications (Ladd, et al., 1977).

In Chapter 8 we briefly mentioned that settlement analyses, to be complete, must also consider the *immediate* or *distortion settlement* of the structure. The procedures for calculating immediate settlement usually involve elastic theory, and one of the biggest problems is to determine or accurately estimate the elastic modulus for the soil. The obvious way would be to take the initial slope of the stress-strain curve, called the *tangent modulus*, as determined in the triaxial test. Or, because the stress-strain curves are so often curved, you could take the *secant modulus*, which is the slope of a straight line drawn from the origin to some predetermined stress level such as 50% of the maximum stress. These definitions are shown in Fig. 11.61. By the way, since the immediate settlement takes place before any consolidation can occur, the triaxial tests should be conducted *undrained*. Thus the modulus, however defined, is called the *undrained modulus* E_u.

However, as shown by many researchers, the undrained modulus is significantly affected by sample disturbance. Most of the time the disturbance tends to reduce the E_u, and thus you would tend to over-predict the immediate settlements in the field. Because of several other factors

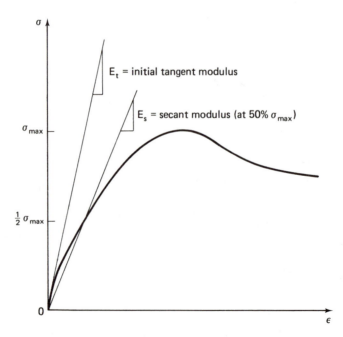

Fig. 11.61 Definitions of the initial tangent modulus and the secant modulus (usually defined at 50% of the maximum stress).

which affect the undrained modulus in laboratory tests (D'Appolonia, Poulos, and Ladd, 1971; Simons, 1974), field loading tests are sometimes used for important projects. Settlements are measured, and the modulus is backcalculated from elastic theory. Load tests have shown that stress level is one very important factor that strongly affects E_u. For example, large-scale loading tests carried out in Norway and Canada (Höeg, et al., 1969; Tavenas, et al., 1974) showed very little settlement since the load was applied rapidly until about one-half the failure load was reached. Then settlements started to accelerate as the load was increased. Thus the backcalculated E_u values were very dependent on the level of the shear stress applied by the surface load.

Because of all the problems with the laboratory determination of E_u and because large-scale field loading tests are expensive, it is common to assume that E_u is somehow related to the undrained shear strength. For example, Bjerrum (1972) said that the ratio E_u/τ_f ranges from 500 to 1500, with τ_f determined by the vane shear test. The lowest value is for highly plastic clays, where the applied load is large compared to the value of $\sigma_p' - \sigma_{vo}'$ (that is, the added stress to the foundation is relatively large). The higher value is for clays of low plasticity, where the added load is relatively small. D'Appolonia, Poulos, and Ladd (1971) reported an average E_u/τ_f of 1200 for load tests at 10 sites, but for the clays of higher plasticity the range was 80 to 400. Simons (1974) found published values ranged from 40 to 3000! These cases plus a few others we have taken from the literature are plotted versus PI in Fig. 11.62 for soft clays. Stiff fissured soils and glacial tills are not included. There is much scatter for PI < 50 but not much data for PI > 50. It seems reasonable to simply use Bjerrum's recommendation (E_u/τ_f of 500 to 1500) and the procedures developed by D'Appolonia, et al. (1971) for estimating immediate settlements of soft clays.

Another factor which strongly affects the undrained shear strength of clays is stress history. We mentioned this factor when we pointed out the difference in behavior between normally consolidated and overconsolidated clays (see, for example, Figs. 11.30 and 11.33). Let's first consider some data showing how the normalized undrained strength τ_f/σ_{vc}' varies with the overconsolidation ratio (OCR). These data are shown for six clays in Fig. 11.63. If you take the ratio of the τ_f/σ_{vc}' ratios, as shown in Fig. 11.64, all these soils fall into a rather narrow band, with only the varved clay somewhat lower. Ladd, et al. (1977) showed that this ratio of ratios is approximately equal to the OCR to the 0.8 power, or

$$\frac{\left(\tau_f/\sigma_{vc}'\right)_{oc}}{\left(\tau_f/\sigma_{vc}'\right)_{nc}} = (\text{OCR})^{0.8} \qquad (11\text{-}10)$$

Relationships such as this can be useful for comparing strength data from different sites or even from the same site.

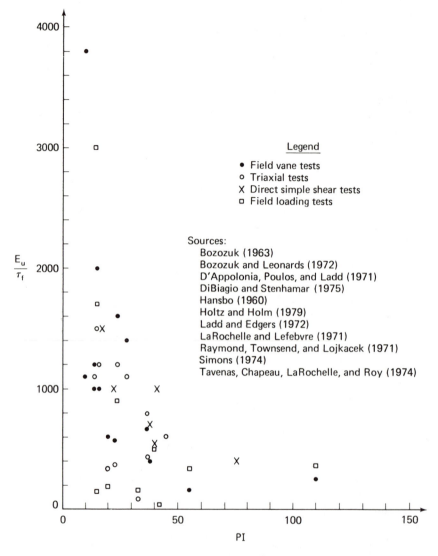

Fig. 11.62 The ratio E_u/τ_f versus plasticity index, as reported by several authors.

Ladd, et al. (1977) also showed how E_u/τ_f varies with OCR, but the relationship is not so simple because, as we mentioned earlier, E_u/τ_f depends so strongly on the level of shear stress. In general, however, it decreases with increasing OCR for a given stress level (Fig. 11.65).

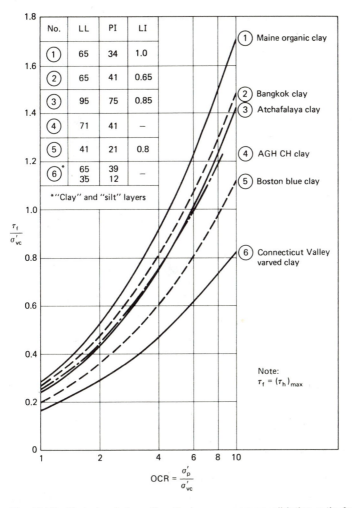

No.	LL	PI	LI
(1)	65	34	1.0
(2)	65	41	0.65
(3)	95	75	0.85
(4)	71	41	–
(5)	41	21	0.8
(6)*	65 35	39 12	–

*"Clay" and "silt" layers

(1) Maine organic clay

(2) Bangkok clay
(3) Atchafalaya clay

(4) AGH CH clay

(5) Boston blue clay

(6) Connecticut Valley varved clay

Note:
$\tau_f = (\tau_h)_{max}$

$\frac{\tau_f}{\sigma'_{vc}}$

$OCR = \frac{\sigma'_p}{\sigma'_{vc}}$

Fig. 11.63 Undrained strength ratio versus overconsolidation ratio from direct-simple shear tests on six clays (after Ladd and Edgers, 1972, and Ladd, et al., 1977).

11.9.13 Special Problems of the Shear Strength of Cohesive Soils

All of the previous discussion has been limited to "well-behaved" cohesive soils. These are the relatively homogeneous marine and fresh water sedimentary clays of low to medium sensitivity which are normally consolidated or only slightly overconsolidated. As you might surmise, there are

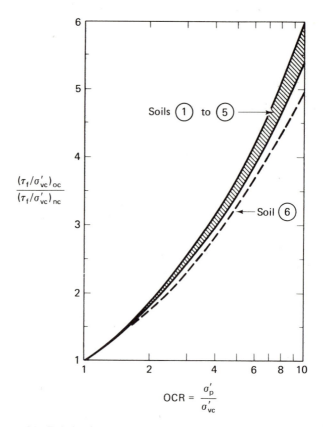

Fig. 11.64   Relative increase in undrained strength ratio with OCR from direct simple shear tests (soils 1 through 6 are identified in Fig. 11.63) (after Ladd, et al., 1977).

many cohesive soil deposits throughout the world that are not "well-behaved." In fact, such soils are probably the rule rather than the exception. In this list are included stiff fissured clays, peats and other organic soils, varved and layered soils, highly sensitive clays, and residual and tropical soils. Often these problem soils possess fissures and other defects that make them difficult to sample and test in the laboratory. They may be very heterogeneous and highly variable even within the confines of a small building site. In situ testing techniques described earlier are a good way to obtain some subsurface information as well as an idea of the statistical spread or variability of the material at the site. In addition to the usual geotechnical literature, two European conferences have been concerned with problem soils—the Geotechnical Conference (Oslo, 1967) and the Seventh European Conference on Soil Mechanics and Foundation Engineering held in Brighton, England, in 1979. The latter conference had as

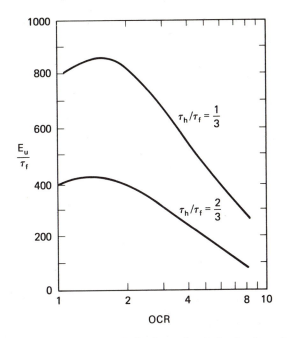

Fig. 11.65   Effect of OCR on E_u/τ_f, from direct simple shear tests on Bangkok clay. (After Ladd and Edgers, 1972, and Ladd, et al., 1977).

its main theme the determination of design parameters for a wide variety of soil conditions. Sometimes the Asian and African regional conferences have sessions on problems associated with residual and tropical soils.

There are some other factors that strongly affect the shear strength of clay soils that are not related to a specific geologic deposit or region of the world. We commonly assume that clays are isotropic—that is, their strength is the same in all directions. It has been known for many years that the undrained strength of many clays is directionally dependent (for example, Hvorslev, 1960). Recently, even c' and ϕ' have been found to be *intrinsically anisotropic* (for example, Saada and Bianchini, 1965). There is also evidence of an *apparent anisotropy* due to the stress system, both during consolidation and during shear. Anisotropy is important because, for stability analyses, the variation of the shear strength with direction along a potential sliding surface significantly affects the calculated safety factor. This variation is shown by Bjerrum (1972), and it is one of the details included in the correction factor to the vane strength (Fig. 11.50).

Another factor included in Bjerrum's vane shear correction factor is the *strain rate effect*. Taylor (1948) showed that the undrained strength of a remolded Boston blue clay increased about 10% per log cycle of time increase in speed of shear (Fig. 11.66a). Bjerrum (1972) showed about the

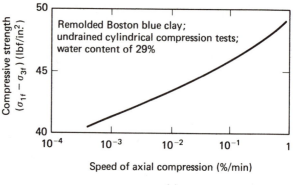

(a)

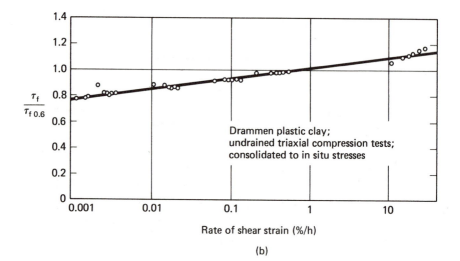

(b)

Fig. 11.66 Effect of rate of loading on the undrained strength of (a) Boston blue clay (after Taylor, 1948); and (b) Drammen, Norway, plastic clay. The strength ratio in these latter tests is with respect to the strength at the NGI standard rate of 0.6% per h (after Bjerrum, 1972).

same increase in CU tests on a Norwegian plastic clay (Fig. 11.66b). Difference between the rate of loading in the laboratory and in the field can sharply affect the undrained shear strength. Ladd, et al. (1977) also discussed this point.

    Finally, we mention briefly the problem of the *residual strength* of soils. When stiff overconsolidated clays have work-softening stress-strain curves like those shown in Fig. 11.24, the ultimate strength at large strains is called the *residual strength* of the soil (Skempton, 1964; 1977). A torsional or ring shear device such as shown in Fig. 10.15a is used to determine the residual strength.

11.10 PORE PRESSURE PARAMETERS

It should now be apparent that when saturated soils are loaded, pore water pressures will develop. In the case of one-dimensional loadings (Chapter 8), the induced pore water pressure is initially *equal* to the magnitude of the applied vertical stress. In three-dimensional or triaxial-type loadings, pore water pressures are also induced, but the actual magnitude will depend on the soil type and its stress history. Of course the rate of loading as well as the soil type determines whether we have drained or undrained loading.

It is often necessary in engineering practice to be able to estimate just how much excess pore water pressure develops in undrained loading due to a given set of stress changes. Note that these stress changes are in terms of *total stresses*, and they can be either hydrostatic (equal all-around) or non-hydrostatic (shear). Because we are interested in how the pore water pressure Δu responds to these changes in total stress, $\Delta\sigma_1$, $\Delta\sigma_2$, and $\Delta\sigma_3$, it is convenient to express these changes in terms of *pore pressure coefficients* or *parameters*, which were first introduced in 1954 by Prof. A. W. Skempton of Imperial College in England.

In general, we can visualize the soil mass as a compressible soil skeleton with air and water in the voids. If we increase the principal stresses acting on a soil element, as in the triaxial test for example, then we will obtain a decrease in volume of the element and an increase in pore pressure. Refer again to Fig. 11.38, which represents the stress conditions in the UU test. Consider what happens when we apply the hydrostatic cell pressure σ_c and prevent any drainage from occurring. If the soil is 100% saturated, then we will obtain a change in pore pressure $\Delta u(= \Delta u_c$ in Fig. 11.38), numerically equal to the change in cell pressure $\Delta\sigma_c(= \sigma_c$ in Fig. 11.38) we just applied. In other words, the ratio $\Delta u/\Delta\sigma_c$ equals 1. If the soil were less than 100% saturated, then the ratio of the induced Δu due to the increase in cell pressure $\Delta\sigma_c$ would be less than 1. It can be shown (see Appendix B-3 for details) that this ratio for the ordinary triaxial test is

$$\frac{\Delta u}{\Delta\sigma_3} = \frac{1}{1 + \dfrac{nC_v}{C_{sk}}} = B \tag{11-11}$$

where $n =$ porosity,
  $C_v =$ compressibility of the voids, and
  $C_{sk} =$ compressibility of the soil skeleton.

For convenience, Prof. Skempton called this ratio B. The pore pressure parameter B expresses the increase in pore pressure in undrained loading due to the increase in hydrostatic or cell pressure.

If the soil is completely saturated with water, then $C_v = C_w$, and for most soils $C_w/C_{sk} \to 0$ since the compressibility of water C_w is so small compared with the compressibility of the soil skeleton. Therefore, for saturated soils, $B = 1$. If the soil is dry, then the ratio of C_v/C_{sk} approaches infinity since the compressibility of air is vastly greater than the soil structure; hence $B = 0$ for dry soils. Partially saturated soils have values of B ranging between 0 and 1. Because in general both C_v and C_{sk} are nonlinear for soils, the relationship between B and the degree of saturation S is also nonlinear, as shown in Fig. 11.67. This relationship will depend on the soil type and stress level, and the exact relationship will have to be determined experimentally.

Equation 11-11 is very useful in the triaxial testing to determine if the test specimen is saturated. The pore pressure response to a small change in cell pressure is measured, and B is calculated. If $B = 1$ or nearly so, then for soft clays the test specimen is saturated. However if the soil skeleton is relatively stiff, then it is possible to have B less than 1 and still have $S = 100\%$ (see Table 11-8). This condition is possible because as C_{sk} gets smaller (a more rigid soil skeleton), the ratio C_w/C_{sk} becomes larger; thus B decreases. Wissa (1969) and Black and Lee (1973) suggest procedures to increase saturation and thereby increase the reliability of pore pressure measurements in undrained tests.

Now let's apply a stress difference or a shear stress to our soil sample (see Fig. 11.38 again for the UU test). In this case, a pore pressure Δu is induced in the specimen due to the change in stress difference $\Delta \sigma = \Delta \sigma_1 - \Delta \sigma_3$, or we can write, as Prof. Skempton did for triaxial compression

TABLE 11-8 Theoretical B-Values for Different Soils at Complete or Nearly Complete Saturation*

Soil Type	$S = 100\%$	$S = 99\%$
Soft, normally consolidated clays	0.9998	0.986
Compacted silts and clays; lightly over consolidated clays	0.9988	0.930
Overconsolidated stiff clays; sands at most densities	0.9877	0.51
Very dense sands; very stiff clays at high confining pressures	0.9130	0.10

*After Black and Lee (1973).

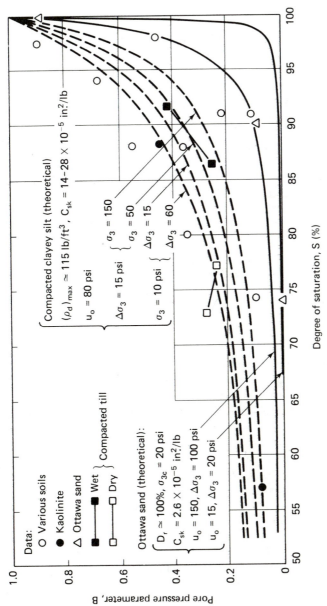

Fig. 11.67 The pore pressure parameter *B* as a function of the degree of saturation for several soils (after Black and Lee, 1973).

601

conditions ($\Delta\sigma_2 = \Delta\sigma_3$),

$$\Delta u = B\tfrac{1}{3}(\Delta\sigma_1 - \Delta\sigma_3) \qquad (11\text{-}12)$$

if the soil skeleton is *elastic*. Since soils in general are not elastic materials, the coefficient for the principal stress difference term is not $1/3$. So Skempton used instead the symbol A for this coefficient. Now we can combine Eqs. 11-11 and 11-12 to take into account the two components of pore pressure: (1) that due to change in average or mean stress and (2) that due to change in shear stress, or

$$\Delta u = B\left[\Delta\sigma_3 + A(\Delta\sigma_1 - \Delta\sigma_3)\right] \qquad (11\text{-}13)$$

Equation 11-13 is the well-known Skempton equation for relating the induced pore pressure to the changes in *total* stress in undrained loading. If $B = 1$ and $S = 100\%$, then we normally write Eq. 11-13 as

$$\Delta u = \Delta\sigma_3 + A(\Delta\sigma_1 - \Delta\sigma_3) \qquad (11\text{-}14)$$

Sometimes it is convenient to write Eq. 11-14 as

$$\Delta u = B\,\Delta\sigma_3 + \overline{A}(\Delta\sigma_1 - \Delta\sigma_3) \qquad (11\text{-}15)$$

where $\overline{A} = BA$.

Equations 11-13 through 11-15 are derived in detail in Appendix B-3. There we show that these equations are true for both triaxial compression ($\Delta\sigma_2 = \Delta\sigma_3$) and triaxial extension ($\Delta\sigma_2 = \Delta\sigma_1$) conditions, although the specific value of A is dependent on the stress path, as discussed in Sec. 11.12.

Like the parameter B, the parameter A also is not a constant; it must be determined for each soil and stress path. The parameter A is very dependent on the strain, the magnitude of σ_2, the overconsolidation ratio, anisotropy, and for natural clays tested in the laboratory, on sample disturbance. Table 11-9 relates the type of clay to different values of the A parameter at failure, A_f in triaxial compression. Of course A can be calculated for the stress conditions at any strain up to failure, as well as at failure.

The Skempton pore pressure coefficients are most useful in engineering practice since they enable us to predict the induced pore pressure if we know or can estimate the change in the total stresses. In the field, the Skempton equations are used, for example, when we want to estimate the pore pressure response during undrained loadings that might be applied by

TABLE 11-9 Values of A_f for Various Soil Types*

Type of Clay	A_f
Highly sensitive clays	$+\frac{3}{4}$ to $+1\frac{1}{2}$
Normally consolidated clays	$+\frac{1}{2}$ to $+1$
Compacted sandy clays	$+\frac{1}{4}$ to $+\frac{3}{4}$
Lightly overconsolidated clays	0 to $+\frac{1}{2}$
Compacted clay-gravels	$-\frac{1}{4}$ to $+\frac{1}{4}$
Heavily overconsolidated clays	$-\frac{1}{2}$ to 0

*After Skempton (1954).

a highway embankment constructed on a very soft clay foundation. Typically, the embankment is constructed more rapidly than the excess pore water pressure can dissipate, and thus we assume that undrained conditions apply. The increase in excess pore pressure can result in instability if the pore pressure gets too high. Consequently, it is important to be able to estimate just how high the pore pressures are likely to get and thereby obtain some idea of how close to failure the embankment might be. If it is too high, stage construction might be utilized; then field monitoring of the pore pressures would be advisable. Skempton's parameters have also been used for the design and construction control of compacted earthfill dams.

EXAMPLE 11.14

Given:

The CU test of Example 11.11.

Required:

A_f.

Solution:

Use Eq. 11-13. Since pore pressures were measured, the specimen must have been saturated. Thus assume $B = 1$. So A at failure is

$$A_f = \frac{\Delta u - \Delta\sigma_3}{\Delta\sigma_1 - \Delta\sigma_3}$$

In an ordinary triaxial compression test, $\Delta\sigma_3 = 0$ since the cell pressure is held constant throughout the test (see Fig. 11.29). From Example 11.11, $\Delta\sigma_{1f} = (\sigma_1 - \sigma_3)_f = 100$ kPa and $\Delta u_f = 88$ kPa. Therefore

$$A_f = \frac{88}{100} = 0.88$$

From Table 11-9 you can see that the clay was probably somewhat sensitive.

As shown by Law and Holtz (1978) and in Appendix B-3, where rotation of principal stresses occurs, it is better to define the pore pressure parameter A in terms of principal stress increments which are independent of the initial stress system. If this is done, then the equations for A for each of the common triaxial stress paths (discussed in Sec. 11.12) are

$$A_{ac} = \frac{\Delta u}{\Delta\sigma_v} \tag{11-16}$$

$$A_{le} = 1 - \frac{\Delta u}{\Delta\sigma_h} \tag{11-17}$$

$$A_{ae} = 1 - \frac{\Delta u}{\Delta\sigma_v} \tag{11-18}$$

$$A_{lc} = \frac{\Delta u}{\Delta\sigma_h} \tag{11-19}$$

It is also shown in Appendix B-3 that

$$A_{ac} = A_{le} \tag{11-20}$$

and

$$A_{ae} = A_{lc} \tag{11-21}$$

You will find these equations useful in Sec. 11.12 (and for the problems at the end of this chapter).

A more general pore pressure equation was proposed by Henkel (1960) to take into account the effect of the intermediate principal stress. It is

$$\Delta u = B(\Delta\sigma_{\text{oct}} + a\,\Delta\tau_{\text{oct}}) \tag{11-22}$$

where

$$\sigma_{oct} = \tfrac{1}{3}(\sigma_1 + \sigma_2 + \sigma_3) \qquad (11\text{-}23)$$

$$\tau_{oct} = \tfrac{1}{3}\sqrt{(\sigma_1 - \sigma_2)^2 + (\sigma_2 - \sigma_3)^2 + (\sigma_3 - \sigma_1)^2} \qquad (11\text{-}24)$$

and a is the *Henkel pore pressure parameter*. Sometimes the Henkel parameter is denoted by the symbol α, and sometimes $a = 3\alpha$. Equation 11-22 is derived in Appendix B-3. Also in B-3, the equations for getting the equivalent Skempton A from Henkel's a parameter for triaxial compression and extension conditions are developed. These relationships are for triaxial compression (AC and LE) conditions

$$A = \frac{1}{3} + a\frac{\sqrt{2}}{3} \qquad (11\text{-}25)$$

For triaxial extension (AE and LC) conditions

$$A = \frac{2}{3} + a\frac{\sqrt{2}}{3} \qquad (11\text{-}26)$$

These equations mean, of course, that $a = 0$ for elastic materials (since $A = \tfrac{1}{3}$ in triaxial compression and $A = \tfrac{2}{3}$ in triaxial extension).

If you have some idea of what the intermediate principal stress is in the field, then you probably should use Eqs. 11-22 through 11-24 to estimate the in situ pore pressures. It is not easy to predict the field pore pressures from laboratory test results, primarily because the pore pressure parameters are very sensitive to sample disturbance. Höeg, et al. (1969), D'Appolonia, et al. (1971), and Leroueil, et al. (1978a and b) provide methods for estimating pore pressures under embankments on soft clays.

11.11 THE COEFFICIENT OF EARTH PRESSURE AT REST FOR CLAYS

As is true for sands, a knowledge of the coefficient of earth pressure at rest, K_o, for a clay deposit is often very important for the design of earth-retaining structures, excavations, and some foundations. In Sec. 11.7, we indicated some typical values of K_o for sands. We said that K_o was empirically related to ϕ' (Eq. 11-6 and Fig. 11.14), and we also mentioned that the coefficient for overconsolidated sand deposits is greater than for normally consolidated sands (Eq. 11-7).

Correlations between K_o and ϕ' have been made for clays by Brooker and Ireland (1965) and others. Their data for normally consolidated clays

Remolded	Undisturbed	Reference
○		Brooker and Ireland (1965)
□		R. Ladd (1965)
⊙	⬢	Bishop (1958)
	◆	Simons (1958)
	▲	Campanella and Vaid (1972)
◎		Compiled by Wroth (1972)
✳		Abdelhamid and Krizek (1976)

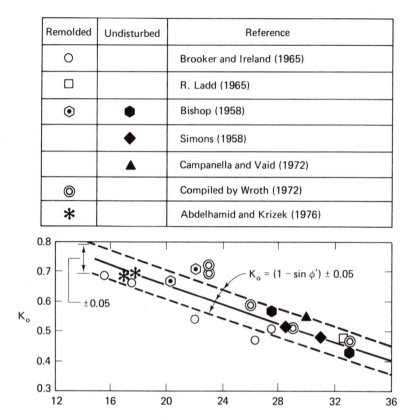

Fig. 11.68 K_o versus ϕ' for normally consolidated clays (after Ladd, et al., 1977).

are shown in Fig. 11.68. Brooker and Ireland (1965) also found a tendency for the normally consolidated K_o to increase with plasticity index. Massarsch (1979) has collected the results from 12 investigations, including the compilation by Ladd, et al. (1977), and they are shown in Fig. 11.69. The intercept of the best-fit line of Fig. 11.69 is very close to the average of K_o for sands shown in Fig. 11.14.

The effect of increasing the overburden stress and subsequent unloading on σ'_h and K_o is shown in Figs. 11.70a and b, respectively. During sedimentation, the effective horizontal stress σ'_h increases in proportion to the increase in effective vertical stress, so K_o is constant. If unloading occurs because of erosion, for example, then there is a hysteresis effect, and the value of K_o increases. Depending on how much unloading actually takes place, it is possible for the lateral stresses to *approach* a state of

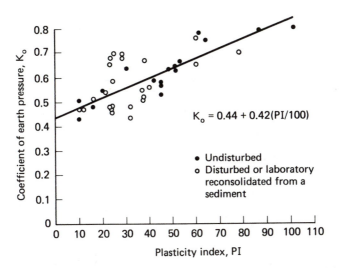

$$K_o = 0.44 + 0.42(PI/100)$$

• Undisturbed
○ Disturbed or laboratory
reconsolidated from a
sediment

Fig. 11.69 Correlation between K_o from laboratory tests and plasticity index PI (after Massarsch, 1979).

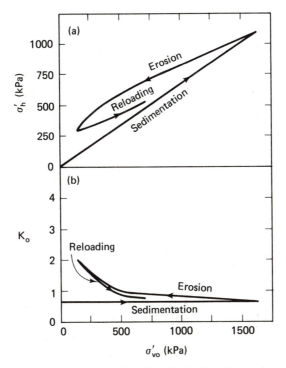

Fig. 11.70 Relationships showing the effect of a changing overburden stress during sedimentation, erosion, and reloading on (a) horizontal stress σ'_h and (b) coefficient of earth pressure at rest, K_o (after Morgenstern and Einstein, 1970).

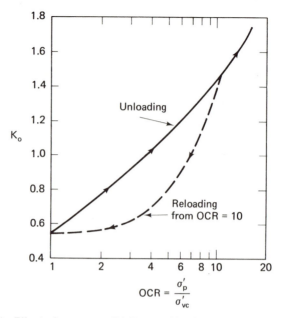

Fig. 11.71 Effect of overconsolidation on K_o of a sensitive clay during unloading and reloading. The data by Campanella and Vaid (1972) was replotted by Ladd, et al. (1977).

failure;* that is, the ratio σ_h'/σ_{vo}' could be 3.0 or 3.5, which corresponds to $\phi' = 30°$ or $35°$ (Eq. 10-14). If there is subsequent reloading, then the K_o tends to decrease, as shown in Fig. 11.70b.

The effect of overconsolidation on the K_o of a sensitive clay is shown in Fig. 11.71. Again, there is some hysteresis when the clay is rebounded from a high OCR. There is limited evidence that the relationship between K_o and OCR depends to some extent on the plasticity of the clay (Fig. 11.72). Ladd, et al. (1977) also determined the exponent h in Eq. 11-7 for several clays during unloading and recompression. For clays with a PI of about 20, a value of $h = 0.4$ is reasonable. Then h decreases slightly as PI increases, with the lowest value of $h = 0.32$ at PI = 80. These values of h are somewhat lower than those for sands (Sec. 11.7). Keep in mind too that all these data are for laboratory consolidated samples. Field behavior is much more erratic, as shown by Massarsch, et al. (1975, Fig. 18). These authors as well as Tavenas, et al. (1975) and Wroth (1975) describe techniques for estimating the K_o in situ in deposits of soft clays. Wroth (1975) also discusses the effects of erosion and a fluctuating ground water

*In terms of lateral earth pressures, this is called a *passive state of failure*, and the stress ratio K_P is called the *coefficient of passive earth pressure*.

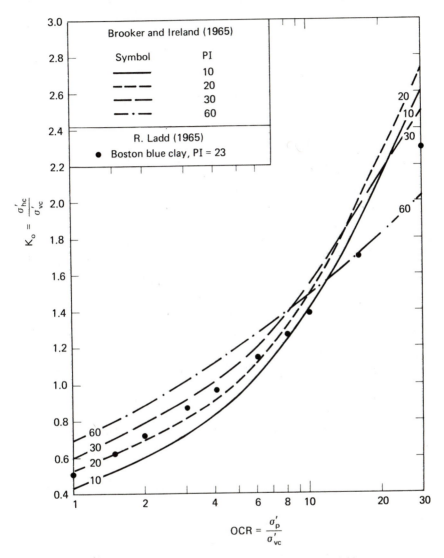

Fig. 11.72 K_o versus OCR for soils of different plasticities. The data by Brooker and Ireland (1965, Fig. 11) was re-plotted by Ladd (1971a).

table on the variation of K_o with depth. Generally, the upper few metres of a soft clay deposit are overconsolidated (the *dry crust*) and K_o can be quite high. Then it will decrease with depth as the OCR decreases, until it is equal to the normally consolidated value when OCR = 1.

11.12 STRESS PATHS DURING UNDRAINED LOADING—NORMALLY CONSOLIDATED CLAYS

We show examples of stress paths for undrained loading of normally consolidated clays in Figs. 10.24, 10.26, 11.34a, and 11.43. Undrained stress paths for overconsolidated clays are shown in Figs. 10.25 and 11.34b. From our comments concerning those figures you should now understand why these stress paths have the shapes they do have. The stress paths we showed for undrained shear were for the most common type of triaxial test used in engineering practice, the axial compression (AC) test. Most of the time, the initial consolidation stresses are *hydrostatic* ($K_o = 1$) because laboratory procedures are simpler. However, a better model for in situ stress conditions would be *non-hydrostatic* consolidation; that is, the axial stress would be different than the cell pressure ($K_o \neq 1$). As we mentioned in Sec. 10.6, there are stress paths other than axial compression that model real engineering design situations. Some of these are shown in Fig. 11.73, along with their laboratory model. Axial compression (AC) models foundation loading such as from an embankment or footing. Lateral extension (LE) models the active earth pressure conditions behind retaining walls. Axial extension (AE) models unloading situations like excavations, and lateral compression (LC) models passive earth pressure conditions such as might occur around an earth anchor.

If you think about it, the ordinary triaxial test is not the best model for the design conditions illustrated in Fig. 11.73. It would be all right for cases (a) and (c) if the foundation or excavation were circular in shape (for example, an oil tank, missile silo, or reactor pit). The more usual case is where one dimension (perpendicular to the page in Fig. 11.73) is very long compared to the others. This is the case for *plane strain*. Examples are long embankments, strip footings, and long retaining walls. In these cases, strictly speaking, the shear strengths should be determined by using plane strain tests (Fig. 10.14b). The laboratory models of Fig. 11.73 can also apply to stress conditions in this test just as well as in the triaxial test. Since the plane strain test is more complicated in several respects than the triaxial test, it is not often used in engineering practice. Triaxial strengths are still commonly obtained for design problems that are obviously plane strain.

Field Situation	Laboratory Model

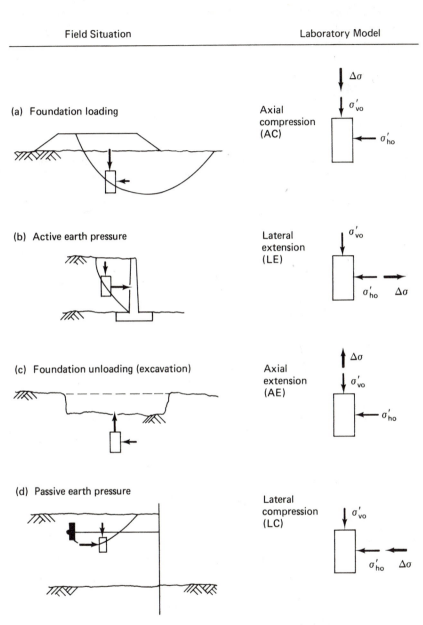

(a) Foundation loading

Axial compression (AC)

$\Delta\sigma$

σ'_{vo}

σ'_{ho}

(b) Active earth pressure

Lateral extension (LE)

σ'_{vo}

σ'_{ho} $\Delta\sigma$

(c) Foundation unloading (excavation)

Axial extension (AE)

$\Delta\sigma$

σ'_{vo}

σ'_{ho}

(d) Passive earth pressure

Lateral compression (LC)

σ'_{vo}

σ'_{ho} $\Delta\sigma$

Fig. 11.73 Some common field stability situations along with their laboratory model.

It is important that you know how to make the computations necessary to plot undrained stress paths; the procedures for doing so are illustrated by the following examples.

EXAMPLE 11.15

Given:

The σ-ϵ and u-ϵ data of Fig. Ex. 11.15a were recorded when the normally consolidated clay of Ex. 11.11 was tested in axial compression.

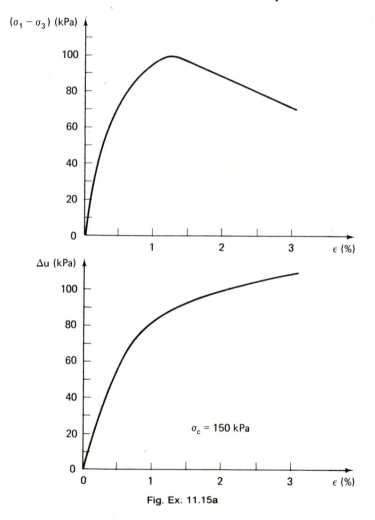

Fig. Ex. 11.15a

Required:

Draw the total and effective stress paths for this test. Determine the Mohr-Coulomb strength parameters.

Solution:

Using Eqs. 10-18 and 10-19, we have to determine p, p', q, and q' for several strains in order to plot the stress paths. Usually five or six points are sufficient. Sometimes, to keep things in order, a table is helpful. Then just fill in the appropriate columns. It may also be helpful to know that

$$\sigma_1 + \sigma_3 = (\sigma_1 - \sigma_3) + 2\sigma_3 \tag{11-27}$$

and

$$\frac{\sigma_1 + \sigma_3}{2} = \frac{\sigma_1 - \sigma_3}{2} + \sigma_3 \tag{11-28}$$

Also, since $\sigma' = \sigma - u$, $p' = p - u$. And finally,

$$\frac{\sigma_1' + \sigma_3'}{2} = \frac{\sigma_1 - \sigma_3}{2} + \sigma_3' \tag{11-29}$$

because $(\sigma_1' - \sigma_3') = (\sigma_1 - \sigma_3)$.

Now just choose the values of $(\sigma_1 - \sigma_3)$ and Δu at several convenient strains, and fill in the table (Table Ex. 11.15) by using the above equations. Note that σ_3 in Example 11.11 was 150 kPa.

TABLE EX. 11.15

	ϵ (%)	$\sigma_1 - \sigma_3$ (kPa)	Δu (kPa)	σ_3' (kPa)	$q = \dfrac{\sigma_1 - \sigma_3}{2}$ (kPa)	$p = \dfrac{\sigma_1 + \sigma_3}{2}$ (kPa)	$p' = \dfrac{\sigma_1' + \sigma_3'}{2}$ (kPa)
	$\frac{1}{4}$	49	35	115	24.5	174.5	139.5
	$\frac{1}{2}$	73	57	93	36.5	186.5	129.5
	$\frac{3}{4}$	86	72	78	43.0	193.0	121.0
	1	94	79.5	68	47.0	197.0	115.0
Failure	$1\frac{1}{4}$	100	88	62	50.0	200.0	112.0
	$1\frac{1}{2}$	96	92	58	48.0	198.0	106.0
	2	89	99	51	44.5	194.5	95.5

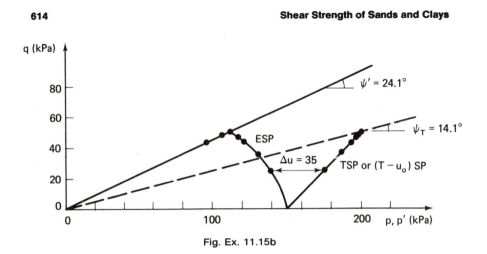

Fig. Ex. 11.15b

Total and effective stress paths are shown in Fig. Ex. 11.15b. The failure lines are also drawn, assuming $a' = a = 0$.

Since $\psi' = 24.1°,$ $\phi' = 26.6°$

and

$\psi_T = 14.1°,$ $\phi_T = 14.5°$

Note that the problem could be solved graphically by plotting the TSP first, then scaling off the corresponding Δu values horizontally to the *left* of the TSP; one point done this way is shown in Fig. Ex. 11.15b.

EXAMPLE 11.16

Given:

A long embankment shown in Fig. Ex. 11.16a is to be constructed rapidly on a deposit of soft organic silty clay in northern Sweden. The soil profile and properties are also shown in Fig. Ex. 11.16a. Assume $K_o = 0.6$. Also assume A before failure is about 0.35; at failure, $A_f = 0.5$ (after Holtz and Holm, 1979).

Required:

Determine the TSP, $(T - u_o)$SP, and ESP for a typical element 5 m below the centerline of the embankment.

Solution:

First, calculate the initial stress conditions for the element. Use Eqs. 7-13,

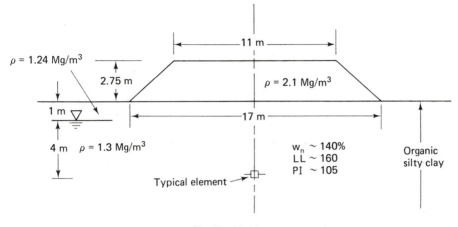

Fig. Ex. 11.16a

7-14, and 7-15.

$$\sigma_{vo} = 1.24(9.81)(1) + 1.30(9.81)(4) = 63 \text{ kPa}$$
$$u_o = 1.0(9.81)(4) \qquad\qquad\quad = 39 \text{ kPa}$$
$$\sigma'_{vo} = \sigma_{vo} - u_o \qquad\qquad\qquad = 24 \text{ kPa}$$
$$\sigma'_{ho} = 0.6\sigma'_{vo}(K_o = 0.6) \qquad\quad = 14 \text{ kPa}$$
$$\sigma_{ho} = \sigma'_{ho} + u_o \qquad\qquad\qquad = 53 \text{ kPa}$$

Second, calculate the $\Delta\sigma$ due to the embankment.

$$\Delta\sigma \text{ at the surface} = 2.1(9.81)(2.75) = 57 \text{ kPa}$$

σ_z at -5 m; use Fig. 8.23,

$$I = 0.45 \times 2 = 0.9$$
$$\sigma_z = 0.9 \times 57 = 51 \text{ kPa}$$

This is $\Delta\sigma_v$ on the typical element.

To determine the increase in horizontal stress $\Delta\sigma_h$, there are equations and some charts available for a limited number of geometries (see, for example, Poulos and Davis, 1974). In this case, assume the increase in horizontal stress is one-third of the increase in vertical stress.

$$\Delta\sigma_h = 0.33(51) = 17 \text{ kPa}$$

Next, use Eqs. 10-18 and 10-19 to determine q, p, and p' for both initial and final conditions. Don't forget the conditions for the $(T - u_o)$ SP. To get the final effective stresses, we need to estimate the induced pore pressures. Use the pore pressure parameter information given. Assume initially that the soil is not stressed to failure; so $A = 0.35$. $B = 1$ below the water table. Use Eq. 11-14.

$$\Delta u = \Delta\sigma_3 + A(\Delta\sigma_1 - \Delta\sigma_3) = 17 + 0.35(51 - 17) = 29 \text{ kPa}$$

If the embankment was overstressing the underlying soil, then the induced Δu would be 34 kPa (because $A = 0.5$). If we used Henkel's pore pressure equations, Eqs. 11-22 and 11-25, we would predict Δu to be about 32 kPa. As we said in Sec. 11.10, predicting in situ pore pressures is not easy.

It is sometimes helpful when calculating stress paths to draw little elements with the appropriate total, total $- u_o$, neutral, and effective stresses indicated (similar to Fig. 11.29). This technique is shown in Fig. Ex. 11.16b both for initial conditions and after loading. Note that stresses on the elements for initial conditions are those we calculated at the beginning of this example. For final stresses, the vertical total stress increased by 51 kPa and the horizontal total stress increased by 17 kPa, as we determined previously from elastic theory. The induced pore pressures shown are those we found from the pore pressure equation. The calculations for p, p', and q for both initial and final conditions are shown below the elements.

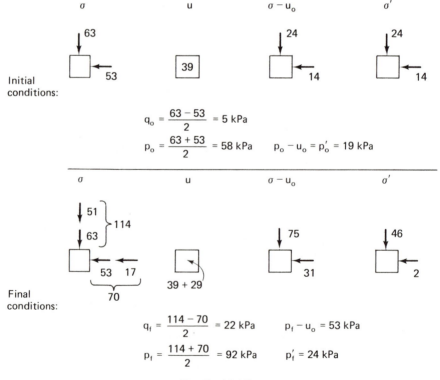

Fig. Ex. 11.16b

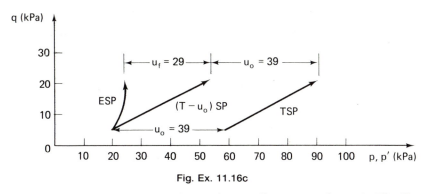

Fig. Ex. 11.16c

Finally, plot the stress paths on the p–q diagram, as shown in Fig. Ex. 11.16c. Sketch the ESP so as to have a shape similar to those shown previously (for example, Figs. 11.34 and 11.43) for normally consolidated clays.

———

The next example is a little more complicated. First, we shall construct the stress paths and determine the strength parameters for an axial compression test; then we shall use the AC test and our knowledge of stress paths to determine the pore pressure response of a lateral extension test. We will see that the effective stress paths for both tests are identical, even though the total stress paths are very different.

———

EXAMPLE 11.17

Given:

Two identical specimens (same w, e, etc.) of a normally consolidated saturated clay were hydrostatically consolidated ($K_o = 1$) and then sheared undrained. In test A, the axial compression (AC) test, the cell pressure was held constant while the axial stress was increased until failure. Specimen B was failed by lateral extension (LE) in which the vertical stress was held constant while the cell pressure was decreased until failure occurred. Stress-strain and pore pressure data for test A are shown in Table Ex. 11.17a.

Required:

    a. Compute and plot the stress-strain and pore pressure-strain curves for test A.
    b. Plot the TSP and ESP for both tests.

c. Determine ϕ' and ϕ_T for both tests.

d. Show that the stress-strain curve for the test A (AC) is identical to that for the test B (LE).

e. Evaluate the pore pressure-strain data for test B from the LE stress paths.

f. Compute the pore pressure parameter $\bar{A}$ for both tests.

TABLE EX. 11.17a Test A (AC test data)*

ϵ (%)	$\Delta\sigma/\sigma_c'$	$\Delta u/\sigma_c'$
0	0	0
1	0.35	0.19
2	0.45	0.29
4	0.52	0.41
6	0.54	0.47
8	0.56	0.51
10	0.57	0.53
12	0.58	0.55

*After Ladd (1964).

Solution:

a. Plot σ-ϵ and Δu-ϵ curves for test A (AC), as shown in Fig. Ex. 11.17a. Note that the data in Table Ex. 11.17a is normalized with respect to the effective consolidation stress σ_c' in the test. We could assume a σ_c' (in whatever units the test was conducted), or we can work everything out in terms of the normalized stresses.

b. As for the previous example, it is helpful to sketch elements showing the total, neutral, and effective stresses for the initial consolidation conditions, during shear, and at failure, as is done in Fig. Ex. 11.17b. Use these stresses to compute the TSP for both tests and the ESP for the AC test. Since at this time we don't know anything about the pore pressures developed in the LE test, we cannot plot its ESP.

Calculations for p, p', and q for test A (AC):
Initial conditions:

$$p_0 = p_0' = \frac{1 + 1}{2} = 1$$

$$q_0 = \frac{1 - 1}{2} = 0$$

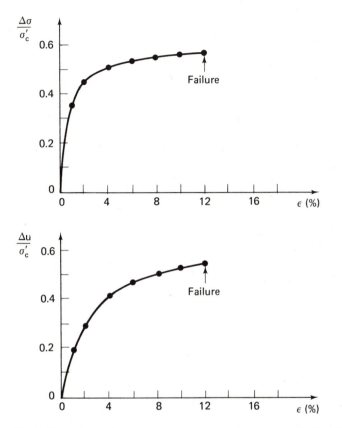

Fig. Ex. 11.17a   Stress-strain and pore pressure-strain curves for the AC test.

At failure:

$$p_f = \frac{1.58 + 1}{2} = 1.29$$

$$p'_f = p_f - \Delta u = 0.74$$

$$\left(\text{Check:} \; p'_f = \frac{1.03 + 0.45}{2} = 0.74 \right)$$

$$q_f = \frac{1.58 - 1}{2} = 0.29$$

$$\left(\text{Check:} \; \frac{1.03 - 0.45}{2} = 0.29 \right)$$

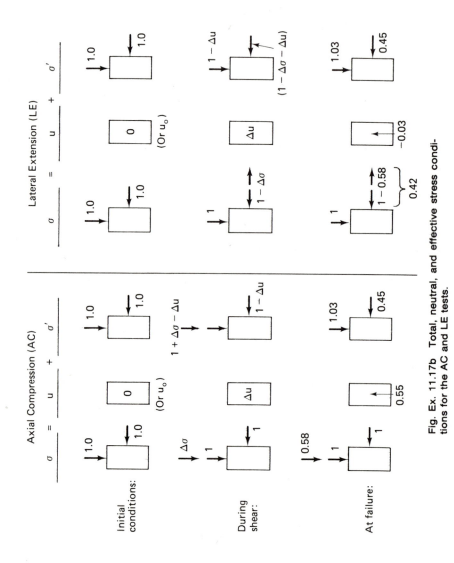

Fig. Ex. 11.17b Total, neutral, and effective stress conditions for the AC and LE tests.

For Test *B* (LE):

$$p_o = p_o' = \frac{1+1}{2} = 1$$

$$q_o = \frac{1-1}{2} = 0$$

$$p_f = \frac{1+0.42}{2} = 0.71$$

$$q_f = \frac{1-0.42}{2} = 0.29$$

Now plot the TSP's for both test *A* (AC) and test *B* (LE). We know that the TSP's will be straight lines inclined at 45° from the stress conditions in both tests since one of the principal stresses remains constant during the test. Therefore we need only calculate and plot the end points q_o, p_o, and q_f on Fig. Ex. 11.17c, and connect these points with straight lines.

Intermediate points for both the TSP and ESP may be calculated from the stress-strain and pore pressure-strain information of Table 11.17a and Fig. Ex. 11.17a. This process is exactly like that shown in Example 11.15. Usually it is easier to do the problem graphically by simply plotting the intermediate *q* values on the TSP ($q = \Delta\sigma/2$) at several conveniently spaced strains. This determines the intermediate *p* values. Then, scale off the Δu values horizontally to determine the intermediate p' values at these same strains. This process, shown in Fig. Ex. 11.15b and Fig. Ex. 11.17c, determines the corresponding ESP.

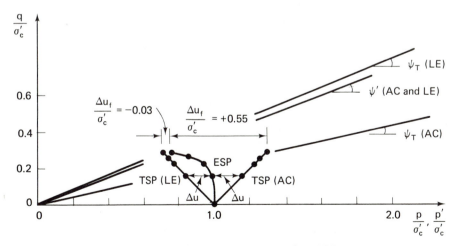

Fig. Ex. 11.17c   Stress paths for the AC and LE tests.

Note that only one ESP is shown for *both* tests. This is so because the effective stress conditions in both tests are the same. Why? Note that during shear the stress difference $\Delta\sigma$, which is equal to $(\sigma_1 - \sigma_3)$, is the same for both tests. Looking at it another way:

For the AC test,

$$\Delta\sigma_{AC} = 1 + \Delta\sigma - 1 = \Delta\sigma$$

For the LE test,

$$\Delta\sigma_{LE} = 1 - 1 + \Delta\sigma = \Delta\sigma$$

Therefore at every strain (including at failure) $\Delta\sigma_{AC} = \Delta\sigma_{LE}$. Thus the stress-strain curves for both tests must be the *same*. So, if we plotted the LE stress-strain curve, it would look exactly like the AC curve shown in Fig. Ex. 11.17a. By the way, this is the answer to part **d**.

If the two specimens have exactly the same stress-strain curve and identical strengths, then the effective stress conditions in the specimens must be identical, both at failure and during loading. This means that the ESP's must also be the same.

Another way of looking at this is that in the LE test, the change in stress difference $\Delta\sigma$ is produced by a change, a decrease, in cell or hydrostatic pressure. When the hydrostatic pressure changes, in an undrained test, only a change in pore pressure results, not a change in effective stresses. If there is no change in effective stresses, then the stress-strain and strength behavior must be the same (Hirschfeld, 1963). The only difference at failure between the tests must be in the amount of pore pressure Δu that develops. If this is true at failure, then it is true throughout the test. Therefore we can construct the pore pressure-strain curve [part **e**. of this example] for the test B (LE) from the stress path plots.

As with test A (AC) the amount of pore pressure developed in test B (LE) is simply the horizontal distance between the TSP and the ESP for that test. Note that for the LE test all values of Δu are *negative*. The

TABLE EX. 11.17b   Test B (LE test data)

ϵ (%)	$\Delta\sigma/\sigma_c'$	$\Delta u/\sigma_c'$
0	0	0
1	0.35	-0.16
2	0.45	-0.16
4	0.52	-0.11
6	0.54	-0.07
8	0.56	-0.05
10	0.57	-0.04
12	0.58	-0.03

constructed pore pressure-strain curve for the LE test is shown in Fig. Ex. 11.17d along with the stress-strain curve for both tests and the pore pressure-strain curve for the AC test. For easy comparison, numerical values of the pore pressure are listed in Table Ex. 11.17b. Fig. Ex. 11.17d and Table Ex. 11.17b are solutions to part **e**.

Now you can see where the effective stress values for the LE test in Fig. Ex. 11.17b came from. Another curious fact about the AC and LE test is that the numerical difference between the two pore pressure curves at a

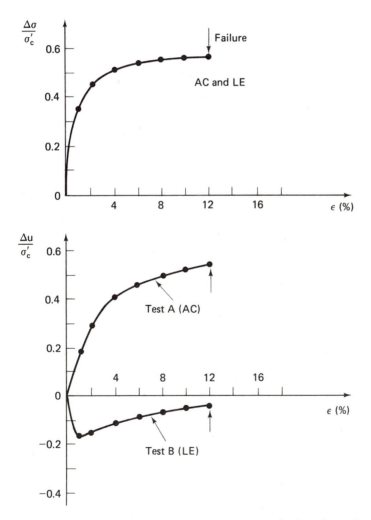

Fig. Ex. 11.17d   Stress-strain and pore pressure-strain data for both tests.

given strain is exactly equal to the principal stress difference at that strain. You can check this statement by using the values of Tables Ex. 11.17a and b or scaling off $\Delta\sigma$ values between the two Δu curves of Fig. Ex. 11.17d. Also the horizontal distance between the two TSP's in Fig. Ex. 11.17c is equal to $\Delta\sigma$ at a given strain.

Now that we know the TSP's and ESP's for both tests, we can compute ϕ' and ϕ_T for the two tests [part **c**]. From Fig. Ex. 11.17c we can measure the angles ψ', $\psi_{T(AC)}$, and $\psi_{T(LE)}$ with a protractor, or we can use Eq. 10-21. Since the clay is normally consolidated, we shall assume that $c' \simeq 0$ and this is why we drew the intercepts a and a' on the p–q diagram to be essentially zero. From Eqs. 10-24 and 10-25 we may readily compute ϕ', $\phi_{T(AC)}$, and $\phi_{T(LE)}$. These values are shown in Table Ex. 11.17c.

TABLE EX. 11.17c   Strength Parameters from Fig. Ex. 11.17c (in degrees)

Angle	Test A (AC)	Test B (LE)
ψ_T	12.5	22
ϕ_T	12.8	23.8
ψ'	21	21
ϕ'	22.6	22.6

f. Let us now compute the pore pressure parameter $\overline{A}$ for both tests. By Eq. 11-15,

$$\overline{A}_f = \frac{\Delta u - \Delta\sigma_3}{\Delta\sigma_1 - \Delta\sigma_3}$$

To obtain the stress changes during the test, it is usually easier to refer to the elements of Fig. Ex. 11.17b, and select the changes in total stress from the initial conditions to the conditions at failure.

For the test A (AC), $\Delta\sigma_3 = 0$ and $\Delta\sigma_1 = \sigma_{1f} - \sigma_{1o} = 1.58 - 1.0 = 0.58$; $\Delta u_f = 0.55$. So,

$$\overline{A}_f = \frac{0.55 - 0}{0.58 - 0} = 0.95$$

For une test B (LE), $\Delta\sigma_1 = 0$ and $\Delta\sigma_3 = \sigma_{3f} - \sigma_{3o} = 0.42 - 1.0 = -0.58$; $\Delta u_f = -0.03$. So,

$$\overline{A}_f = \frac{-0.03 - (-0.58)}{0 - (-0.58)} = \frac{0.55}{0.58} = 0.95$$

If this is confusing, it might be easier to use Eq. 11-17 for the LE test,

$$\overline{A}_{le} = 1 - \frac{\Delta u}{\Delta\sigma_h} = 1 - \frac{\Delta u}{\Delta\sigma_3} = 1 - \frac{-0.03}{-0.58} = 0.95$$

(Of course we knew from Eq. 11-20 that $\overline{A}_{le}$ should equal $\overline{A}_{ac}$.) The term

$\Delta\sigma_h$ is negative because it is decreased during the LE test (refer again to Fig. Ex. 11.17b).

What conclusions can we draw from this example? First, both the axial compression and lateral extension tests have identical stress-strain curves and their compressive strengths $\Delta\sigma_f$ are the same. If the stress-strain curves are the same, then they have the same E modulus. They also have the same ESP. However they have markedly different TSP's and markedly different pore pressure responses, but A_f (and thus $\overline{A}_f$) is the same for both tests. We can summarize these observations as follows:

Same $\Delta\sigma$ and $\Delta\sigma_f$
Same σ-ϵ curves and E modulus
Same ESP
Same ϕ'
Same A_f (and $\overline{A}_f$)
Different TSP
Different ϕ_T
Different Δu

In Example 11.17 we showed the stress conditions and plotted the stress paths for the AC and LE tests, where you will note that the principal stresses at failure had the same orientation as they did at the beginning of the test. For the axial extension (AE) and lateral compression (LC) tests (see Figs. 10.22 and 11.73 for a review of these tests), the principal stresses *rotate* during shear, and the stress paths go *below* the horizontal axis. In this case, q becomes negative. If we went through a similar exercise as we did in Example 11.17, we would reach the same conclusions as for the AC and LE tests: they have the same strength, ESP, $\overline{A}_f$, and ϕ', but different TSP and Δu. The stress conditions for the AE and LC tests are shown in Fig. 11.74; you might compare these stresses with those shown in Fig. Ex. 11.17b and see what is meant by the rotation of principal stresses. Figure 11.75 then shows typical test results from AE and LC tests. The stress paths for both tests are shown in Fig. 11.76.

The difference between the AC-LE and the AE-LC tests is really a function of the intermediate principal stress σ_2. Note that for the first two types of tests we assume that $\sigma_2 = \sigma_3$, and there is no rotation of principal stresses from the beginning of the test until failure. On the other hand, for the AE-LC tests $\sigma_2 = \sigma_1$, and a rotation of principal stresses occurs. This rotation would be even more dramatic if, for initial conditions, we had different vertical stresses than horizontal stresses: that is, if $\sigma_{vo} \neq \sigma_{ho} = \sigma_{cell}$. For this initial condition, $\sigma_{vo} = \sigma_{1o}$ and $\sigma_{ho} = \sigma_{3o} = \sigma_{cell}$. For both the AE and LC tests, the horizontal stress at failure becomes the major principal stress, as shown in Fig. 11.74.

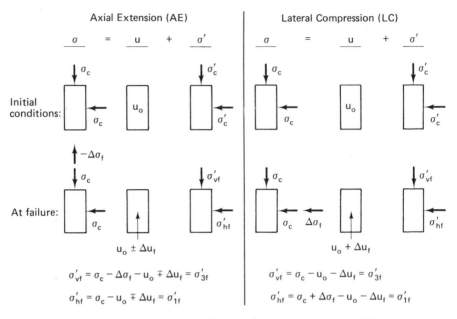

Fig. 11.74 Stress conditions for the axial extension (AE) and lateral compression (LC) tests. Note that the major principal stress is now horizontal for both these tests at failure.

Some actual test data on natural clays is shown in Figs. 11.77 and 11.78. These results verify the assertions made above that the ESP, σ-ϵ, and A_f responses of AC and LE, and AE and LC tests are essentially the same for saturated soils. The effective stress and σ-ϵ behavior is determined *only* by the sign and magnitude of the principal stress difference, $\Delta\sigma = \sigma_v - \sigma_h$, and is independent of the particular shape of the total stress path (Bishop and Wesley, 1975).

Note that the ESP for the AE and LC tests in Figs. 11.77 and 11.78 did not cross the AE-TSP as it did in Fig. 11.76. This means that the induced pore pressure in these tests did not go slightly negative, in contrast to the behavior of Fig. 11.76. The specific ESP characteristics for any given soil must be determined by laboratory tests.

The angle of inclination of the failure planes as determined according to the Mohr failure hypothesis (discussed in Sec. 10.4) is different for the AE and LC tests because of the rotation of principal stresses. We may determine this angle by using the pole method. This procedure is shown in Fig. 11.79 for the AC and AE tests; similar results would be found for the LE and LC tests. In summary, then, for:

For AC and LE, no rotation of σ_1 and σ_3: $\alpha_f = 45° + \phi'/2$
For AE and LC, with rotation of σ_1 and σ_3: $\alpha_f = 45° - \phi'/2$

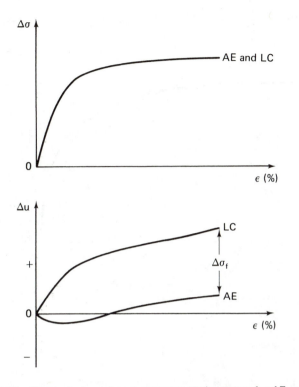

Fig. 11.75 Stress-strain and pore pressure-strain curves for AE and LC tests on a normally consolidated clay (after Hirschfeld, 1963).

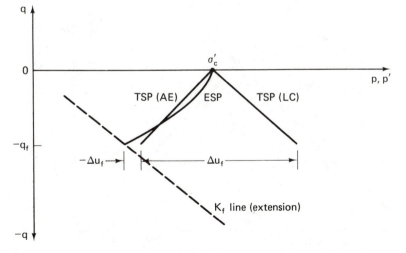

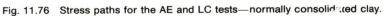

Fig. 11.76 Stress paths for the AE and LC tests—normally consolidated clay.

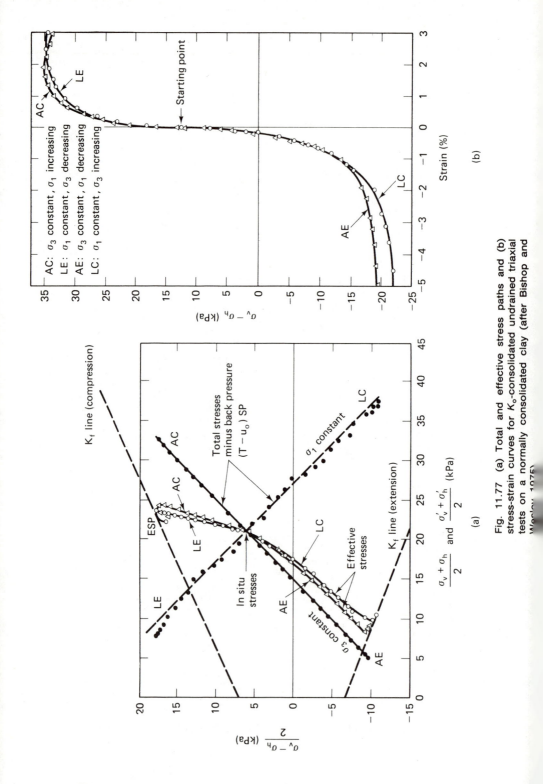

Fig. 11.77 (a) Total and effective stress paths and (b) stress-strain curves for K_o-consolidated undrained triaxial tests on a normally consolidated clay (after Bishop and Wesley, 1975).

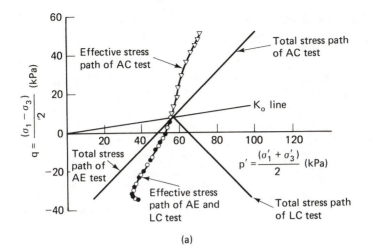

(a)

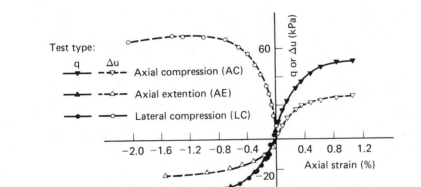

(b)

Specimen	K_o	Test Type	w_n (%)	w_n (%)	$\left(\dfrac{\sigma_1 - \sigma_3}{2}\right)_{max}$ (kPa)	A_f^*
Kars clay:						
195-22-5	0.75	AC	71.5	70.4	51.2	0.39
195-22-7	0.75	LC	73.5	72.0	34.9	0.73
195-22-3	0.75	AE	71.5	70.3	34.5	0.73

$^*A_f$ is the pore pressure parameter at failure based on expressions in Table B-3-2.

Fig. 11.78 (a) Total and effective stress paths and (b) stress-strain and pore pressure-strain response of K_o-consolidated undrained triaxial tests on undisturbed samples of Leda clay from Kars, Ontario (after Law and Holtz, 1978).

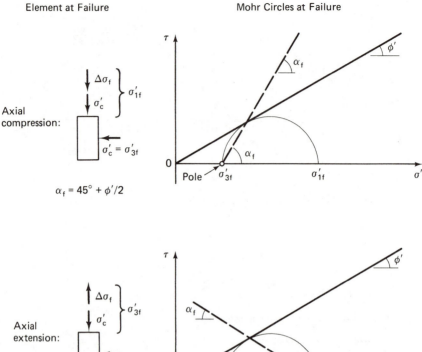

Fig. 11.79 Angle of inclination of the failure plane for AC and AE tests.

11.13 STRESS PATHS DURING UNDRAINED LOADING—OVERCONSOLIDATED CLAYS

All of the previous section on undrained stress paths concerned the behavior of normally consolidated clays. For overconsolidated clays, the principles are the same but the shapes of the stress paths are different because the developed pore pressures are different. Examples of stress paths for axial compression tests on overconsolidated clays are shown in Figs. 10.25 and 11.34b. Knowing how the excess pore water pressures develop along with the shapes of the total stress paths for the various types of tests, you can readily construct the ESP's for overconsolidated clays.

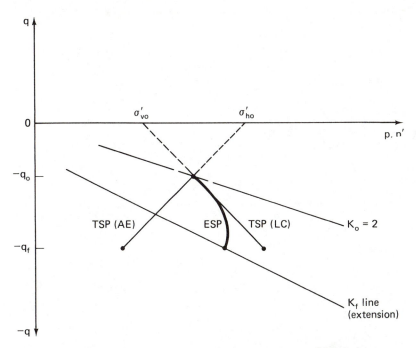

Fig. 11.80.   AE and LC stress paths for an overconsolidated clay.

As discussed in Sec. 11.11, overconsolidated clays usually have a K_o much greater than one. Therefore the stress paths for overconsolidated clays in situ (or for samples reconsolidated to in situ stresses in the laboratory) will start from below the hydrostatic ($K_o = 1$) axis, as shown in Fig. 10.25. Figure 11.80 shows how the stress paths for AE and LC tests on an overconsolidated clay might appear.

EXAMPLE 11.18

Given:

Consolidated undrained triaxial compression tests are conducted on an overconsolidated clay with preconsolidation stress σ'_p of 800 kPa, which is equivalent to an OCR of 10. The results are shown in Fig. Ex. 11.18a. Another CU test is co. `cted on the same clay at the same OCR and thus the same σ'_c. In the latte. st, the lateral stress is not held constant, but is increased at the same time as the axial stress is increased so that $\Delta\sigma_3 = 0.2\,\Delta\sigma_1$. (See Fig. Ex. 11.18b.) Assume that the test results on this clay

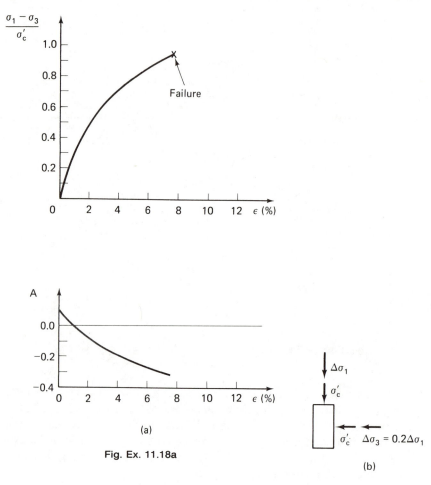

(a)

Fig. Ex. 11.18a

(b)

Fig. Ex. 11.18b

shown in Fig. Ex. 11.18a are valid for all ways of changing the boundary stresses in compression, that is, both σ_1 and σ_3 increasing during the test.

Required:

Predict the behavior of the second CU test.

    a. Calculate the quantities and fill in the columns of Table Ex. 11.18 for 0, 0.5, 2.5, 5, and 7.5% strain.

    b. Draw the TSP and the ESP for this test.

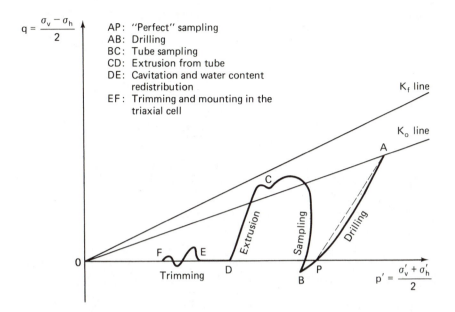

Fig. 11.81 Stress paths during sampling of a normally
consolidated clay (after Ladd and Lambe, 1963).

Procedures for evaluating sample disturbance and correcting the
measured shear strength are suggested by Ladd and Lambe (1963) and
Ladd, et al. (1977).

Next, we shall consider the case of foundation loading, for example,
a highway embankment constructed on a soft clay foundation. Let us
assume that the clay is very nearly 100% saturated and is normally
consolidated. This case, as shown by Fig. 11.73a, may be modeled by axial
compression stress conditions. Strictly speaking, as mentioned previously,
the loading should be plane strain ($\epsilon_2 = 0$) for a long embankment, but we
shall use the common triaxial test, with which you are familiar, for
illustrative purposes. The stress paths for this case are shown in Fig. 11.82
(compare with Fig. 10.26).

Let's look a little more closely at these stress paths and their engineering implications. For this normally consolidated clay, the K_o is less
than 1 (about 0.6), so that the initial stress conditions in the ground are
plotted as point A on the figure. In a foundation *loading*, the horizontal
stresses probably increase slightly, as we assumed in Example 11.16, but
for this case we will assume that they are essentially constant. Then, the
$(T - u_o)$SP is the straight line AC. The total stresses represented by point
C are applied at the end of construction. The induced pore pressures are
positive, of course, for a normally consolidated clay, and so we will have

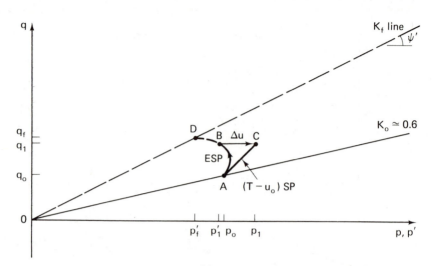

(a)

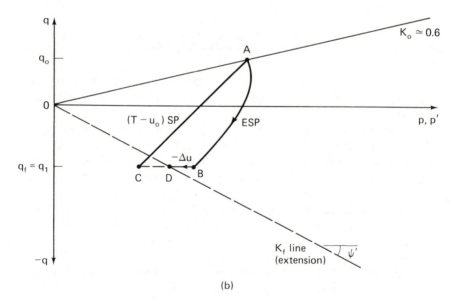

(b)

Fig. 11.82   Stress paths for (a) foundation loading and (b)
foundation excavation of normally consolidated clay.

the typically shaped ESP hooking off to the left, as is illustrated by curve *AB*. The distance *BC*, then, is numerically equal to the excess pore pressure induced by the embankment loading. Note that the shear stress on a typical element under the embankment increases from its initial value of q_o to q_1. Had loading continued to the level of q_f, the ESP would have intersected the K_f line and failure would have occurred.

For this example, let's assume that we were good designers, that we correctly estimated the in situ shear strength of the soil, and that no failure occurred. Then we are at point *B* on the ESP at the *end of construction*, the most critical design condition for foundation loadings on normally consolidated clays. Why is this? Well, look at what happens after we reach point *B*. The applied loading is constant thereafter (assuming no additional construction occurs), the clay starts to consolidate, and the excess pore water pressure that was caused by the load dissipates. This excess pore pressure is represented by the distance *BC*. Thus the ESP proceeds along line *BC*. Ultimately at $u = 100\%$, all the excess pore pressure will be dissipated and our element will be at point *C* in equilibrium under the embankment load. It will still have a shear stress q_1 acting on it, and $p = p' = p_1$. Since there is no excess pore water pressure remaining in the element, the total stresses will equal the effective stresses at point *C*. Now you can see why point *B* at the end of construction was the most critical for this case. Point *B* was the closest point to the failure line K_f. After that, because of consolidation, the foundation soil became stronger with time (safer) until at point *C* we were at the farthest point from the K_f line for this particular loading situation. That is why the end of construction is the most critical for foundation loading of normally consolidated clays. The engineering lesson here is that if you make it through the end of construction period for this type of loading, then conditions become *safer* with time.

For the foundation loading of an overconsolidated clay, the TSP and ESP would look something like the paths shown in Figs. 10.25 and 11.34b. As the negative excess pore pressure dissipates, the stresses on the element move closer to the K_f line, which means that the long-term conditions are actually the *least* safe after dissipation of the pore pressure has occurred. But in most cases, we are so far from the K_f line anyway that long term conditions are usually not critical.

EXAMPLE 11.19

Given:

The embankment of Example 11.16. Triaxial compression tests indicate $\phi' = 23°$ and $c' = 7$ kPa.

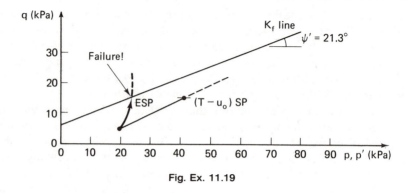

Fig. Ex. 11.19

Required:

Construct the K_f line and determine whether the embankment will be stable.

Solution:

From Eqs. 10-24 and 10-25, $\psi' = 21.3°$ and $a' = 6.4$ kPa. Draw the K_f line on the p-q diagram (Fig. Ex. 11.19). Since the ESP would intersect the K_f line before the final design loads could be applied, failure would occur. At that time, q would be approximately 15 kPa.

Another important engineering situation concerns an excavation for a foundation in normally consolidated clay. This situation is illustrated in Fig. 11.73 as an example of axial extension. We already know from Fig. 11.76 what the TSP and ESP look like for this case, they are also in Fig. 11.82b. Since the vertical stress decreases during an excavation, the total stress path goes from the initial conditions at point A to point C. As with the case of foundation loading, the horizontal stresses may also decrease slightly, but for illustration purposes, we shall assume that they remain essentially unchanged. Since negative pore pressures occur due to unloading the ESP must lie to the right of the $(T - u_o)$SP. For the case shown with unloading from q_o down to q_1, the ESP then follows curve AB, and point B represents conditions at the end of construction. For this case, failure did not occur, and we are safe at the end of construction. Now, the excess pore pressure starts to dissipate—it is negative in this case, and now it starts to become more and more positive, following line BDC. At point C, of course, all the excess negative pore pressure would be dissipated and the total stresses would equal the effective stresses. But, this would never occur because when the ESP reached point D, it would intersect the K_f line in extension and failure would occur. Therefore the long-term conditions are the more critical for the case of an excavation in normally consolidated clays. In contrast to the case of foundation loading, just because you get through construction without a failure doesn't mean that you are free of a possible failure. No, the excavation will become less and less safe with time. Field measurements (for example, Lambe and Whitman, 1969) have shown that the rate of dissipation of this negative pore pressure occurs relatively fast, much faster than in the case of foundation loading. Therefore the engineering implication for this case is to get that excavation filled and the clay loaded as fast as possible. Otherwise you risk a failure occurring at some time, perhaps only a few weeks after completion of the excavation. This is another example of the long-term conditions being more critical than the end of construction conditions.

These examples illustrate the value of the stress path method. You can construct similar TSP and ESP diagrams for the other cases shown in Fig. 11.73, for both normally and overconsolidated clays, and see what the critical design situations are. Some of the critical conditions for stability are summarized in Table 11-10 (Ladd, 1971b).

TABLE 11-10 Critical Conditions for the Stability of Saturated Clays*

Foundation Loading:		
Soil Type:	Soft (NC) clay	Stiff (highly OC) clay
Critical condition:	UU case (no drainage)	Probably UU case but check CD case (drainage with equilibrium pore pressures)
Remarks:	Use $\phi = 0$, $c = \tau_f$ with appropriate corrections (Sec. 11.9).	Stability usually not a major problem
Excavation or Natural Slope:		
Soil type:	Soft (NC) clay	Stiff (highly OC) clay
Critical condition:	Could be either UU or CD case	CD case (complete drainage)
Remarks:	If soil is very sensitive, it may change from drained to undrained conditions.	Use effective stress analysis with equilibrium pore pressures. If clay is fissured, c' and perhaps ϕ' may decrease with time.

*After Ladd (1971b).

PROBLEMS

11-1. A granular material is observed being dumped from a conveyor belt. It formed a conical pile with about the same slope angle, 1.8 horizontal to 1 vertical. What can you say about the material?

11-2. A battery filler is filled with a medium rounded sand in the densest state possible. Every effort is made to keep the sand saturated. A transparent tube allows observation of the water level in the battery filler. What will happen to the water level, if anything, as the bulb is squeezed very hard? Why? Would it matter if the sand were loose? Explain.

11-3. You are climbing up a large sand dune west of Yuma, Arizona. The slope angle is 33°. In what compass direction are you traveling? Don't forget the declination!

11-4. The principal stress ratio for a drained test at failure was 4.60. What was the probable relative density of the sand?

11-5. Derive Eq. 11-3.

11-6. A direct shear test is conducted on a fairly dense sample of Franklin Falls sand from New Hampshire. The initial void ratio was 0.668. The shear box was 76 mm square, and initially the height of the specimen was 11 mm. The following data were collected during shear. Compute the data needed and plot the usual curves for this type of test.

Time Elapsed (min)	Vertical Load (kN)	Horizontal Displacement (mm)	Thickness Change (mm)	Horizontal Load (N)
0	2.25	8.89	3.56	0
0.5	(constant)	8.82	3.54	356
1		8.63	3.52	721
2		8.44	3.51	1014
3		7.92	3.53	1428
4		7.18	3.59	1655
5		6.38	3.63	1770
6		5.49	3.65	1744

(After Taylor, 1948.)

11-7. A conventional triaxial compression test is conducted on a sample of dense sand from Ft. Peck Dam, Montana. The initial area of the test specimen was 10 cm^2 and its initial height was 70 mm. Initial void ratio was 0.605. The following data were observed during shear. First, calculate the average area of the specimen, assuming it is a right circular cylinder at all times during the test. Then make the calculations necessary to plot the axial stress versus axial strain and volumetric strain versus axial strain curves for this test. Assuming $c' = 0$, what is ϕ'?

Time Elapsed (min)	Chamber Pressure (kPa)	Strain Dial, Giving ΔH (mm)	Buret, Giving ΔV (cc)	Axial Load (N)
0	30	200	2.00	0
	(constant)	205	1.91	41
		210	1.86	84
45		224	1.92	144
		240	2.13	177
90		278	2.80	207
		319	3.66	218
		359	4.56	221
240		402	5.40	218
		508	7.30	202
460		603	8.09	183

(After Taylor, 1948.)

11-8. The results of two CD triaxial tests at different confining pressures on a medium dense, cohesionless sand are summarized in the table below. The void ratios of both specimens were approximately the

Test No. 1 ($\sigma_c = 100$ kPa)			Test No. 2 ($\sigma_c = 3000$ kPa)		
Axial Strain (%)	$(\sigma_1 - \sigma_3)$ (kPa)	Volumetric Strain (%)	Axial Strain (%)	$(\sigma_1 - \sigma_3)$ (kPa)	Volumetric Strain (%)
0	0	0	0	0	0
1.71	325	− 0.10	0.82	2090	− 0.68
3.22	414	+ 0.60	2.50	4290	− 1.80
4.76	441	+ 1.66	4.24	5810	− 2.71
6.51	439	+ 2.94	6.00	6950	− 3.36
8.44	405	+ 4.10	7.76	7760	− 3.88
10.4	370	+ 5.10	9.56	8350	− 4.27
12.3	344	+ 5.77	11.4	8710	− 4.53
14.3	333	+ 6.33	13.2	8980	− 4.71
16.3	319	+ 6.70	14.9	9120	− 4.84
18.3	318	+ 7.04	16.8	9140	− 4.92
20.4	308	+ 7.34	18.6	9100	− 4.96
			20.5	9090	− 5.01

(After A. Casagrande.)

same at the start of the test. Plot on one set of axes the principal stress difference versus axial strain and volumetric strain (Eq. 11-4) versus axial strain for both tests. Estimate the initial tangent modulus of deformation, the "50%" secant modulus, and the strain at failure for each of these tests.

11-9. For the two tests of Problem 11-8, determine the angle of internal friction of the sand at (a) peak compressive strength, (b) at ultimate compressive strength, and (c) at 5% axial strain. Comments?

11-10. A sand is hydrostatically consolidated in a triaxial test apparatus to 420 kPa and then sheared with the drainage valves open. At failure, $(\sigma_1 - \sigma_3)$ is 1046 kPa. Determine the major and minor principal stresses at failure and the angle of shearing resistance. Plot the Mohr diagram. (This problem should be followed by the next one.)

11-11. The same sand as in Problem 11-10 is tested in a direct shear apparatus under a normal pressure of 420 kPa. The sample fails when a shear stress of 280 kPa is reached. Determine the major and minor principal stresses at failure and the angle of shearing resistance. Plot the Mohr diagram. Explain the differences, if any, of these values with those obtained in the preceding problem.

11-12. Indicate the orientations of the major principal stress, the minor principal stress, and the failure plane of the tests in Problems 11-10 and 11-11.

11-13. A granular soil is tested in direct shear under a normal stress of 300 kPa. The size of the sample is 7.62 cm in diameter. If the soil to be tested is a dense sand with an angle of internal friction of 42°, determine the size of the load cell required to measure the shear force with a factor of safety of 2 (that is, the capacity of the load cell should be twice that required to shear the sand).

11-14. The stresses induced by a surface load on a *loose* horizontal sand layer were found to be $\sigma_v = 4.62$ kPa, $\tau_v = 1.32$ kPa, $\sigma_h = 2.90$ kPa, and $\tau_h = -1.32$ kPa. By means of Mohr circles, determine if such a state of stress is safe. Use Eq. 10-11 for the definition of factor of safety.

11-15. If the same stress conditions as in Problem 11-14 act on a very dense gravelly sand, is such a state safe against failure?

11-16. The effective normal stresses acting on the horizontal and vertical planes in a silty gravel soil are 1.91 MPa and 3.18 MPa, respectively. The shear stress on these planes is ± 0.64 MPa. For these conditions, what are the magnitude and direction of the principal stresses? Is this a state of failure?

11-17. A sample of dense sand tested in a triaxial CD test failed along a well-defined failure plane at an angle of 66° with the horizontal. Find the effective confining pressure of the test if the principal stress difference at failure was 100 kPa.

11-18. A dry loose sand is tested in a vacuum triaxial test in which the pore air pressure of the sample is lowered below gage pressure to within about 95% of −1 atm. Compute the principal stress difference and the major principal stress ratio at failure.

11-19. For the data shown in Fig. 11.4a, what is (a) the principal stress difference and (b) the principal stress ratio at an axial strain of 10% for an effective confining pressure of 2 MPa?

11-20. For the conditions given in Problem 11-19, plot the Mohr circle.

11-21. Do Problems 11-19 and 11-20 for the data shown in Fig. 11.5a.

11-22. A sample of Sacramento River sand has a critical confining pressure of 1000 kPa. If the sample is tested at an effective confining pressure of 1500 kPa, describe its behavior in drained and undrained shear. Show results in the form of unscaled Mohr circles.

11-23. For the sand of Problem 11-22, describe the behavior in drained and undrained shear in a triaxial test if the effective confining pressure is 750 kPa.

11-24. A drained triaxial test is performed on a sand with $\sigma'_{3c} = \sigma'_{3f} = 500$ kPa. At failure, $\tau_{max} = 660$ kPa. Find σ'_{1f}, $(\sigma_1 - \sigma_3)_f$, and ϕ'.

11-25. If the test of Problem 11-24 had been conducted undrained, determine $(\sigma_1 - \sigma_3)_f$, ϕ', ϕ_{total}, and the angle of the failure plane in the specimen. $\Delta u_f = 100$ kPa.

11-26. If the test of Problem 11-25 were conducted at an initial confining pressure of 1000 kPa, estimate the principal stress difference and the induced pore water pressure at failure.

11-27. Assume the sand of Problem 11-24 is Sacramento River sand at a void ratio of 0.8. If the initial volume of the specimen was 70 cm³, what change in volume would you expect during shear?

11-28. What volume change would you expect during the test of Problem 11-25?

11-29. A silty sand is tested consolidated-drained in a triaxial cell where both principal stresses at the start of the test were 500 kPa. If the axial pressure at failure is 1.63 MPa while the horizontal pressure remains constant, compute the angle of shearing resistance and the theoretical orientation of the failure plane with respect to the horizontal.

11-30. The silty sand of Problem 11-29 was inadvertently tested consolidated-undrained, but the laboratory technician noticed that the pore pressure at failure was 290 kPa. What was the principal stress difference at failure?

11-31. If the consolidation pressure in the CU test of Problem 11-30 were 1000 kPa instead of 500 kPa, estimate the pore pressure at failure.

11-32. A sample of sand failed when $(\sigma_1 - \sigma_3)$ was 900 kPa. If the hydrostatic consolidation stress were 300 kPa, compute the angle of shearing resistance of the sand. What else can you say about the sand?

11-33. If the sample of Problem 11-32 were sheared undrained and the induced pore pressure at failure were 200 kPa, estimate the principal stress difference at failure. What would be the angle of shearing resistance in terms of total stresses?

11-34. A sample of sand at the field density is known to have a $(\sigma_1/\sigma_3)_{max}$ of 4.0. If such a specimen is hydrostatically consolidated to 1210 kPa in a triaxial test apparatus, at what effective confining pressure σ'_{3f} will the sample fail if the vertical stress is held constant? (This is a lateral extension test.)

11-35. Two CD stress path triaxial tests are conducted on identical

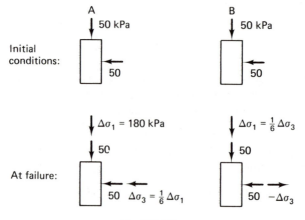

Fig. P11-35

samples of the same sand. Both specimens are initially consolidated hydrostatically to 50 kPa; then each specimen is loaded as shown in Fig. P11-35. Specimen A failed when the applied $\Delta\sigma_1$ was 180 kPa. Make the necessary calculations to plot (a) the Mohr circles at failure for both tests, and (b) the stress paths for both tests. (c) Determine ϕ' for the sand. (After C. W. Lovell.)

11-36. Plot a graph of σ_1'/σ_3', versus ϕ'. (Aren't you sorry you didn't do this sooner? It would have been helpful for solving some of these problems.) What range of values of ϕ' should be used?

11-37. Estimate the shear strength parameters of a fine (beach) sand (SP). Estimate the minimum and maximum void ratios.

11-38. A subrounded to subangular sand has a D_{10} of about 0.1 mm and a uniformity coefficient of 3. The angle of shearing resistance measured in the direct shear test was 47°. Is this reasonable? Why?

11-39. Estimate the ϕ' values for (a) a well-graded sandy gravel (GW) at a density of 2.1 Mg/m^3; (b) a poorly graded silty sand with a field density of 1.55 Mg/m^3; (c) an SW material at 100% relative density; and (d) a poorly graded gravel with an in situ void ratio of 0.4.

11-40. The results of a series of CD triaxial tests on a medium dense, cohesionless sand are summarized in the table below. The void ratios for all the test specimens were approximately the same at the start of the test. Plot the strength circles and draw the Mohr failure envelope for this series of tests. What angle of internal friction should be used in solving stability problems in which the range of normal stresses is (a) 0–500 kPa; (b) 1000–1500 kPa; (c) 3–6 MPa; and (d) 0–6 MPa?

Test No.	Confining Pressure (kPa)	Compressive Strength (kPa)
1	100	480
2	400	1870
3	997	4080
4	1880	7050
5	2990	10 200
6	3850	12 690

(After A. Casagrande.)

11-41. Estimate the values of the coefficient of earth pressure at rest, K_o, for the four soils of Problem 11-39.

11-42. If the sands of Problem 11-41 had been preloaded, would your estimate of K_o be any different? If so, would it be higher or lower? Why?

11-43. Estimate K_o for sands 1, 4, 5, 6, 8, and 10 in Table 11-2 for relative densities of 40% and 85%.

11-44. For future reference, place a scale of K_o on the ordinate of Fig. 11.13. You should probably also indicate a range of values of K_o.

11-45. Explain the difference between liquefaction and cyclic mobility.

11-46. The Peacock diagram (Fig. 11.10) has been used to predict the pore pressure response of undrained tests on sands, based on the volume changes observed at failure in drained tests. At a given void ratio a sample consolidated at an effective confining pressure less than $\sigma'_{3\,crit}$ would be expected to offer *more* resistance to liquefaction (since it should have a dilative tendency and therefore develop negative pore water pressure) than a sample consolidated at a confining pressure higher than $\sigma'_{3\,crit}$ (as this one should tend to decrease in volume during shear). This is contrary to what has been found in the laboratory in cyclic triaxial tests. Explain the apparent contradiction.

11-47. Figure 11.18b shows that at the tenth cycle, the change in pore water pressure is about 66 kPa just at the beginning of the application of the principal stress difference. Yet, at a quarter of a cycle later (as well as slightly before) the pore water pressure is just about equal to the effective confining pressure. At this time the principal stress difference is *zero*! Explain this observation. (It will help if you understand the answer to Problem 11-46.)

11-48. A large power plant is to be constructed at a site immediately adjacent to the Ohio River. The soils at the site consist of 50 m of loose to medium dense granular materials, and the ground water table is near the ground surface. Since there are several potential earthquake source areas that could influence the site, list some measures that could be taken to protect the foundation of this important structure from liquefaction and/or cyclic mobility.

11-49. We stated in Secs. 10.5 and 11.9 that the unconsolidated-drained test was meaningless because it could not be properly interpreted. Why is this so? Discuss in terms of laboratory tests as well as possible practical applications.

11-50. A CD axial compression triaxial test on a normally consolidated clay failed along a clearly defined failure plane of 57°. The cell pressure during the test was 200 kPa. Estimate ϕ', the maximum σ_1'/σ_3', and the principal stress difference at failure.

11-51. Suppose an identical specimen of the same clay as in Problem 11-50 was sheared undrained, and the induced pore pressure at failure was 85 kPa. Determine the principal stress difference, total and effective principal stress ratios, ϕ', ϕ_{total}, A_f and α_f for this test.

11-52. A series of *drained* direct shear tests were performed on a saturated clay. The results, when plotted on a Mohr diagram, gave $c' = 10$ kPa and $\tan\phi' = 0.5$. Another sample of this clay was consolidated to an effective pressure of 100 kPa. An *undrained* direct shear test was performed, and the measured value of τ_{ff} was 60 kPa. What was the pore water pressure at failure? Was the sample normally consolidated? Why?

11-53. The following information was obtained from laboratory tests on specimens from a completely saturated sample of clay:

(a) The sample had in the past been precompressed to at least 200 kPa.
(b) A specimen tested in direct shear under a normal stress of 600 kPa, with complete drainage allowed, showed a shearing strength of 350 kPa.
(c) A specimen which was first consolidated to 600 kPa, and then subjected to a direct shear test in which no drainage occurred, showed a shearing strength of 175 kPa.

Compute ϕ' and ϕ_T for the undrained case. Sketch the Mohr envelopes which you would expect to obtain from a series of undrained and drained tests on this clay. (After Taylor, 1948.)

11-54. Triaxial tests were performed on undisturbed samples from the same depth of organic clay whose preconsolidation load, determined from consolidation tests, was in the range 90 to 160 kPa. The principal stresses at failure of two CD tests were

Test No. 1:	$\sigma_3 = 200$ kPa,	$\sigma_1 = 704$ kPa
Test No. 2:	$\sigma_3 = 278$ kPa,	$\sigma_1 = 979$ kPa

Data from one CU test on the same clay are shown below. The effective consolidation pressure was 330 kPa and the sample was loaded in axial compression.

Stress Difference (kPa)	Strain (%)	Pore Pressure (kPa)
0	0	0
30	0.06	15
60	0.15	32
90	0.30	49
120	0.53	73
150	0.90	105
180	1.68	144
210	4.40	187
240	15.50	238

(a) Plot the Mohr circles at failure and determine ϕ' from the CD tests for the normally consolidated portion of the failure envelope.

(b) For the CU test, plot curves of principal stress difference and pore pressure versus strain.

(c) On a p-q diagram, plot the stress paths for the CD and CU tests. What is the OCR of the normal stresses on the failure plane at failure?

(d) Assuming that the single CU test for which data are given is representative for CU tests run at pressures well above the preconsolidation stress: (a) What is ϕ in terms of total stresses above the effects of preconsolidation? (b) What is ϕ' determined by the CU test above the effects of preconsolidation? (After A. Casagrande.)

11-55. In Problem 11-54, failure in the CU test was assumed to have occurred when the maximum principal stress difference was reached. Calculate and plot the principal effective stress ratio versus strain for this test. What is the maximum σ_1'/σ_3'? Is there any difference in ϕ' for the two failure criteria? Hint: Study Fig. 11.35.

11-56. A CU triaxial test is performed on a cohesive soil. The effective consolidation stress was 750 kPa. At failure, the principal stress difference was 1250 kPa, and the major effective principal stress was 1800 kPa. Compute Skempton's pore pressure coefficient A at failure.

11-57. Suppose another specimen of the soil in the preceding problem developed a major effective principal stress of 2200 kPa at failure. What would Skempton's pore pressure coefficient A at failure be, if $\sigma_c' = 900$ kPa?

11-58. Two samples of a slightly overconsolidated clay were tested in triaxial compression, and the following data at failure were obtained. The preconsolidation stress for the clay was estimated from oedometer tests to be about 400 kPa.

Specimen	X (kPa)	Y (kPa)
σ'_c	75	750
$(\sigma_1 - \sigma_3)_f$	265	620
Δu_f	-5	$+450$

(a) Determine the Skempton pore pressure parameter A at failure for both tests.

(b) Plot the Mohr circles at failure for both total and effective stresses.

(c) Estimate ϕ' in the normally consolidated range, and c' and ϕ' for the overconsolidated range of stresses.

11-59. Two identical specimens of soft saturated normally consolidated clay were consolidated to 150 kPa in a triaxial apparatus. One specimen was sheared drained, and the principal stress difference at failure was 300 kPa. The other specimen was sheared undrained, and the principal stress difference at failure was 200 kPa. Determine (a) ϕ' and ϕ_{total}; (b) u_f in the undrained specimen; (c) A_f in the undrained specimen; and (d) the theoretical angle of failure planes for both specimens.

11-60. A clay sample is hydrostatically consolidated to 1.0 MPa and then sheared undrained. The $(\sigma_1 - \sigma_3)$ at failure was also equal to 1 MPa. If drained tests on identical samples gave $\phi' = 22°$, evaluate the pore pressure at failure in the undrained test and compute Skempton's A parameter.

11-61. An undrained triaxial compression test was performed on a saturated sample of normally consolidated clay. The consolidation pressure was 100 kPa. The specimen failed when the principal stress difference was 85 kPa and the induced pore water pressure was 67 kPa. A companion undrained test was performed on an identical sample of the same clay, but at a consolidation pressure of 250 kPa. What maximum principal stress difference would you expect at failure for this second test specimen? What are ϕ' and

ϕ_{total}? Predict the angle of the failure planes for the two undrained tests. Determine A_f for this clay.

11-62. The following data were obtained from a CU test with pore pressures measured on an undisturbed specimen of sandy silt. The consolidation pressure was 850 kPa and the specimen was sheared in axial compression.

Principal Stress Difference (kPa)	Strain (%)	Induced Pore Pressure (kPa)
0	0	0
226	0.11	81
415	0.25	187
697	0.54	323
968	0.99	400
1470	2.20	360
2060	3.74	219
2820	5.78	− 009
3590	8.41	− 281
4160	11.18	− 530
4430	13.93	− 703
4310	16.82	− 767
4210	19.71	− 789

(a) Plot curves of principal stress difference and pore pressures versus strain. Plot on one sheet.
(b) Plot the stress paths on a p-q diagram.
(c) What is the maximum effective principal stress ratio developed in this test? Is it the same as the maximum obliquity for this specimen?
(d) Is there any difference in ϕ' as determined when the principal stress difference or the principal effective stress ratio is a maximum?
(After A. Casagrande.)

11-63. Typical consolidated-drained behavior of saturated normally consolidated samples of Ladd's (1964) simple clay are shown in Fig. P11-63. You are to conduct another axial compression CD triaxial test on the same clay with the effective consolidation stress equal to 100 kPa. For this test estimate (a) the water content and (b) the principal stress difference at an axial strain of 5%. (After C. W. Lovell.)

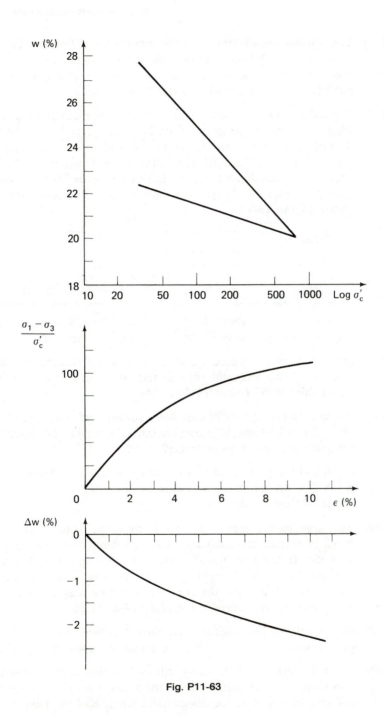

Fig. P11-63

11-64. The consolidation behavior of the simple clay of Problem 11-63 is shown in Fig. P11-63. Estimate the water content of a sample of this clay at an OCR of 10, if the maximum consolidation stress is 500 kPa instead of 800 kPa. (After C. W. Lovell.)

11-65. Triaxial compression tests were run on specimens from a large undisturbed block sample of clay. Data are given below. Tests 1 through 4 were run so slowly that complete drainage may be assumed. In tests 5 through 8, no drainage was permitted. Plot the Mohr failure envelopes for this soil. Determine the Mohr-Coulomb strength parameters in terms of both total and effective stresses. (After Taylor, 1948.)

Test No.:	1	2	3	4	5	6	7	8
$(\sigma_1 - \sigma_3)_f$, kPa	447	167	95	37	331	155	133	119
σ'_{3f}, kPa	246	89	36	6				
σ_c, kPa					481	231	131	53

What can you say about the probable in situ OCR and K_o of this clay? Is it possible to estimate the E_u and τ_f of this soil?

11-66. Strength tests conducted on samples of a stiff overconsolidated clay gave lower strengths for CD tests than for CU tests. Is this reasonable? Why? (After Taylor, 1948.)

11-67. An undisturbed sample of clay has a preconsolidation load of 500 kPa. In which of the following triaxial tests would you expect the compressive strength to be larger? Why?

(a) A CD test performed at a chamber pressure of 10 kPa.
(b) A CU test performed at a chamber pressure of 10 kPa.
(After A. Casagrande.)

11-68. An unconfined compression test is performed on a dense silt. Previous drained triaxial tests on similar samples of the silt gave $\phi' = 35°$. If the unconfined compressive strength was 475 kPa, estimate the height of capillary rise in this soil above the ground water table. Hint: Find the effective confining pressure acting on the specimen. Draw elements similar to Fig. 11.44.

11-69. Estimate the in situ value of K_o of the silt of Problem 11-68. Is this value reasonable in terms of the correlation shown in Fig. 11.69?

11-70. Another specimen of the dense silt of Problem 11-68 is tested in unconfined compression. Assume the average pore size of the silt is 2 μm, and estimate the compressive strength of the sample.

11-71. What would happen if the specimen of Problem 11-68 was prepared in a loose state, then sheared? What would be its unconfined compressive strength?

11-72. The results of unconfined compression tests on a sample of clay in both the undisturbed and remolded states are summarized below. Determine the compressive strength, the initial tangent modulus of deformation, and the secant modulus of deformation at 50% of the compressive strength for both the undisturbed and remolded specimens. Determine the sensitivity of the clay. For the solution of a practical stability problem involving this clay in the undisturbed state, what shear strength would you use if no change in water content occurs during construction? (After A. Casagrande.)

Undisturbed State		Remolded State	
Axial Strain (%)	$\Delta\sigma$ (kPa)	Axial Strain (%)	$\Delta\sigma$ (kPa)
0	0	0	0
1	33	1	7
2	61	2	11
4	109	4	23
6	133	6	32
8	149	8	40
12	160	12	47
16	161	16	50
20	161	20	51

11-73. (a) Show that Eq. 11-8 (in Example 11.12) is correct for *undrained* triaxial or unconfined compression tests. (b) Derive a similar expression for the area of the specimen in a *drained* triaxial test. Hint: $A_s = f(A_o, H_o, \epsilon, \Delta V)$.

11-74. In each of the following cases state which test, X or Y, should show the greater shearing strength. Except for the difference stated below, the two tests are the same type in each case (triaxial, direct shear, etc.) and for identical clay samples.
   (a) The tests are run with no drainage allowed, and test Y is run much faster than test X.
   (b) Sample Y is preconsolidated to a larger pressure than sample X; the pressures during the tests are alike for the two cases.
   (c) Neither sample is preconsolidated; test X is allowed to drain during shear and test Y is not allowed to drain.
   (d) Both samples are highly overconsolidated; test X is not allowed to drain and test Y is allowed to drain.

(e) Test Y is on a sample that is essentially in the undisturbed state, and test X is on a specimen with appreciably disturbed structure but with the same void ratio as Y.
(After Taylor, 1948.)

11-75. List the advantages and disadvantages of each of the field tests listed in Table 11-6 for determining the undrained shear strength of cohesive soils.

11-76. Which of the tests in Table 11-6 are appropriate to measure the undrained shear strength for (a) a building foundation and (b) a cut slope for a highway in each of the following five cases:
  (i) Sensitive Scandinavian clay.
  (ii) Organic marine clay from the U.S. Gulf Coast.
  (iii) Stiff fissured clay till from the midwest United States.
  (iv) Canadian fiberous peat.
  (v) Heavily overconsolidated swelling clay from New Mexico.

11-77. Estimate the maximum expected value of the pore pressure parameter B for the following soils:

  (a) Compacted glacial till at $S = 90\%$.
  (b) Soft saturated normally consolidated Boston blue clay.
  (c) Soil (a) at $S = 100\%$.
  (d) Stiff overconsolidated clay at $S = 99\%$.
  (e) Loose Ottawa sand at $S = 95\%$ and 100%.
  (f) Compacted clayey silt at $S = 90\%$ and subjected to high confining pressures.
  (g) Dense Ottawa sand at $S = 99\%$ and 100%.

11-78. A 2 m thick fill is constructed at the surface of the soil profile of Example 7.5. If the clay is slightly overconsolidated, estimate the change in pore pressure at point A of Fig. Ex. 7.5.

11-79. A soil sample is taken from the midpoint of the clay layer of Example 7.5, that is, from a depth of 6 m. If the pore pressure parameter A_u for unloading is 0.90, estimate the effective vertical and horizontal stresses acting on the sample just before testing in the laboratory. Assume ϕ' for the clay is 25°. Hint: Draw elements with stresses similar to Fig. 11.38, and use the definition of stress *increments* in Appendix B-3. (After G. A. Leonards.)

11-80. What would your answer to Problem 11-79 be if you used Eq. 11-22 instead of 11-13?

11-81. A sample of normally consolidated clay is removed from -10 m below the ground surface. The effective vertical overburden stress is 250 kPa, and K_o is 0.8. If the pore pressure parameter due to sampling is 0.7, estimate the change in pore pressure in the sample when it is removed from the clay layer. What effective stresses act on the specimen after extrusion from the sample tube?

11-82. Show that Δu in Example 11-16 is about 32 kPa, as predicted by Eqs. 11-22 and 11-25.

11-83. Prepare a listing of those relationships in Chapter 11 and elsewhere that can be used to predict a soil parameter or property (namely, ϕ, c, K_o) when some other property (w, PI, etc.) is known. List the page and figure number, ordinate, abscissa, and variables of the graph, if any. (This listing will be helpful for solving the next four problems.)

11-84. For the data shown in Fig. 8.5, estimate the unconfined compressive strength and the sensitivity of this soil. Typical values for the clay are LL = 88, PL = 43, and PI = 45.

11-85. The data presented in Fig. 8.15b are for a black fissured organic silty clay or clayey silt. At a depth of 6 m, estimate the expected value or range of values of the undrained modulus.

11-86. A cohesive soil with a liquidity index of 1 has a natural water content of 50%. Estimate as many soil parameters as you can. Include, if you can, compressibility and rate of compression parameters as well as those related to shear strength.

11-87. The medium gray silty clay of Fig. 8.18b at a depth of 20 m had an LL of 38 and a PL of 23. Estimate the following parameters for this soil: (a) coefficient of earth pressure at rest; (b) effective angle of internal friction; (c) ratio of τ_f/σ'_{vo}; (d) activity; (e) sensitivity; and (f) the undrained Young's modulus. Are there any inconsistencies in the values you obtained? If so, discuss the possible reasons.

11-88. A normally consolidated clay has a ϕ' of 30°. Two identical specimens of this clay are consolidated to 200 kPa in a triaxial cell. Predict the maximum and minimum possible axial stresses in the specimens for a constant cell pressure. Hint: The first test is an axial compression test, the second test is an axial extension test. What assumptions are necessary to solve this problem?

11-89. The effective stresses at failure for three identical triaxial specimens of an overconsolidated clay are shown in Fig. P11-89. Plot the Mohr circles at failure and determine ϕ' and c'. Determine the theoretical angle of inclination of the failure planes in each test

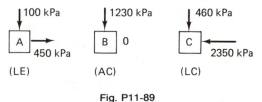

Fig. P11-89

specimen, and show these on a small sketch. Also sketch the effective stress paths for the three tests. (After C. W. Lovell.)

11-90. Three identical specimens (same e, w) of a clay are normally consolidated and sheared consolidated-drained (CD) in both compression and extension. The stresses at failure for the three specimens are as shown in Fig. P11-90.

(a) Plot the Mohr circles at failure, and determine ϕ' and ϕ_{total}.

(b) Determine the inclination of the predicted failure planes (from the Mohr failure hypothesis). Sketch the failed specimens, showing their failure planes.

(c) Sketch the three stress paths.

(After C. W. Lovell.)

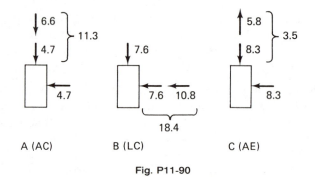

Fig. P11-90

11-91. A series of conventional triaxial compression tests were conducted on three identical specimens of a saturated clay soil. Test results are tabulated below.

Specimen	σ_c (kPa)	$(\sigma_1 - \sigma_3)_f$ (kPa)	Δu_f (kPa)
A	100	170	40
B	200	260	95
C	300	360	135

(a) Sketch the total and effective stress paths for each test, and determine the Mohr-Coulomb strength parameters in terms of both total and effective stresses. (b) Estimate the theoretical angle of the failure planes for each specimen. (c) Do you believe this clay is normally or overconsolidated? Why?

11-92. Assume that the induced pore pressures at failure for Problem 11-91 were: specimen A, -15 kPa; specimen B, -40 kPa; and specimen C, -80 kPa; and that everything else was the same. Now do parts (a) and (b) above, and then answer part (c).

11-93. An axial compression CU test has been performed on an undisturbed specimen of 100% saturated organic clay. The data for the test are given in Problem 11-54. A lateral extension test is to be performed on an identical specimen at the same consolidation pressure and with the same time of consolidation and time of loading as in the axial compression test.

(a) Plot the total and effective stress paths. Determine the curve of pore pressure versus (1) principal stress difference and (2) axial strain that you would predict theoretically for the lateral extension test.

(b) On the p-q diagram, draw the line corresponding to zero induced pore pressure and the line along which the magnitude of the induced negative pore pressure is equal to the principal stress difference.

(c) What is A_f for both the AC and LE test?

(After A. Casagrande and R. C. Hirschfeld.)

11-94. The following data were obtained from a conventional triaxial compression test on a saturated ($B = 1$), normally consolidated simple clay (Ladd, 1964). The cell pressure was held constant at 10 kPa, while the axial stress was *increased* to failure (axial compression test).

ϵ_{axial} (%)	$\Delta\sigma_{axial}$ (kPa)	Δu (kPa)
0	0	0
1	3.5	1.9
2	4.5	2.8
4	5.2	3.5
6	5.4	3.9
8	5.6	4.1
10	5.7	4.3
12	5.8 failure	4.4

(a) Plot the $\Delta\sigma$ and Δu versus axial strain curves. Determine A_f.

(b) Plot the total and effective stress paths for the AC test.

(c) What is ϕ'? (Assume $c' = 0$ for normally consolidated clay.)

A lateral extension (LE) test was conducted on an identical sample of the same clay (same e, w). In this test, the axial or vertical stress was held constant at 10 kPa, while the cell pressure was *decreased* to 4.2 kPa, at which time the specimen failed.

(d) Plot both the total and effective stress paths for the LE test.
(e) Determine u_f, σ'_{1f}, σ'_{3f} and A_f for this test.
(f) Find ϕ_{total} for both the AC and the LE tests.
(g) Find the theoretical inclinations (from the Mohr failure hypothesis) of the failure planes in each test. Sketch the specimen at failure, indicating the effective stresses at failure and the failure plane inclination.

11-95. A conventional triaxial compression (AC) test was conducted on a saturated sample of overconsolidated clay, and the following data, normalized with respect to the effective confining pressure, were obtained.

ϵ_{axial} (%)	$\Delta\sigma/\sigma'_c$	$\Delta u/\sigma'_c$
0	0	0
0.5	0.57	+ 0.07
1	0.92	+ 0.05
2	1.36	− 0.03
4	1.77	− 0.22
6	1.97	− 0.35
8	2.10	− 0.46
10	2.17	− 0.52
12	2.23	− 0.58
14	2.28	− 0.62
16	2.33 failure	− 0.67

A lateral extension (LE) test was conducted on an identical specimen of the same clay. While the vertical stress was maintained constant, the cell pressure was decreased until failure occurred at the same principal stress difference as the AC specimen ($\Delta\sigma/\sigma'_c = 2.33$). From your knowledge of stress paths and soil behavior, determine (a) the effective and total stress paths for both tests and (b) the pore pressure versus strain response of the LE test. (c) Can the Mohr-Coulomb strength parameters be determined? Why? (After C. W. Lovell.)

11-96. A K_o consolidated-undrained triaxial compression ($\sigma_{cell} =$ constant) test was conducted on an undisturbed specimen of sensitive Swedish clay. The initial conditions were as shown in Fig. P11-96a. The stress-strain and pore pressure response of the specimen is shown in Fig. P11-96b.

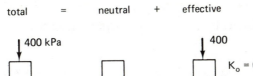

total   =   neutral   +   effective

Fig. P11-96a

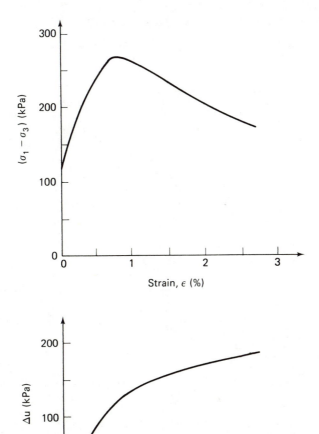

Fig. P11-96b

(a) Find the stress conditions at failure and symbolically show the total, neutral, and effective stresses (like the "initial conditions" shown above).

(b) Sketch the total and effective stress paths.

(c) Plot A versus ϵ. What is A_f? What are ϕ' and ϕ_T?

11-97. If an LE test were conducted on a sample of Swedish clay identical to that tested in Problem 11-96, predict the pore pressure versus strain response of the clay. What is u_f and $\overline{A}_f$? What is ϕ_T?

11-98. The data shown in Fig. P11-98 were obtained from several CU tests on a saturated clay which has an OCR of 10 and a precon-

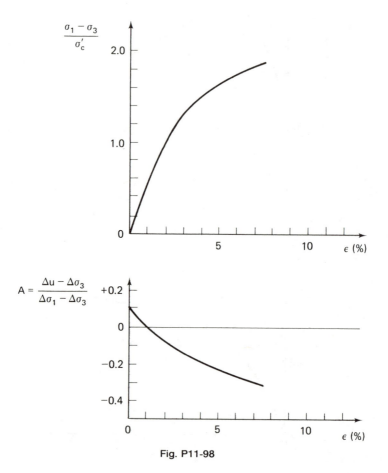

Fig. P11-98

solidation stress of 800 kPa. It is assumed that these results are valid for all compression stress paths on this clay. You are going to run a special stress path test on this clay. After consolidation at σ'_{vo}, the cell pressure will be increased in such a way that $\Delta\sigma_3 = 0.2\,\Delta\sigma_1$ until failure occurs. For this special stress path test, fill in the table below and plot the total and effective stress paths. (After C. W. Lovell.)

ϵ (%)	$\Delta\sigma_1$ (kPa)	$\Delta\sigma_3$ (kPa)	σ_1 (kPa)	σ_3 (kPa)	Δu (kPa)	A
0						
0.5						
2.5						
5.0						
7.5						

11-99. A series of CU compression tests on a simple clay (Ladd, 1964) provided the following test results:

ϵ_{axial} (%)	$2\tau_f/\sigma'_c$	A
0	0	—
1	0.35	0.53
2	0.45	0.64
3	0.50	0.72
4	0.52	0.76
6	0.54	0.88
8	0.56	0.92
10	0.57	0.93
12 failure	0.58	0.945

(a) In an axial compression test, if $\sigma'_c = 200$ kPa, determine q_f, p_f, and p'_f. (b) Find ϕ' and c'. A special lateral extension stress path test was conducted on this clay in which the decrease in lateral stress was exactly equal to the increase in axial stress; that is, $-\Delta\sigma_3 = \Delta\sigma_1$. For this case, if $\sigma'_c = 400$ kPa, determine $\Delta\sigma_1$, q, p, p', and Δu when (c) the axial strain is 4% and (d) at failure. (After C. W. Lovell.)

11-100. Figure P11-100 shows normalized data from an axial compression (AC) triaxial test and a lateral compression (LC) triaxial test on saturated simple clay (Ladd, 1964). Make the appropriate calculations, and plot the complete total and effective stress paths for both tests. What are the Mohr-Coulomb strength parameters? Determine A_f for each test.

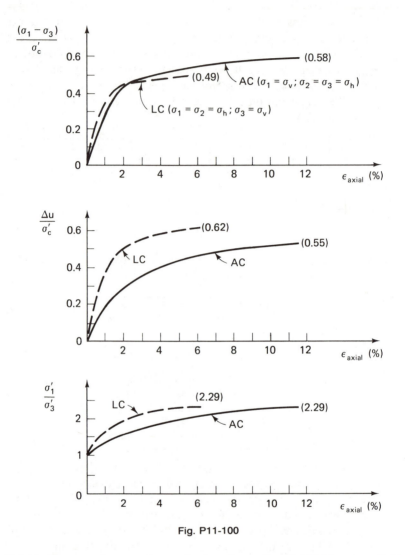

Fig. P11-100

11-101. Two specimens of a soft clay from the Skå-Edeby test field in Sweden were reconsolidated to their initial in situ effective stress conditions and then sheared to failure. One specimen was loaded in axial compression (AC), while the other was failed by axial extension (AE). The normalized stress-strain and pore pressure strain data for both tests are shown in Fig. P11-101 (after Zimmie, 1973). Pertinent specimen data is given in the accompanying table. (a) On a p-q diagram, sketch the total, total $- u_o$, and effective stress paths for both tests. (b) Determine ϕ' and ϕ_{total} in both

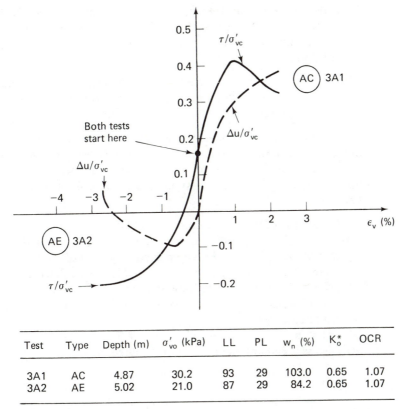

Fig. P11-101

Test	Type	Depth (m)	σ'_{vo} (kPa)	LL	PL	w_n (%)	K_o^*	OCR
3A1	AC	4.87	30.2	93	29	103.0	0.65	1.07
3A2	AE	5.02	21.0	87	29	84.2	0.65	1.07

*Assumed.

Note: Tests were conducted with a back pressure of 200 kPa. In situ pore water pressure is approximately 400 kPa.

compression and extension. (c) Calculate the Skempton pore pressure parameter A at failure for both tests. (d) Show in a sketch the predicted theoretical angles of the failure planes for the two specimens.

11-102. Are the values given and calculated for K_o, τ_f/σ'_{vo}, ϕ', etc., for the Skå-Edeby clay of Problem 11-101 reasonable in terms of the simple correlations with PI, LI, etc., given in this chapter?

11-103. For the oil tank problem in Chapter 8 (Problem 8-46), plot the complete total, total $- u_o$, and effective stress paths due to construction and filling of the tank for an element under the centerline of the tank and at the midpoint of the clay layer. Assume that

K_o at the site is 0.7 and that the average value of the A parameter before failure is 0.4; assume $A_f = 0.5$. Make reasonable estimates of the strength parameters, and estimate the factor of safety against failure.

11-104. What is the maximum safe height of the embankment for Examples 11.16 and 11.19? Plot a graph of factor of safety versus height of the embankment.

11-105. How would you recommend the shear strength be determined for the following design situations? Your answer can include both laboratory and field tests or, in some cases, no tests but some other design approach that may be appropriate. Be as specific as you can.

(a) Long-term stability of a compacted clay earth dam.
(b) Stability of a hydraulic fill sand dam under seismic loading.
(c) End of construction of a compacted clay earthfill dam.
(d) Foundation on a soft saturated normally consolidated clay.
(e) Shallow foundation on a loose dry sand.
(f) End of construction of an excavation in soft normally consolidated clay.
(g) Cut slope in an overconsolidated stiff fissured clay.
(h) Highway embankment on a stiff fissured clay.

appendix a

Application of the SI System of Units to Geotechnical Engineering[*]

A.1 INTRODUCTION

Within the scientific and engineering community, there has always been some confusion as to the proper system of units for physical measurements and quantities. Many systems have been advanced during the last few centuries and some, such as the Imperial or British Engineering system, the so-called "metric system," and a few hybrids, have achieved popular usage. Recently, with the growth of international cooperation and trade, it has become increasingly apparent that one, single, commonly accepted system of units would be not only convenient but also of tremendous practical value.

Although the field of geotechnical engineering may not claim the greatest confusion in the use of units, it undoubtedly ranks near the top of all fields in the number of different systems in common usage. Laboratory engineers, following their counterparts in the physical sciences, have attempted to use some sort of metric system, usually the cgs (centimetre-gram-second) system, for the simple laboratory tests. They, with ease, apply the mks (metre-kilogram-second) system to measurements of pressure and stress in consolidation and triaxial tests, and, with some impunity, they use British Engineering units for compaction tests. As any teacher of soil mechanics can testify, the confusion to the uninitiated is tremendous. At least, practicing geotechnical engineers in North America have been somewhat consistent in the use of the British Engineering system for laboratory and field densities, stress measurements, etc., although they commonly alternate between pounds per square foot, kips per square foot,

---

[*]This appendix has been adapted from an article written by R. D. Holtz at Northwestern University, November 1969. See also Holtz (1980).

tons per square foot, and pounds per square inch, depending on how they or their clients feel about the subject. Fortunately, 1 ton-force/ft^2 is within 2% of 1 kg-force/cm^2, a common laboratory unit for stress and pressure, and the foundation engineer utilizing consolidation test data can convert directly with small error. Strictly speaking, using force as a basic unit is incorrect; mass should be the basic unit, with force derived according to Newton's second law of motion. Use of the kilogram as a unit of force is one of the difficulties with the so-called "metric system," a modified version of the mks system, which was common among continental European engineers. At least they tried to keep the distinction between mass and force by calling the kilogram-force a *kilopond* (kp).

A modernized version of the metric system has been developed over the past 30 years. The system is known as *SI*, which stands for "Le Système International d'Unités" ("The International System of Units"); it is described in detail in the ASTM (1966) *Metric Practice Guide* and in the *ASTM Standard for Metric Practice* (1980), Designation E 380-79, available in the back of every current ASTM Annual Book of Standards. The system may eventually become the common, and perhaps the only legal, system used in the United States, Canada and a few other countries still using the Imperial or British Engineering system. In fact, Great Britain itself converted completely to SI in 1972, and Australia and New Zealand followed suit shortly thereafter. Most European countries already have de facto conversion to SI, especially in engineering practice.

A.2 THE SI METRIC SYSTEM

The SI metric system is a fully coherent and rationalized system. It is founded on seven basic units for *length* (metre or meter), *mass* (kilogram), *time* (second), *electric current* (ampere), *thermodynamic temperature* (kelvin), *luminous intensity* (candela), and *amount of substance* (mole). All these basic units have precise definitions, names, and symbols. Units for all other physical quantities can be derived in terms of these basic units. Sometimes the derived quantities are given specific names, such as the *newton* for force and the *watt* for power. The derived unit of force replaces the kilogram-force (kgf) of the mks system, so that the name of the unit indicates that it is a unit of force, not mass. A great advantage is that *one and only one unit exists fo, each physical quantity*, and all other mechanical quantities such as velocity, force, and work can be derived from the basic units. In addition, the SI units for force, energy, and power are *independent* of the nature of the physical process, whether mechanical, electrical, or chemical.

As previously mentioned, a major advantage of SI is that it is a fully coherent system, which means that a product or quotient of any two unit quantities is a unit of the resulting quantity. For example, unit length squared should be unit area, and unit force should be unit mass times unit acceleration. Obviously, many of the engineering units in common use (for example, acre, lb-force, kg-force) are not coherent units. Also, units which might be related to basic units by powers of 10 are *not* consistent within the SI system. A good example is the litre, or liter, which is a cubic decimetre. The equivalent volume of the litre has been defined as exactly 10^{-3} m^3 (1000 cm^3). Additional advantages of SI include the use of unique and well-defined symbols and abbreviations and the convenient decimal relation between multiples and submultiples of the basic units.

In the next two sections of this appendix we describe in detail the SI units of particular interest in geotechnical engineering and present appropriate conversion factors for some of the common mks and British Engineering units. Since you are likely to encounter just about anything in your engineering practice, it is important that you know how to convert between these systems and SI, and that you have some feel for physical quantities in both sets of units.

A.3 BASIC AND DERIVED SI METRIC UNITS

The three *base units* of interest to geotechnical engineers are *length*, *mass*, and *time*. The SI units for these quantities are the *metre*, m, the *kilogram*, kg, and the *second*, s. Temperature, which might also be of interest, is expressed in *kelvins* (K), although the system does allow for use of the degree Celsius (°C), which has the same interval. Electric current is expressed in *amperes* (A). Supplementary units include the *radian* and *steradian*, the units of plane and solid angles, respectively.

As mentioned, these basic SI units have precise physical definitions. For example, contrary to a popular misconception, the metre is *not* the distance between two bars in Paris, but rather it has been defined as exactly equal to a certain number of wavelengths of radiation corresponding to a specific transition level of krypton 86. The standard kilogram is equal to the mass of the international prototype kilogram, a cylinder of platinum-iridium alloy preserved in a vault at Le Bureau International des Poids et Mesures at Sèvres, France. Similar standard kilograms can also be found at the National Bureau of Standards near Washington, D.C. The second has been defined as the duration of a certain number of periods of the radiation corresponding to a specific transition state of cesium 133.

Derived units geotechnical engineers use include those listed in Table A-1.

Prefixes are used to indicate multiples and submultiples of the basic and derived units. SI prefixes are listed in Table A-2.

The prefixes should be applied to indicate orders of magnitude of the basic or derived units and to reduce redundant zeros so that numerical values lie between 0.1 and 1000. They should *not* be applied to the

TABLE A-1

Quantity	Unit	SI Symbol	Formula
acceleration	metre per second squared	m/s^2	—
area	square metre	m^2	—
area	hectare	ha	$hm^2 = 10^4 \, m^2$
density	kilogram per cubic metre	kg/m^3	—
force	newton	N	$kg \cdot m/s^2$
frequency	hertz	Hz	$1/s$
moment or torque	newton metre	$N \cdot m$	$kg \cdot m^2/s^2$
power	watt	W	J/s
pressure	pascal	Pa	N/m^2
stress	pascal	Pa	N/m^2
unit weight	newton per cubic metre	N/m^3	$kg/s^2 \cdot m^2$
velocity	metre per second	m/s	—
voltage	volt	V	W/A
volume	cubic metre	m^3	—
volume	litre	L	$dm^3 = 10^{-3} \, m^3$
work (energy)	joule	J	$N \cdot m$

TABLE A-2

Factor	Prefix	Symbol
10^{18}	exa	E
10^{15}	peta	P
10^{12}	tera	T
10^{9}	giga	G
10^{6}	mega	M
10^{3}	kilo	k
10^{2}	hecto	h
10^{1}	deka	da
10^{-1}	deci	d
10^{-2}	centi	c
10^{-3}	milli	m
10^{-6}	micro	μ
10^{-9}	nano	n
10^{-12}	pico	p
10^{-15}	femto	f
10^{-18}	atto	a

denominator of compound units (kilogram is an exception since kg is a basic unit in the SI system). Note that *spaces*, not commas, should be used to separate groups of zeros (a concession to the Europeans to persuade them to stop using a comma as a decimal point!).

To maintain the coherence of the system, it is recommended that *only basic units be used to form derived units*. For example, the unit of force, the newton, is derived according to Newton's second law, $F = Ma$, where the mass M is in kilograms and the acceleration a is in m/s^2, all basic units. For derived combinational units such as pressure or stress (pascals or newtons per square metre), multiples and submultiples of the basic metric units (in this case metres) should be avoided. For example, N/cm^2 and N/mm^2 are wrong; the appropriate prefix should be used with the numerator to indicate larger or smaller quantities, for example, kN/m^2 or MN/m^2 (for kilonewtons per square metre or meganewtons per square metre).

A.4 SI UNITS OF INTEREST TO GEOTECHNICAL ENGINEERS AND THEIR CONVERSION FACTORS

Length. You should already be familiar with the SI unit for length (the metre, m). (By the way, this is the ASTM recommended spelling.) Useful SI length multiples and submultiples are the kilometre (km), millimetre (mm), micrometre (μm), and nanometre (nm). Conversion factors for the common British Engineering and mks systems are:

1 inch, in.	= 25.4 mm = 0.0254 m
1 foot, ft	= 0.3048 m
1 yard, yd	= 0.9144 m
1 mile (U.S. statute)	= 1.609×10^3 m = 1.609 km
1 mile (nautical)	= 1.852×10^3 m = 1.852 km
1 angstrom, Å	= 1×10^{-10} m = 0.1 nm
1 mil	= 2.54×10^{-5} m = 0.0254 mm = 25.4 μm

Good SI practice suggests that multiple and submultiple metric units be used in increments of 1000, for example, mm, m, km. Use of the centimetre, especially for lengths under 300 mm, should be avoided.

Mass. You may recall from physics that the inertia or mass (SI unit: kilogram, kg) of a physical object is a measure of the property which controls the response of that object to an applied force. It is convenient to measure the mass in terms of the acceleration of an object produced by a unit force, as related by Newton's second law of motion. Thus a unit force causes 1 kg mass to accelerate 1 m/s^2. The mass then is an appropriate measure of the amount of matter an object contains. The mass remains the same even if the object's temperature, shape, or other physical attributes change. Unlike weight, which is discussed later, the mass of an object does not depend on the local gravitational attraction, and thus it is also independent of the object's location in the universe.

Among all the SI units, the kilogram is the only one whose name, for historical reasons, contains a prefix. The names of multiples and submultiples of the kilogram are formed by attaching prefixes to the word *gram* rather than to *kilogram*. In other words, 10^{-6} kg is not a micro-kilogram, but a milligram = 10^{-3} g. Similarly, 1000 kg is not 1 kilo-kilogram but is equivalent to 1 megagram (Mg); 1000 kg is also the metric ton (t), sometimes spelled "tonne" to avoid confusion with the British ton = 2000 lb. ASTM recommends that the metric ton be restricted to commercial usage and that the term *tonne* be avoided altogether. Practical units of mass in engineering practice are the megagram (Mg), the kilogram (kg), and gram (g), the latter two units being primarily used in laboratory work.

Some useful relationships and conversion factors are:

> 1 pound mass, 1bm (avoirdupois) = 0.4536 kg
> 1 British (short) ton = 2000 lbm   = 907.2 kg
> 1 gram, g   = 10^{-3} kg
> 1 metric ton, t   = 10^3 kg = 10^6 g = 1 Mg
> 1 slug (1 lb-force/ft/s^2)   = 14.59 kg

Time. Although the second (s) is the basic SI time unit, minutes (min), hours (h), days (d), etc., may be used where convenient, even though they are not decimally related. (Maybe some day we will even have a decimal time system; see Carrigan, 1978.)

Force. As mentioned, the SI unit of force is derived from $F = Ma$, and it is termed the *newton* (N), which is equal to 1 kg$\cdot$m/s^2. Conversion factors for common engineering force units are:

> 1 lb-force   = 4.448 N
> 1 British ton-force   = 8.896×10^3 N = 8.896 kN
> 1 kg-force = 1 kp   = 9.807 N
> 1 kip = 1000 lb-force   = 4.448×10^3 N = 4.448 kN
> 1 metric ton-force = 1000 kg-force = 9.807×10^3 N = 9.807 kN
> 1 dyne (g$\cdot$cm/s^2)   = 10^{-5} N = 10 μN

It is obvious that the numbers in newtons for such items as column loads would be very large indeed and consequently somewhat awkward. Therefore, consistent with the rules for application of prefixes, it is simple to adjust these rather large numbers to more manageable quantities for engineering work. The common prefixes would be kilo (10^3), mega (10^6), and giga (10^9), so that engineering forces would be kilonewtons, kN, meganewtons, MN, and giganewtons, GN. (The symbol for mega is M, to avoid confusion with the symbol for milli, m.) Thus, since 1 ton-force is 8.9 kN, 1000 tons would be 8.9 MN.

Some useful relationships of these prefixes are:

1 kilonewton, kN	$= 10^3$ newton		$= 1000$ N	
1 meganewton, MN	$= 10^6$ newton	$= 10^3$ kN	$= 1000$ kN	
1 figanewton, FN	$= 10^8$ newton	$= 10^5$ kN	$= 10^2$ MN	$= 100$ MN
1 giganewton, GN	$= 10^9$ newton	$= 10^6$ kN	$= 10^3$ MN	$= 1000$ MN
3 giganewtons	$= 30$ figanewtons	$= 1$ boxafiganewtons*		
14.4 giganewtons	$= 1$ grossafiganewtons			

*This unit is only a constant prior to opening the box.

The correct unit to express the *weight* of an object is the newton since the weight is the gravitational force that causes a downward acceleration of the object. Or, weight W equals Mg, where M is the mass of the object and g is the acceleration due to gravity. You will recall that the acceleration due to gravity varies with latitude and elevation and, in fact, SI recommends that weight be avoided and that mass be used instead. If weight must be used, it is suggested that the location and gravitational acceleration also be stated. However, for most ordinary engineering purposes, the difference in acceleration (about 0.5%) can be neglected, and as long as we express the weight in newtons, the units will be consistent.

Another problem with weight is that it is commonly used when we really mean the mass of an object. For example, in the laboratory when we "weigh" an object on a laboratory balance, we really are comparing two masses, the mass of the unknown object with objects of known mass. Even scales or balances which displace linear springs are calibrated by using objects of known mass.

Further ambiguity occurs, of course, because common units of mass such as the pound or kilogram are often used in engineering practice as a unit of force. If pound is used as a unit of force, then depending on the resulting accelerations, different mass units are defined. For example, if a 1 lb-force causes an acceleration of 1 ft/s^2, then the mass is 1 lb-force·s^2/ft, which is called a *slug*. In other words, 1 lb-force $= 1$ slug $\times$ 1 ft/s^2. Using slugs as units of mass avoids the confusion with pounds-mass, and this unit has been commonly used in aerodynamics and fluid mechanics.

If we wanted to use instead a pound-mass system, we could define a unit of force called the *poundal*, where 1 poundal $= 1$ lbm $\times$ 1 ft/s^2. Poundals are apparently used only in physics books.

EXAMPLE A.1

Given:

A force of 1 lb acts on an object weighing 1 lb.

Required:

Find the resulting acceleration.

Solution:

From Newton's second law,

$$F = Ma = \left(\frac{W}{g}\right)a$$

or

$$a = \frac{Fg}{W} = \frac{(1\ \text{lbf})(32.17\ \text{ft/s}^2)}{1\ \text{lbf}} = 32.17\ \text{ft/s}^2$$

EXAMPLE A.2

Given:

The object in Example A.1, which weighs 1 lbf.

Required:

Find its mass when a 1 lbf causes an acceleration of 1 ft/s^2.

Solution:

$$F = Ma = \left(\frac{W}{g}\right)a$$

or

$$M = \frac{W}{g} = \frac{1\ \text{lbf}}{32.17\ \text{ft/s}^2} = 0.031\frac{\text{lbf} \cdot \text{s}^2}{\text{ft}} = 0.031\ \text{slug}$$

EXAMPLE A.3

Given:

Neil Armstrong weighs 150 lb on earth.

Required:

How much does he weigh on the surface of the moon?

Solution:

First, we have to calculate Mr. Armstrong's mass on earth. Unless he had health problems during the voyage, his mass will be the same on the moon.

$$M = \frac{W}{g} = \frac{150 \text{ lbf}}{32.17 \text{ ft/s}^2} = 4.66 \frac{\text{lbf} \cdot \text{s}^2}{\text{ft}}, \text{ or } 4.66 \text{ slugs}$$

Since 1 slug = 14.59 kg, his mass is 68.03 kg. Another way to calculate his mass is to convert his weight to newtons; then divide by g.

$$W = 150 \text{ lbf} \left(\frac{4.448 \text{ N}}{1 \text{ lbf}} \right) = 667.20 \text{ N or } 667.2 \frac{\text{kg} \cdot \text{m}}{\text{s}^2}$$

$$M = \frac{W}{g} = \frac{667.2 \text{ kg} \cdot \text{m/s}^2}{9.807 \text{ m/s}^2} = 68.03 \text{ kg}$$

Next, we have to either ask an astronomer or look up in the *Handbook of Chemistry and Physics* (1977) or some other reference the gravitational acceleration on the surface of the moon. We find that $g_{moon} = 1.67 \text{ m/s}^2$. Thus,

$$W_{moon} = Mg_{moon} = 68.03 \text{ kg} (1.67 \text{ m/s}^2) = 113.62 \text{ N}$$

Or, since 4.448 N = 1 lbf,

$$W_{moon} = 113.62 \text{ N} \left(\frac{1 \text{ lbf}}{4.448 \text{ N}} \right) = 25.54 \text{ lbf}$$

Check: On earth, 667 N $\left(\frac{1.67}{9.81} \right) = 113.6$ N on the moon.

———————

See how confusing the old British Engineering system can be? However, if you think this is bad, wait until you try to convert densities and unit weights!

Stress and Pressure. The SI unit for stress and pressure is the *pascal* (Pa), which is exactly equal to 1 newton per square metre (N/m^2).

There has been some objection, especially in Europe, to the use of the pascal as the basic unit of stress and pressure because it is so small. The Germans and French, for example, often use the *bar*, which is exactly 10^5 Pa. However the pascal is more logical since it is a coherent unit; that is, equations involving the pascal with other SI units can be written without coefficients of proportionality being required.

Conversion factors for some common engineering units are:

1 psi (lb-force/in.^{2})	= 6.895×10^3 Pa or 6.895 kPa
1 atm at STP*	= 1.013×10^5 Pa or 101.3 kPa
1 kg-force/cm^2	= 9.807×10^4 Pa or 98.07 kPa
1 metric ton-force/m^2	= 9.807×10^3 Pa or 9.807 kPa
1 bar	= 1×10^5 Pa or 100 kPa
1 ksi (kip/in.^{2})	= 6.895×10^6 Pa or 6.895 MPa
1 British ton-force/ft^2	= 95.76×10^3 Pa or 95.76 kPa
1 lb-force/ft^2	= 47.88 Pa

*Standard temperature and pressure, not a motor oil additive or Soil Test Probe.

It is obvious that the pascal is a small unit, but as with SI force units, it is easy to add prefixes to make the large numbers more manageable. Thus, 1 psi in the above table is more conveniently expressed as 6.9 kPa (kN/m^2) than as 6.9×10^3 Pa. For ordinary triaxial testing of soils, for example, hydrostatic cell pressures rarely exceed 200 or 300 psi (1379 or 2068 kPa). Or, if all the pressures in a test series are in this range, it might be convenient to use 1.4 or 2.1 MPa. And, as with other systems of units, a rounded or even interval may be more convenient; for example, in this case, 1.5 and 2.0 MPa.

Similar examples could be given for engineering stresses. Either kilopascals or megapascals, kPa or MPa, or kilo- or meganewtons per square metre, kN/m^2 or MN/m^2, will become commonly used for foundation stresses, lateral earth pressures, allowable bearing values, etc. In the laboratory, force is measured by a proving ring or load cell and then converted to stress (for example, in the unconfined compression or direct shear tests), so the computational process will be no more complicated than it is now. Similarly, with electrical pressure transducers, a calibration factor must be used to convert millivolts (mV) output to pressure in whatever units are used.

A convenient approximation, part of which is already in use in geotechnical engineering practice, is the following:

$$1 \text{ British ton-force/ft}^2 \simeq 1 \text{ kg-force/cm}^2 \simeq 1 \text{ atmosphere}$$
$$\simeq 10 \text{ metric ton-force/m}^2 \simeq 100 \text{ kPa} = 100 \text{ kN/m}^2$$

The error involved is between 2% and 4%, which is certainly less than ordinary engineering accuracy requirements.

EXAMPLE A.4

Given:

The pressure or stress is 100 kPa.

Required:

Convert this pressure or stress to (a) psi (lb-force/in².), (b) ksi (kips/in².), (c) tsf (British ton-force/ft²), (d) kg-force/cm², (e) bar, (f) metric ton-force/m², (g) mm of mercury, (h) ft of water, and (i) m of water.

Solution:

A simple way to convert from one set of units to another is to set up an equation with the equivalents in either the numerator or denominator of the equation so that the appropriate cancellations occur.

a. $p = 100 \text{ kPa} = 100 \dfrac{\text{kN}}{\text{m}^2} \left(\dfrac{1 \text{ lbf}}{4.448 \text{ N}} \right) \left(\dfrac{1000 \text{ N}}{\text{kN}} \right) \left(\dfrac{0.0254 \text{ m}}{1 \text{ in.}} \right)^2$

$= 14.5 \text{ psi}$

Note: The exact conversion value is $14.503\,773\,77 \text{ kN/m}^2$, which comes about if you use the exact value for $1 \text{ lbf} = 4.448\,221\,615\,260\,5 \text{ N}$. 1 in. is exactly equal to 0.0254 m.

b. $p = 100 \text{ kPa}$

$= 100 \dfrac{\text{kN}}{\text{m}^2} \left(\dfrac{1 \text{ lbf}}{4.448 \text{ N}} \right) \left(\dfrac{1000 \text{ N}}{\text{kN}} \right) \left(\dfrac{1 \text{ kip}}{1000 \text{ lbf}} \right) \left(\dfrac{0.0254 \text{ m}}{1 \text{ in.}} \right)^2$

$= 0.0145 \text{ ksi}$

Again, as in part (a), the exact conversion value is slightly different.

c. $p = 100 \text{ kPa}$

$= 100 \dfrac{\text{kN}}{\text{m}^2} \left(\dfrac{1 \text{ lbf}}{4.448 \text{ N}} \right) \left(\dfrac{1000 \text{ N}}{\text{kN}} \right) \left(\dfrac{1 \text{ tonf}}{2000 \text{ lbf}} \right) \left(\dfrac{0.3048 \text{ m}}{1 \text{ ft}} \right)^2$

$= 1.04 \text{ tonf/ft}^2$

d. $p = 100 \text{ kPa} = 100 \dfrac{\text{kN}}{\text{m}^2} \left(\dfrac{1 \text{ lbf}}{9.807 \text{ N}} \right) \left(\dfrac{1000 \text{ N}}{\text{kN}} \right) \left(\dfrac{\text{m}}{100 \text{ cm}} \right)^2$

$= 1.02 \text{ kgf/cm}^2$

Note: The exact conversion for kgf to N is 9.806 65.

e. $p = 100 \text{ kPa} = 100 \text{ kPa} \left(\dfrac{1 \text{ bar}}{10^5 \text{ Pa}} \right) \left(\dfrac{10^3 \text{ Pa}}{1 \text{ kPa}} \right) = 1 \text{ bar}$

f. $p = 100 \text{kPa} = 100 \dfrac{\text{kN}}{\text{m}^2} \left(\dfrac{1 \text{ kgf}}{9.807 \text{ N}} \right) \left(\dfrac{1000 \text{ N}}{\text{kN}} \right) \left(\dfrac{1 \text{ tonf}}{1000 \text{ kgf}} \right)$

$= 10.2 \text{ metric tonf/m}^2$

g. For p in mm of mercury, we need to remember or look up the density of Hg. It is 13.6 g/cm^3. Also recall from hydrostatics that $p = \gamma z$

$= \rho gz$, where z is the depth of the fluid. Thus for pressure in cm of mercury, $z = p/\rho g$. So

$$z = 100\frac{kN}{m^2} \left(\frac{1000\ N}{kN}\right)\left(\frac{cm^3}{13.6\ g}\right)\left(\frac{1000\ g}{kg}\right)\left(\frac{m}{100\ cm}\right)^3$$

$$\cdot\left(\frac{s^2}{9.807\ m}\right)\left(\frac{1000\ mm}{m}\right)$$

$$= 750\ mm\ Hg$$

h. Again, use $z = p/\rho g$

$$z = 100\frac{kN}{m^2}\left(\frac{1000\ N}{kN}\right)\left(\frac{m^3}{1000\ kg}\right)\left(\frac{s^2}{9.807\ m}\right)\left(\frac{1\ ft}{0.3048\ m}\right)$$

$$= 33.5\ ft\ of\ water$$

i. $z = 100\dfrac{kN}{m^2}\left(\dfrac{1000\ N}{kN}\right)\left(\dfrac{m^3}{1000\ kg}\right)\left(\dfrac{s^2}{9.807\ m}\right)$

$$= 10.2\ m\ of\ water$$

Density and Unit Weight. Density is defined as mass per unit volume. Its units in the SI metric system are kilograms per cubic metre, kg/m^3. In many cases, it may be more convenient to express density in megagrams per cubic metre, Mg/m^3. Conversions from the common laboratory and field densities are:

$$1\ lb\text{-}mass/ft^3 = 16.018\ kg/m^3$$
$$1\ g/cm^3 \quad = 10^3\ kg/m^3 = 1\ Mg/m^3 = 1\ t/m^3$$

You will recall that the density of water, ρ_w, is exactly 1.000 g/cm^3 at 4°C, and the variation is relatively small over the range of temperatures encountered in ordinary engineering practice. Therefore it is usually sufficiently accurate to take $\rho_w = 10^3\ kg/m^3 = 1\ Mg/m^3$, which simplifies phase computations considerably. It is also useful to know that 1000 kg/m^3 is equal to 62.4 lb-mass/ft^3.

Typical densities that might be encountered in geotechnical practice are 1.2 Mg/m^3 (74.8 lb/ft^3), 1.6 Mg/m^3 (100 lb/ft^3), and 2.0 Mg/m^3 (125 lb/ft^3). Ranges of different densities are also listed in Table 2-1. The commonly used density for concrete, 150 lb/ft^3, is almost exactly 2.4 Mg/m^3.

You should note that all mass and volume ratios common in geotechnical engineering practice are not affected by the use of SI units. For example, void ratio or water content of any given soil still has the same numerical value.

Unit weight or weight per unit volume is still the common measurement in geotechnical engineering practice. However, since weight should be avoided in technical work for all the reasons discussed earlier, then unit weight also should be avoided. ASTM now recommends that density be used in place of unit weight. If you must convert from density to unit weight, then simply use $\gamma = \rho g$, which means you will have to consider the appropriate value for the acceleration due to gravity. The "standard" value of g is 9.807 m/s² (32.17 ft/s²), which as mentioned previously, can be used with sufficient accuracy for ordinary engineering work for most places on this earth. If you ever have a job on the moon or some other planet, then you must use the local value for g. Keep in mind, also, to be very careful which "pounds" you are working with, lbf or lbm, in these conversions.

EXAMPLE A.5

Given:

The density of water is 1000 kg/m³.

Required:

The density of water in (a) g/cm³ and (b) lb/ft³.

Solution:

Set up an equation as follows:

$$\text{a. } 1000\frac{kg}{m^3} = 1000\frac{kg}{m^3}\left(\frac{1000\ g}{1\ kg}\right)\left(\frac{1\ m}{100\ cm}\right)^3 = 1\frac{g}{cm^3}$$

$$\text{b. } 1000\frac{kg}{m^3} = 1000\frac{kg}{m^3}\left(\frac{1\ lbm}{0.4536\ kg}\right)\left(\frac{0.3048\ m}{1\ ft}\right)$$

$$= 62.43\frac{lbm}{ft^3}$$

Another way to do part **b** is to recall that 1 lbm/ft³ = 16.018 kg/m³; so

$$1000\frac{kg}{m^3} = 1000\frac{kg}{m^3}\left(\frac{1\ lbm/ft^3}{16.018\ kg/m^3}\right)$$

$$= 62\ 43\ \frac{lbm}{ft^3}$$

EXAMPLE A.6

Given:

Density of water, $\rho_w = 1000 \text{ kg/m}^3$.

Required:

Convert this density to unit weight in (a) SI and (b) British Engineering units.

Solution:

a. *SI units*: We know that $\gamma = \rho g$; so

$$\gamma = 1000\frac{\text{kg}}{\text{m}^3}\left(9.807\frac{\text{m}}{\text{s}^2}\right) = 9807\frac{\text{kg}\cdot\text{m}}{\text{m}^3\cdot\text{s}^2}$$

Recall that $1 \text{ N} = 1\dfrac{\text{kg}\cdot\text{m}}{\text{s}^2}$

$$\therefore \quad \gamma = 9807\frac{\text{N}}{\text{m}^3} = 9.807\frac{\text{kN}}{\text{m}^3}$$

b. *British Engineering units*: From Example A.5, we know that

$$1000\frac{\text{kg}}{\text{m}^3} = 62.43\frac{\text{lbm}}{\text{ft}^3}$$

$$\gamma = 62.43\frac{\text{lbm}}{\text{ft}^3}\left(32.17\frac{\text{ft}}{\text{s}^2}\right) = 2008\frac{\text{lbm}\cdot\text{ft}}{\text{s}^2\cdot\text{ft}^3}$$

If lbf are used, from part (a),

$$\gamma = 9.8\frac{\text{kN}}{\text{m}^3}\left(\frac{1000 \text{ N}}{1 \text{ kN}}\right)\left(\frac{1 \text{ lbf}}{4.448 \text{ N}}\right)\left(\frac{0.3048 \text{ m}}{1 \text{ ft}}\right)$$

$$= 62.4\frac{\text{lbf}}{\text{ft}^3}$$

This is the commonly used value for the unit weight of water.

EXAMPLE A.7

Given:

A soil has a dry density of 1.7 Mg/m^3.

Required:

Convert this density into unit weights, in terms of both (a) SI and (b) British Engineering units.

Solution:

a. *SI units*:

$$\rho_d = 1.7 \text{ Mg/m}^3 = 1700 \text{ kg/m}^3$$
$$\gamma = \rho g$$
$$\gamma = 1700 \frac{\text{kg}}{\text{m}^3} \left(9.81 \frac{\text{m}}{\text{s}^2} \right) = 16\,677 \frac{\text{N}}{\text{m}^3} = 16.7 \frac{\text{kN}}{\text{m}^3}$$

b. *British Engineering units*, in terms of lbm:

$$\rho = 1700 \frac{\text{kg}}{\text{m}^3} \left(\frac{1 \text{ lbm/ft}^3}{16.018 \text{ kg/m}^3} \right) = 106.13 \frac{\text{lbm}}{\text{ft}^3}$$
$$\gamma = \rho g$$
$$\gamma = 106.13 \frac{\text{lbm}}{\text{ft}^3} \left(32.17 \frac{\text{ft}}{\text{s}^2} \right) = 3414 \frac{\text{lbm} \cdot \text{ft}}{\text{s}^2 \cdot \text{ft}^3}$$

In terms of lbf: From part (a),

$$\gamma = 16.7 \frac{\text{kN}}{\text{m}^3} \left(\frac{1000 \text{ N}}{1 \text{ kN}} \right) \left(\frac{1 \text{ lbf}}{4.448 \text{ N}} \right) \left(\frac{0.3048 \text{ m}}{1 \text{ ft}} \right)$$
$$= 106.3 \text{ lbf/ft}^3$$

The latter value in terms of lbf is, of course, the more familiar figure.

Geostatic Stress. For computations of geostatic stresses, the unit weights of the various soil layers can be easily replaced by the ρg of the layers. The usual formula

$$\sigma_v = \sum_{i=1}^{n} \gamma_i z_i$$

then becomes

$$\sigma_v = \sum_{i=1}^{n} \rho_i g z_i \qquad \text{(7-14c)}$$

where σ_v = total vertical stress at some depth,
ρ_i = density of each layer,
z_i = thickness of each layer, and
g = acceleration of gravity.

If ρg is a constant throughout the depth h, then

$$\sigma_v = \rho g h \tag{7-14b}$$

By analogy, computation of the static pore water pressure u_o at some depth h_w below the ground water table is

$$u_o = \rho_w g h_w \tag{7-15}$$

where ρ_w = the density of water ($1 \ Mg/m^3$).

Similarly, to obtain the effective vertical overburden stress, the effective or buoyant density ρ' for each layer below the ground water table can be used or, perhaps more simply, $\sigma'_{vo} = \sigma_{vo} - u_o$.

Dimensional analysis of these equations for stress shows that if the densities are expressed in Mg/m^3, then stresses automatically result in kPa, or

$$\left(\frac{Mg}{m^3}\right)\left(\frac{m}{s^2}\right) m = 1000\frac{kg \cdot m}{s^2 \cdot m^2} = 1000\frac{N}{m^2} = 1 \ kPa$$

Several examples of geostatic stress computations using SI units can be found in Chapter 7.

appendix b-1

Derivation of Laplace's Equation*

As mentioned in Sec. 7.9, a flow net is actually a graphical solution of Laplace's equation, Eq. 7-24. The assumptions necessary for the derivation of this equation are:

$S = 100\%$

e = constant [i.e., no consolidation (Chapter 8) or compression of the medium occurs]

k is isotropic

Darcy's law (Eq. 7-5) is valid

Consider the flow of water into an element with dimensions dx and dy (Fig. B-1.1). Two-dimensional flow is assumed here for simplicity; you could do the exact same thing in three dimensions, but it would just be more complicated. The term $(\partial v_x/\partial x)\, dx$ indicates the *rate of change* in velocity v_x in the x-direction; similarly, $(\partial v_y/\partial y)\, dy$ is the rate of change in v_y in the y-direction. From continuity, we know that q = constant = VA_{in} = VA_{out}. So

$$VA_{in} = v_x\, dy + v_y\, dx$$

$$VA_{out} = \left(v_x + \frac{\partial v_x}{\partial x} dx \right) dy + \left(v_y + \frac{\partial v_y}{\partial y} dy \right) dx$$

If we set these two equations equal, we get

$$\frac{\partial v_x}{\partial x} dx\, dy + \frac{\partial v_y}{\partial y} dx\, dy = 0$$

*Chapter 7.

681

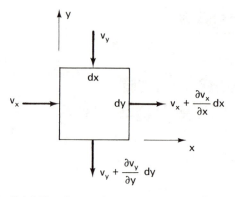

Fig. B-1.1 Flow into and out of an element dx by dy.

or

$$\frac{\partial v_x}{\partial x} + \frac{\partial v_y}{\partial y} = 0 \qquad \text{(B-1-1)}$$

since dx and dy cannot be zero.

From Darcy's law (Eqs. 7-2 and 7-5), $v = ki = k\Delta h/L$. Thus we can write for our element:

$$v_x = k_x \frac{\partial h}{\partial x}, \qquad v_y = k_y \frac{\partial h}{\partial y}$$

Substituting these terms into Eq. B-1-1 we obtain

$$k_x \frac{\partial^2 h}{\partial x^2} + k_y \frac{\partial^2 h}{\partial y^2} = 0$$

Since k was assumed to be isotropic, $k_x = k_y$. So we have

$$\frac{\partial^2 h}{\partial x^2} + \frac{\partial^2 h}{\partial y^2} = 0 \qquad \text{(7-24)}$$

which is *Laplace's equation in two dimensions*. For the equation in three dimensions, simply add the term $\partial^2 h/\partial z^2$ to Eq. 7-24.

appendix b-2

Derivation and Solution of Terzaghi's One-Dimensional Consolidation Theory[*]

B-2.1 ASSUMPTIONS

To develop the Terzaghi one-dimensional consolidation theory, we need to assume the following:

1. The clay is homogeneous and 100% saturated.
2. Drainage is provided at both the top and bottom of the compressible layer.
3. Darcy's law (Eq. 7-5) is valid.
4. The soil grains and water are incompressible.
5. Compression and flow are one dimensional.
6. The small load increment applied produces essentially no change in thickness (that is, small strains), and k and a_v remain constant.
7. There is a *unique* linear relationship between the volume change Δe and the effective stress $\Delta\sigma'$. In other words, $de = -a_v\,d\sigma'$ and a_v is assumed constant over the increment of applied stress. This important assumption also implies that there is *no secondary compression*.

B-2.2 DERIVATION

Now let us borrow a little element from Fig. 9.1 f and enlarge it in Fig. B-2.1. Our element exists at a depth z below the top of the compressible layer, has thickness dz, and has an area dx times dy. The volume change of the element is the difference between the amount of flow in and

*Chapter 9.

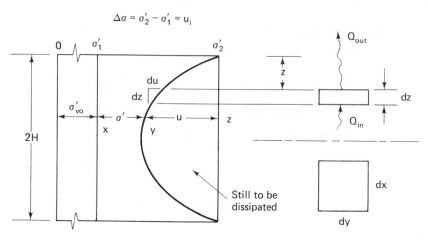

Fig. B-2.1 Soil layer undergoing compression, similar to Fig. 9.1f.

out of the element. Since consolidation under these conditions is directly dependent on the escape of pore water from the soil voids, we may develop the consolidation equation by considering the continuity of flow in our element. The hydraulic gradient i_z at the top of our element is given by

$$i_z = \frac{\text{head loss}}{\text{distance}} = \frac{\partial}{\partial z}\left(\frac{u}{\rho_w g}\right) = \frac{1}{\rho_w g}\frac{\partial u}{\partial z} \qquad \text{(B-2-1)}$$

The corresponding hydraulic gradient at the bottom of our element dz is given by

$$i_{z+dz} = \frac{1}{\rho_w g}\frac{\partial u}{\partial z} + \frac{1}{\rho_w g}\frac{\partial^2 u}{\partial z^2}dz \qquad \text{(B-2-2)}$$

From Darcy's law, $dQ = kiadt$, we may compute the quantity of flow dQ in time dt out of the top of our element by

$$dQ_{\text{out}} = k\frac{1}{\rho_w g}\frac{\partial u}{\partial z}dz\,dx\,dy\,dt \qquad \text{(B-2-3)}$$

Likewise we may compute the quantity of flow in time dt at the bottom into the element by

$$dQ_{\text{in}} = k\frac{1}{\rho_w g}\left(\frac{\partial u}{\partial z} + \frac{\partial^2 u}{\partial z^2}\right)dz\,dx\,dy\,dt \qquad \text{(B-2-4)}$$

We can now compute the volume change from the difference in rates of flow, $Q_{\text{out}} - Q_{\text{in}}$. Also we assume the area $dx\,dy$ to be a unit area. Therefore

$$\text{volume change} = dQ_{\text{out}} - dQ_{\text{in}} = -\frac{k}{\rho_w g}\frac{\partial^2 u}{\partial z^2}1\,dz\,dt \qquad \text{(B-2-5)}$$

The volume change may also be determined from the laboratory oedometer test. Remember, from Chapter 8, that we would obtain a laboratory curve similar to Fig. 8.4, which we again show as Fig. B-2.2. From Eqs. 8-5a and b, the coefficient of compressibility a_v is

$$a_v = -\frac{de}{d\sigma'} = \frac{e_1 - e_2}{\sigma_2' - \sigma_1'} \tag{B-2-6}$$

To be correct, we should write these equations in terms of effective stresses. From Fig. B-2.2, you can see that the slope of the e-σ' curve is negative, and you know that e_1 is numerically larger than e_2.

From Eq. 8-4, $s = \Delta e H_o / (1 + e_o)$, or in terms of our element in Fig. B-2.1 and the e-σ' relationship in Fig. B-2.2, we obtain

$$s = \Delta\,dz = \frac{-de}{1 + e_1}\,dz \tag{B-2-7}$$

where e_1 corresponds to the initial void ratio e_o. From Eq. B-2-6, $-de = a_v\,d\sigma'$. Therefore

$$\Delta\,dz = \frac{a_v\,d\sigma'}{1 + e_1}\,dz \tag{B-2-8}$$

Now, from our discussion in Chapter 9, we know that as the excess pore

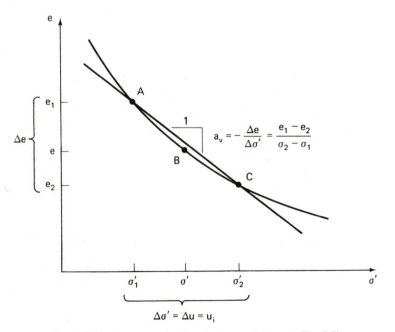

Fig. B-2.2 Laboratory compression curve (same as Fig. 9.2).

water pressure dissipates, the effective stress in the soil skeleton increases. This is shown schematically in Figs. 9.1c and f. Thus we can write that $\Delta\sigma' = -\Delta u$, because any *change* in effective stress is numerically equal to the negative of the change in excess pore water pressure. This relationship is true, of course, as long as the total stress does not change. Now, Eq. B-2-8 can be written as

$$\Delta\,dz = -\frac{a_v\,du}{1 + e_1}\,dz \qquad (\text{B-2-9})$$

and since $du = (\partial u/\partial t)\,dt$, Eq. B-2-9 becomes

$$\Delta\,dz = \frac{-a_v}{1 + e_1}\,\frac{\partial u}{\partial t}\,dt\,dz \qquad (\text{B-2-10})$$

By equating the volume change obtained in Eq. B-2-5 and the volume change in Eq. B-2-10, we have

$$-\frac{k}{\rho_w g}\,\frac{\partial^2 u}{\partial z^2}\,dz\,dt = -\frac{a_v}{1 + e_1}\,\frac{\partial u}{\partial t}\,dt\,dz \qquad (\text{B-2-11})$$

We can collect the soil properties terms as in Eq. 9-3,

$$c_v = \frac{k}{\rho_w g}\,\frac{1 + e_1}{a_v} \qquad (\text{9-3})$$

where c_v is called the *coefficient of consolidation* since it governs the consolidation process. Note that it has units of $L^2 T^{-1}$. We thus obtain

$$c_v \frac{\partial^2 u}{\partial z^2} = \frac{\partial u}{\partial t} \qquad (\text{9-2})$$

Equation 9-2 is the Terzaghi *one-dimensional consolidation equation*. If we assume c_v is a constant with respect to time and position, then Eq. 9-2 is a second-order partial differential equation with constant coefficients. There are a variety of ways to solve such equations; some are mathematically exact, others are only approximate. For example, Harr (1966) presents an approximate solution using the method of finite differences. Taylor (1948), following Terzaghi (1925), provides a mathematically rigorous solution in terms of a Fourier series expansion. The development that follows is adapted from Taylor (1948) and Leonards (1962).

B-2.3 MATHEMATICAL SOLUTION

The boundary and initial conditions for the case of one-dimensional consolidation are as follows:

1. There is complete drainage at the top and bottom of the compressible layer.

2. The initial *excess* hydrostatic pressure $\Delta u = u_i$ is equal to the applied increment of stress at the boundary, $\Delta \sigma$.

These boundary and initial conditions can be written

when $z = 0$ and when $z = 2H$, $u = 0$

when $t = 0$, $u = u_i = \Delta \sigma = (\sigma_2' - \sigma_1')$

The general solution of Eq. 9-2, when the initial excess pore pressure u_i is a function of the depth z, is

$$u = \sum_{n=1}^{n=\infty} \left(\frac{1}{H} \int_{z=0}^{z=2H} u_i \sin \frac{n\pi z}{2H} dz \right) \sin \frac{n\pi z}{2H} \exp\left(\frac{-c_v t n^2 \pi^2}{4H^2} \right) \quad \text{(B-2-12)}$$

When u_i is a constant or varies linearly with depth, the solution becomes

$$u = (\sigma_2' - \sigma_1') \sum_{n=0}^{\infty} \underbrace{\frac{4}{(2n+1)\pi} \sin\left(\frac{2n+1}{2} \pi \frac{z}{H} \right)}_{Z}$$

$$\times \exp - \left[\frac{(2n+1)^2}{4} \pi^2 \underbrace{\frac{k(1+e_1)}{\underbrace{a_v \rho_w g}_{c_v}} \frac{t}{H^2}}_{T} \right] \quad \text{(B-2-13)}$$

where $\sigma_1' =$ initial effective stress, $\sigma_2' = \sigma_1' + \Delta \sigma$, and $n = 0, 1, 2, 3, \ldots$. The solution provides the instantaneous value of the pore water pressure u at any specified time and point in the soil mass. The only part of Eq. B-2-13 that is a function of the soil properties is c_v.

You can see that the solution is in terms of two dimensionless quantities, Z and T or as we wrote in Eq. 9-4,

$$u = (\sigma_2' - \sigma_1') \sum_{n=0}^{\infty} f_1(Z) f_2(T) \quad \text{(9-4)}$$

You will recall that the dimensionless quantity T is called the *time factor*, and it is related to c_v (Eqs. 9-5 and 9-6) by

$$T = c_v \frac{t}{H_{dr}^2} = \frac{k(1+e_o)}{a_v \rho_w g} \frac{t}{H_{dr}^2} \quad \text{(B-2-14)}$$

In this equation H_{dr} is the longest drainage path a drop of water has to follow in a compressible soil deposit to get to a free draining boundary. In Fig. B-2.1 you can see that the height of a doubly drained layer is $2H$. Therefore the drainage path H_{dr} is equal to H. If we had only a singly drained layer, we would only consider the top half of Fig. B-2.1, and again the drainage path would be the height of H.

The *consolidation ratio* U_z relates the change in volume at depth z and time t to the ultimate volume change at depth z, or

$$U_z = \frac{\text{volume change at depth } z \text{ and time } t}{\text{ultimate volume change at depth } z} \qquad \text{(B-2-15)}$$

The change in volume, of course, means a change in void ratio, or as we wrote in Eq. 9-7,

$$U_z = \frac{e_1 - e}{e_1 - e_2} \qquad \text{(9-7)}$$

The changes in void ratio can be related to the stress increment through the coefficient of consolidation a_v. These relationships are shown in Fig. B-2.2. Because in one-dimensional consolidation, the initial excess hydrostatic (pore) pressure is equal to the increment of applied stress, Eq. 9-7 becomes

$$U_z = \frac{\sigma' - \sigma_1'}{\sigma_2' - \sigma_1'} = \frac{\sigma' - \sigma_1'}{\Delta\sigma'} = \frac{u_i - u}{u_i} = 1 - \frac{u}{u_i} \qquad \text{(9-8)}$$

Now we can write our solution to the consolidation equation (Eq. 9-4) as in Eq. 9-9, or

$$U_z = 1 - \sum_{n=0}^{\infty} f_1(Z)f_2(T) \qquad \text{(9-9)}$$

This equation is shown graphically in Fig. 9.3, and we explain in Chapter 9 how to use this figure to obtain the amount of consolidation at any depth and time in the consolidating layer (see Examples 9.1 and 9.2).

Generally in engineering practice we are interested in the volume change of the entire soil layer. So we want the *average degree or percent consolidation U*, which is defined as

$$U(\%) = \frac{\text{total volume change at time } t}{\text{ultimate total volume change}} \times 100(\%) \qquad \text{(B-2-16)}$$

For one-dimensional compression, the change in volume is, of course, equal to the change in height of the layer. To obtain the average degree of consolidation over the entire layer we have to find the area under the curve corresponding to a given time factor in Fig. 9.3; this is shown in Fig. 9.5. Mathematically, $U(\%)$ = average value of U_z, or

$$U(\%) = \frac{\sum U_z \, dz}{2H} = \frac{1}{2H} \int_0^{2H} U_z \, dz \qquad \text{(B-2-17)}$$

or from Fig. B-2.1,

$$U(\%) = \frac{\int_0^{2H} xy}{\int_0^{2H} xz} = \frac{\int_0^{2H} \left[(\sigma_2' - \sigma_1') - u \right] dz}{(\sigma_2' - \sigma_1')2H} \times 100 \qquad \text{(B-2-18)}$$

Rewriting,

$$U(\%) = \frac{100}{2H(\sigma_2' - \sigma_1')} \int_0^{2H} \left[(\sigma_2' - \sigma_1') - u \right] dz \qquad \text{(B-2-19)}$$

or

$$U(\%) = \frac{100}{2H(\sigma_2' - \sigma_1')} \left[\int_0^{2H} (\sigma_2' - \sigma_1') \, dz - \int_0^{2H} u \, dz \right] \qquad \text{(B-2-20)}$$

Substituting the value of u from Eq. B-2-13 into Eq. B-2-20 and integrating, we obtain:

$$U(\%) = \frac{100}{2H(\sigma_2' - \sigma_1')} \left[(\sigma_2' - \sigma_1')2H - (\sigma_2' - \sigma_1') \sum_{n=0}^{\infty} \frac{4}{(2n+1)\pi} \right.$$

$$\times (-1) \cos \frac{(2n+1)\pi}{2H} z \left(\frac{2H}{(2n+1)\pi} \right)$$

$$\left. \times \exp \left. -\left(\frac{(2n+1)^2 \pi^2}{4} T \right) \right|_0^{2H} \right] \qquad \text{(B-2-21)}$$

Putting in the limits, we obtain

$$U(\%) = 100 \left\{ 1 - \sum_{n=0}^{\infty} \frac{4}{(2n+1)^2 \pi^2} (-1)(-1-1) \right.$$

$$\left. \times \exp -\left[\frac{(2n+1)^2 \pi^2}{4} T \right] \right\} \qquad \text{(B-2-22)}$$

or

$$U(\%) = 100 \left\{ 1 - \sum_{n=0}^{\infty} \frac{8}{(2n+1)^2 \pi^2} \exp -\left[\frac{(2n+1)^2 \pi^2}{4} T \right] \right\}$$

$$\text{(B-2-23)}$$

This solution is for the special case of constant or linear initial hydrostatic excess pressure and is valid for all values of U. Solutions for other initial pore pressure distributions are provided by Taylor (1948) and Leonards (1962), but the differences are negligible for practical purposes. The summation indicated by Eq. B-2-23 can be carried out once and for all and tabulated (Table 9-1) or shown graphically (Fig. 9.5). Casagrande (1938) and Taylor (1948) give the following approximations for Eq. B-2-23, which are useful to know:

For $U < 60\%$,

$$T = \frac{\pi}{4} U^2 \qquad \text{(9-10)}$$

For $U > 60\%$,

$$T = 1.781 - 0.933 \log(100 - U\%) \qquad (9\text{-}11)$$

For values of $U > 60\ \%$, the series in Eq. B-2-23 converges extremely rapidly so that only the first term is significant. Therefore, letting $n = 0$, Eq. B-2-23 becomes

$$U(\%) = 100\left[1 - \frac{8}{\pi^2}\exp\left(-\frac{\pi^2}{4}T \right) \right] \qquad (B\text{-}2\text{-}24)$$

Rearranging, Eq. B-2-24 gives Eq. 9-11.

appendix b-3

Pore Pressure Parameters*

B-3.1 DERIVATION OF SKEMPTON'S PORE PRESSURE EQUATION

The pore pressure parameters (Sec. 11.10), first defined by Skempton (1954), relate the change in pore water pressure to the change in *total* stress during undrained loading.

First, let's derive Eq. 11-11. This can be done in several ways. One simple way is to assume for a start that we have a triaxial specimen in equilibrium with the cell pressure σ_c acting on it. Assume for the moment that the soil skeleton is elastic and isotropic, and that there are both air and water in the voids (that is, $S < 100\%$). Now, when we apply a small change in the cell pressure $\Delta\sigma_c$ to the sample, by Terzaghi's principle of effective stress (Eq. 7-13), the *change* in effective stress is

$$\Delta\sigma_c' = \Delta\sigma_c - \Delta u$$

The volume change ΔV caused by this change in stress is

$$\Delta V - C_{sk}V_o(\Delta\sigma_c') = -C_{sk}V_o(\Delta\sigma_c - \Delta u)$$

where C_{sk} is the compressibility of the soil skeleton and V_o is the original volume of the sample.

As mentioned in Chapter 8, the mineral grains themselves are relatively incompressible, so any decrease in the volume of the soil skeleton results in a decrease in volume of the voids, or

$$\Delta V = -V_v C_v \Delta u = -nV_o C_v \Delta u \qquad \text{(B-3-1)}$$

where n is the porosity, and C_v is the compressibility of the pore fluid

*Chapter 11.

(air + water). If $S = 100\%$, then $C_v = C_w$, the compressibility of water. If we allow no drainage to occur, then these two changes in volume must be equal, or

$$- nV_oC_v \Delta u = -C_{sk}V_o(\Delta\sigma_c - \Delta u)$$

Solving for the ratio $\Delta u / \Delta\sigma_c$, we obtain Eq. 11-11.

$$\frac{\Delta u}{\Delta\sigma_c} = \frac{1}{1 + \dfrac{nC_v}{C_{sk}}} = B \tag{11-11}$$

where $\Delta\sigma_c = \Delta\sigma_3$. We discussed in Sec. 11.10 the values of B for different soils and test conditions (see Table 11-8). A more general way to obtain Eq. 11-11 is shown later in this appendix.

We can follow a similar development for the change in pore pressure due to the change in the principal stress difference or shear stress in our triaxial test specimen in order to derive Eqs. 11-13 through 11-15. Assume that the soil skeleton still behaves elastically; then the volume change caused by the change in effective stresses is

$$\Delta V = -C_{sk}V_o\frac{1}{3}(\Delta\sigma_1' + \Delta\sigma_2' + \Delta\sigma_3')$$

The symbols were previously defined. For the common triaxial compression test, $\Delta\sigma_2 = \Delta\sigma_3$, so

$$\Delta V = -C_{sk}V_o\frac{1}{3}(\Delta\sigma_1' + 2\Delta\sigma_3')$$

The coefficient $1/3$ comes about because for elastic isotropic materials the volume change is due to the *average* of the changes in the three principal stresses. Now add and subtract $3\Delta\sigma_3$ to the right-hand side of the equation, and invoke Terzaghi's principle of effective stress. We then obtain

$$\Delta V = -C_{sk}V_o\frac{1}{3}(\Delta\sigma_1 - \Delta\sigma_3 + 3\Delta\sigma_3 - 3\Delta u)$$

As before, the decrease in voids is

$$\Delta V = -nV_oC_v \Delta u \tag{B-3-1}$$

For undrained conditions, the two volumes must be equal. Solving for Δu and noting that

$$B = \frac{1}{1 + \dfrac{nC_v}{C_{sk}}} \tag{11-11}$$

we obtain

$$\Delta u = B\left[\Delta\sigma_3 + \frac{1}{3}(\Delta\sigma_1 - \Delta\sigma_3)\right] \tag{B-3-2}$$

Note that the coefficient $1/3$ for the stress difference term is for elastic materials and triaxial compression conditions. If we make a similar derivation for *triaxial extension* conditions ($\Delta\sigma_2 = \Delta\sigma_1$), we get

$$\Delta u = B\left[\Delta\sigma_3 + \frac{2}{3}(\Delta\sigma_1 - \Delta\sigma)\right] \qquad \text{(B-3-3)}$$

(Note that you have to add and subtract $2\Delta\sigma_3$ in this case.) Thus for elastic soil skeletons, the pore pressure parameter in extension is twice that in compression.

Since soils in general are inelastic materials, Skempton (1954) replaced the two constants in Eqs. B-3-2 and B-3-3 by the coefficient A, so that

$$\Delta u = B\left[\Delta\sigma_3 + A(\Delta\sigma_1 - \Delta\sigma_3)\right] \qquad \text{(11-13)}$$

Often it is convenient to write Eq. 11-13 as

$$\Delta u = B\,\Delta\sigma_3 + \overline{A}(\Delta\sigma_1 - \Delta\sigma_3) \qquad \text{(11-15)}$$

where $\overline{A} = BA$. For saturated soils, we usually write Eq. 11-13 as

$$\Delta u = \Delta\sigma_3 + A(\Delta\sigma_1 - \Delta\sigma_3) \qquad \text{(11-14)}$$

Other convenient ways to write the pore pressure equation (11-13) are given by Skempton (1954). For triaxial compression conditions,

$$\Delta u = B\left[\frac{1}{3}(\Delta\sigma_1 + 2\Delta\sigma_3) + \frac{3A-1}{3}(\Delta\sigma_1 - \Delta\sigma_3)\right] \qquad \text{(B-3-4)}$$

And for triaxial extension conditions

$$\Delta u = B\left[\frac{1}{3}(2\Delta\sigma_1 + \Delta\sigma_3) + \frac{3A-2}{3}(\Delta\sigma_1 - \Delta\sigma_3)\right] \qquad \text{(B-3-5)}$$

These equations show that if soils behaved as perfectly elastic materials (that is, $A = 1/3$ in compression and $A = 2/3$ in extension), then the pore pressure would depend only on the average change in principal stress, which is the first part of Eqs. B-3-4 and B-3-5.

B-3.2 DEFINITION OF $\Delta\sigma_1$ AND $\Delta\sigma_3$ FOR ROTATION OF PRINCIPAL STRESSES

Law and Holtz (1978) showed that contradictory definitions of the pore pressure parameter A exist in the literature because of the lack of a consistent definition of principal stress increment for cases where the principal stresses rotate. They proposed the following system, to take care of any possible ambiguities when the principle stresses rotate 90°.

In this system, $\Delta\sigma_1$ and $\Delta\sigma_3$ are called the *major and minor principal stress increments*, respectively. A principal stress increment is defined as the maximum or minimum normal stress increment imposed on a given stress system. The sign convention is positive for compression and negative for tension. $\Delta\sigma_1$ is the *algebraically largest* normal component of a given system of stress increments, and $\Delta\sigma_3$ is the *algebraically smallest* normal component of that system.

The advantage of this system is that the stress increment is not connected to the original stress. Thus the direction of $\Delta\sigma_1$ is *independent* of the direction of the original or final σ_1, and so is $\Delta\sigma_3$. This point is illustrated in Table B-3-1, which shows some combinations of $\Delta\sigma_1$ and $\Delta\sigma_3$ being applied to typical existing stress systems represented by σ_1 and σ_3.

TABLE B-3-1 Examples Using the Proposed New Definition of Principal Stress Increments (units of stress are arbitrary, and axisymmetry in stress system is assumed)*

Initial Stress System	Stress Increment	Final Stress State	$\Delta\sigma_1$		$\Delta\sigma_3$	
			Magnitude	Direction	Magnitude	Direction
(5, 3, 5) +	(4, 4) =	(9, 3, 9)	4	V*	0	H*
(5, 3, 5) +	(2, 2) =	(3, 3, 3)	0	H	-2	V
(5, 3, 5) +	(4, 1, 1, 4) =	(1, 2, 1)	-1	H	-4	V

*V = Vertical; H = Horizontal.

B-3.3 FORMULAS FOR PORE PRESSURE PARAMETERS FOR DIFFERENT STRESS PATH TESTS

To aid in calculating the correct value of the parameter A, Law and Holtz (1978) derived the appropriate expressions for A for the four types of triaxial stress path tests, AC, AE, LC, and LE (Secs. 10.6 and 11.12). These

*After Law and Holtz (1978).

TABLE B-3-2 Definition of Principal Stress Increments and Formulas for Pore Pressure Parameters for Various Types of Triaxial Tests*

Test Type	$\Delta\sigma_1$	$\Delta\sigma_2$	$\Delta\sigma_3$	Formula for A	Equation
Compression test:					
Axial compression, AC	$\Delta\sigma_v$	0	0	$A_{ac} = \Delta u/\Delta\sigma_v$	11-16
Lateral extension, LE	0	$\Delta\sigma_h$	$\Delta\sigma_h$	$A_{le} = 1 - \Delta u/\Delta\sigma_h$	11-17
Extension test:					
Axial extension, AE	0	0	$\Delta\sigma_v$	$A_{ae} = 1 - \Delta u/\Delta\sigma_v$	11-18
Lateral compression, LC	$\Delta\sigma_h$	$\Delta\sigma_h$	0	$A_{lc} = \Delta u/\Delta\sigma_h$	11-19

*After Law and Holtz (1978).

are shown in Table B-3-2. The derivation of these expressions is shown in the following example.

EXAMPLE B-3.1

Given:

An axial extension (AE) triaxial test is conducted on a saturated clay.

Required:

Determine the correct formula for the pore pressure parameter A.

Solution:

In the AE test, the lateral (cell) pressure remains constant while the axial stress is decreased. Therefore

$$\Delta\sigma_1 = \Delta\sigma_2 = 0, \qquad \Delta\sigma_3 = \Delta\sigma_v$$

According to the definition of principal stress increments proposed by Law and Holtz (1978), $\Delta\sigma_v$ is negative since it decreases. Thus it is algebraically the smallest component of the stress increment. Substituting these definitions for $\Delta\sigma_1$ and $\Delta\sigma_3$ into Eq. 11-13 (assume $B = 1$), we obtain

$$A_{ae} = 1 - \frac{\Delta u}{\Delta\sigma_v} \tag{11-18}$$

This formula is the same as shown for the AE test in Table B-3-2.

B-3.4 PROOF THAT $A_{ac} = A_{le}$ AND $A_{ae} = A_{lc}$

It was shown by example in Sec. 11.12 that the pore pressure parameter A was the same in axial compression (AC) as in lateral extension (LE). It was inferred that A in axial extension (AE) was identical to A in lateral compression (LC). The statements are true even though these sets of tests have different total stress paths. The proof of this contention was given by Law and Holtz (1978).

We first define $p' = (\sigma_1' + \sigma_3')/2$ as the average of the major and minor effective stresses and $q = (\sigma_1 - \sigma_3)/2$ as half the principal stress difference (Sec. 10.6). We can express the slope at any point on the effective stress path in a p'-q diagram as

$$\left(\frac{dq}{dp'}\right) = \frac{d(\sigma_1 - \sigma_3)}{d(\sigma_1 + \sigma_3 - 2u)}$$

For the axial compression case $d\sigma_1 = d\sigma_v$ and $d\sigma_3 = 0$. Hence

$$\left(\frac{dq}{dp'}\right)_{ac} = \frac{1}{1 - 2A_{ac}}$$

For the lateral extension case $d\sigma_1 = 0$ and $d\sigma_3 = d\sigma_l$. Hence

$$\left(\frac{dq}{dp'}\right)_{le} = \frac{-1}{1 - 2(1 - A_{le})} = \frac{1}{1 - 2A_{le}}$$

Since both tests have the same effective stress paths (see, for example, Example 11.17) then

$$\left(\frac{dq}{dp'}\right)_{ac} = \left(\frac{dq}{dp'}\right)_{le}$$

Hence

$$A_{ac} = A_{le} \tag{11-20}$$

Similarly, we can show that

$$A_{ae} = A_{lc} \tag{11-21}$$

B-3.5 DERIVATION OF THE HENKEL PORE PRESSURE EQUATION AND COEFFICIENTS*

Assume an element of soil in equilibrium with stresses σ_1, σ_2, and σ_3 on it. When we apply stress increments $\Delta\sigma_1$, $\Delta\sigma_2$, and $\Delta\sigma_3$ to the element, an excess pore pressure Δu and a resulting change in effective stresses

*After Scott, 1963, and Perloff and Baron, 1976.

occurs. So,

$$\Delta\sigma_1' = \Delta\sigma_1 - \Delta u, \qquad \Delta\sigma_2' = \Delta\sigma_2 - \Delta u, \qquad \Delta\sigma_3' = \Delta\sigma_3 - \Delta u$$

Assume for now that the soil skeleton is elastic and isotropic. Thus it has a bulk modulus $K_{sk} = E/3(1 - 2\nu)$. Since the definition of bulk modulus is the volumetric effective stress $\frac{1}{3}(\sigma_1' + \sigma_2' + \sigma_3')$ divided by the volumetric strain $\Delta V/V_o$,

$$K_{sk} = \frac{\frac{1}{3}(\sigma_1' + \sigma_2' + \sigma_3')}{\Delta V/V_o} = \frac{E}{3(1 - 2\nu)}$$

Rearranging, the volumetric strain of the soil skeleton is

$$\frac{\Delta V}{V_o} = \epsilon_1 + \epsilon_2 + \epsilon_3 = C_{sk}\left(\frac{\sigma_1' + \sigma_2' + \sigma_3'}{3}\right)$$

where $C_{sk} = 1/K_{sk}$ is called the *compressibility of soil skeleton*, and ϵ_1, ϵ_2, and ϵ_3 are principal strains. Since E and ν are difficult to determine for a real soil, the general coefficient C_{sk} is more practical (Scott, 1963). Now, if we state this equation in terms of total stress changes and pore pressure, then we have

$$\frac{\Delta V}{V_o} = C_{sk}\left(\frac{\Delta\sigma_1 + \Delta\sigma_2 + \Delta\sigma_3}{3} - \Delta u\right)$$

This equation states that the volumetric strain is a function only of the change in mean effective stress for a *linearly elastic* material (or, in fact, for any non-dilative, no-volume-change-during-shear material). However soils do change volume due to the change in shear stress, and this is accounted for by an empirical correction factor, $D|\Delta\tau_{oct}|$, where $|\Delta\tau_{oct}|$ is the absolute value of the *increment* in τ_{oct}. Thus we have

$$\frac{\Delta V}{V_o} = C_{sk}\left[\Delta\sigma_{oct} - u\right] + D|\Delta\tau_{oct}| \qquad \text{(B-3-6)}$$

because, by definition from continuum mechanics,

$$\sigma_{oct} = \frac{\sigma_1 + \sigma_2 + \sigma_3}{3} \qquad \text{(11-23)}$$

and

$$\tau_{oct} = \frac{1}{3}\sqrt{(\sigma_1 - \sigma_2)^2 + (\sigma_2 - \sigma_3)^2 + (\sigma_3 - \sigma_1)^2} \qquad \text{(11-24)}$$

As Perloff and Baron (1976) point out, since τ_{oct} is a nonlinear function of the principal stress differences, we cannot in general calculate it directly from the stress increments. Instead we must determine $\Delta\tau_{oct}$ from the difference $(\tau_{oct})_2 - (\tau_{oct})_1$.

Now, as we did in Section B-3.1, let us look at what happens to the voids. The volumetric strain in the voids is

$$\frac{\Delta V_v}{V_v} = -C_v \Delta u \qquad \text{(B-3-1)}$$

where C_v is the compressibility of the voids, and V_v, the volume of voids, is nV_0. If $S = 100\%$, then $C_v = C_w$, the compressibility of water. And if there is no change in volume permitted (that is, undrained conditions prevail), then setting Eq. B-3-1 equal to Eq. B-3-6 and solving for Δu, we have

$$\Delta u = \frac{1}{1 + n\left(\dfrac{C_v}{C_{sk}}\right)}\left[(\Delta\sigma_{oct}) + \frac{D}{C_{sk}}|\Delta\tau_{oct}|\right] \qquad \text{(B-3-7)}$$

Since soils are not linearly elastic materials, as before we use empirical coefficients which are to be determined by experiment,

$$B = \frac{1}{1 + \dfrac{nC_v}{C_{sk}}} \qquad \text{(11-11)}$$

and

$$a = \frac{D}{C_{sk}} \qquad \text{(B-3-8)}$$

So Eq. B-3-5 becomes

$$\Delta u = B(\Delta\sigma_{oct} + a\,\Delta\tau_{oct}) \qquad \text{(11-22)}$$

The coefficient a is the *Henkel pore pressure* parameter.

Although this derivation for Eq. 11-22 is rather elegant mathematically, it may be easier to simply write the equation as

$$\Delta u = B\frac{\Delta\sigma_1 + \Delta\sigma_2 + \Delta\sigma_3}{3}$$
$$+ \frac{a}{3}\sqrt{(\Delta\sigma_1 - \Delta\sigma_2)^2 + (\Delta\sigma_2 - \Delta\sigma_3)^2 + (\Delta\sigma_3 - \Delta\sigma_1)^2} \qquad \text{(B-3-9)}$$

This latter formulation is more consistent with the definition of principal stress increments presented in Sec. B-3.2. With this definition, a systematic separation of the stress increments from the initial and final stress states is possible.

Equations 11-13 and B-3-9 are useful since they allow the separation of pore pressure effects observed in soils into two components, that due to (1) the change in mean or average stress, and (2) the change in shear stress.

The Henkel parameter a is, like the Skempton parameter A, nonlinear and must be determined for each stress path. It is also very dependent on strain, on the magnitude of σ_2, on the overconsolidation ratio, and on material properties such as anisotropy. The parameters a and B are for *general* changes in total stress. They enable the engineer to predict the pore pressure if the changes in the total stresses are known or can be estimated, therefore they can be very useful in engineering practice.

Sometimes in the geotechnical literature the Henkel parameters are denoted by the symbol α, where $\alpha = a/3$. In this case, Eq. B-3-9 would be

$$\Delta u = B\frac{\Delta\sigma_1 + \Delta\sigma_2 + \Delta\sigma_3}{3}$$
$$+ \alpha\sqrt{(\Delta\sigma_1 - \Delta\sigma_2)^2 + (\Delta\sigma_2 - \Delta\sigma_3)^2 + (\Delta\sigma_3 - \Delta\sigma_1)^2} \tag{B-3-10}$$

This is the way Henkel (1960) originally wrote his equation, but with the symbol a for α. Thus Henkel's original a or α was one-third our a. Later Henkel and Wade (1966) suggested the notation used herein, along with Eq. 11-22.

It is often useful to be able to convert between the Henkel parameter a and the Skempton parameter A. For the special case of triaxial compression (AC), $\sigma_2 = \sigma_3$ and $S = 100\%$ ($B = 1$), we have

$$\Delta\sigma_{oct} = \frac{1}{3}(\Delta\sigma_1 + 2\Delta\sigma_3)$$

and

$$\Delta\tau_{oct} = \frac{\sqrt{2}}{3}(\Delta\sigma_1 - \Delta\sigma_3)$$

so (Eq. 11-22)

$$\Delta u = \frac{1}{3}(\Delta\sigma_1 + 2\Delta\sigma_3) + a\frac{\sqrt{2}}{3}(\Delta\sigma_1 - \Delta\sigma_3)$$

but since $\Delta\sigma_2 = \Delta\sigma_3 = 0$ (constant cell pressure) and $\Delta\sigma_1 = \Delta\sigma_v$,

$$\Delta u = \left(\frac{1}{3} + a\frac{\sqrt{2}}{3}\right)\Delta\sigma_v$$

From Eq. 11-13 and for triaxial compression conditions in Table B-3-2, we know that $A_{ac} = \Delta u/\Delta\sigma_v$. Therefore

$$A_{ac} = \frac{1}{3} + a\frac{\sqrt{2}}{3} \tag{11-25a}$$

For the lateral extension (LE) test, $\sigma_2 = \sigma_3$, and Eq. 11-22 becomes

$$\Delta u = \frac{1}{3}(\Delta\sigma_1 + 2\Delta\sigma_3) + a\frac{\sqrt{2}}{3}(\Delta\sigma_1 - \Delta\sigma_3)$$

But since $\Delta\sigma_1 = 0$ and $\Delta\sigma_2 = \Delta\sigma_3 = \Delta\sigma_h$,

$$\Delta u = \frac{2}{3}\Delta\sigma_h - a\frac{\sqrt{2}}{3}\Delta\sigma_h = \frac{2}{3} - \frac{\sqrt{2}}{3}\Delta\sigma_h$$

From Eq. 11-13 and Table B-3-2, we know that $A_{le} = 1 - \Delta u/\Delta\sigma_h$. Therefore

$$A_{le} = 1 - \frac{2}{3} - a\frac{\sqrt{2}}{3} = \frac{1}{3} + a\frac{\sqrt{2}}{3} \qquad (11\text{-}25\text{b})$$

which is the same as Eq. 11-25a. This result should not be unexpected since we have already shown that $A_{ac} = A_{le}$ (Eq. 11-20).

For the case of axial extension (AE), $\sigma_2 = \sigma_1$, and Eq. 11-22 becomes

$$\Delta u = \frac{1}{3}(2\Delta\sigma_1 + \Delta\sigma_3) + a\frac{\sqrt{2}}{3}(\Delta\sigma_1 - \Delta\sigma_3)$$

But since $\Delta\sigma_1 = \Delta\sigma_2 = 0$ and $\Delta\sigma_3 = \Delta\sigma_v$,

$$\Delta u = \frac{1}{3}\Delta\sigma_v - a\frac{\sqrt{2}}{3}\Delta\sigma_v = \frac{1}{3} - a\frac{\sqrt{2}}{3}\Delta\sigma_v$$

From Eq. 11-13 and Table B-3-2, we know that $A_{ae} = 1 - \Delta u/\Delta\sigma_v$. Therefore

$$A_{ae} = 1 - \frac{1}{3} - a\frac{\sqrt{2}}{3} = \frac{2}{3} + a\frac{\sqrt{2}}{3} \qquad (11\text{-}26\text{a})$$

For the lateral compression (LC) test, $\sigma_2 = \sigma_1$, so Eq. 11-22 becomes

$$\Delta u = \frac{1}{3}(2\Delta\sigma_1 + \Delta\sigma_3) + a\frac{\sqrt{2}}{3}(\Delta\sigma_1 - \Delta\sigma_3)$$

Since $\Delta\sigma_1 = \Delta\sigma_2 = \Delta\sigma_h$ and $\Delta\sigma_3 = 0$, we have

$$\Delta u = \left(\frac{2}{3} + a\frac{\sqrt{2}}{3}\right)\Delta\sigma_h$$

From Eq. 11-13 and Table B-3-2, we know that $A_{lc} = \Delta u/\Delta\sigma_h$. Therefore

$$A_{lc} = \frac{2}{3} + a\frac{\sqrt{2}}{3} \qquad (11\text{-}26\text{b})$$

As expected (Eq. 11-21), $A_{ae} = A_{lc}$.

Note that for elastic materials, $A_{ac} = A_{le} = 1/3$ and $A_{ae} = A_{lc} = 2/3$, and $a = 0$. In general, since $A_{ac} \neq A_{lc}$, then the a parameters are not necessarily the same for the two cases, primarily because the compressibility of the soil skeleton C_{sk} is not the same in compression as in extension.

References

ABDELHAMID, M.S., AND KRIZEK, R.J. (1976) "At Rest Lateral Earth Pressure of a Consolidating Clay," *Journal of the Geotechnical Engineering Division*, ASCE, Vol. 102, No. GT7, pp. 721–738.

ABELEV, Y.M. (1957) "The Stabilization of Foundations of Structures on Loess Soils," *Proceedings of the Fourth International Conference on Soil Mechanics and Foundation Engineering*, London, Vol. I, pp. 259–263.

ABOSHI, H. (1973) "An Experimental Investigation on the Similitude in the Consolidation of a Soft Clay, Including the Secondary Creep Settlement," *Proceedings of the Eighth International Conference on Soil Mechanics and Foundation Engineering*, Moscow, Vol. 4.3, p. 88.

AL-HUSSAINI, M.M. (1977) "Contribution to the Engineering Soil Classifications of Cohesionless Soils," *Final Report*, Miscellaneous Paper S-77-21, U.S. Army Engineer Waterways Experiment Station, Vicksburg, Mississippi, 61 pp.

AL-HUSSAINI, M.M., AND TOWNSEND, F.C. (1975) "Investigation of K_o Testing in Cohesionless Soils," *Technical Report S-75-11*, U.S. Army Engineer Waterways Experiment Station, Vicksburg, Mississippi, 70 pp.

ALPAN, I. (1967) "The Empirical Evaluation of the Coefficient K_o and K_{oR}," *Soil and Foundation*, Vol. VII, No. 1, pp. 31–40.

AMERICAN ASSOCIATION FOR STATE HIGHWAY AND TRANSPORTATION OFFICIALS (1978) *Standard Specifications for Transportation Materials and Methods of Sampling and Testing*, 12th Ed., Washington, D.C. Part I, Specifications, 828 pp.; Part II, Tests, 998 pp.

AMERICAN SOCIETY OF CIVIL ENGINEERS (1978) *Soil Improvement—History, Capabilities, and Outlook*, Committee on Placement and Improvement of Soils, Geotechnical Engineering Division, ASCE, 182 pp.

AMERICAN SOCIETY FOR TESTING AND MATERIALS (1980) "Natural Building Stones; Soil and Rock," *Annual Book of ASTM Standards*, Part 19, Philadelphia, 634 pp.

AMERICAN SOCIETY FOR TESTING AND MATERIALS (1966) *ASTM Metric Practice Guide*, 2nd Ed., 46 pp.

ARAVIN, V.I., AND NUMEROV, S. (1955) "Seepage Computations for Hydraulic Structures," *Stpoitel'stvu i Arkhitekture*, Moscow, (referenced in Harr, 1962).

ATTERBERG, A. (1905) "Die Rationelle Klassifikation der Sande und Kiese," *Chemiker-Zeitung*, Vol. 29, pp. 195–198.

ATTERBERG, A. (1911) "Lerornas Förhallande till Vatten, deras Plasticitetsgränser och Plasticitetsgrader," ("The Behavior of Clays with Water, their Limits of Plasticity and their Degrees of Plasticity,") *Kungliga Lantbruksakademiens Handlingar och Tidskrift*, Vol. 50, No. 2, pp. 132–158; also in *Internationale Mitteilungen für Bodenkunde*, Vol. 1, pp. 10–43 ("Über die Physikalische Bodenuntersuchung und über die Plastizität der Tone").

AZZOUZ, A.S., KRIZEK, R.J. AND COROTIS, R.B. (1976) "Regression Analysis of Soil Compressibility," *Soils and Foundations*, Vol. 16, No. 2, pp. 19–29.

BAGUELIN, F., JÉZÉQUEL, J.F., AND SHIELDS, D.H. (1978) *The Pressuremeter and Foundation Engineering*, Trans Tech Publications, Clausthal, Germany, and Aedermannsdorf, Switzerland, 617 pp.

VON BANDAT, H.F. (1962) *Aerogeology*, Gulf Publishing Company, Houston, Texas, 350 pp.

BARDEN, L., AND MCGOWN, A. (1973) "Microstructural Disturbance in Soft Clays Resulting from Site Investigation Sampling," *Proceedings of the International Symposium on Soil Structure*, Gothenburg, Sweden, p. 213.

BARRETT, R.J. (1966) "Use of Plastic Filters in Coastal Construction," *Proceedings of the Tenth International Conference on Coastal Engineering*, Tokyo, pp. 1048–1067.

BEGEMANN, H.K.S.PH. (1953) "Improved Methods of Determining Resistance to Adhesion by Sounding through a Loose Sleeve Placed Behind the Cone," *Proceedings of the Third International Conference on Soil Mechanics and Foundation Engineering*, Zurich, Vol. I, pp. 213–217.

BESKOW, G. (1935) "Soil Freezing and Frost Heaving with Special Application to Roads and Railroads," *The Swedish Geological Society*, Series C, No. 375, 26th Year Book No. 3; translated by J.O. Osterberg, Northwestern University, 1947, 145 pp.

BISHOP, A.W. (1958) "Test Requirements of Measuring the Coefficient of Earth Pressure at Rest," *Proceedings of the Conference on Earth Pressure Problems*, Brussels, Vol. I, pp. 2–14.

BISHOP, A.W., AND HENKEL, D.J. (1962) *The Measurement of Soil Properties in the Triaxial Test*, Edward Arnold Ltd., London, 2nd Ed., 228 pp.

BISHOP, A.W., AND WESLEY, L.D. (1975) "Triaxial Apparatus for Controlled Stress Path Testing," *Géotechnique*, Vol. XXV, No. 4, pp. 657–670.

BJERRUM, L. (1954) "Geotechnical Properties of Norwegian Marine Clays," *Géotechnique*, Vol. IV, No. 2, pp. 49–69.

BJERRUM, L. (1967) "Engineering Geology of Norwegian Normally Consolidated Marine Clays as Related to Settlements of Buildings," *Géotechnique*, Vol. XVII, No. 2, pp. 81–118.

BJERRUM, L. (1972) "Embankments on Soft Ground," *Proceedings of the ASCE Specialty Conference on Performance of Earth and Earth-Supported Structures*, Purdue University, Vol. II, pp. 1–54.

BJERRUM, L., AND SIMONS, N.E. (1960) "Comparison of Shear Strength Characteristics of Normally Consolidated Clays," *Proceedings of the ASCE Research Conference on the Shear Strength of Cohesive Soils*, Boulder, pp. 711–726.

BLACK, D.K., AND LEE, K.L. (1973) "Saturating Laboratory Samples by Back Pressure," *Journal of the Soil Mechanics and Foundations Division*, ASCE, Vol. 99, No. SM1, pp. 75–93.

BOUSSINESQ, J. (1885) *Application des Potentiels à L'Étude de L'Équilibre et due Mouvement des Solides Élastiques*, Gauthier-Villars, Paris.

BOZOZUK, M. (1963) "The Modulus of Elasticity of Leda Clay from Field Measurements," *Canadian Geotechnical Journal*, Vol. I, No. 1, pp. 43–51.

BOZOZUK, M., AND LEONARDS, G.A. (1972) "The Gloucester Test Fill," *Proceedings of the ASCE Specialty Conference on Performance of Earth and Earth-Supported Structures*," Purdue University, Vol. I, Part 1, pp. 299–317.

BROMS, B.B., AND FORSSBLAD, L. (1969) "Vibratory Compaction of Cohesionless Soils," *Proceedings of the Specialty Session No. 2 on Soil Dynamics*, Seventh International Conference on Soil Mechanics and Foundation Engineering, Mexico City, pp. 101–118.

BROOKER, E.W., AND IRELAND, H.O. (1965) "Earth Pressures at Rest Related to Stress History," *Canadian Geotechnical Journal*, Vol. II, No. 1, pp. 1–15.

BRUMUND, W.F., JONAS, E., AND LADD, C.C. (1976) "Estimating In Situ Maximum Past Preconsolidation Pressure of Saturated Clays from Results of Laboratory Consolidometer Tests," *Special Report 163*, Transportation Research Board, pp. 4–12.

CADLING, L., AND ODENSTAD, S. (1950) "The Vane Borer," *Proceedings No. 2*, Royal Swedish Geotechnical Institute, pp. 1–88.

CAMPANELLA, R.G., AND VAID, Y.P. (1972) "A Simple K_o-Triaxial Cell," *Canadian Geotechnical Journal*, Vol. 9, No. 3, pp. 249–260.

CARRIGAN, R.A. (1978) "Decimal Time," *American Scientist*, Vol. 66, No. 3, pp. 305–313.

CASAGRANDE, A. (1932a) Discussion of "A New Theory of Frost Heaving," by A.C. Benkelman and F.R. Ohlmstead, *Proceedings of the Highway Research Board*, Vol. 11, pp. 168–172.

CASAGRANDE, A. (1932b) "Research on the Atterberg Limits of Soils," *Public Roads*, Vol. 13, No. 8, pp. 121–136.

CASAGRANDE, A. (1932c) "The Structure of Clay and Its Importance in Foundation Engineering," *Journal of the Boston Society of Civil Engineers*, April; reprinted in *Contributions to Soil Mechanics 1925–1940*, BSCE, pp. 72–113.

CASAGRANDE, A. (1936a) "Characteristics of Cohesionless Soils Affecting the Stability of Slopes and Earth Fills," *Journal of the Boston Society of Civil Engineers*, January; reprinted in *Contributions to Soil Mechanics 1925–1940*, BSCE, pp. 257–276.

CASAGRANDE, A. (1936b) "The Determination of the Pre-Consolidation Load and Its Practical Significance," Discussion D-34, *Proceedings of the First International Conference on Soil Mechanics and Foundation Engineering*, Cambridge, Vol. III, pp. 60–64.

CASAGRANDE, A. (1937) "Seepage Through Dams," *Journal of the New England*

Water Works Association, Vol. LI, No. 2; reprinted in *Contributions to Soil Mechanics 1925–1940*, BSCE, pp. 295–336.

CASAGRANDE, A. (1938) "Notes on Soil Mechanics—First Semester," Harvard University (unpublished), 129 pp.

CASAGRANDE, A. (1948) "Classification and Identification of Soils," *Transactions*, ASCE, Vol. 113, pp. 901–930.

CASAGRANDE, A. (1950) "Notes on the Design of Earth Dams," *Journal of the Boston Society of Civil Engineers*, October; reprinted in *Contributions to Soil Mechanics 1941–1953*, BSCE, pp. 231–255.

CASAGRANDE, A. (1958) "Notes on the Design of the Liquid Limit Device," *Géotechnique*, Vol. VIII, No. 2, pp. 84–91.

CASAGRANDE, A. (1975) "Liquefaction and Cyclic Deformation of Sands, a Critical Review," *Proceedings of the Fifth Panamerican Conference on Soil Mechanics and Foundation Engineering*, Buenos Aires; reprinted as Harvard Soil Mechanics Series, No. 88, 27 pp.

CASAGRANDE, A., AND FADUM, R.E. (1944) Closure to "Application of Soil Mechanics in Designing Building Foundations," *Transactions*, ASCE, Vol. 109, p. 467.

CASTRO, G. (1969) "Liquefaction of Sands," PhD Thesis, Harvard University; reprinted as Harvard Soil Mechanics Series, No. 81, 112 pp.

CASTRO, G. (1975) "Liquefaction and Cyclic Mobility of Saturated Sands," *Journal of the Geotechnical Engineering Division*, ASCE, Vol. 101, No. GT6, June, pp. 551–569.

CASTRO, G., AND POULOS, S.J. (1977) "Factors Affecting Liquefaction and Cyclic Mobility," *Journal of the Geotechnical Engineering Division*, ASCE, Vol. 103, No. GT6, pp. 501–516.

CATERPILLAR TRACTOR CO. (1977) *Caterpillar Performance Handbook*, Form AEKQ 3313, 8th Ed., Chapter 14, Peoria, Illinois, p. 6.

CEDERGREN, H.R. (1977) *Seepage, Drainage, and Flow Nets*, 2nd Ed., John Wiley & Sons, Inc., New York, 534 pp.

CLEVENGER, W.A. (1958) "Experiences With Loess as a Foundation Material," *Transactions*, ASCE, Vol. 123, pp. 151–169.

COLLINS, K., AND McGOWN, A. (1974) "The Form and Function of Microfabric Features in a Variety of Natural Soils," *Géotechnique*, Vol. XXIV, No. 2, pp. 223–254.

COULOMB, C.A. (1776) "Essai sur une application des règles de Maximus et Minimis à quelques Problèmes de Statique, relatifs à l'Architecture," *Mémoires de Mathématique et de Physique, Présentés a l'Académie Royale des Sciences, par divers Savans, et lûs dans ses Assemblées*, Paris, Vol. 7, (volume for 1773 published in 1776), pp. 343–382.

D'APPOLONIA, D.J., WHITMAN, R.V., AND D'APPOLONIA, E.D. (1969) "Sand Compaction with Vibratory Rollers," *Journal of the Soil Mechanics and Foundations Division*, ASCE, Vol. 95, No. SM1, pp. 263–284.

D'APPOLONIA, D.J., LAMBE, T.W., AND POULOS, H.G. (1971) "Evaluation of Pore Pressures Beneath an Embankment," *Journal of the Soil Mechanics and Foundations Division*, ASCE, Vol. 97, No. SM6, pp. 881–897.

D'APPOLONIA, D.J., POULOS, H.G., AND LADD, C.C. (1971) "Initial Settlement of Structures on Clay," *Journal of the Soil Mechanics and Foundations Division*, ASCE, Vol. 97, No. SM10, pp. 1359–1377.

DALLAIRE, G. (1976) "Filter Fabrics: Bright Future in Road and Highway Construction," *Civil Engineering*, ASCE, Vol. 46, No. 5, pp. 61–65.

D'ARCY, H. (1856) "Les Fontaines Publiques de la Ville de Dijon," Dalmont, Paris.

DAWSON, R.F. (1944) Discussion of "Relation of Undisturbed Sampling to Laboratory Testing," by P.C. Rutledge, *Transactions*, ASCE, Vol. 109, pp. 1190–1193.

DE MELLO, V.F.B. (1971) "The Standard Penetration Test," State of the Art Paper, *Proceedings of the Fourth Panamerican Conference on Soil Mechanics and Foundation Engineering*, Vol. I, pp. 1–86.

DIBERNARDO, A. (1979) "The Effect of Laboratory Compaction on the Compressibility of a Compacted Highly Plastic Clay," MSCE Thesis, Purdue University, 187 pp.

DIBIAGIO, E., AND STENHAMAR, P. (1975) "Prøvefylling til Brudd på Bløt Leire," *Proceedings of the Seventh Scandinavian Geotechnical Meeting*, Copenhagen, Polyteknisk Forlag, pp. 173–185.

DUNCAN, J.M., AND BUCHIGNANI, A.L. (1976) "An Engineering Manual for Settlement Studies," *Geotechnical Engineering Report*, University of California at Berkeley, 94 pp.

EDEN, W.J. (1971) "Sampler Trials in Overconsolidated Sensitive Clay," *Sampling of Soil and Rock*, ASTM Special Technical Publication No. 483, pp. 132–142.

EDEN, W.J., AND KUBOTA, J.K. (1962) "Some Observations on the Measurement of Sensitivity of Clays," *Proceedings of the American Society for Testing and Materials*, Vol. 61, pp. 1239–1249.

EIDE, O., AND HOLMBERG, S. (1972) "Test Fills to Failure on the Soft Bangkok Clay," *Proceedings of the ASCE Specialty Conference on Performance of Earth and Earth-Supported Structures*, Purdue University, Vol. I, Part 1, p. 163.

ESOPT (1974) *Proceedings of the European Symposium on Penetration Testing*, Stockholm, Swedish Council for Building Research, Vols. 1, 2.1, 2.2, and 3.

FINN, W.D.L., PICKERING, D.J., AND BRANSBY, P.L. (1971) "Sand Liquefaction in Triaxial and Simple Shear Tests," *Journal of the Soil Mechanics and Foundations Division*, ASCE, Vol. 97, No. SM4, pp. 639–659.

FLAATE, K., AND PREBER, T. (1974) "Stability of Road Embankments," *Canadian Geotechnical Journal*, Vol. 11, No. 1, pp. 72–88.

FOSTER, C.R., AND AHLVIN, R.G. (1954) "Stresses and Deflections Induced by a Uniform Circular Load," *Proceedings of the Highway Research Board*, Vol. 33, pp. 467–470.

GARCIA-BENGOCHEA, I., LOVELL, C.W., AND ALTSCHAEFFL, A.G. (1979) "Pore Distribution and Permeability of Silty Clays," *Journal of the Geotechnical Engineering Division*, ASCE, Vol. 105, No. GT7, pp. 839–856.

GIBBS, H.J. (1969) Discussion, *Proceedings of the Specialty Session No. 3 on Expansive Soils and Moisture Movement in Partly Saturated Soils*, Seventh International Conference on Soil Mechanics and Foundation Engineering, Mexico City.

GRIM, R.E. (1959) "Physico-Chemical Properties of Soils: Clay Minerals," *Journal of the Soil Mechanics and Foundations Division*, ASCE, Vol. 85, No. SM2, pp. 1–17.

GROMKO, G.J. (1974) "Review of Expansive Soils," *Journal of the Geotechnical Engineering Division*, ASCE, Vol. 100, No. GT6, pp. 667–687.

Handbook of Chemistry and Physics (1977) CRC Press, Cleveland, 58th Ed.

HANSBO, S. (1957) "A New Approach to the Determination of the Shear Strength of Clay by the Fall-Cone Test," *Proceedings No. 14*, Swedish Geotechnical Institute, 47 pp.

HANSBO, S. (1960) "Consolidation of Clay with Special Reference to Influence of Vertical Sand Drains," *Proceedings No. 18*, Swedish Geotechnical Institute, pp. 45–50.

HANSBO, S. (1975) *Jordmateriallära*, Almqvist & Wiksell Förlag AB, Stockholm, 218 pp.

HARR, M.E. (1962) *Groundwater and Seepage*, McGraw-Hill Book Company, New York, 315 pp.

HARR, M.E. (1966) *Foundations of Theoretical Soil Mechanics*, McGraw-Hill Book Company, New York, 381 pp.

HARR, M.E. (1977) *Mechanics of Particulate Media*, McGraw-Hill Book Company, New York, 543 pp.

HAZEN, A. (1911) Discussion of "Dams on Sand Foundations," by A.C. Koenig, *Transactions*, ASCE, Vol. 73, pp. 199–203.

HENKEL, D.J. (1960) "The Shear Strength of Saturated Remoulded Clays," *Proceedings of the ASCE Research Conference on Shear Strength of Cohesive Soils*, Boulder, pp. 533–554.

HENKEL, D.J., AND WADE, N.H. (1966) "Plane Strain Tests on a Saturated Remolded Clay," *Journal of the Soil Mechanics and Foundations Division*, ASCE, Vol. 92, No. SM6, pp. 67–80.

HILF, J.W. (1961) "A Rapid Method of Construction Control for Embankments of Cohesive Soils," *Engineering Monograph No. 26*, revised, U.S. Bureau of Reclamation, Denver, 29 pp.

HIRSCHFELD, R.C. (1963) "Stress-Deformation and Strength Characteristics of Soils," Harvard University (unpublished), 87 pp.

HÖEG, K., ANDERSLAND, O.B., AND ROLFSEN, E.N. (1969) "Undrained Behaviour of Quick Clay Under Load Tests at Åsrum," *Géotechnique*, Vol. XIX, No. 1, pp. 101–115.

HOGENTOGLER, C.A., AND TERZAGHI, C. (1929) "Interrelationship of Load, Road, and Subgrade," *Public Roads*, Vol. 10, No. 3, pp. 37–64.

HOLDEN, J.C. (1974) "Penetration Testing in Australia," *Proceedings of the European Symposium on Penetration Testing* (ESOPT), Vol. 1, pp. 155–162.

HOLM, G., AND HOLTZ, R.D. (1977) "A Study of Large Diameter Piston Samplers," *Proceedings of the Specialty Session No. 2*, Ninth International Conference on Soil Mechanics and Foundation Engineering, Tokyo, p. 77; also in *Proceedings of the International Symposium on Soft Clay*, Bangkok, p. 381.

HOLTZ, R.D. (1980) "SI Units in Geotechnical Engineering," *Geotechnical Testing Journal*, ASTM, Vol. 3, No. 2.

HOLTZ, R.D., AND BROMS, B.B. (1972) "Long-Term Loading Tests at Skå-Edeby, Sweden," *Proceedings of the ASCE Specialty Conference on Performance of Earth and Earth-Supported Structures*, Purdue University, Vol. I, Part 1, pp. 435–464.

HOLTZ, R.D., AND HOLM, G. (1979) "Test Embankment on an Organic Silty Clay," *Proceedings of the Seventh European Conference on Soil Mechanics and Foundation Engineering*, Brighton, England, Vol. 3, pp. 79–86.

HOLTZ, W.G. (1959) "Expansive Clays—Properties and Problems," *Quarterly of the Colorado School of Mines*, Vol. 54, No. 4, pp. 89–125.

HOLTZ, W.G., AND GIBBS, H.J. (1956) "Engineering Properties of Expansive Clays," *Transactions*, ASCE, Vol. 121, pp. 641–677.

HOUGH, B.K. (1969) *Basic Soils Engineering*, 2nd Ed., The Ronald Press Company, New York, 634 pp.

HOWARD, A.K. (1977) "Laboratory Classification of Soils—Unified Soil Classification System," *Earth Sciences Training Manual No. 4*, U.S. Bureau of Reclamation, Denver, 56 pp.

HVORSLEV, M.J. (1949) *Subsurface Exploration and Sampling of Soils for Civil Engineering Purposes*, U.S. Army Engineer Waterways Experiment Station, Vicksburg, Mississippi, 521 pp; reprinted by the Engineering Foundation, 1962.

HVORSLEV, M.J. (1960) "Physical Components of the Shear Strength of Saturated Clays," *Proceedings of the ASCE Research Conference on the Shear Strength of Cohesive Soils*, Boulder, pp. 169–173.

INTERNATIONAL SOCIETY FOR SOIL MECHANICS AND FOUNDATION ENGINEERING (1977) "List of Symbols, Units, and Definitions," Subcommittee on Symbols, Units, and Definitions, *Proceedings of the Ninth International Conference on Soil Mechanics and Foundation Engineering*, Tokyo, Vol. 3, pp. 156–170.

JÁKY, J. (1944) "The Coefficient of Earth Pressure at Rest," (in Hungarian), *Magyar Mérnök és Épitész Egylet Közdönye* (Journal of the Society of Hungarian Architects and Engineers), Vol. 78, No. 22, pp. 355–358.

JÁKY, J. (1948) "Earth Pressure in Silos," *Proceedings of the Second International Conference on Soil Mechanics and Foundation Engineering*, Rotterdam, Vol. I, pp. 103–107.

JANBU, N., AND SENNESET, K. (1973) "Field Compressometer—Principles and Applications," *Proceedings of the Eighth International Conference on Soil Mechanics and Foundation Engineering*, Moscow, Vol. 1.1, pp. 191–198.

JOHNSON, A.W., AND SALLBERG, J.R. (1960) "Factors that Influence Field Compaction of Soils," *Bulletin 272,* Highway Research Board, 206 pp.

JONES, D.E., AND HOLTZ, W.G. (1973) "Expansive Soils—the Hidden Disaster," *Civil Engineering*, ASCE, Vol. 43, No. 8, pp. 49–51.

KARLSSON, R. (1977) "Consistency Limits," in cooperation with the Laboratory Committee of the Swedish Geotechnical Society, Swedish Council for Building Research, Document D6, 40 pp.

KAUFMAN, R.I., AND SHERMAN, W.C., JR., (1964) "Engineering Measurements for Port Allen Lock," *Journal of the Soil Mechanics and Foundations Division*, ASCE, Vol. 90, No. SM5, pp. 221–247; also in *Design of Foundations for Control of Settlement*, ASCE, pp. 281–307.

KENNEY, T.C. (1959) Discussion of "Geotechnical Properties of Glacial Lake Clays," by T.H. Wu, *Journal of the Soil Mechanics and Foundations Division*, ASCE, Vol. 85, No. SM3, pp. 67–79.

KENNEY, T.C. (1964) "Sea-Level Movements and the Geologic Histories of the Post-Glacial Marine Soils at Boston, Nicolet, Ottawa, and Oslo," *Géotechnique*, Vol. XIV, No. 3, pp. 203–230.

KENNEY, T.C. (1976) "Formation and Geotechnical Characteristics of Glacial-Lake Varved Soils," *Laurits Bjerrum Memorial Volume—Contributions to Soil Mechanics*, Norwegian Geotechnical Institute, Oslo, pp. 15–39.

KOVACS, W.D., EVANS, J.C., AND GRIFFITH, A.H. (1977) "Towards a More Standardized SPT," *Proceedings of the Ninth International Conference on Soil Mechanics and Foundation Engineering*, Tokyo, Vol. 2, pp. 269–276.

LADD, C.C. (1964) "Stress-Strain Behavior of Saturated Clay and Basic Strength Principles," *Research Report R64-17*, Department of Civil Engineering, Massachusetts Institute of Technology, 67 pp.

LADD, C.C. (1971a) "Settlement Analyses for Cohesive Soils," *Research Report R71-2*, Soils Publication 272, Department of Civil Engineering, Massachusetts Institute of Technology, 107 pp.

LADD, C.C. (1971b) "Strength Parameters and Stress-Strain Behavior of Saturated Clays," *Research Report R71-23*, Soils Publication 278, Department of Civil Engineering, Massachusetts Institute of Technology, 280 pp.

LADD, C.C. (1972) "Test Embankment on Sensitive Clay," *Proceedings of the ASCE Specialty Conference on Performance of Earth and Earth-Supported Structures*, ASCE, Purdue University, Vol. I, Part 1, pp. 103 and 107.

LADD, C.C. (1975) "Foundation Design of Embankments Constructed on Connecticut Valley Varved Clays," *Research Report R75-7*, Geotechnical Publication 343, Department of Civil Engineering, Massachusetts Institute of Technology, 438 pp.

LADD, C.C., AND LAMBE, T.W. (1963) "The Strength of Undisturbed Clay Determined from Undrained Tests," *Laboratory Shear Testing of Soils*, ASTM Special Technical Publication No. 361, pp. 342–371.

LADD, C.C., AND LUSCHER, U. (1965) "Engineering Properties of the Soils Underlying the M.I.T. Campus," *Research Report R65-68*, Soils Publication 185, Department of Civil Engineering, Massachusetts Institute of Technology.

LADD, C.C., AND EDGERS, L. (1972) "Consolidated-Undrained Direct-Simple Shear Tests on Saturated Clays," *Research Report R72-92*, Soils Publication 284, Department of Civil Engineering, Massachusetts Institute of Technology, 245 pp.

LADD, C.C., AND FOOTE, R. (1974) "A New Design Procedure for Stability of Soft Clays," *Journal of the Geotechnical Engineering Division*, ASCE, Vol. 100, No. GT7, pp. 763–786.

LADD, C.C., FOOTE, R., ISHIHARA, K., SCHLOSSER, F., AND POULOS, H.G. (1977) "Stress-Deformation and Strength Characteristics," State-of-the-Art Report, *Proceedings of the Ninth International Conference on Soil Mechanics and Foundation Engineering*, Tokyo, Vol. 2, pp. 421–494.

LADD, R.S. (1965) "Use of Electrical Pressure Transducers to Measure Soil Pressure," *Research Report R65-48*, Soils Publication 180, Department of Civil Engineering, Massachusetts Institute of Technology, 79 pp.

LADD, R.S. (1977) "Specimen Preparation and Cyclic Stability of Sands," *Journal of the Soil Mechanics and Foundations Division*, ASCE, Vol. 103, No. GT2, pp. 535–547.

LADE, P.V., AND LEE, K.L. (1976) "Engineering Properties of Soils," *Report*, UCLA-ENG-7652, 145 pp.

LAMBE, T.W. (1951) *Soil Testing for Engineers*, John Wiley & Sons, Inc., New York, 165 pp.

LAMBE, T.W. (1953) "The Structure of Inorganic Soil," *Proceedings*, ASCE, Vol. 79, Separate No. 315, 49 pp.

LAMBE, T.W. (1958a) "The Structure of Compacted Clay," *Journal of the Soil Mechanics and Foundations Division*, ASCE, Vol. 84, No. SM2, 1654-1 to 1654-34.

LAMBE, T.W. (1958b) "The Engineering Behavior of Compacted Clay," *Journal of the Soil Mechanics and Foundations Division*, ASCE, Vol. 84, No. SM2, pp. 1655-1 to 1655-35.

LAMBE, T.W. (1964) "Methods of Estimating Settlement," *Journal of the Soil Mechanics and Foundations Division*, ASCE, Vol. 90, No. SM5, pp. 43–67; also in *Design of Foundations for Control of Settlement*, ASCE, pp. 47–72.

LAMBE, T.W. (1967) "Stress Path Method," *Journal of the Soil Mechanics and Foundations Division*, ASCE, Vol. 93, No. SM6, pp. 309–331.

LAMBE, T.W., AND MARR, W.A. (1979) "Stress Path Method: Second Edition," *Journal of the Geotechnical Engineering Division*, ASCE, Vol. 105, No. GT6, pp. 727–738.

LAMBE, T.W., AND WHITMAN, R.V. (1969) *Soil Mechanics*, John Wiley & Sons, Inc., New York, 553 pp.

LAMBRECHTS, J.R., AND LEONARDS, G.A. (1978) "Effects of Stress History on Deformation of Sand," *Journal of the Geotechnical Engineering Division*, ASCE, Vol. 104, No. GT11, pp. 1371–1387.

LAROCHELLE, P., AND LEFEBVRE, G. (1971) "Sampling Disturbance in Champlain Clays," *Sampling of Soil and Rock*, ASTM Special Technical Publication No. 483, pp. 143–163.

LAROCHELLE, P., TRAK, B., TAVENAS, F., AND ROY, M. (1974) "Failure of a Test Embankment on a Sensitive Champlain Clay Deposit," *Canadian Geotechnical Journal*, Vol. 11, No. 1, pp. 142–164.

LAW, K.T., AND HOLTZ, R.D. (1978) "A Note on Skempton's *A* Parameter with Rotation of Principle Stresses," *Géotechnique*, Vol. XXVIII, No. 1, pp. 57–64.

LEE, K.L. (1965) "Triaxial Compressive Strength of Saturated Sands Under Seismic Loading Conditions," PhD Dissertation, University of California at Berkeley, 521 pp.

LEE, K.L., AND SEED, H.B. (1967) "Drained Strength Characteristics of Sands," *Journal of the Soil Mechanics and Foundations Division*, ASCE, Vol. 93, No. SM6, pp. 117–141.

LEE, K.L., AND SINGH, A. (1971) "Compaction of Granular Soils," *Proceedings of the Ninth Annual Symposium on Engineering Geology and Soils Engineering*, Boise, Idaho, pp. 161–174.

LEONARDS, G.A., Ed. (1962) *Foundation Engineering*, McGraw-Hill Book Company, New York, 1136 pp.

LEONARDS, G.A. (1973) Discussion of "The Empress Hotel, Victoria, British Columbia: Sixty-five Years of Foundation Settlements," *Canadian Geotechnical Journal*, Vol. 10, No. 1, pp. 120–122.

LEONARDS, G.A. (1976) "Estimating Consolidation Settlements of Shallow Foundations on Overconsolidated Clays," *Special Report 163*, Transportation Research Board, pp. 13–16.

LEONARDS, G.A. (1977) Discussion to Main Session 2, *Proceedings of the Ninth International Conference on Soil Mechanics and Foundation Engineering*, Tokyo, Vol. 3, pp. 384–386.

LEONARDS, G.A., AND ALTSCHAEFFL, A.G. (1964) "Compressibility of Clay," *Journal of the Soil Mechanics and Foundations Division*, ASCE, Vol. 90, No. SM5, pp. 133–156; also in *Design of Foundations for Control of Settlement*, ASCE, pp. 163–185.

LEONARDS, G.A., CUTTER, W.A., AND HOLTZ, R.D. (1980) "Dynamic Compaction of Granular Soils," *Journal of the Geotechnical Engineering Division*, ASCE, Vol. 106, No. 1, pp. 35–44.

LEONARDS, G.A., AND GIRAULT, P. (1961) "A Study of the One-Dimensional Consolidation Test," *Proceedings of the Fifth International Conference on Soil Mechanics and Foundation Engineering*, Paris, Vol. I, pp. 116–130.

LEONARDS, G.A., AND RAMIAH, B.K. (1959) "Time Effects in the Consolidation of Clay," *Papers on Soils—1959 Meeting*, American Society for Testing and Materials, Special Technical Publication No. 254, pp. 116–130.

LEROUEIL, S., TAVENAS, F., TRAK, B., LaROCHELLE, P., AND ROY, M. (1978a) "Construction Pore Pressure in Clay Foundations Under Embankments. Part I: The Saint-Alban Test Fills," *Canadian Geotechnical Journal*, Vol. 15, No. 1, p. 56.

LEROUEIL, S., TAVENAS, F., MIEUSSENS, C., AND PEIGNAUD, M. (1978b) "Construction Pore Pressures in Clay Foundations Under Embankments. Part II: Generalized Behaviour," *Canadian Geotechnical Journal*, Vol. 15, No. 1, pp. 66–82.

LIU, T.K. (1970) "A Review of Engineering Soil Classification Systems," *Special Procedures for Testing Soil and Rock for Engineering Purposes*, 5th Ed., ASTM Special Technical Publication 479, pp. 361–382.

LOOS, W. (1936) "Comparative Studies of the Effectiveness of Different Methods for Compacting Cohesionless Soils," *Proceedings of the First International Conference on Soil Mechanics and Foundation Engineering*, Cambridge, Vol. III, pp. 174–179.

LOWE, J., III (1974) "New Concepts in Consolidation and Settlement Analysis," *Journal of the Geotechnical Engineering Division*, ASCE, Vol. 100, No. GT6, pp. 574–612.

LOWE, J., III, ZACCHEO, P.F., AND FELDMAN, H.S. (1964) "Consolidation Testing with Back Pressure," *Journal of the Soil Mechanics and Foundations Division*,

ASCE, Vol. 90, No. SM5, pp. 69–86; also in *Design of Foundations for Control of Settlement*, ASCE, pp. 73–90.

LUKAS, R.G. (1980) "Densification of Loose Deposits by Pounding," *Journal of the Geotechnical Engineering Division*, ASCE, Vol. 106, No. GT4, pp. 435–446.

MACDONALD, A.B., AND SAUER, E.K. (1970) "The Engineering Significance of Pleistocene Stratigraphy in the Saskatoon Area, Saskatchewan, Canada," *Canadian Geotechnical Journal*, Vol. 7, No. 2, pp. 116–126.

MANNHEIM, B.R.S. (1979) "A New Conception of the Formation of the Kola Peninsula Apatite Deposit: The Coprogenic Impact Theory-CIT," *The Journal of Irreproducible Results*, Society for Basic Irreproducible Research, Vol. 25, No. 1, pp. 6–7.

MASSARSCH, K.R. (1979) "Lateral Earth Pressure in Normally Consolidated Clay," *Proceedings of the Seventh European Conference on Soil Mechanics and Foundation Engineering*, Brighton, England, Vol. 2, pp. 245–250.

MASSARSCH, K.R., HOLTZ, R.D., HOLM, B.G., AND FREDRICKSSON, A. (1975) "Measurement of Horizontal In Situ Stresses," *Proceedings of the ASCE Specialty Conference on In Situ Measurement of Soil Properties*, Raleigh, North Carolina, Vol. I, pp. 266–286.

McGOWN, A. (1973) "The Nature of the Matrix in Glacial Ablation Tills," *Proceedings of the International Symposium on Soil Structure*, Gothenburg, Sweden, p. 95.

MEEHAN, R.L. (1967) "The Uselessness of Elephants in Compacting Fill," *Canadian Geotechnical Journal*, Vol. IV, No. 3, pp. 358–360.

MÉNARD, L. (1956) "An Apparatus for Measuring the Strength of Soils in Place," MSCE Thesis, University of Illinois.

MÉNARD, L. (1975) "The Ménard Pressuremeter," *Les Éditions Sols-Soils*, No. 26, pp. 7–43.

MÉNARD, L., AND BROISE, Y. (1975) "Theoretical and Practical Aspects of Dynamic Consolidation," *Géotechnique*, Vol. XXV, No. 1, pp. 3–18.

MESRI, G. (1973) "Coefficient of Secondary Compression," *Journal of the Soil Mechanics and Foundations Division*, ASCE, Vol. 99, No. SM1, pp. 123–137.

MESRI, G. (1975) Discussion of "New Design Procedures for Stability of Soft Clays," *Journal of the Geotechnical Engineering Division*, ASCE, Vol. 101, No. GT4, pp. 409–412.

MESRI, G., AND GODLEWSKI, P.M. (1977) "Time- and Stress-Compressibility Interrelationship," *Journal of the Geotechnical Engineering Division*, ASCE, Vol. 103, No. GT5, pp. 417–430.

MILLIGAN, V. (1972) Discussion of "Embankments on Soft Ground," *Proceedings of the ASCE Specialty Conference on Performance of Earth and Earth-Supported Structures*, Purdue University, Vol. III, pp. 41–48.

MILLS, W.T., AND DeSALVO, J.M. (1978) "Soil Compaction and Proofrolling," *Soils*, newsletter of Converse, Ward, Davis, and Dixon, Inc., Pasadena, California, and Caldwell, New Jersey, Autumn, pp. 6–7.

MITCHELL, J.K. (1968) "In-Place Treatment of Foundation Soils," *Placement and Improvement of Soil to Support Structures*, ASCE, pp. 93–130; also in *Journal of the Soil Mechanics and Foundations Division*, ASCE, Vol. 96, No. SM1, (1970) pp. 73–110.

MITCHELL, J.K. (1976) *Fundamentals of Soil Behavior*, John Wiley & Sons, Inc., New York, 422 pp.

MITCHELL, J.K., AND GARDNER, W.S. (1975) "In Situ Measurement of Volume Change Characteristics," State-of-the-Art Report, *Proceedings of the ASCE Specialty Conference on In Situ Measurement of Soil Properties*, Raleigh, North Carolina, Vol. II, p. 333.

MOH, Z.C., BRAND, E.W., AND NELSON, J.D. (1972) "Pore Pressures Under a Bund on Soft Fissured Clay," *Proceedings of the ASCE Specialty Conference on Performance of Earth and Earth-Supported Structures*, Purdue University, Vol. I, Part 1, pp. 245–246.

MOHR, O. (1887) "Über die Bestimmung und die graphische Darstellung von Trägheitsmomenten ebener Flächen," *Civilingenieur*, columns 43–68; also in *Abhandlungen aus dem Gebiete der technischen Mechanik*, 2nd Ed., W. Ernst u. Sohn, Berlin, pp. 90 and 109 (1914).

MOHR, O. (1900) "Welche Umstände bedingen die Elastizitätsgrenze und den Bruch eines Materiales?" *Zeitschrift des Vereines Deutscher Ingenieure*, Vol. 44, pp. 1524–1530; 1572–1577.

MORGENSTERN, N.R., AND EISENSTEIN, Z. (1970) "Methods of Estimating Lateral Loads and Deformation," *Proceedings of the ASCE Specialty Conference on Lateral Stresses in the Ground and Design of Earth-Retaining Structures*, Cornell University, pp. 51–102.

MULILIS, J.P., SEED, H.B., AND CHAN, C.K. (1977) "Effects of Sample Preparation on Sand Liquefaction," *Journal of the Soil Mechanics and Foundations Division*, ASCE, Vol. 103, No. GT2, pp. 91–108.

NEWMARK, N.M. (1935) "Simplified Computation of Vertical Pressures in Elastic Foundations," *University of Illinois Engineering Experiment Station Circular 24*, Urbana, Illinois, 19 pp.

NEWMARK, N.M. (1942) "Influence Charts for Computation of Stresses in Elastic Foundations," *University of Illinois Engineering Experiment Station Bulletin*, Series No. 338, Vol. 61, No. 92, Urbana, Illinois, reprinted 1964, 28 pp.

OSTERBERG, J.O. (1957) "Influence Values for Vertical Stresses in a Semi-infinite Mass Due to an Embankment Loading," *Proceedings of the Fourth International Conference on Soil Mechanics and Foundation Engineering*, London, Vol. I, pp. 393–394.

OSTERBERG, J.O. (1963) "Current Practice in Foundation Design-II," *Chicago Soil Mechanics Lecture Series*, ASCE and Illinois Institute of Technology, pp. 6–22.

OSTERMAN, J. (1960) "Notes on the Shearing Resistance of Soft Clays," *Acta Polytechnica Scandinavica*, Series Ci-2, (AP 263 1959), pp. 1–22.

PARK, T.K., AND SILVER, M.L. (1975) "Dynamic Triaxial and Simple Shear Behavior of Sand," *Journal of the Geotechnical Engineering Division*, ASCE, Vol. 101, No. GT6, pp. 513–529.

PARSONS, A.W., KRAWCZYK, J., AND CROSS, J.E. (1962) "An Investigation of the Performance on an $8\frac{1}{2}$-Ton Vibrating Roller for the Compaction of Soil," Road Research Laboratory, Laboratory Note No. LN/64/AWP. JK. JEC.

PAVLOVSKY, N.N. (1956) *Collected Works*, Akad. Nauk USSR, Leningrad.

PECK, R.B., HANSON, W.E., AND THORNBURN, T.H. (1974) *Foundation Engineering*, 2nd Ed., John Wiley & Sons, Inc., New York, 514 pp.

PENNER, E. (1963) "Sensitivity in Leda Clay," *Nature*, Vol. 197, No. 4865, pp. 347–348.

PERLOFF, W.H., AND BARON, W. (1976) *Soil Mechanics—Principles and Applications*, The Ronald Press Company, New York, pp. 359–361.

POULOS, H.G., AND DAVIS, E.H. (1974) *Elastic Solutions for Soil and Rock Mechanics*, John Wiley & Sons, Inc., New York, 411 pp.

PROCTOR, R.R. (1933) "Fundamental Principles of Soil Compaction," *Engineering News-Record*, Vol. 111, Nos. 9, 10, 12, and 13.

PUSCH, R. (1973) General Report on "Physico-Chemical Processes which Affect Soil Structure and Vice Versa," *Proceedings of the International Symposium on Soil Structure*, Gothenburg, Sweden, Appendix p. 33.

QUIGLEY, R.M., AND THOMPSON, C.D. (1966) "The Fabric of Anisotropically Consolidated Sensitive Marine Clay," *Canadian Geotechnical Journal*, Vol. III, No. 2, pp. 61–73.

RAJU, A.A. (1956) "The Preconsolidation Pressure in Clay Soils," MSCE Thesis, Purdue University, 41 pp.

RAYMOND, G.P., TOWNSEND, D.L., AND LOJKACEK, M.J. (1971) "The Effects of Sampling on the Undrained Soil Properties of the Leda Soil," *Canadian Geotechnical Journal*, Vol. 8, No. 4, pp. 546–557.

RAYMOND, G.P., AND WAHLS, H.E. (1976) "Estimating One-Dimensional Consolidation, Including Secondary Compression of Clay Loaded from Overconsolidated to Normally Consolidated State," *Special Report 163*, Transportation Research Board, pp. 17–23.

REED, M.A., LOVELL, C.W., ALTSCHAEFFL, A.G., AND WOOD, L.E. (1979) "Frost Heaving Rate Predicted from Pore Size Distribution," *Canadian Geotechnical Journal*, Vol. 16, No. 3, pp. 463–472.

RUTLEDGE, P.C. (1944) "Relation of Undisturbed Sampling to Laboratory Testing," *Transactions*, ASCE, Vol. 109, pp. 1162–1163.

SAADA, A.S., AND BIANCHINI, G.F. (1975) "Strength of One Dimensionally Consolidated Clays," *Journal of the Geotechnical Engineering Division*, ASCE, Vol. 101, No. GT11, pp. 1151–1164.

SANGLERAT, G. (1972) *The Penetrometer and Soil Exploration*, Elsevier, Amsterdam, 464 pp.

SCHMERTMANN, J.H. (1955) "The Undisturbed Consolidation Behavior of Clay," *Transactions*, ASCE, Vol. 120, pp. 1201–1233.

SCHMERTMANN, J.H. (1970) "Suggested Method for Screw-Plate Load Test," *Special Procedures for Testing Soil and Rock for Engineering Purposes*, 5th Ed., ASTM Special Technical Publication 479, pp. 81–85.

SCHMERTMANN, J.H. (1975) "Measurement of In Situ Shear Strength," State-of-the-Art Report, *Proceedings of the ASCE Specialty Conference on In Situ Measurement of Soil Properties*, Raleigh, North Carolina, Vol. II, pp. 57–138.

SCHMIDT, B. (1966) Discussion of "Earth Pressures at Rest Related to Stress History," *Canadian Geotechnical Journal*, Vol. III, No. 4, pp. 239–242.

SCHMIDT, B. (1967) "Lateral Stresses in Uniaxial Strain," *Bulletin No. 23*, Danish Geotechnical Institute, pp. 5–12.

SCHWAB, E.F. (1976) "Bearing Capacity, Strength, and Deformation Behaviour of Soft Organic Sulphide Soils," PhD Thesis, Institutionen för Jord- och Bergmekanik, Kungliga Tekniska Högskolan, Stockholm, 368 pp.

SCOTT, R.F. (1963) *Principles of Soil Mechanics*, Addison-Wesley Publishing Company, Inc., Reading, Massachusetts, pp. 267–275.

SEED, H.B. (1959) "A Modern Approach to Soil Compaction," *Proceedings of the Eleventh California Street and Highway Conference*, reprint No. 69, The Institute of Transportation and Traffic Engineering, University of California at Berkeley, p. 93.

SEED, H.B. (1964) Lecture Notes, CE 271, Seepage and Earth Dam Design, University of California at Berkeley (by W.D. Kovacs, April 13).

SEED, H.B. (1979) "Soil Liquefaction and Cyclic Mobility Evaluation for Level Ground During Earthquakes," *Journal of the Geotechnical Engineering Division*, ASCE, Vol. 105, No. GT2, pp. 201–255.

SEED, H.B., AND CHAN, C.K. (1959) "Structure and Strength Characteristics of Compacted Clays," *Journal of the Soil Mechanics and Foundations Division*, ASCE, Vol. 85, No. SM5, pp. 87–128.

SEED, H.B., AND IDRISS, I.M. (1967) "Analysis of Soil Liquefaction: Niigata Earthquake," *Journal of the Soil Mechanics and Foundations Division*, ASCE, Vol. 93, No. SM3, pp. 83–108.

SEED, H.B., AND LEE, K.L. (1966) "Liquefaction of Saturated Sands During Cyclic Loading," *Journal of the Soil Mechanics and Foundations Division*, ASCE, Vol. 92, No. SM6, pp. 105–134.

SEED, H.B., AND LEE, K.L. (1967) "Undrained Strength Characteristics of Cohesionless Soils," *Journal of the Soil Mechanics and Foundations Division*, ASCE, Vol. 93, No. SM6, pp. 333–360.

SEED, H.B., AND PEACOCK, W.H. (1971) "Test Procedures for Measuring Soil Liquefaction Characteristics," *Journal of the Soil Mechanics and Foundations Division*, ASCE, Vol. 97, No. SM8, pp. 1099–1119.

SEED, H.B., AND WILSON, S.D. (1967) "The Turnagain Heights Landslide, Anchorage, Alaska," *Journal of the Soil Mechanics and Foundations Division*, ASCE, Vol. 93, No. SM4, pp. 325–353.

SEED, H.B., WOODWARD, R.J., AND LUNDGREN, R. (1962) "Prediction of Swelling Potential for Compacted Clays," *Journal of the Soil Mechanics and Foundations Division*, ASCE, Vol. 88, No. SM4, pp. 107–131.

SEEMEL, R.M. (1976) "Plastic Filter Fabrics Challenging the Conventional Granular Filter," *Civil Engineering*, ASCE, Vol. 46, No. 3, pp. 57–59.

SELIG, E.T., AND YOO, T.-S. (1977) "Fundamentals of Vibratory Roller Behavior," *Proceedings of the Ninth International Conference on Soil Mechanics and Foundation Engineering*, Tokyo, Vol. 2, pp. 375–380.

SIMONS, N.E. (1958) Discussion of "Test Requirements for Measuring the Coefficient of Earth Pressure at Rest," *Proceedings of the Conference on Earth Pressure Problems*, Brussels, Vol. III, pp. 50–53.

SIMONS, N.E. (1974) "Normally Consolidated and Lightly Overconsolidated Cohesive Materials," General Report, *Proceedings of the Conference on Settlement of Structures*, Cambridge University, British Geotechnical Society, pp. 500–530.

SKEMPTON, A.W. (1953) "The Colloidal Activity of Clays," *Proceedings of the Third International Conference on Soil Mechanics and Foundation Engineering*, Vol. I, pp. 57–61.

SKEMPTON, A.W. (1954) "The Pore-Pressure Coefficients A and B," *Géotechnique*, Vol. IV, pp. 143–147.

SKEMPTON, A.W. (1960) "Effective Stress in Soils, Concrete and Rocks," *Pore Pressure and Suction in Soils*, Butterworths, London, p. 5.

SKEMPTON, A.W. (1964) "Long-Term Stability of Clay Slopes," *Géotechnique*, Vol. XIV, No. 2, pp. 77–101.

SKEMPTON, A.W. (1977) "Slope Stability of Cuttings in Brown London Clay," *Proceedings of the Ninth International Conference on Soil Mechanics and Foundation Engineering*, Tokyo, Vol. 3, pp. 261–270.

SKEMPTON, A.W., AND NORTHEY, R.D. (1952) "The Sensitivity of Clays," *Géotechnique*, Vol. III, No. 1, pp. 30–53.

SKEMPTON, A.W., AND HENKEL, D.J. (1953) "The Post-Glacial Clays of the Thames Estuary at Tillbury and Shellhaven," *Proceedings of the Third International Conference on Soil Mechanics and Foundation Engineering*, Zurich, Vol. I, pp. 302–308.

SODERMAN, L.G., AND KIM, Y.D. (1970) "Effect of Groundwater Levels on Stress History of the St. Clair Clay Till Deposit," *Canadian Geotechnical Journal*, Vol. 7, No. 2, pp. 173–187.

STEWARD, J.E., WILLIAMSON, R., AND MOHNEY, J. (1977) "Guidelines for Use of Fabrics in Construction and Maintenance of Low-Volume Roads," *Internal Report*, U.S. Forest Service, Portland, Oregon; also published as *Report No. FHWA-TS-78-2-5* by the Federal Highway Administration, Washington, D.C., 171 pp.

TAYLOR, D.W. (1948) *Fundamentals of Soil Mechanics*, John Wiley & Sons, Inc., New York, 700 pp.

TAVENAS, F.A., CHAPEAU, C., LAROCHELLE, P., AND ROY, M. (1974) "Immediate Settlements of Three Test Embankments on Champlain Clay," *Canadian Geotechnical Journal*, Vol. 11, No. 1, pp. 109–141.

TAVENAS, F.A., BLANCHETTE, G., LEROUEIL, S., ROY, M., AND LAROCHELLE, P. (1975) "Difficulties in the In Situ Determination of K_0 in Soft Sensitive Clays," *Proceedings of the ASCE Specialty Conference on In Situ Measurement of Soil Properties*, Raleigh, North Carolina, Vol. I, pp. 450–476.

TERZAGHI, K. (1922) "Der Grundbruch an Stauwerken und seine Verhütung," *Die Wasserkraft*, Vol. 17, No. 24, pp. 445–449; reprinted in *From Theory to Practice in Soil Mechanics*, John Wiley & Sons, Inc., New York, pp. 114–118.

TERZAGHI, K. (1925) *Erdbaumechanik auf Bodenphysikalischer Grundlage*, Franz Deuticke, Leipzig und Wein, 399 pp.

TERZAGHI, K. (1927) "Concrete Roads—A Problem in Foundation Engineering," *Journal of the Boston Society of Civil Engineers*, May; reprinted in *Contributions to Soil Mechanics 1925–1940*, BSCE, pp. 57–58.

TERZAGHI, K. (1943) *Theoretical Soil Mechanics*, John Wiley & Sons, Inc., New York, 510 pp.

TERZAGHI, K., AND PECK, R.B. (1967) *Soil Mechanics in Engineering Practice*, 2nd Ed., John Wiley & Sons, Inc., New York, pp. 729.

TURNBULL, W.J. (1950) "Compaction and Strength Tests on Soil," presented at Annual Meeting, ASCE, January, as cited by Lambe, T.W., and Whitman, R.V. (1969) *Soil Mechanics*, John Wiley & Sons, Inc., New York, p. 517.

TURNBULL, W.J., AND FOSTER, C.R. (1956) "Stabilization of Materials by Compaction," *Journal of the Soil Mechanics and Foundations Division*, ASCE, Vol. 82, No. SM2, pp. 934-1 to 934-23; also in *Transactions*, ASCE, Vol. 123, (1958), pp. 1–26.

U.S. ARMY CORPS OF ENGINEERS (1970) "Laboratory Soils Testing," *Engineer Manual, EM 1110-2-1906*.

U.S. ARMY ENGINEER WATERWAYS EXPERIMENT STATION (1960) "The Unified Soil Classification System," *Technical Memorandum No. 3-357*. Appendix A, Characteristics of Soil Groups Pertaining to Embankments and Foundations, 1953; Appendix B, Characteristics of Soil Groups Pertaining to Roads and Airfields, 1957.

U.S. BUREAU OF RECLAMATION (1974) *Earth Manual*, 2nd Ed., Denver, 810 pp.

U.S. NAVY (1971) "Soil Mechanics, Foundations, and Earth Structures," *NAVFAC Design Manual DM-7*, Washington, D.C.

WALLACE, G.B., AND OTTO, W.C. (1964) "Differential Settlement at Selfridge Air Force Base," *Journal of the Soil Mechanics and Foundations Division*, ASCE, Vol. 90, No. SM5, pp. 197–220; also in *Design of Foundations for Control of Settlement*, ASCE, pp. 249–272.

WARSHAW, C.M., AND ROY, R. (1961) "Classification and a Scheme for the Identification of Layer Silicates," *Geological Society of America Bulletin*, Vol. 72, pp. 1455–1492.

WESTERGAARD, H.M. (1938) "A Problem of Elasticity Suggested by a Problem in Soil Mechanics: A Soft Material Reinforced by Numerous Strong Horizontal Sheets," in *Contributions to the Mechanics of Solids, Stephen Timoshenko 60th Anniversary Volume*, Macmillan, New York, pp. 268–277.

WILSON, S.D. (1970) "Suggested Method of Test for Moisture-Density Relations of Soils Using Harvard Compaction Apparatus," *Special Procedures for Testing Soil and Rock for Engineering Purposes*, 5th Ed., ASTM Special Technical Publication 479, pp. 101–103.

WINELAND, J.D. (1975) "Borehole Shear Device," *Proceedings of the ASCE Specialty Conference on In Situ Measurement of Soil Properties*, Raleigh, North Carolina, Vol. I, pp. 511–522.

WISSA, A.E.Z. (1969) "Pore Pressure Measurement in Saturated Stiff Soils," *Journal of the Soil Mechanics and Foundations Division*, ASCE, Vol. 95, No. SM4, pp. 1063–1073.

WROTH, C.P. (1972) "General Theories of Earth Pressures and Deformations," General Report, *Proceedings of the Fifth European Conference on Soil Mechanics and Foundation Engineering, Madrid, Vol. II, pp. 33–52*.

WROTH, C.P. (1975) "In Situ Measurement of Initial Stresses and Deformation Characteristics," *Proceedings of the ASCE Specialty Conference on In Situ Measurement of Soil Properties*, Raleigh, North Carolina, Vol. II, pp. 181–230.

YODER, E.J., AND WITCZAK, M.W. (1975) *Principles of Pavement Design*, John Wiley & Sons, Inc., New York, 711 pp.

YONG, R.N., AND SHEERAN, D.E. (1973) "Fabric Unit Interaction and Soil Behaviour," *Proceedings of the International Symposium on Soil Structure*, Gothenburg, Sweden, pp. 176–183.

YONG, R.N., AND WARKENTIN, B.P. (1975) *Soil Properties and Behaviour*, Elsevier Scientific Publishing Co., New York, 449 pp.

ZIMMIE, T.F. (1973) "Soil Tests on a Clay from Skå-Edeby, Sweden," Norwegian Geotechnical Institute, Internal Report No. 50306-3, 21 pp.

Index